EARTHQUAKE RESEARCH AND ANALYSIS – SEISMOLOGY, SEISMOTECTONIC AND EARTHQUAKE GEOLOGY

Edited by **Sebastiano D'Amico**

INTECHOPEN.COM

Earthquake Research and Analysis – Seismology, Seismotectonic and Earthquake Geology
Edited by Sebastiano D'Amico

Published by InTech
Janeza Trdine 9, 51000 Rijeka, Croatia

Edition 2014

Publishing Process Manager Igor Babic

Technical Editor Teodora Smiljanic

Cover Designer InTech Design Team

Additional hard copies can be obtained from orders@intechopen.com

Earthquake Research and Analysis – Seismology, Seismotectonic and Earthquake Geology,
Edited by Sebastiano D'Amico
p. cm.
ISBN 978-953-307-991-2

INTECH

open science | open minds

Contents

Preface

This book presents different aspects of earthquake research. In includes syntheses of recent works combined with new results and interpretations, applicable to diverse branches of earthquake seismology and geology.

The destructive potential of earthquakes depends on their magnitude and the placement of the hypocenter. Worldwide efforts, such as the Global Seismographic Network of seismological and geophysical sensors and data centers, provides information that is essential to identify earthquake locations, understand the physics of the earthquakes and faulting mechanisms, as well as studies on paleoseismicity and earthquakes in the human history. These types of studies play a key role in mitigating earthquake hazards and planning emergency response. Geological studies on tectonic landforms and advances in the theoretical and applied physics of seismology have led to the creation of numerical models for seismic activity, improvement of design methods and testing practices for earthquake-resistant construction and seismic retrofit, as well as observation and monitoring. The study of earthquakes combines science, technology and expertise in infrastructure and engineering in an effort to minimize human and material losses when their occurrence is inevitable.

The chapters in this book are devoted to various aspects of earthquake research and analysis, such as seismology, seismotectonic and earthquake geology.

The first chapter presents a study on ways to find out about the seismic past of a region. This is really important because identification and reliable description of past earthquakes is of great importance for the seismic assessment in a particular area and it represents a reliable projection of future earthquakes. Even though the paper focuses on, but is not limited to, German attempts, it offers a clear picture of historical earthquakes and their reception history, ancient and early modern attempts to collect earthquake information, early earthquake catalogue compilation and open source ways to collect earthquake information.

Chapters 2 and 3 discuss the trace of earthquakes through some examples of seismites from Turkey and through studies on paleoearthquakes in Central Asia. The authors describe sedimentary beds disturbed by seismic shaking known as seismites. The main goal of chapter 2 is to determine the different deformation structures in Lake Van deposits, interpret the triggering mechanism and discuss the importance of these structures in regional tectonic. Chapter 3 focuses on strong paleoearthquakes along the Talas-Fergana Fault to improve the understanding of the intracontinental deformation of the Tien Shan mountain belt as a whole, and the occurrence of strong earthquakes along the largest wrench structure in Central Asia.

Chapter 4 presents some study on paleoseismic investigation. The authors analyzed the paleoseismic evidence for earthquake rupture at four excavated sites along the fault responsible for the destructive earthquake in Taiwan in 1999 (Mw=7.6). They suggest that "the evolutionary path of thrust-related folds develops from fault-propagation folding to breakthrough fault-propagation folding, and finally to fault-bend folding".

Chapter 5 deals with the generation mechanism of giant earthquakes in subduction zones. The authors discuss a numerical simulation approach in order to set up a model in subduciton zones. They were able to quantitatively explain the features of the 2011 M9 earthquake in Japan using a hierarchical asperity model.

Chapter 6 also deals with earthquakes in subduction zones by investigating the process and mechanisms of slow-earthquake migration. The chapter presents a comparison of results obtained by observation and numerical simulations. Furthermore, the authors discuss a new method for detecting changes around important asperities before megathrust earthquake.

Chapter 7 is an example of a seismology paper and deals with the determination of focal depth of small and moderate earthquakes using teleseismic depth phases, which is crucial information for many studies in seismology. The chapter shows several examples and comparisons among the focal depth obtained using teleseismic phase as well as other results obtained using other methods, and the consistency in the comparisons is good.

Chapter 8 investigates the crustal structure of the Azorean plateau region, describing the results obtained by using 1-D and 3-D inversions. It illustrates the use of several techniques well-known in earthquake seismology and shows the results of intense seismic and volcanic activity.

Chapter 9 deals with identifications of structure in non-tectonic areas inferred using seismogram analysis of earthquakes recorded in Taiwan.

Chapter 10 presents results on the tectonic background of the Wenchuan earthquake area. Due to the complicated tectonic environment of northwestern Sichuan, several different results are presented. The paper sheds light on the tectonic process of the area and tries to clarify several hypotheses.

Chapter 11 reports statistical studies applied to seismology. In particular, two different approaches are used to investigate non-linear analysis. Fractal approach to characterize the time-clustering phenomenon is also investigated. The goal of the paper is to propose a general methodology to obtain better estimates of the earthquake hazard.

Chapters 12 and 13 analyze complex signals associated with geoelectric activity and methods of electrical conductivity determination respectively, focusing on two different areas of the globe. In particular, chapter 12 reports some complexity studies of geoelectric signals during a two-year period, from June 1, 1994 to May 31, 1996, in two sites (Acapulco and Coyuca stations) located in southern Mexico. Chapter 13 focuses on the EM image beneath southwest Japan. A comparative study with the northeast Japan area is made.

Chapter 14 covers the hydrological changes associated with some of the big earthquakes that occurred in the Southern Apennines (Italy) during the XX century. The seismic events that occurred in 1930, 1980 and 1984 are investigated and the types of hydrological effects are also reported for each earthquake.

Chapter 15 presents the important issue of rock-fluid interactions along seismogenic faults. In this paper, the authors explain the interactions at the fault by analyzing clay minerals within the fossil seismogenic fault along the subduction interface.

Chapter 16 is a nice example of a mathematical approach to derive inequalities for comparing the change of pore pressure with that of frictional coefficient during fault slip.

Chapter 17 discusses the contribution of seismic frictional heat to the total earthquake energy budget. The author describes the ESR (electron spin resonance) technique and shows the scanning ESR microscopic technique for sequential high-resolution measurements for the detection of seismic frictional heat. The case of the Nojima fault rocks in Japan is presented in detail.

Chapter 18 describes a methodology for constraining the structural lineaments in active tectonic areas by integrating morphological and morphometrical data derived by DEMs (Digital Elevation Models) processing, with different geophysical data, as local seismicity and ground deformation data. Furthermore, validation of the lineaments extracted from DEM is carried out by looking over geological and geomorphological maps of literature, available aerial photo and field surveys reports. Three different case study areas are presented.

Chapter 19 presents a description of studying seismic signals associated to volcanic earthquakes. In particular, the pattern recognition (PR) technique is applied as a valid tool for the volcano-seismic monitoring. This chapter briefly introduces fundamental concepts regarding seismic volcanic signals and PR systems. Moreover, it raises the important issue of transferring prototype academic results into deployed technology.

We can conclude that the book collects different approaches which are really important to better understand such a complex phenomenon and which are also critical in terms of civil protection and associated risk assessment. This represents the ultimate goal since such seismic activity often has a significant impact in economic and human terms.

Finally, I would like to express my special thanks to Mr. Igor Babic and Ms. Ivana Lorkovic. Last but not least, I would like to thank the whole staff of InTech Open Access Publishing, especially Mr. Igor Babic, for their professional assistance and technical support during the publishing process that has led to the realization of this book.

Sebastiano D'Amico
Research Officer III
Physics Department
University of Malta
Malta

1

Earthquakes in History – Ways to Find out About the Seismic Past of a Region

Friedrich Barnikel[1] and Mark Vetter[2]
[1]*Educational Coordinator for Geography, City of Munich*
[2]*Department of Geography, University of Erlangen-Nuremberg*
Germany

1. Introduction

The onset of the 21st century has brought a new public awareness of natural hazards. Recent catastrophic events like the 2004 tsunami in Asia or the 2005 flooding of New Orleans have made it not only to the headlines in news publications around the globe but have also contributed to a more profound desire to accumulate knowledge about natural hazards in general among people all over the world. Earthquake research belongs to the most fascinating (albeit problematic) topics in the field of natural hazard research. Hardly any other hazard claims more lives, destroys more values and can lead to catastrophic after-effects (as can be seen when looking at the 2011 earthquake with resulting tsunami and nuclear disaster in Japan). In most cases scientists are not able to forecast when and where an earthquake may take place, but only the approximate region and the probability, not the precise date and the magnitude. Geoscientists therefore by and large concentrate on assessing and mapping regions that experienced earthquakes in the past.

As a consequence, the description of past earthquakes is of utmost importance for a reliable projection of future earthquakes. Several ways exist to sum up details about historical earthquakes, the most important of which was the analysis of written documents in pre-measurement times (see Fig. 1). From the 19th century onwards the installation of seismometers has considerably improved the situation and nowadays the analysis of a new earthquake relies almost completely on measurements.

This paper strives to offer a short introduction to the following aspects:

- Historical earthquakes and their reception history
- Ancient attempts to forecast and measure earthquakes
- Early modern attempts to collect written information on earthquake events
- Early modern earthquake catalogues in Europe
- The application of early seismometers
- The expansion of a seismometer network
- Open source ways to collect earthquake information

The paper will concentrate on (but not limit itself to) German attempts in the aforementioned aspects, including a short overview of the current state of research. Of special importance in this context is the Bavarian BASE-project, which serves as a role model for a modern and complete collection of historical earthquake data

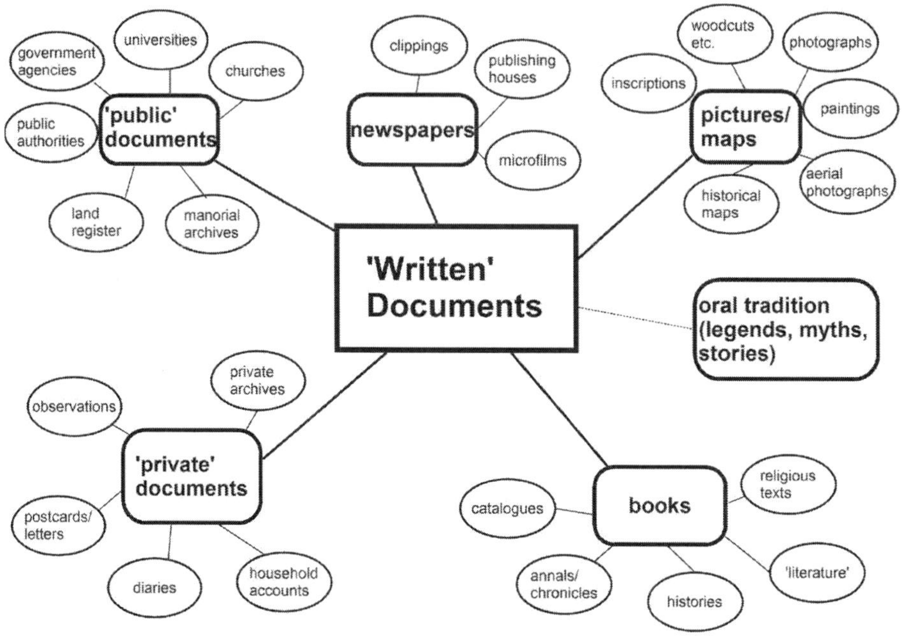

Fig. 1. A selection of sources for written documents for the analysis of historical earthquakes (F. Barnikel).

2. Historical earthquakes and aspects of their reception history

Bolt et al. (1975) point out the different causes of earthquakes as follows: they list (in that order) tectonic earthquakes, the principle of elastic rebound, the dilatancy in crustal rocks, nuclear earthquakes, reservoir-induced earthquakes and volcanic earthquakes. Even if some of these causes can be discarded for ancient times, the event of an earthquake has shaken the trust of ancient populations in a *terra firma* at all times.

Consequently, it is no wonder that the first record of an earthquake dates back to the year 1831 BC. It is mentioned in the Chinese *Chronicle on Bamboo*: "Mount Taishan quaked." Even older, but considered not very reliable by the Chinese historians, is a reference to the year 2221 BC: "While the San-Miao was going to be destroyed, the Earth quaked, fountains sprang" (quoted after Xie, 1988). But the oldest 'known' (and completely unspecified) earthquake on the other hand was, as Seyfart, with a piety not typical for his time, noted in the 18th century, "undoubtedly the one triggered by God on the third day of the creation through the power of the fire inside the globe" (quoted after Fréchet, 2008).

Approximately one thousand years later the Chinese began their continuous listing of quake events, which lasts until today. Some of the descriptions are thus detailed that current research is able to reconstruct the approximate magnitude of the event. Chinese scholars first explained earthquakes with other forms of disasters, like floodings, droughts or the plague. Among the most popular lores from ancient times is the well known ancient Japanese belief that linked earthquakes to the fish species of brown bullhead (*namazu*), which were said to move their tails to and fro. Other mythological creatures include frogs (China), a snake (Philippines) or the God Poseidon himself (Greece; for the history of earthquake reception cf. Bolt et al., 1995 or Zeilinga de Boer & Sanders, 2005).

In Asia among the most severe earthquakes in China are the events in 110 AD (Dian), in 1290 (Chihli) and the 1556 event in Shaanxi. In nearby Japan some of the gravest events include the 869 AD Sanriku Earthquake, the 1293 Kamakura event, an earthquake 1498 in Honshu, and the 1707 Hoei and the 1896 Meiji events (all events before the introduction of modern seismometers). Srivastava & Das (1988) quote archaeological evidence pointing at an earthquake c. 1730 ± 100 BC in Rajasthan. The earliest earthquake records for Indochina date back to 642 BC (Prachuab, 1988). The case lies different with New Zealand for example, where no older written sources exist. But Eiby mentions that "Maori oral tradition records an important earthquake in about 1460" (Eiby, 1988), an interesting reminder of ancient ways of transmitting information.

The oldest specific date for an earthquake event in Europe and the Middle East dates back to about 1274 - 1234 BC, when the Assyrians noted an earthquake in Nineveh. But it was the Greeks who made the first attempts to write about explanations for earthquakes and their origin in Europe and the adjoining regions. Thales of Milet was the first to think that land would drift on the surface of the oceans and that the waves might cause earthquakes. In the 6th century BC Anaximenes thought that falling stones inside the globe caused earthquakes, Anaxagoras, one hundred years later, believed fire was the reason for it (for a beautiful illustration of a similar concept, see Fig. 2). In 464 BC a strong earthquake is thought to be one reason for the outbreak of the Peloponnese War, when the poor helots revolted against the rulers, of whom many soldiers, women and boys (as future soldiers) had died as a result of the event (Zeilinga de Boer & Sanders, 2005). Aristoteles wrote about a central fire inside the Earth and thus was one of the first to acknowledge the energy inside the Earth in the 4th c. BC. A lot of events were listed by the geographer Strabo and known to the more educated Greeks. In 226 BC a very notable earthquake event destroyed the Colossus of Rhodes, one of the ancient wonders of the earth. Other severe events in historic times include the 365 AD Crete event and other earthquakes 856 in Corinth, 1303 in Crete, and 1481 in Rhodes.

Among the earliest (and most colourful) descriptions of earthquakes in the Near and Middle East belong some bible texts (as aforementioned), although they cannot be dated precisely. One example is Zechariah 14:4, which describes in detail a horizontal displacement close to an epicentre: "In that day His feet will stand on the Mount of Olives, which is in front of Jerusalem on the east; and the Mount of Olives will be split in its middle from east to west by a very large valley, so that half of the mountain will move toward the north and the other half toward the south." Other early and unspecific descriptions include Genesis 19:24f. ("Then the Lord rained upon Sodom and upon Gomorrah brimstone and fire from the Lord out of heaven; And he overthrew those cities, and all the plain, and all the inhabitants of the cities, and that which grew upon the ground"), 2 Samuel 22:8 ("Then the earth shook and trembled; the foundations of heaven moved and shook, because He was wroth") and Jeremiah 10:10 ("But the Lord is the true God; He is the living God and an everlasting King. At His wrath the earth shall tremble, and the nations shall not be able to abide His indignation"), which all describe the fury in the wrath of God. Some grave events in the region include: 526 AD in Antioch, 551 in Beirut, 749 in the Levant, 847 in Damascus, 856 in Qumis, 893 in Armenia and in Iran the same year, 1042 in Syria, 1138 in Aleppo, 1157 in Syria, 1170 in Aleppo, 1202 in Syria, 1268 in Cilicia, and 1509 in Istanbul, 1667 in Shamakhi, 1727 in Tabriz, and 1780 in Iran.

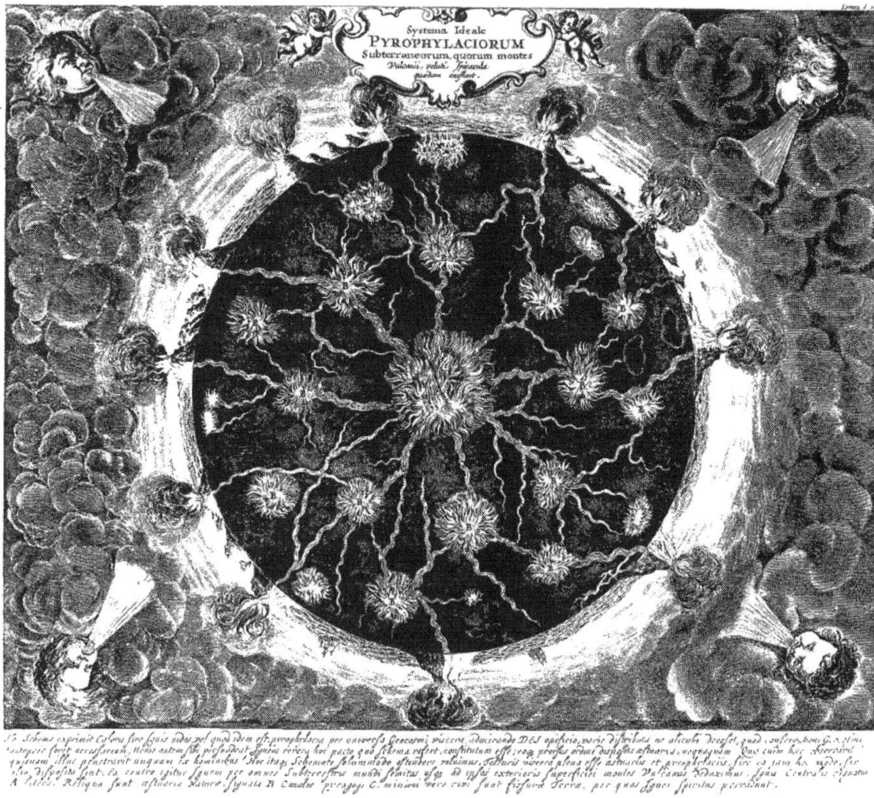

Fig. 2. The interior of the earth according to the perception of Athanasius Kirchner in the year 1678 with subterranean lakes, rivers and pools of fire.

Seneca (4 BC - 65 AD) assumed movements in the air inside the Earth as reasons for earthquakes. More severe events in Italy took place in the second millennium AD, e.g. 1169 in Sicily, 1348 in Friuli, 1693 in Sicily, 1694 in Irpinia, 1783 in Calabria, and 1857 close to Naples. The latter event led the British scientist Robert Mallet to lay down the foundations of modern seismology and to begin with the compilation of an earthquake catalogue, which in the end listed 6800 events with location and effects, just a few years after the first earthquake statistics put together by K.E.A. Hoff in 1840 (Bolt et al., 1995). Other European events of that time include the 1356 Basel event and the earthquakes in Lisbon in 1531 and 1755. It was especially the 18th century that brought a large number of new scientific

developments. Isaac Newton delivered a scientific theory for the explanation of seismic waves. Several earthquake events 1750 in London led to a number of papers presented to the Royal Society. The 1755 earthquake of Lisbon, which was perceived over large parts of Western Europe, triggered off further inquiries.
From other parts of the world the 1575 Valdivia and the 1868 Arica events in South America are among the most notable. The Geological Survey of India, founded in 1857, and later the U.S. Geological Survey (founded in the year 1879) began to list and assess earthquakes worldwide and thus contributed enormously to the knowledge about earthquakes. The desire to assess and list historical earthquakes within the framework of systematic and scientific progress has gathered its enormous momentum within the past 150 years foremost in countries around the Pacific Ring of Fire, especially the USA.

3. Ancient attempts to forecast and measure earthquakes

Ambraseys (2009) notes that

> *historical evidence shows that earthquake prediction was a serious preoccupation for the early soothsayer, astrologer, or prophet, and there are many recorded instances in history of earthquakes having been forecast (Ambraseys, 2009).*

The earliest known seismoscope, which was designed to substitute sheer soothsaying, was made in China by Chang Heng in 132 AD. Its mechanism remains unknown, although it is said that his device actually measured some earthquake four hundred miles away which was otherwise unnoticed at the site of the seismoscope. It took science a very long way until Jean de Hautefeuille led the European movement to invent a device that would measure earthquakes in 1703. Nicholas Cirrillo used pendulums to measure a series of earthquakes in Naples in the year 1731. James Forbes was probably the first scientist to attempt explicitly to give a seismological instrument a "long" period, something which was finally achieved by Emil Wiechert at the turn of the century, who invented probably the oldest type of seismograph still in use today (Dewey & Byerly, 1969). But the direction, intensity and duration of seismic waves had already been recorded by Luigi Palmieri in as early as 1856 (Bolt et al., 1995).
The forecast of earthquakes has always been highly desired. From the year 373 BC in Greece stories exist that describe the reaction of rats and centipedes that flee from a severe earthquake just hours before the event. Modern studies concentrate more on electrical resistivity of rocks, electrotelluric fields, electrochemical potential or electrical conductivity of water (Sidorin, 2002). Fossil quakes may help with the forecast if a region is earthquake prone, archaeoseismology plays an important role in this context. Tree ring analyses may result in the reconstruction of historical events, just as sediments in lakes might prove ground liquefaction (Bolt et al., 1995), and there are many more fields in which palaeoseismology may contribute historical and prehistorical records. Reicherter et al. (2009) sum up the current knowledge on earthquake ground effects for hazard assessment.

4. Earthquakes in mediaeval and modern Europe

Mediaeval earthquakes were seen as a continuation of ancient natural conditions and have found their way even into literary works of the Renaissance, like those by Shakespeare, who mentions earthquakes for example in Henry IV, Part I ("I say the earth did shake when I was born").

Ambraseys (2009) in his exhausting catalogue of earthquake events for the Mediterranean and the Middle East lists archaeological data, epigraphs and inscriptions as important sources for the analysis of past earthquake events. But he prefers the literary sources and states:

It is symptomatic of cultural changes since the First World War that, as instrumental, electronic or other mechanical reporting of events has grown, and news is increasingly disseminated by radio and television, a parallel decline is visible in both the volume and the quality of documentary and descriptive accounts of earthquakes in the twentieth century (Ambraseys, 2009).

Kozák & Ebel (1996), in addition to that, suggest not to forget pictorial sources when assessing earthquake data (confer fig. 3):

The depictions of historic earthquakes provide some macroseismic information for reevaluating the intensities of the portrayed events. Furthermore, the depictions may be used to infer other macroseismic information, such as ground acceleration levels, soil amplification or liquefaction, and the amount of tsunami damage. In some cases, an analysis of the depictions could indicate the need to reclassify the sizes or locations of some historic earthquakes (Kózak & Ebel, 1996).

For his catalogue Ambraseys writes about the European/Occidental sources that "for Classical, Roman and Byzantine times almost all the sources are well known, and they are relatively limited in number and mostly published" (Ambraseys, 2009). He writes that on top of that Arabic sources "have generally been identified and published", most of them being narrative histories. However, he claims, "little or no archival material survives from this early period". Ottoman sources are often connected to the cost for the repair or reconstruction of structures affected by the shock. Venetian sources, on the other hand, have a long period of observations, chiefly from coastal regions of the north-eastern part of the Mediterranean, thus adding reliable historical written material to earthquake catalogues.

During the second part of the twentieth century several scientists in developed countries realised that a new approach with respect to historical earthquakes had to be tried. In France, for example, it was Jean Vogt, who understood that historical seismicity needed a revision through a return *ad fontes* (cf. Vogt, 1979). He began to collect the original sources as far as available: periodicals, newspapers, administrative, notarial and family archives, and he did not stop short of libraries and archives of neighbouring countries. With regards to the need to identify smaller earthquake events, especially in regions with a low seismicity, Fréchet (2008) remarks:

Often, seismic catalogues concentrate only on the largest damaging earthquakes in a region, neglecting valuable information on foreshocks and aftershocks and on smaller events.

And he carries on by asking for more enthusiasm from the people involved in hazard mitigation also with regards to the financial aspect:

For each event, it is necessary to make exhaustive use of all existing catalogues in order to identify the least trace of earthquake, aftershock, and background seismicity. Once an event date and location is known approximately, it is usually straightforward to search for original descriptions in newspapers, periodicals, etc., for the last three centuries at last (Fréchet, 2008).

Fig. 3. Earthquake events in Rossanna and Constantinople, according to Hermann Gall, in the year 1556.

5. Criticism of the sources

But of course we are well aware of the fact that the quoting of "original" texts is often unprecise or even faulty. Many of the written sources have already been shortened, translated, paraphrased or compiled, thus distorting the information of the underlying primary sources. On top of that, some compilations may include mistakes like wrong datings. Sometimes compilations are a mixture of reliable and questionable data (e.g. Pfister, 1988; Glade et al., 2001 on that topic), as is especially the case with the "long" German catalogues of Central European earthquakes collected by Reindl (1903a, 1903b; Gießberger, 1922; Sieberg, 1940, see also Fig. 4).

An illustrative collection of problems when dealing with historical written sources is given by Kárník (1988):

> *In some cases storm effects, landslides or subsidences are reported as seismic phenomena. Another source of error is the wrong transcription of names of localities, or the case of some localities having identical or very similar names; as a consequence, earthquake epicentres have been moved to wrong places. 'New' earthquakes can originate simply by listing twice an event reported with the date given in different sources according to the Julian or the Gregorian calendar. Another source of similar manufacturing of earthquakes are errors in transcribing the dates, e.g. Jan.-June, VI-XI, etc. An opposite phenomenon may occur because of a long period of war, foreign occupation of a country, plague, or other reason for which the records either were not made or were destroyed, which results in an artificial interval of quiescence. It is imperative to work with the original reports as much as possible, but this is not easy because some old sources are not accessible to an investigator or have been destroyed or require special knowledge of language.*

The plausibility of the data can only be elicited by a thorough assessment of the data itself and a comparison of the data in question with established catalogues. In general, the co-operation of historians and geoscientists proves very valuable for the assessment of historical hazard data (Alexandre, 1990; Coeur et al., 1998), even if the majority of the historical documents only provide binominal data compared to the more valuable censored data of mostly younger documents (cf. Stedinger & Cohn, 1986).

As usual with historical data, several aspects regarding the quality of written sources need of course to be kept in mind (cf. e.g. Pfister & Hächler, 1991; Coeur et al., 1998; Fliri, 1998; Pfister, 1999; Glaser et al., 2002). The documents may differ greatly with regard to terminology, detail, educational background of the author etc. Many sources are not capable of providing all the data expected and needed for a thorough assessment of earthquakes. But the more different sources (or -later- archives) can be included, the more precise earthquake data will become overall. Excellent examples of the analysis of historical data are the publications of Galadini et al. (2001) about the Veronese earthquake of 1117, the work of Mucciarelli & Stucchi (2001) about disaster scenarios (or, for the general approach, e.g. Hammerl & Lenhardt, 1997 and Gisler et al., 2004). An up-to-date overview on German historical approaches is for example given by Grünthal (2004).

The past few hundred years have seen quite a large number of descriptive and parametric catalogues of historical earthquakes as a result of long-lasting archive work with written sources. Among the larger and exhaustive catalogues are the writings of Bonito (1691), Coronelli (1693), Seyfart (1756), whose catalogue is considered the first 'modern' catalogue (Fréchet 2008), Hoff (1840), and Milne (1911), as Ambraseys (2009) points out. In France for example major catalogue projects were undertaken from the 19th c. onwards, e.g. by Alexis Perrey (cf. Perrey, 1841), later came the global catalogue of Ferdinand Montessus de Ballore, completed in 1907 (Montessus de Ballore, 1904-1907), a process which is typical for most of the developed countries in Europe.

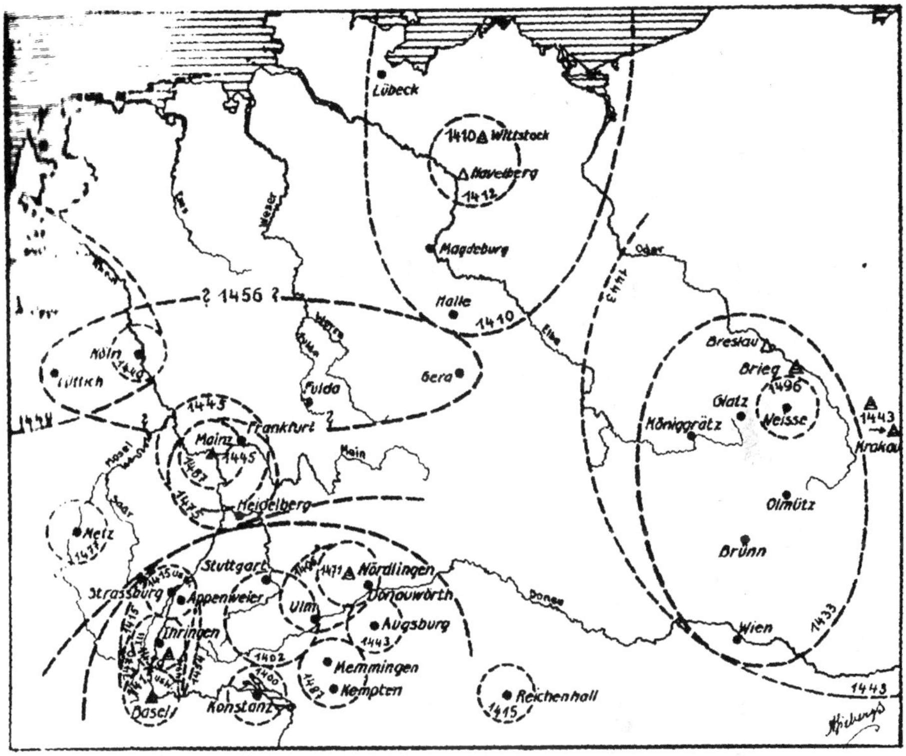

Fig. 4. Earthquakes in central Europe in the 15th century, according to Sieberg (1940).

6. Early modern attempts to collect written information on quake events, the case of Bavaria

Earthquakes which can be perceived in Bavaria are, in most cases, not "Bavarian" earthquakes *sensu stricto*. "Real" Bavarian earthquakes have usually been less severe so far, stronger events took place in the neighbouring countries Switzerland or Austria, sometimes also in Italy. Some of the earthquake events perceived in Northern Bavaria have occurred in what is today the Czech Republic. Kárník (1988) remarks that "the important role of historical data is most evident in regions of medium or low seismicity", just as is the case in Bavaria. In order to find out about the epicentres of historic earthquakes, the simplest way is to screen and assess historical documents, especially written documents.

Since a thorough assessment of historical documents (the value of which has been proven many times, foremost in the field of hydrology, cf. Stedinger & Cohn, 1986; Alexandre, 1987; Baker, 1987; to name only a few of the eminent publications on that topic) filed in the numerous archives all over the state of Bavaria would have proven too time-consuming and costly by far (cf. Coeur et al., 1998; Barnikel & Becht, 2004), a new way had to be found to gather as much relevant information as possible in a shorter period of time. Especially local archives, which have found to be among the most productive and important sources for documents describing hazardous natural events from the 1800s and older (cf. Barnikel & Becht, 2003) cannot be screened by natural scientists alone. Eventually, people with access to local documents, like librarians or historians, need to be encouraged to contribute to the growing data base as a kind of open-source catalogue.

7. The BASE-project as example for historical analysis

The old Bavarian earthquake catalogue lists four events from the 14th until the 17th century, the earliest one being the 1390 event in Bad Reichenhall. BASE adds 232 events from the 4th century until the 17th century, the oldest one being a (questionable) earthquake recorded in Memmingen in 369 AD. The past few centuries are of course much better represented in the data base than the older ones. The renaissance of natural sciences after the 1500s led to a more profound occupation with natural hazards. More people noted earthquakes and reported them. As a not surprising result we have more detailed information about quakes for the past few centuries than for the time before 1500. More than half of all events filed in the BASE-catalogue date back to the 18th and 19th centuries, whereas the 10th and 15th centuries are, astonishingly enough, only sparsely represented (see Kárník's remark above).
A more problematic part of the assessment was the inclusion of earthquakes mentioned only in maps of important publications (especially Sieberg, 1940, whose compilation is problematic enough - see also Fig. 4). Those maps are in general quite speculative and only in very few cases specific. But in order to get a complete picture of the seismic situation in Bavaria it was necessary to include events which were shown as relevant for Bavaria in these maps. This resulted in the inclusion of at least 232 different earthquakes (project BASE II).
Thus, in a first step all existing data about earthquakes felt in Bavaria needed to be collected. This step was limited to already published data in 27 crucial publications over the past two centuries (Perrey, 1844; Boegner, 1847; Volger, 1857; Credner, 1884; Gümbel, 1889; Langenbeck, 1892, Günther, 1897; Günther n.d.; Gümbel, 1898; Brunhuber, 1903; Reindl, 1903a, 1903b; Günther & Reindl, 1904; Reindl, 1905a, 1905b; Credner, 1907; Reindl, 1907; Heritsch, 1908; Messerschmitt, 1907; Gießberger, 1922; Sieberg, 1940; Sponheuer, 1952; Schmedes, 1979; Leydecker & Brüning, 1988; Wolf & Wolf, 1989; Bachmann & Schmedes, 1993; Schmedes et al., 1993). The data were filed in a specially designed data base (Fig. 5), which was modified from the one successfully used in the HANG-project about natural hazards in the Alps (Barnikel, 2004). A second, future step will be the inclusion of (validated) contributions made by citizens all over Bavaria who share an interest in earthquakes and have access to local publications or documents which may have been denied wider distribution in journals or other scientific publications and are, therefore, largely unknown.
Germany's standard earthquake catalogue was first published by Leydecker (1986) in the year 1986 (in parts based on the catalogue published by Grünthal, for example in 1988) and issued on behalf of the German Federal Institute for Geosciences and Natural Resources. It claimed full coverage for all quakes from intensity MSK-1964 IV (Sponheuer, 1965) upwards and listed about 1900 events with either their epicentres in Germany or with macroseismic

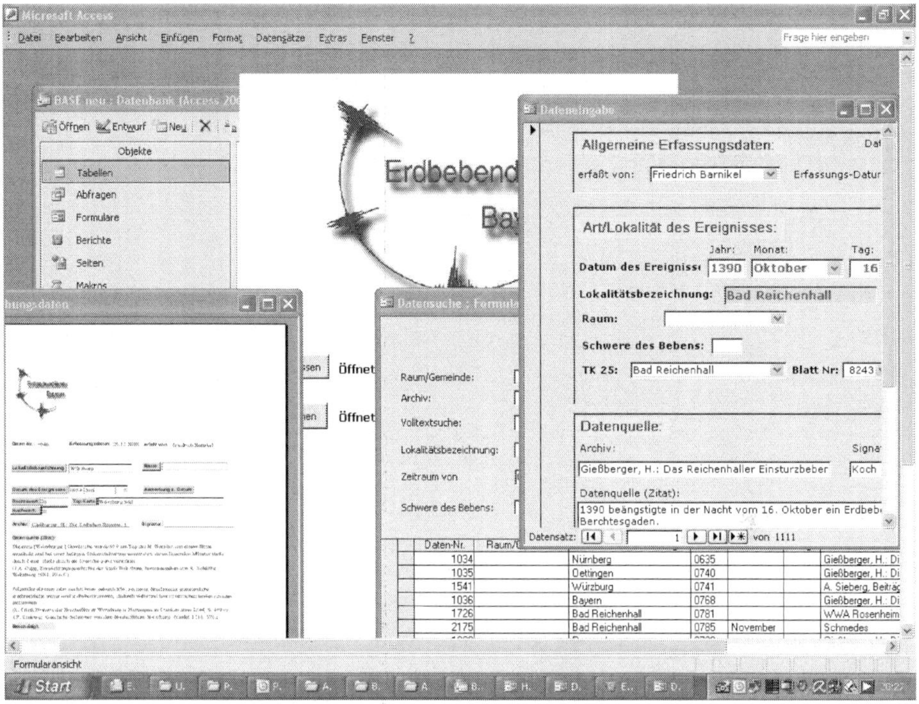

Fig. 5. BASE data bank, screenshot.

effects on Germany, but originating in neighbouring countries. Today it lists more than 2500 events relevant for Germany before 1906. Leydecker not only shaped the German earthquake catalogue, he also, together with his late colleague van Gils, issued the European earthquake catalogue with more than 8500 events (intensity IV and more) for the covered countries until 1906 (van Gils & Leydecker 1991). The latest addition of interest is the earthquake catalogue for southeastern Europe (Shebalin et al., 1998) with approximately 2000 events before 1906. Other important catalogues for earthquakes relevant for Bavaria are the Swiss catalogue (Swiss Seismological Service, 2002), the Austrian catalogue (Austrian Central Institute for Meteorology and Geodynamics, 2006) and the latest catalogue for central and northern Europe by Grünthal & Wahlström (2003). Between all catalogues slight discrepancies can be found, most of them regarding the precise location of the epicentre, its coordinates, the intensity and the corresponding radius. Even the Bavarian earthquake catalogue, although based on the German catalogue, shows some minor differences to the German catalogue.

Consequently all existing earthquake catalogues with possible relevance for Bavaria had to be screened in order to find more earthquake events to be included in the BASE data bank (Grünthal, 1988; Leydecker, 1986ff; Swiss Seismological Service, 2002; Shebalin et al., 1998; Austrian Central Institute for Meteorology and Geodynamics, 2006) and the Kövesligethy formula (see Equation 1) was used to determine the intensity of earthquakes from neighbouring countries. As a result all earthquakes with a calculated intensity of $I \geq 3$ (with $\alpha = 0{,}001$) for Bavaria were included.

$$I = Io - 3 \cdot \log(R/h) - 1.3 \cdot \alpha \cdot (R-h) \quad (1)$$

8. The application of early seismometers

In Bavaria the recording and scientific assessment of earthquakes is principally undertaken by the Department for Geo- and Environmental Sciences at the Ludwig-Maximilians-University of Munich and the Bavarian Environment Agency (together forming the Bavarian Seismological Service). Instrumental recording of earthquakes in Bavaria began with the 1000kg Wiechert seismometer in Munich-Bogenhausen in 1905. Today the Geophysical Observatory in Fürstenfeldbruck is the data centre for a modern digital seismological network in Bavaria.

An interesting finding when comparing pre-instrumental data with the seismographs from Bogenhausen is the difference of the geographical distribution of the earthquakes before 1905 and after. Before 1905 most earthquakes felt in Bavaria took place in either the Alps or the northern fringe of the Bohemian Forest. But a surprisingly high number of events happened all over central Europe with no apparent connection to the more active seismic regions. After 1905 we find quite a different picture, which obviously is connected to the installation of seismological stations in Bavaria, that were able to record also smaller earthquakes, "invisible" in the past. A large number of events was recorded for the Ries crater around Nördlingen, although most epicentres are still found in the Alps or the northern Bohemian Forest (cf. Barnikel & Geiss, 2008, who compare the two periods).

But even the introduction of seismometers does not mean highly precise earthquake data from that moment onwards.

Batlló (2008) points out that

> *in the early XX century, fundamental concepts of seismic source physics [...] were yet to be discovered, as well as the benefits of computer technology and digital signal processing.*

9. An open source project to collect earthquake information

In order to complete the existing data base of historical earthquakes in Bavaria or relevant for Bavaria, the Bavarian Environment Agency has teamed up with the Department of Geography at the University of Munich, which offers expertise in historical assessment of natural hazards (see Barnikel & Becht, 2004). The project BASE (Bavaria's Seismicity in Historical Documents), which is currently running in its ninth year, works with historical written documents and all information about earthquakes relevant for Bavaria stored in them. The first two parts of the project (BASE I and II) dealt with the inclusion of data from already published literature, the current parts (BASE-NET and BASE20) are aiming at the inclusion of earthquake data from original written documents (letters, postcards, etc.) and the setting up of an internet website to enable interested and informed citizens of Bavaria to contribute to the data base in the future.

Of special importance in this context, when working with historical written sources, is always the exact quotation of the written texts to ensure accessibility for later evaluations. BASE links the data on the event to other researched material, like pictures or maps. The data was then compared with the data in the German and European catalogues and, if necessary, adjusted (with regards to time, coordinates etc.). Important for the data files are descriptions of the impact an earthquake had on the society. The more details about damages or destructions we get, the more accurate our intensity estimations will later become. Therefore, all the data have been graded after the new EMS-98-Intensity-Scale (Grünthal, 1998), which allows a direct comparison of data from different centuries and of data from different documentary sources.

Some examples of historical written data may illustrate the particular problems and chances of working with this kind of sources in the BASE-project: "In the year 740 AD the Earth trembled so much in the Swabian Countries for almost a year that many monasteries and churches collapsed" (after Gießberger, 1922). This text was found in a manuscript from 1723, therefore at least a second-hand source. Even if authors often used to simply copy older texts without changing them, a description of an earthquake almost a thousand years later is highly problematic (see Barnikel, 2004, who deals at length with this topic). The nature of the information is also doubtful. The destruction of numerous churches and monasteries (some of them certainly wooden structures as were common in these times) sounds improbable. In addition to that, no other known source recorded that many devastating earthquakes in that year in Bavaria. The validity of this source, as a result, remains very poor.

Several sources were found for a suspected earthquake in 841 AD in Würzburg (after Boegner, 1847; Gießberger, 1922; Sieberg, 1940; the mentioned sources date back to e.g. 1578, 1644, 1692 and 1756 respectively): "AD 841 an earthquake hit the town of Würzburg about twenty times and with it came terrible hail and a great storm", depicting the ancient belief that earthquakes were connected to atmospheric events. Judging from the text the event in question sounds more like a heavy storm which shook the buildings and inhabitants of Würzburg and is nowadays considered to be a fake. These "original" sources, used in the secondary sources exploited by the BASE-project, are of course hardly precise and valid, since they cannot be considered as contemporary (real contemporary sources being very rare and far between). Often sources like these make use of data already written down in older documents, thus only quoting other sources and thereby adding "new" evidence where there is none (cf. the exemplary analysis of a series of earthquakes in southern Germany found to be fakes by Grünthal & Fischer, 2001), or just mix up the dates, as is the case with earthquakes listed after either the Julian or the Gregorian calendar (as for example Grünthal & Wahlström (2003) point out; also see chapter 5).

Much more precise and valid in general are understandably younger documents, one example being an earthquake in 1889: "On February 22nd, 2 o'clock and 40 minutes in the afternoon, a heavy blow sounded, which was accompanied by a short rolling sub-surface, so that the windows clattered in many houses. This blow seemed to move from W to E. In the lower part of town the same was felt so heavily that the inhabitants of the surrounding streets ran terrified onto the streets." This information appears not only much more reliable, it is also very precise in terms of date, time, place, process and effect. It is especially this kind of source we need to specify an earthquake.

The BASE-project so far was able to collect a total of 516 events which were perceived and, consequently, recorded in Bavaria. Astonishing enough is the fact that about 76% of these events could not be linked to a specific date or place from the earthquake catalogues of the surrounding areas (the European earthquake catalogue by van Gils & Leydecker (1991) or the German earthquake catalogue by Leydecker (1986ff.), although quite a few events must

be considered fakes, especially when just mentioned by a few (or only one) sources. Only 122 earthquake epicentres in addition to the 24 events already listed in the Bavarian earthquake catalogue could be specified. Most of these events took place in either Switzerland (23%) or Austria (16%), just under 10% in the German state of Baden-Württemberg, the Czech Republic or Italy respectively. The large number of uncertain epicentres for earthquakes felt in Bavaria is nevertheless puzzling, but so is the fact that both, the German and the European catalogue, list a significant different number of German quakes with the same intensity span for the time period up to 1905. The European catalogue mentions 1019 events in Germany before 1906, the German catalogue 1821.
In total a number of 1673 references to earthquakes in Bavaria for the time up to 1905 have been collected. For these references date, time, location, quotation, details about the earthquake itself, damages and other crucial information are listed in the data base. The exceptions are of course those earthquakes which have been extracted from maps, where no further information from within Bavaria could be found. These earthquakes are attributed to their origin outside Bavaria and are listed in the data base under the names of Bavarian cities and towns on the maps.

10. Outlook from the BASE-Project

As a result, data from the BASE-project are useful with regard to several aspects (cf. Barnikel et al. 2009):

- The data is compiled in a catalogue which in the future can be accessed by every user via internet. It is not only a valuable tool for specialists, it also helps the public understand geodynamics better.
- The catalogue can be enlarged and improved by citizens, thus including more man-power in scientific research and showing the public that everybody can contribute to the betterment of science and society, as an open-source project it will be a kind of Web 2.0-try to link science and public (project BASE-NET).
- This step will produce new pieces of information from areas which have not produced earthquake reports so far. Hopefully this will help to judge the tentative records only based on maps or mathematic calculations so far.
- The data will help us to calibrate existing catalogues, especially the Bavarian catalogue, and should serve as a model for the calibration of other existing catalogues.
- The list of known earthquakes will be prolonged significantly and so provide a basis for future risk modelling.

Another step, as mentioned above, will be the inclusion of (validated) contributions made by citizens all over Bavaria who share an interest in earthquakes and have access to local publications or documents which may have been denied wider distribution in journals or other scientific publications and are, therefore, largely unknown. Of course a *caveat* must be kept in mind, namely that "the tendency of some chroniclers to attract attention by exaggerating of manufacturing information is known" (Kárník, 1988).
The figure of unknown earthquake-epicentres remains a problem that needs to be solved. The inclusion of data from local archives is, consequently, of utmost importance. The future presentation of the BASE-DB on the web (www.erdbeben-in-bayern.de) is an important step to reach this goal and serves as an example for other catalogues. The inclusion of the original text sources is crucial in this respect, because it allows later adjustment and

validation. But a future assimilation of the existing catalogues from the different European countries will also be important. The BASE-project collects, in addition to that, written data for events after 1905 in a second step. A comparison between the written sources and instrumental data from the seismometre may prove useful for the calibration of older written data. Another step should be an examination, how reliability scales (cf. Papadopoulos et al., 2000) could be used to classify the reliability of the historical earthquakes collected by BASE and the implementation of indicative magnitude values (cf. Sibol et al., 1987; Cavallini & Rebez, 1996; Papazachos & Papaioannou, 1997) could be valuable. In the end the catalogue should then be ready to be used in further scientific studies about earthquakes in southern Germany.

In general, earthquake catalogues need to be screened for double entries and uncertain data. In Europe a lot of promising steps have already been undertaken in that direction. Further historical studies can enormously contribute to a more precise analysis of the seismic situation of a region. The cooperation with historians might prove useful as is the microfilming and scanning of historical seismograms worldwide (cf. Lee & Benson, 2008). Especially since the teleseismic instrumental recording dates back to 1889, as Stein et al. (1988) point out. And the expansion of the now existing Global Digital Network of seismic stations will support a much more reliable assessment of earthquake events in the future (cf. Beck, 1996). Especially in regions with low seismicity a data sample covering at least several centuries is highly recommended for the understanding of earthquake generating processes, even if "historical information is normally not suitable for statistical processing because its homogeneity can be rarely guaranteed" (see Kárník, 1988). A result can be a seismic risk map including all sorts of entries, as the impressive work of Tyagunov et al. (2006) shows, and catalogues that are consistent, complete and, in the best of cases, merged across political boundaries (Bayliss & Burton, 2007).

11. Acknowledgment

The authors want to thank the Bavarian Environment Agency for funding the BASE-project.

12. References

Alexandre, P. (1987). Le Climat en Europe au Moyen Age, In: *Recherches d'histoire et de sciences sociales,* Vol. 24, ISSN 0249-5619

Alexandre, P. (1990). *Les séismes en Europe occidentale de 394 à 1259 – Nouveau catalogue critique.* Communications Observatoire Royal de Belgique, Série Géophysique, Brussels, Belgium

Ambraseys, N. (2009). *Earthquakes in the Mediterranean and Middle East,* Cambridge University Press, ISBN 978 0 521 87292 8, Cambridge, UK

Austrian Central Institute for Meteorology and Geodynamics (Ed.) (2006). *Katalog des Österreichischen Erdbebendienstes der ZAMG,* Avaiable from http://www.zamg.ac.at

Bachmann, C. & Schmedes, E. (1993). Ein Schadensbeben in Neuhausen, Landkreis Landshut am 7. Februar 1822 - eine Zeitungsente, In: Z. *angew. Geol.,* Vol.39, No.2, (1993), pp. 106-107, ISSN 0044-2259

Baker, V.R. (1987). Paleoflood hydrology and extraordinary flood events, In: *Journal of Hydrology,* Vol.96, pp. 79-99, ISSN 0022-1694

Barnikel, F. & Becht, M. (2003). A historical analysis of hazardous events in the Alps - the case of Hindelang (Bavaria, Germany). In: *Natural Hazards and Earth System Sciences,* Vol.3., No.6, pp. 625-635, ISSN 1561-8633

Barnikel, F. (2004). Analyse von Naturgefahren im Alpenraum anhand historischer Quellen am Beispiel der Untersuchungsgebiete Hindelang und Tegernseer Tal, Bayern, In: *Göttinger Geographische Abhandlungen,* Vol. 111, Göttingen, Germany, 210 pp., ISSN 0341-3780

Barnikel, F. & Becht, M. (2004). Möglichkeiten einer Bewertung des regionalen Gefährdungspotentials auf der Basis historischer Naturgefahrenforschungen. In: Geomorphologische und hydrologische Naturgefahren in Mitteleuropa, Becht, M. & Damm, B. [Eds.] *Zeitschrift für Geomorphologie,* N.F., Supplement Vol. 135, pp.1-10, ISSN 0372-8854

Barnikel, F. & Geiss, E. (2008). The BASE-Project - An open-source catalogue for earthquakes in Bavaria, Germany. In: *Natural Hazards and Earth System Sciences,* Vol.8, No.6, pp. 1395-1401, ISSN 1561-8633

Barnikel, F.; Vetter, M.; Geiss, E.; Frank, C. & Witossek, G. (2009). Bavaria's Seismicity in Historical Documents, Extended Abstracts, *6th European Congress on Regional Geoscientific Cartography and Information Systems (München),* pp. 312-315, München, Germany, June 9-12, 2009

Battló, J.; Stich, D. & Macià, R. (2008). Quantitative Analysis of Early Seismograph Recordings. In: *Historical Seismology (modern Approaches in Solid Earth Sciences),* Fréchet, J., M. Meghraoui & M. Stucchi (Eds.), 385-402, Springer, ISBN 978-1-4020-8221-4, Heidelberg, Germany

Bayliss, T.J. & Burton, P.W. (2007). A New Earthquake Catalogue for Bulgaria and the Conterminous Balkan High hazard Region, In: *Natural Hazards and Earth System Sciences,* Vol.7, pp. 345-359, ISSN 1561-8633

Beck, S. (1996). Introduction to the Topical Section on Historic Earthquakes, *Pure and Applied Geophysics,* Vol.146, No.1, pp. 3-4, ISSN 0033-4553

Boegner, J. (1847). *Das Erdbeben und seine Erscheinungen - Nebst einer chronologischen Uebersicht der Erderschütterungen im mittleren Deutschland vom 8. Jahrhundert bis auf die neueste Zeit und ihres Zusammenhanges mit vulkanischen Erscheinungen in entfernten Ländern,* Frankfurt a.M., Germany

Bolt, B.A.; Horn, W.L.; Macdonald, G.A. & Scott, R.F. (1975): *Geological Hazards,* Cambridge University Press, ISBN 3-540-06948-8, Cambridge, UK

Bonito, M. (1691). *Terra tremante,* Naples, Italy

Brunhuber, A. (1903). Zwei Erdbeben im Gebiete der Oberpfalz, *Berichte des naturwissenschaftlichen Vereins zu Regensburg,* Vol.9, No.2, pp. 1-14

Cavallini, F. & Rebez, A. (1996). Representing earthquake intensity-magnitude relationship with a nonlinear function, *Bulletin of the Seismological Society of America,* Vol.86, No.1A, pp. 73-78, ISSN 0037-1106

Coeur, D.; Lang, M.; Naulet, R.; Burnet, R. & Strazzeri, D. (1998). Histoire et connaissance des phénomenes naturels extremes. *Ingénieres-EAT,* pp. 15-26

Coronelli, P. (1693). Epitome cosmografica: delle cause de' tremuoti, Cologne, Germany

Credner, H. (1884). Die erzgebirgisch-vogtländischen Erdbeben während der Jahre 1878 bis Anfang 1884. *Zeitschrift für Naturwissenschaften,* Vol.57, pp. 1-29

Credner, H. (1907). Die Sächsischen Erdbeben während der Jahre 1904 bis 1906, Aus den Berichten der mathematisch-physischen Klasse der Königlich Sächsischen Gesellschaft der Wissenschaften zu Leipzig, Vol.59, pp. 333-355

Dewey, J. & Byerly, P. (1969). Earthquake Monitoring. *Bulletin of the Seismological Society of America,* Vol.59, No.1, pp 183-227, ISSN 0037-1106

Eiby, G.A. (1988). Documenting New Zealand Earthquakes. *Historical Seismograms and Earthquakes of the World,* Lee, W.H.K.; H. Meyers & Shimazaki, K. (Eds.), 232-240, ISBN 0-12-440870-2, San Diego, USA

Fliri, F. (1998). *Naturchronik von Tirol,* Wagner, ISBN 3-7030-0313-8, Innsbruck, Austria

Fréchet, J. (2008). Past and Future of Historical Seismicity Studies in France. In: *Historical Seismology (modern Approaches in Solid Earth Sciences),* Fréchet, J., M. Meghraoui & M. Stucchi (Eds.), 131-145, Springer, ISBN 978-1-4020-8221-4, Heidelberg, Germany

Galadini, F.; Galli, P.; Molin, D. & Ciurletti, G. (2001). Searching for the source of the 1117 earthquake in northern Italy: A multidisciplinary approach. In: *The Use of Historical Data in Natural Hazard Assessments, Advances in Natural and Technological Hazards Research,* Glade, T., Albini, P. and Francés, F. (Eds.), 3-27, Kluwer, ISBN 0792371542, Dordrecht, Netherlands

Gießberger, H. (1922). Die Erdbeben Bayerns I. Teil. *Abhandlungen der Bayer. Akad. der Wissenschaften Mathematisch-Physikalische Klasse,* Vol.29, No.6

Gisler, M.; Fäh, D. & Kästli, P. (2004). Historical Seismicity in Central Switzerland. *Eclogae geo. Helv.,*Vol. 97, pp. 221-236, ISSN 0112-9402

Glade, T., Albini, P. & Francés, F. (Eds.) (2001). *The Use of Historical Data in Natural Hazard Assessments. Advances in Natural and Technological Hazards Research,* Kluwer, ISBN 0792371542, Dordrecht, Netherlands

Glaser, R.; Jacobeit, J.; Deutsch, M. & Stangl, H. (2002). Hochwasser als historisches Phänomen. *Rundgespräche der Kommission für Ökologie,* Vol. 24, pp. 15-30, ISSN 0938-5851

Grünthal, G. (1988). Erdbebenkatalog des Territoriums der Deutschen Demokratischen Republik und angrenzender Gebiete von 823 bis 1984, *Veröffentlichungen des Zentralinstituts für Physik der Erde,* Vol. 99, ISSN 0514-8790

Grünthal, G. (Ed.) (1998). European Macroseismic Scale 1998. *Cahiers du Centre Européen du Géodynamique et de Séismologie,* Vol.15, ISSN 1340-4202

Grünthal, G. & Fischer, J. (2001). Eine Serie irrtümlicher Schadenbeben im Gebiet zwischen Nördlingen und Neuburg an der Donau vom 15. bis zum 18. Jahrhundert. *Mainzer naturwissenschaftliches Archiv* Vol. 39, pp. 15-32, ISSN 0542-1535

Grünthal, G. & Wahlström, R. (2003). An earthquake catalogue for central, northern and northwestern Europe based on MW magnitudes, Avaible from http://www-app1.gfz-potsdam.de/catalog/index.html

Grünthal, G. (2004). The history of historical earthquake research in Germany. *Annals of Geophysics,* Vol. 47, No. 2/3, pp. 631-643, ISSN 1593-5213

Gümbel, C.W. (1889). Das Erdbeben vom 22. Februar 1889 in der Umgegend von Neuburg a.D. *Aus den Sitzungsberichten der mathematisch-physikalischen Classe der k. bayer. Akad. der Wiss.,* Vol. 19, No. 1, pp. 79-108

Gümbel, C.W. (1898): Ueber die in den letzten Jahren in Bayern wahrgenommenen Erdbeben. *Aus den Sitzungsberichten der mathematisch-physikalischen Classe der k. bayer. Akad. der Wiss.*, Vol. 28., pp. 3-18

Günther, S. (1897). Das bayerisch-böhmische Erdbeben vom Jahre 1329, *Jahresbericht der Geographischen Gesellschaft*, Vol. 17, pp. 76-88

Günther, S. (n. d.) : Münchener Erdbeben- und Prodigienlitteratur in älterer Zeit, *Jahrbuch für Münchener Geschichte*, Vol. 4, No. 15, pp. 234-256

Günther, S. & Reindl, J. (1904). Seismologische Untersuchungen. *Aus den Sitzungsberichten der mathematisch-physikalischen Classe der k. bayer. Akad. der Wiss.*, Vol. 33, No. 4, pp. 631-671

Hammerl, C. & Lenhardt, W. (1997). *Erdbeben in Österreich*, Leykam Verlag, ISBN 3701173346, Wien, Austria

Heritsch, F. (1908). Über das Mürztaler Erdbeben vom 1. Mai 1885, *Mitteilungen der Erdbeben-Komission der Kaiserlichen Akademie der Wissenschaften in Wien*, Vol. 32

Hoff, K.E. von (1840). Geschichte der natürlichen Veränderungen der Erdoberfläche, Chronik der Erdbeben und Vulkanausbrüche, Gotha, Germany

Kárník, V. (1988). Compilation and Processing of Historical Data: Summary Remarks. In: *Historical Seismograms and Earthquakes of the World*, Lee, W.H.K.; H. Meyers & Shimazaki, K. (Eds.), 139-141, ISBN 0-12-440870-2, San Diego, USA

Kozák, J. & Ebel, J.E. (1996). Macroseismic Information from Historical Pictorial Sources, *Pure and Applied Geophysics*, Vol. 146, No. 1, pp. 103-111, ISSN 0033-4553

Langenbeck, R. (1892). Die Erdbebenerscheinungen in der oberrheinischen Tiefebene und ihrer Umgebung. In: *Geographische Abhandlungen aus den Reichslanden Elsass-Lothringen*, Gerland, G. (Ed.), Stuttgart, pp. 1-120, 1892.

Lee, W.H.K. & Benson, R.B. (2008). Making Non-Digitally-Recorded Seismograms Accessible Online for Studying Earthquakes. In: *Historical Seismology (modern Approaches in Solid Earth Sciences)*, Fréchet, J., M. Meghraoui & M. Stucchi (Eds.), 402-412, Springer, ISBN 978-1-4020-8221-4, Heidelberg, Germany

Leydecker, G., (1986). Erdbebenkatalog für die Bundesrepublik Deutschland mit Randgebieten für die Jahre 1000-1981, *Geol. Jb.*, Vol. E 36, 83 pp., ISSN 0016-7800, avaible from http://www.bgr.bund.de

Leydecker, G. & Brüning, H.J. (1988). Ein vermeintliches Schadenbeben im Jahre 1046 im Raum Höxter und Holzminden in Norddeutschland – Über die Notwendigkeit des Studiums der Quellen historischer Erdbeben, *Geol. Jb.*, Vol. E 42, pp. 119-125, ISSN 0016-7800

Messerschmitt, J.B. (1907). Die Erdbeben in Bayern 1905 bis 1907, *Die Erdbebenwarte*, Vol. 4, pp. 94-96

Milne, J. (1911). Catalogue of Destructive Earthquakes AD 7 to AD 1899, *British Association for the Advancement of Science, Seismology Reports*, Vol. 80, pp. 694-740

Montessus de Ballore, F. de (1904-1907). Ephémérides sismiques et volcaniques, *Ciel et Terre*, pp. 24-28

Mucciarelli, M. & Stucchi, M. (2001). Expeditious seismic damage scenarios based on intensity data from historical earthquakes. In: *The Use of Historical Data in Natural Hazard Assessments, Advances in Natural and Technological Hazards Research*, Glade, T., Albini, P. and Francés, F. (Eds.), 3-27, Kluwer, ISBN 0792371542, Dordrecht, Netherlands

Papadopoulos, G.; Vassilopoulou, A & Plessa, A. (2000). Historical Earthquakes and Tsunamis in the Corinth Rift, Central Greece, *Institute of Geodynamics - National Observatory of Athens*, Vol. 12, pp. 9-120

Papazachos, C. & Papaioannou, Ch. (1997). The macroseismic field of the Balkan area, *Journal of Seismology*, Vol. 1, ISSN 1735-1669

Perrey, A. (1841). Tremblements de terre dans les différentes siècles et aux différentes époques de l'année, *Comptes Rendus hebdomadaires des séances de l'Académie des Sciences*, Vol. 12, pp. 1185-1187

Perrey, A. (1844). *Mémoire sur les Tremblements de Terre resentis en France, en Belgique et en Hollande,* Bruxelles, Belgium

Pfister, C. (1988). *Das Klima der Schweiz von 1525-1860 und seine Bedeutung in der Geschichte von Bevölkerung und Landwirtschaft.* Academia Helvetica 6, Haupt, ISBN 3-258-03956-9, Bern, Switzerland

Pfister, C. & Hächler, S. (1991). Überschwemmungskatastrophen im Schweizer Alpenraum seit dem Spätmittelalter, *Würzburger Geographische Arbeiten*, Vol. 80, pp. 127-148, ISSN 0510-9833

Pfister, C. (1999). *Wetternachhersage. 500 Jahre Klimavariationen und Naturkatastrophen*, Haupt, ISBN 3-258-05696-X, Bern, Switzerland

Prachuab, S. (1988). Historical Earthquakes of Thailand, Burma, and Indochina. In: *Historical Seismograms and Earthquakes of the World*, Lee, W.H.K.; H. Meyers & Shimazaki, K. (Eds.), 253-266, ISBN 0-12-440870-2, San Diego, USA

Reicherter, K.; Michetti, A.M. & Silva, P.G. (Eds.) (2009). Palaeoseismology: Historical and Prehistorical Records of Earthquake Ground Effects for Seismic Hazard Assessment, *Geological Society Special Publication*, Vol. 316, ISSN 0305-8719

Reindl, J. (1903a). Beiträge zur Erdbebenkunde von Bayern, *Aus den Sitzungsberichten der mathematisch-physikalischen Klasse der Kgl. Bayer. Akademie der Wissenschaften*, Vol. 33, No. 1, pp. 171-203

Reindl, J. (1903b). Die Erdbeben der geschichtlichen Zeit im Königreiche Bayern, *Die Erdbebenwarte*, Vol. 2, pp. 235-243

Reindl, J. (1905a). Die Erdbeben Nordbayerns, *Abhandlungen d. Naturh. Gesellsch.*, Vol. 15, No. 3., pp. 3-46

Reindl, J. (1905b). Ergänzungen und Nachträge zu v. Gümbels Erdbebenkatalog, *Aus den Sitzungsberichten der mathematisch-physikalischen Klasse der Kgl. Bayer. Akademie der Wissenschaften*, Vol. 35, No. 1, pp. 31-68

Reindl, J. (1907). Die Erdbeben Bayerns im Jahre 1904, *Die Erdbebenwarte*, Vol. 4, pp. 178-184

Schmedes, E. (1979). Die seismische Aktivität im Raum Bad Reichenhall, *Geol. Jb.*, Vol. C 22, pp. 91-101, ISSN 0016-7800

Schmedes, E.; Loibl, R. & Gebrande, H. (1993). Ein Schadensbeben in Regensburg am 8. Februar 1062 - eine Fehlinterpretation historischer Quellen, *Z. angew. Geol.*, Vol. 39, No. 2, pp. 103-105, ISSN 0044-2259

Seyfart, F. (1756). *Allgemeine Geschichte der Erdbeben.* Frankfurt, Germany

Shebalin, N.V.; Leydecker, G.; Mokrushina, N.G.; Tatevossian, R.E.; Erteleva, O.O. & Vassiliev, V.YU. (1998). Earthquake Catalogue for Central and Southeastern Europe 342 BC - 1990 AD, Final Report to Contract ETNU - CT 93 – 0087. Avaible from http://www.bgr.bund.de

Sibol, M.; Bollinger, G. & Birch, J. (1987). Estimation of magnitudes in central and eastern North America using intensity and felt area, *Bulletin of the Seismological Society of America*, Vol. 77, No. 5, pp. 1635-1654, ISSN 0037-1106

Sidorin, A Ya. (2002). Search for Earthquake Precursors in Multidisciplinary Data Monitoring of Geophysical and Biological Parameters, *Natural Hazards and Earth System Sciences*, Vol. 3, pp. 153-158, ISSN 1561-8633

Sieberg, A., (1940). Beiträge zum Erdbebenkatalog Deutschlands und angrenzender Gebiete für die Jahre 58 bis 1799, *Mitteilungen des Deutschen Reichs-Erdbebendienstes*, Vol. 2, 111 pp.

Sponheuer, W. (1952). Erdbebenkatalog Deutschlands und der angrenzenden Gebiete für die Jahre 1800 bis 1899, Deutsche Akademie der Wissenschaften, *Mitteilungen des Deutschen Erdbebendienstes*, Vol. 3

Sponheuer, W. (1965). Bericht über die Weiterentwicklung der seismischen Skala, *Veröffentlichungen des Institutes für Geodynamik*, Vol. 8, ISSN 0448-942X

Srivastava, H.N. & Das, S.K. (1988). Historical Seismicity and Earthquake Catalogues for the Indian Region. *Historical Seismograms and Earthquakes of the World*, Lee, W.H.K.; H. Meyers & Shimazaki, K. (Eds.), 335-348, ISBN 0-12-440870-2, San Diego, USA

Stedinger, J.R. & Cohn, T.A. (1986). Flood frequency analysis with historical and paleoflood information, *Water Resources Research*, Vol. 22, No. 5, pp. 785-793, ISSN 0043-1397

Stein, S.; Okal, E.A. & Wiens, D.A. (1988): Application of Modern Techniques to Analysis of Historical Earthquakes. *Historical Seismograms and Earthquakes of the World*, Lee, W.H.K.; H. Meyers & Shimazaki, K. (Eds.), 85-104, ISBN 0-12-440870-2, San Diego, USA

Swiss Seismological Service (Ed.) (2002). ECOS - Earthquake Catalog of Switzerland, ECOS Report to PEGASOS, Version 31.3.2002, avaible from: http://www.seismo.ethz.ch

Tyagunov, S.; Grünthal, G.; Wahlström, R.; Stempniewski, L. & Zschau, J. (2006). Seismic Risk Mapping for Germany, *Natural Hazards and Earth System Sciences*, Vol. 6, pp. 573-586, ISSN 1561-8633

Van Gils, J.M. & Leydecker, G., 1991. Catalogue of European earthquakes with intensities higher than 4, Bundesanstalt für Geowissenschaften und Rohstoffe BGR, avaible from: http://www.bgr.bund.de

Vogt, J. (1979). Les tremblements de terre en France, *Mémoire du Bureau de Recherches Géologiques et Minières*, Vol. 96, ISSN 0071-8246

Volger, O. (1857). *Untersuchungen über das Phänomen der Erdbeben in der Schweiz, Erster Theil: Chronik der Erdbeben in der Schweiz*, Gotha, Germany

Wolf, P. & Wolf, H. (1989). Das Erdbeben in Regensburg von 1062 – Wirklichkeit oder wissenschaftliches Phantom? *Die Oberpfalz*, Vol. 77, No. 2, pp. 35-43, ISSN 0342-9873

Xie, Y. (1988): Historical materials of Chinese Earthquakes and Their Seismological Analysis. In: *Historical Seismograms and Earthquakes of the World*, Lee, W.H.K.; H. Meyers & Shimazaki, K. (Eds.), 162-170, ISBN 0-12-440870-2, San Diego, USA

Zeilinga de Boer, J. & Sanders, D.T. (2005). *Earthquakes in Human History*, Princeton University Press, ISBN 9780691050706, Princeton, USA

2

The Traces of Earthquake (Seismites): Examples from Lake Van Deposits (Turkey)

Serkan Üner, Çetin Yeşilova and Türker Yakupoğlu
Yüzüncü Yıl University, Geological Engineering Department Van Turkey

1. Introduction

The soft sediment deformation structures formed by liquefaction or fluidization in the unconsolidated and cohesionless sediments during deposition or later (Lowe, 1975; Owen, 1996). These structures composed of overpressure of sediments, storm waves, sudden oscillation of groundwater or seismic shakings (Allen, 1982; Owen, 1987, 1996; Molina et al., 1998).

All kinds of earthquake induced soft sediment deformation structures are called as "seismite" (Seilacher, 1969). Seismites frequently observed in lacustrine deposits alike other depositional environments (Sims, 1975; Hempton et al., 1983; Seilacher, 1984; Davenport & Ringrose, 1987; Ringrose, 1989; Mohindra & Bagati, 1996; Alfaro et al., 1997; Calvo et al., 1998; Rodriguez-Pascua et al., 2000; Bowman et al., 2004; Neuwerth et al., 2006; Moretti & Sabato, 2007). Seismites can occur with seismic tremor (M≥5) (Fukuoka, 1971; Kuribayashi & Tatsuoka, 1975; Atkinson, 1984; Ambraseys, 1988) and they use for determination of the location and density of seismic activity (Sims, 1975; Weaver, 1976; Hempton et al., 1983; Talwani & Cox, 1985; Scott & Price, 1988; Ringrose, 1989).

The purposes of this study are; to determine the types of deformation structures in Lake Van deposits, to interpret the triggering mechanism and to discuss the importance of these structures in regional tectonic.

2. Method of the study

This study completed in six steps: (1) to determine the locations of soft sediment deformation structures, (2) to prepare the measured sedimentological sections according to facies properties and depositional subenvironments, (3) to measure the dimension and geometry (shape, symmetry and depth) of deformation structures and to determine the lateral continuity of deformed layers, (4) to detect the liquefaction potential of deposits by the help of sieve analysis, (5) to investigate the active faults and earthquake records (M≥5) at surrounding area, and (6) to match the all data with previous studies.

3. Geological settings

Lake Van is the largest soda lake of the world which has 607 km^3 volume and 451 metres depth (Kempe et al., 1978). The lake was formed at least 500 kyr ago (Litt et al., 2009). Lake

Van Basin exist at Eastern Anatolia Plateau as the product of Middle Miocene collision of Eurasia and Arabian plates (Şengör & Kidd, 1979; Şengör & Yılmaz, 1981; Keskin et al., 1998) (Fig. 1).

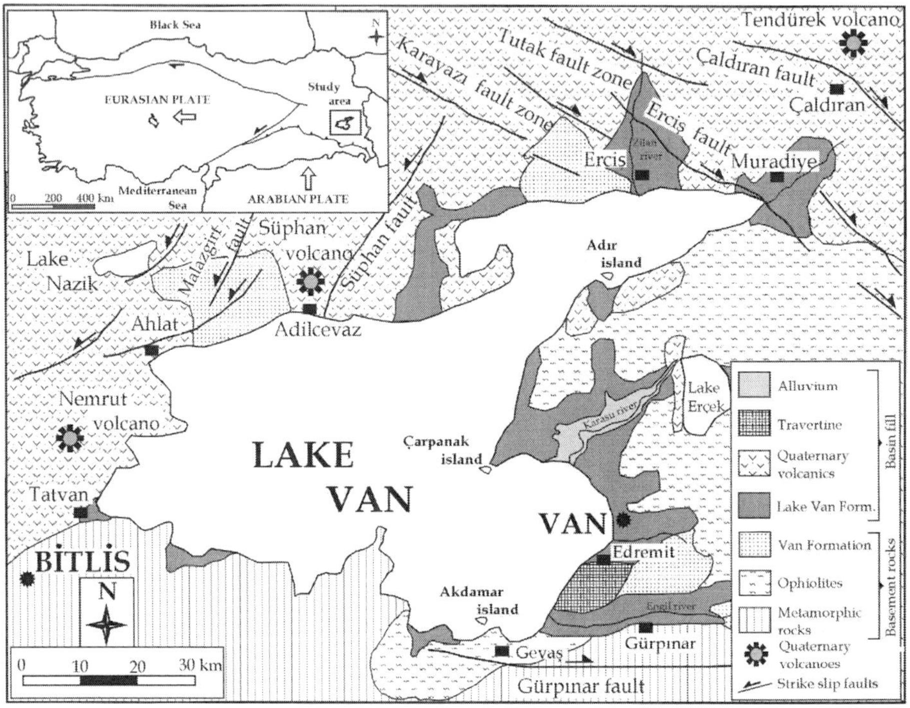

Fig. 1. Simplified geological map showing the active faults (Modified from Kurtman et al., 1978; Bozkurt, 2001; Koçyiğit et al., 2001).

Neotectonic period was started in Pliocene for the Eastern Anatolia Plateau and Lake Van Basin (Koçyiğit et al., 2001). This period is represented by N-S compressional regime. That compressional regime creates NW-SE trending dextral and NE-SW trending sinistral strike-slip faults (Şaroğlu & Yılmaz, 1986; Bozkurt, 2001; Koçyiğit et al., 2001) (Fig. 1). The region has a number of active faults that create earthquakes (M≥5). The Çaldıran Earthquake (1976) is the most known among these with their impact (M_s= 7.2) (Table 1).

Lake Van Basin stays on basement units which are Bitlis Metamorphic Complex, Upper Cretaceous Ophiolites and Oligocene-Miocene turbidites (Van formation). The basin fill consists of Quaternary volcanic rocks (from Nemrut and Süphan volcanoes), contemporaneous lacustrine deposits, Late Quaternary travertines and alluvium (Fig. 2). Generally, lacustrine sediments locate in east and north of Lake Van (Fig.1). These sediments were deposited during the period of highest lake level (+105 m) in 115000 years ago (Kuzucuoğlu et al. 2010). The deformation structures (seismites) are observed in these lacustrine deposits.

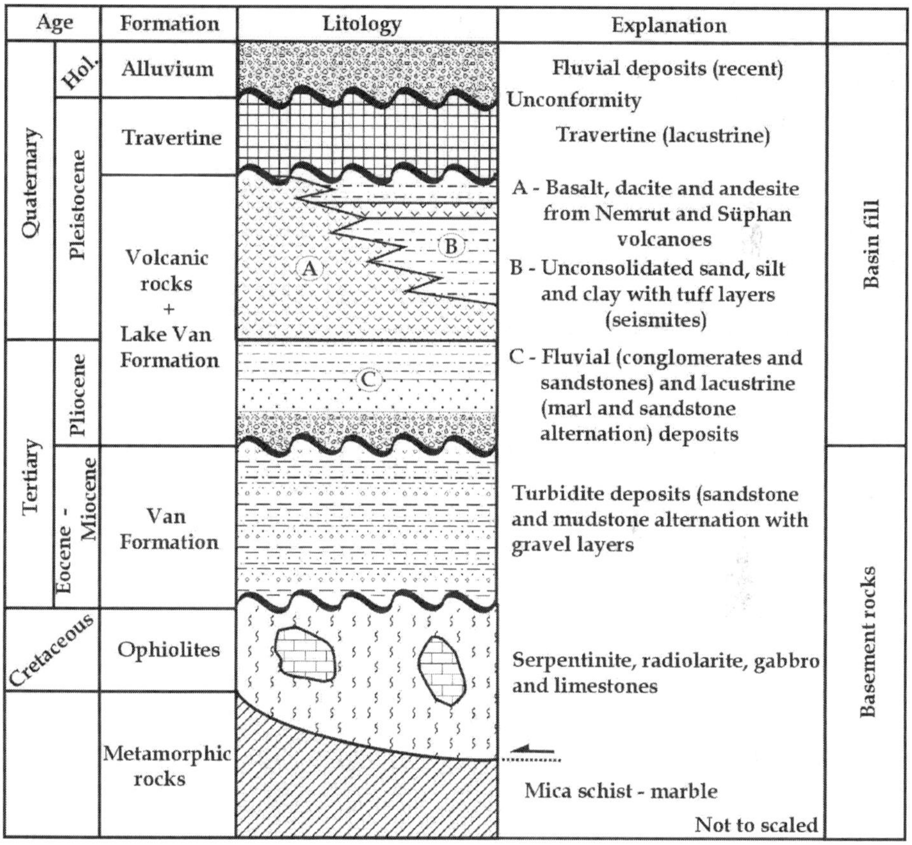

Fig. 2. Generalized stratigraphic columnar section of the study area (from Aksoy, 1988; Acarlar et al., 1991).

Date	Lat.	Long.	Depth (km)	Mag. (M_s)	Date	Lat.	Long.	Depth (km)	Mag. (M_s)
851	40.00	44.60	-	**5.2**	1941	39.45	43.32	20	**5.9**
856	40.00	44.60	-	**5.3**	1945	38.41	43.76	60	**5.2**
858	40.00	44.60	-	**5.2**	1945	38.00	43.00	30	**5.2**
1840	39.70	44.40	-	**6.8**	1945	38.63	43.33	10	**5.4**
1857	38.40	42.10	-	**6.7**	1966	38.14	42.52	28	**5.2**
1869	38.40	42.10	-	**5.0**	1966	38.10	42.50	50	**5**
1871	38.50	43.40	-	**5.5**	1968	38.15	42.85	53	**5**
1881	38.50	43.30	-	**5.0**	1972	38.23	43.86	46	**5**
1884	38.40	42.10	-	**6.1**	1976	38.61	43.20	56	**5.2**
1891	39.15	42.50	-	**5.5**	1976	39.17	43.95	33	**7.2**
1894	38.50	43.30	-	**5.0**	1976	39.09	43.71	49	**5.2**
1900	38.50	43.30	-	**5.0**	1976	39.18	43.71	46	**5.2**
1902	39.00	43.30	-	**5.0**	1976	39.31	43.66	53	**5.2**
1903	39.10	42.50	30	**6.2**	1977	39.35	43.48	24	**5**
1907	39.10	42.50	30	**5.2**	1977	39.29	43.62	46	**5.2**
1907	39.10	42.50	30	**5.4**	1977	39.27	43.70	39	**5.3**
1908	38.00	44.00	30	**6**	1977	39.13	43.90	34	**5**
1913	38.38	42.23	10	**5.5**	1977	39.31	43.53	38	**5.2**
1915	38.80	42.50	30	**5.7**	1979	39.12	43.91	44	**5.2**
1924	38.00	43.00	30	**5.2**	1988	38.50	43.07	49	**5.6**
1929	38.00	42.00	30	**5.2**	2000	38.41	42.95	48	**5.5**

Table 1. Earthquake records with magnitude 5 and higher occured in the study area (from Utkucu, 2006; KOERI, 2009), (Lat. = Latitude, Long. = Longitude, Mag. = Magnitude)

4. Deformation structures (Seismites)

Soft sediment deformation structures are observed in horizontally bedded, sandy, silty and clayey lacustrine deposits of Lake Van. Deformation structures exist in different levels of these shallow water deposits with the other sedimentary structures as cross-beds and wave ripples. Soft sediment deformation structures are classified differently according to morphologic properties or occurrence processes of the structure (Rossetti, 1999; Dramis & Blumetti, 2005; Neuwerth et al., 2006; Taşgın & Türkmen, 2009).

In this study, soft sediment deformation structures, observed in lacustrine deposits of Lake Van, were classified as contorted structures (simple-complex convolute bedding and ball-pillow structures), load structures (flame structures) and water escape structures (dish and pillar structures).

4.1 Contorted structures

Two types of contorted deformation structures exist in lacustrine deposits. These are simple-complex convolute bedding and ball-pillow structures.

4.1.1 Simple and complex convolute bedding

Simple and complex convolute structures are observed frequently in sandy and silty lacustrine sediments of Lake Van. These structures consist of little anticline or syncline like convolutions. The dimensions of these structures access up to 130 cm wide and 70 cm high. Simple convolute beds occur from one folded layer (Fig. 3a), while complex structures are composed of an outer trough and irregular inner laminates (Fig. 3b).

Simple and complex convolute structures may occur by overpressure, seismic shaking or storm waves. The convolute structures are bounded by undeformed horizontal beds in lacustrine sediments of Lake Van. This undeformed beds support the seismic origin suggestion (Cojan & Thiry, 1992). Additionally, the existence of folds at the centre of the complex convolute structures display the repeated tectonic activities (Bhattacharya & Bandyopadhyay, 1998).

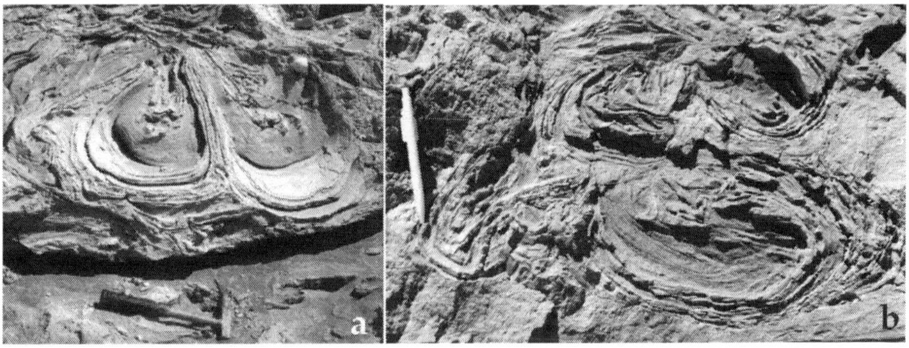

Fig. 3. (a) Simple convolute bedding and (b) complex convolute bedding in lacustrine deposits of Lake Van.

4.1.2 Ball and pillow structures

These structures are observed in sandy and silty deposits of Lake Van. They are composed of spherical or semi-spherical sand bodies in silty deposits (Fig. 4). This sand balls are covered by a silty crust. They have laminations at the inner part of the structures. Some structures are remaining connected to the overlying bed or the others are completely isolated from the bed. These structures in Lake Van sediments are very similar to presented in previous works (Hempton et al., 1983; Allen, 1986; Rossetti, 1999).

Ball and pillow structures occur with liquefaction of unconsolidated sediments. Because of the seismic tremors, liquefied sand size sediments are merged each other and create a ball-like structure (Montenat et al., 1987; Ringrose, 1989; Cojan & Thiry, 1992; Rodriguez-Pascua et al., 2000).

4.2 Load structures

The load balance of the unconsolidated sediments may change by landslide, rock fall or seismic waves. The load structures are formed by that load changing. These structures contain the load marks, pseudo-nodules and flame structures. Merely flame structures exist in lacustrine deposits of Lake Van.

Fig. 4. Ball and pillow structures observed in silty and sandy lacustrine deposits of Lake Van.

4.2.1 Flame structures

Flame structures are observed in sandy and silty deposits of Lake Van. These structures are formed by penetration of silty sediments to sandy deposits. These structures have different dimensions in lacustrine deposits of Lake Van (Fig. 5). They access up to 80 cm wide and 70 cm high. Generally flame structures comprise over pressure, but they can also be formed by seismic tremors (Visher & Cunningham, 1981; Dasgupta, 1998).

4.3 Water escape structures

Water escape structures are formed by sudden movement of pore-water to the upper level of deposits. Dish and pillar structures are formed by that mechanism in lacustrine sediments of Lake Van.

4.3.1 Dish and pillar structures

Dish and pillar structures are frequently observed in sandy and silty deposits of Lake Van. They consist of water movement in unconsolidated sediments due to sudden over pressure or seismic waves. The movement of pore water composes dish-like structures with folding of layers. These dish-like structures are seperated with vertical channels, called as pillars (Fig. 6).

Dish and pillar structures are observed in different sizes in lacustrine deposits of Lake Van. The dimensions of these structures access up to 100 cm wide and 50 cm high. The shape of dish structures may change depending on amount of pressure, movement velocity of pore water and the degree of consolidation. These structures in Lake Van deposits are very similar to presented in previous works (Lowe & LePiccolo 1974; Lowe 1975; Neuwerth et al., 2006). Dish and pillar structures may occur with seismic shakings (Plaziat & Ahmamou, 1998; Moretti et al., 1999).

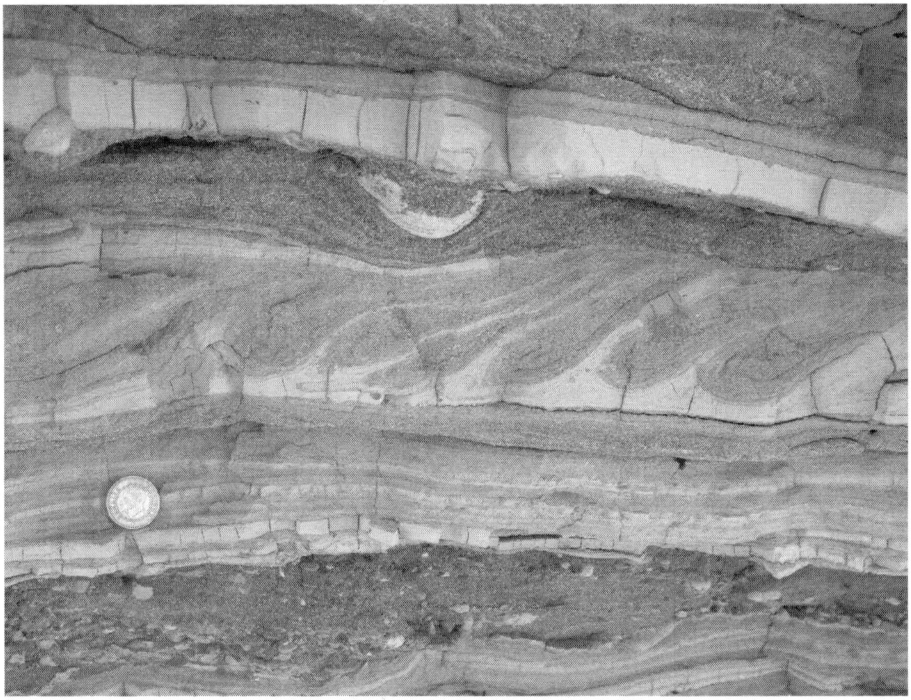

Fig. 5. Flame structures observed in lacustrine deposits of Lake Van.

5. Trigger mechanism

The most known occurence of soft sediment deformation structures are overpressure of sediments (Lowe & LoPiccolo, 1974; Lowe, 1975), storm waves (Molina et al., 1998; Alfaro et al., 2002) and seismic shakings (Seilacher, 1969; Lowe, 1975; Sims, 1975; Rossetti, 1999; Vanneste et al., 1999; Jones & Omoto, 2000; Rodriguez-Pascua et al., 2000; Bowman et al., 2004). Deformation structures in lacustrine deposits of Lake Van were evaluated in the light of these trigger mechanism. There is not any evidence or data about overpressure of sediments or the effect of storm waves. Therefore, seismic shaking mechanism were investigated in detailed.

Seismic waves may form deformation structures (seismites) in unconsolidated sediments becuse of changing of pore water pressure, existence of impermeable layers in sequence and

Fig. 6. Dish and pillar structures occured by the movement of pore water in sediments.

Fig. 7. Deformation structures observed among the undeformed parallel layers at different levels.

heterogeneity of grain size. Whenever, deformation structures in Lake Van deposits were evaluated for seismic origin; (1) grain size of the deformed sediments stay in liquefaction range (Port and Harbour Research Institute of Japan, 1997), (2) deformation structures are frequently observed in different levels of sequence which dissociated with undeformed, parallel beds (Fig. 7), (3) shapes, dimensions, geometry, sedimentologic and geotechnic properties of deformation structures are very similar to presented in previous works (Sims, 1975; Rossetti, 1999; Vanneste et al., 1999; Jones & Omoto, 2000; Bowman et al., 2004), (4) the region is very active for earthquakes (M≥5) and (5) soft sediment deformation structures in lacustrine deposits of Lake Van provide all criteria for the called as seismite (Sims, 1975; Obermeier, 1998; Rossetti, 1999).

6. Conclusion

In this study, the shapes, dimensions and locations of soft sediment deformation structures and facies properties and depositional environments of Quaternary aged lacustrine deposits of Lake Van are investigated. According to these features, deformation structures are classified into three parts as contorted structures, load structures and water escape structures.

Earthquake records show the tectonic activity (M≥5) of Lake Van and surrounding area. This data suggest that, the earthquakes should effect the lacustrine deposits in time of deposition (Late Quaternary). The deformation structures are frequently observed in different levels of lacustrine deposits. These deformed layers are the evidence of the repeated tectonic activity (M≥5) that effect the Lake Van deposits.

The relationship between the earthquake moment magnitude and the distance from epicenter to liquefaction locations appeal to the geologists. This distance may be more than 100 km in big earthquakes (M>7). According to locations and the distribution of soft sediment deformation structures, these structures should be formed by more than one faults activities.

7. Acknowledgment

The authors are grateful to Erman Özsayın, Alkor Kutluay and Ali Özvan for their fruitful discussions, help and suggestions to improve this chapter.

8. References

Acarlar, M., Bilgin, A.Z., Elibol, E., Erkan, T., Gedik, İ., Güner, E., Hakyemez, Y., Şen, A.M., Uğuz, M.F. & Umut, M., 1991. Van Gölü doğusu ve kuzeyinin jeolojisi, *MTA Report* No. 9469, Ankara, Turkey.

Aksoy, E., 1988. Van ili doğu-kuzeydoğu yöresinin stratigrafisi ve tektoniği. *PhD thesis, Fırat University*, Elazığ, Turkey.

Alfaro, P., Delgado, J., Estevez, A., Molina, J.M., Moretti, M. & Soria, J.M., 2002. Liquefaction and fluidization structures in Messinian storm deposits (Bajo Segura Basin, Betic Cordillera, southern Spain). *International Journal Earth Science*, 91, 505- 513.

Alfaro, P., Moretti, M. & Soria, J.M., 1997. Soft-sediment deformation structures induced by earthquakes (seismites) in Pliocene lacustrine deposits (Guadix-Baza Basin, Central Betic Cordillera). *Eclogae Geologicae Helvetiae*, 90, 531–540.

Allen, J.R.L., 1982. Sedimentary structures: their character and physical basis. *Developments in Sedimentology 30*, Elsevier, Amsterdam.

Allen, J.R.L., 1986. Earthquake magnitude-frequency, epicentral distance and soft-sediment deformation in sedimentary basins. *Sedimentary Geology*, 46, 67– 75.

Ambraseys, N.N., 1988. Engineering seismology. *Earthquake Engineering and Structural Dynamics*, 17 (1), 1–105.

Atkinson, G., 1984. Simple computation of liquefaction probability for seismic hazard applications. *Earthquake Spectra*, 1, 107–123.

Bhattacharya, H.N. & Bandyopadhyay, S., 1998. Seismites in a Proterozoic tidal succession, Singhbhum, Bihar, India. *Sedimentary Geology*, 119, 239–252.

Bowman, D., Korjenkov, A. & Porat, N., 2004. Late-Pleistocene seismites from Lake Issyk-Kul, The Tien Shan range, Kyrghyztan. *Sedimentary Geology*, 163, 211 -228.

Bozkurt, E., 2001. Neotectonics of Turkey-a synthesis. *Geodinamica Acta*, 14, 3-30.

Calvo, J.P., Rodriguez-Pascua, M.A., Martin-Velasquez, S., Jimenez, S. & De Vicente, G., 1998. Microdeformation of lacustrine laminite sequences from Late Miocene formations of SE Spain: an interpretation of loop bedding. *Sedimentology*, 45, 279–292.

Cojan, I. & Thiry, M., 1992. Seismically-induced deformation structures in Oligocene shallow marine and eolian coastal sands (Paris Basin). *Tectonophysics*, 206, 79-89.

Dasgupta, P., 1998. Recumbent flame structures in the Lower Gondwana rocks of the Jharia Basin, India- a plausible origin. *Sedimentary Geology*, 119, 253-261.

Davenport, C.A. & Ringrose, P.S., 1987. Deformation of Scottish Quaternary sediment sequences by strong earthquake motions. In: *Deformation of Sediments and Sedimentary Rocks*, M.E. Jones and R.M.F. Preston (eds.), Geological Society Special Publication, 29, Blackwell, Oxford, pp. 299–314.

Dramis, F. & Blumetti, A.M., 2005. Some considerations concerning seismic geomorphology and paleoseismology. *Tectonophysics*, 408, 177– 191.

Fukuoka, M., 1971. Memories of earthquake and foundations. *Bridges and Foundations*, 5, No. 10.

Hempton, M. R., Dunne, L. A. & Dewey, J. F., 1983. Sedimentation in an active strike-slip basin, Southeastern Turkey. *Journal of Geology*, 91, 401-412.

Jones, A.P. & Omoto, K., 2000. Towards establishing criteria for identifying trigger mechanisms for soft-sediment deformation: a case study of Late Pleistocene lacustrine sands and clays, Onikobe and Nakayamadaira Basins, northeastern Japan. *Sedimentology*, 47, 1211– 1226.

Kempe, S., Khoo, F. & Gürleyik, Y., 1978. Hydrography of Lake Van and its drainage area. In: *The Geology of Lake Van*, E.T. Degens and F. Kurtman (eds.), The Mineral Research and Exploration Institute of Turkey (MTA) Publication No.169, pp. 30-44.

Keskin M., Pearce J.A. & Mitchell J.G., 1998. Volcano-stratigraphy and geochemistry of collision-related volcanism on the Erzurum–Kars Plateau, northeastern Turkey. *Journal of Volcanology and Geothermal Research*, 85, 355–404.

Koçyiğit, A., Yılmaz, A., Adamia, S. & Kuloshvili, S., 2001. Neotectonics of East Anatolian Plateau (Turkey) and Lesser Caucasus: implication for transition from thrusting to strike-slip faulting. *Geodinamica Acta*, 14, 177-195.

KOERI (Boğaziçi University Kandilli Observatory and Earthquake Research Institute), 2009. *Turkey Earthquake Catalog*, 25 Mayıs 2009, http://www.koeri.boun.edu.tr/sismo/mudim/ katalog. asp,.

Kuribayashi, E. & Tatsuoka, F., 1975. Brief review of liquefaction during earthquakes in Japan. *Soils and Foundations, Japanese Society of Soil Mechanics and Foundation Engineering*, 15, No. 4, 81-91.

Kurtman, F., Akkuş, M.F. & Gedik, A., 1978. The geology and oil potential of the Muş-Van region. In: *The Geology of Lake Van*, E.T. Degens and F. Kurtman (eds.), The Mineral Research and Exploration Institute of Turkey (MTA) Publication, No.169, pp. 124-133.

Kuzucuoğlu, C., Christol, A., Mouralis, D., Doğu, A.F., Akköprü, E., Fort, M., Brunstein, D., Zorer, H., Fontugne, M., Karabıyıkoğlu, M., Scaillet, S., Reyss, J.L. & Guillou, H., 2010. Formation of the Upper Pleistocene terraces of Lake Van (Turkey). *Journal of Quaternary Science*, 25(7), 1124-1137.

Litt, T., Krastel, S., Sturm, M., Kipfer, R., Örçen, S. & Çağatay, M.N., 2009. Van Gölü Sondaj Projesi 'PALEOVAN', Uluslararası Bilimsel Kıta Sondaj Programı (ICDP): Yaklaşan Derin Sondaj Seferi ve Bilimsel Hedefler. *62. Turkish Geology Kurultai*, Proceeding Books, Ankara, s. 718-719.

Lowe, D. R., 1975. Water escape structures in coarse-grained sediments. *Sedimentology*, 22, 157-204.

Lowe, D.R. & LoPiccolo, R.D., 1974. The characteristics and origins of dish and pillar structures. *Journal of Sedimentary Petrology*, 44, 484–501.

Mohindra, R. & Bagati, T.N., 1996. Seismically induced soft-sediment deformation structures (seismites) around Sumdo in the lower Spiti valley (Tethys Himalaya). *Sedimentary Geology*, 101, 69-83.

Molina, J.M., Alfaro, P., Moretti, M. & Soria, J.M., 1998. Soft-sediment deformation structures induced by cyclic stress of storm waves in tempestites (Miocene, Guadalquivir basin, Spain). *Terra Nova*, 10, 145–150.

Montenat, C., d'Estevou, O.P. & Masse, P., 1987. Tectonic–sedimentary characteristics of the Betic Neogene basins evolving in a crustal transcurrent shear zone (SE Spain). *Bulletin des Centre de Recherches Exploration-Production of Elf-Aquitaine*, 11, 1–22.

Moretti, M., Alfaro, P., Caselles, O. & Canas, J.A., 1999. Modelling seismites with a digital shaking table. *Tectonophysics*, 304, 369–383.

Moretti, M. & Sabato, L., 2007. Recognition of trigger mechanisms for soft-sediment deformation in the Pleistocene lacustrine deposits of the Sant 'Arcangelo Basin (Southern Italy): seismic shock vs. overloading. *Sedimentary Geology*, 196, 31-45.

Neuwerth, R., Suter, F., Guzman, C.A. & Gorin, G.E., 2006. Soft-sediment deformation in a tectonically active area: The Plio-Pleistocene Zarzal Formation in the Cauca Valley (Western Colombia). *Sedimentary Geology*, 186, 67–88.

Obermeier, S.F., 1998. Liquefaction evidence for strong earthquakes of Holocene and latest Pleistocene ages in the states of Indiana and Illinois, USA. *Engineering Geology*, 50, 227–254.

Owen, G., 1987. Deformation processes in unconsolidated sands. In: *Deformation of Sediments and Sedimentary Rocks*, M.E., Jones and R.M.F., Preston (eds.), Geological Society Special Publication, 29, pp. 11–24.

Owen, G., 1996. Experimental soft-sediment deformation: structures formed by the liquefaction of unconsolidated sands and some ancient examples. *Sedimentology*, 43, 279-293.

Plaziat, J.C. & Ahmamou, M., 1998. Les differents mecanismes a l'origine de la diversite des seismites, leur identification dans le Pliocene du Saiss de Fes et de Meknes (Maroc) et leur signification tectonique. *Geodinamica Acta*, 11/4, 183- 203.

Port Harbour Research Institute of Japan (1997). *Handbook on liquefaction remediation of reclaimed land.* A.A. Balkema, Roterdam.

Ringrose, P.S., 1989. Paleoseismic (?) liquefaction event in late Quaternary lake sediment at Glen Roy, Scotland. *Terra Nova*, 1, 57-62.

Rodriguez-Pascua, M.A., Calvo, J.P., De Vicente, G. & Gómez-Gras, D., 2000. Soft sediment deformation structures interpreted as seismites in lacustrine sediments of the Prebetic Zone, SE Spain, and their potential use as indicators of earthquake magnitudes during the Late Miocene. *Sedimentary Geology*, 135, 117-135.

Rossetti, D.F., 1999. Soft-sediment deformational structures in late Albian to Cenomanian deposits, Sao Luis Basin, northern Brazil: evidences for paleoseismicity. *Sedimentology*, 46, 1065-1081.

Scott, B. & Price, S., 1988. Earthquake-induced structures in young sediments. *Tectonophysics*, 147, 165-170.

Seilacher, A., 1969. Fault-graded beds interpreted as seismites. *Sedimentology*, 13, 155- 159.

Seilacher, A., 1984. Sedimentary structures tentatively attributed to seismic events. *Marine Geology*, 55, 1-12.

Sims, J. D., 1975. Determining earthquake recurrence intervals from deformational structures in young lacustrine sediments. *Tectonophysics*, 29, 141-152.

Şaroğlu, F. & Yılmaz, Y., 1986. Doğu Anadolu'da neotektonik dönemdeki jeolojik evrim ve havza modelleri. *Bulletin of The Mineral Research and Exploration*, 107, 73-94.

Şengör A.M.C. & Kidd W.S.F., 1979. Post-collisional tectonics of the Turkish-Iranian Plateau and a comparison with Tibet. *Tectonophysics*, 55, 361-376.

Şengör, A.M.C. & Yılmaz, Y, 1981. Tethyan evolution of Turkey: a plate tectonic approach. *Tectonophysics*, 75, 181-241.

Talwani, P. & Cox, J., 1985. Paleoseismic evidence for recurrence of earthquakes near Charleston, South Carolina. *Science*, 229, 379-381.

Taşgın, C.K. & Türkmen, İ., 2009. Analysis of soft-sediment deformation structures in Neogene fluvio-lacustrine deposits of Çaybağı formation, eastern Turkey. *Sedimentary Geology*, 218, 16-30.

Utkucu, M., 2006. Implications for the water level change triggered moderate (M ≥ 4.0) earthquakes in Lake Van basin, Eastern Turkey. *Journal of Seismology*, 10, 105-117.

Vanneste, K., Meghraoui, M. & Camelbeeck, T., 1999. Late Quaternary earthquake-related soft-sediment deformation along the Belgian portion of the Feldbiss Fault, Lower Rhine Graben system. *Tectonophysics*, 309. 57-79.

Visher, G.S. & Cunningham, R.D., 1981. Convolute laminations - a theoretical analysis: example of Pennsylvanian sandstone. *Sedimentary Geology*, 28, 175-189.

Weaver, J.D., 1976. Seismically-induced load structures in the basal coal measures, South Wales. *Geological Magazine*, 113, 535-543.

3

Parameters of the Strong Paleoearthquakes Along the Talas-Fergana Fault, the Kyrgyz Tien Shan

A.M. Korjenkov[1,2], D. Rust[3], A. Tibaldi[4] and S.V. Abdieva[2]
[1]Institute of Physics of the Earth, Russian Academy of Science
[2]Kyrgyz-Russian Slavic University
[3]University of Portsmouth
[4]University of Milan-Bicocca
[1]Russia
[2]Kyrgyzstan
[3]United Kingdom
[4]Italy

1. Introduction

1.1 Geological-tectonic structure and evolution of the Talas-Fergana Fault in the Cenozoic

The Talas-Fergana Fault (TFF) is the largest strike-slip structure in the central Asia. It forms an obliquely-oriented boundary between the north-eastern and south-western parts of the Tien Shan (Fig. 1). The last one includes the Fergana depression, Chatkal-Kurama mountain system and Alay valley. A wide belt of latitudinal oriented ranges, which are located between the Kazakh platform and Tarim basin, represents the north-eastern Tien Shan.

Many scientists were engaged and are still interested in studying the problem of recent movements along the Talas-Fergana Fault (K.E. Abdrakhmatov, T.P. Belousov, V.S. Burtman, O.K. Chediya, A. Khodzhaev, V.I. Knauf, A.M. Korjenkov, N.P. Kostenko, V.N. Krestnikov, I.N. Lemzin, V.I. Makarov, P. Molnar, V.A. Nikolaev, V.N. Ognev, A.V. Peive, V.I. Popov, G.N. Pshenin, E.Ya. Rantcman, S.V. Ruzhentsev, V.M. Sinitsyn, N.M. Sinitsyn, S.F. Skobelev, D.V. Shtange, A.L. Strom, L.D. Sulerzhitsky, A.I. Suvorov, V.G. Trifonov, N.N. Verzilin, R.E. Wallace, and others. Most researchers interpret the TFF as a right-lateral strike-slip fault active since Palaeozoic time. Right-lateral movements rejuvenated in the late Cenozoic because of crustal shortening linked to the India-Eurasia collision. The tectonic movements along the intracontinental strike-slip faults contribute to absorb part of the regional crustal shortening, thus strike-slip motions along the TFF are necessary for the complete assessment of the active deformation of the Tien Shan orogen.

Our focus is to improve the understanding of the intracontinental deformation of the Tien Shan mountain belt as a whole and the occurrence of strong earthquakes along the whole length of the TFF. The aim of the work is an attempt to reveal features of relief occurred during strong paleoearthquakes along the Talas-Fergana Fault, TFF fault segmentation, length of the seismogenic ruptures, energy and age of ancient catastrophes. Mentioned data are critical for complete seismic hazard assessment for a territory with absence of materials on historical seismicity.

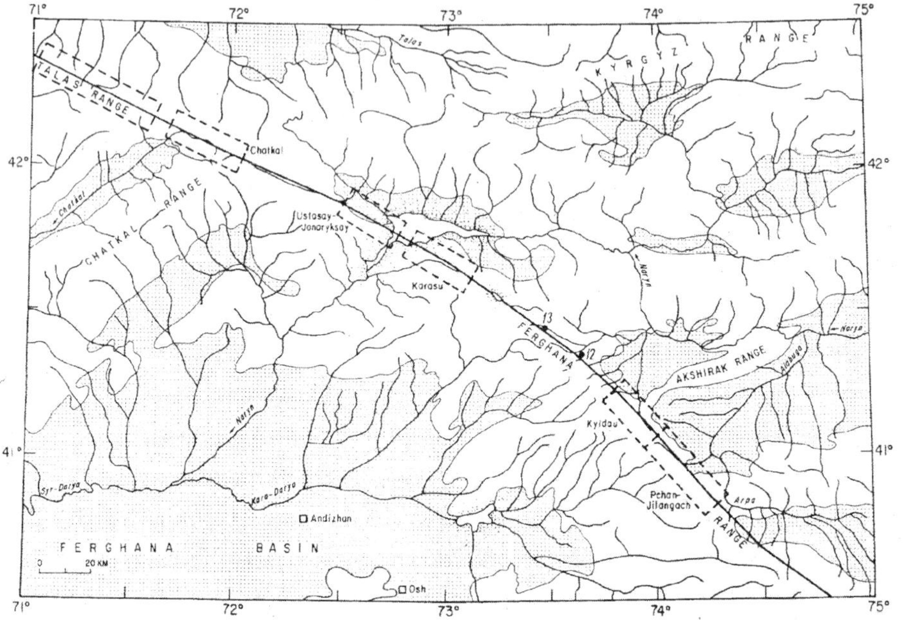

Fig. 1. The map of the Talas-Fergana Fault's line and adjacent territories (modified after Burtman et al., 1996). Dashed rectangulars shows studied portions of the fault. Sedimentary basins are indicated by regular dotty filling. Irregular dotty area shows lakes and reservoirs. Numbers along the fault's line are observation points N12 and 13 from Burtman et al. (1996).

Many authors (Khodzhaev, 1985; Burtman et al., 1987, 1996; Trifonov et al., 1990, 1992; Abdrakhmatov and Lemzin, 1991; Korjenkov, 1993, 2006; Korjenkov et al., 2006, 2009, 2010 and others) were occupied also by a detailed paleoseismological study of the TFF zone. Some of them (Burtman et al., 1987, 1996; Trifonov et al., 1990, 1992; Rust et al., 2008; Korjenkov et al., 2009, 2010) collected samples for the radiocarbon dating. Because the organic material has deposited later than the formation of the upslope facing scarp, displacing channels of gullies and watersheds, the radiocarbon dates (Table 1) point on minimum ages of the events which led to relief forms' displacement along the fault zone.

All features pointing on seismic-rupturing character of the upslope facing scarp, developed along the fault zone, are testifying that the Talas-Fargana Fault is "alive" until present. As related to its morphologic-kinematic characteristics, most of scholars believe that the fault is right-lateral strike-slip fault's structure, they point on amplitude of displacement along it from hundreds meters to 12-14 kilometers during Cenozoic time (Ranzman and Pshenin, 1963; Trifonov et al., 1990 and others).

Last summary of materials of previous investigations along the Talas-Fergana fault is cited in papers by Korjenkov (2006), Korjenkov et al. (2006, 2007, 2009, 2010), Rust et al. (2008).

Field station	River valley or name of the fault segment	Value of the horizontal displacement, m	Radiocarbon age, years
1	Dzhilangach	19	3970 ±40
2	Dzhilangach	40 ±3 40 ±3 40 ±3	1940 ±50 2630 ±70 2740 ±70
3	Dzhilangach	45 ±3	1720 ±70
4	Chitty-Western	40 40	4590 ±100 15800 ±1300
5	Birguzy	35 ±5	3030 ±90
6	Birguzy	27	3740 ±600
7	Pchan	90 ±3	3150 ±40
8	Pchan	21–24 21–24 21–24	2180 ±120 2280 ±70 2540 ±70
9	Pchan	25 ±1	2640 ±600
10	Kyldau	23–24	2320 ±40
11	Kyldau	125 ±25 60 ±25	3670 ±80 3670 ±80
12	Urumbash	17–20	1510 ±60
13	Keklikbel	10–12	1240 ±60
14	Dzhanaryksay	14 ±2	1440 ±30
15	Chatkal	17–20	1450 ±40
16	Chatkal	20 40	1350 ±60 1350 ±60
17	Chatkal	20 20 40	1150 ±40 2020 ±50 2020 ±50
18	Chatkal	20 40	1220 ±50 1220 ±50

Table 1. Radiocarbon dates of the samples collected from the displaced gullies along the TFF (by Trifonov et al., 1990, 1992 и Burtman et al., 1996)

2. Methodology

Besides traditional route field investigations, forestalling by interpretation of air-photos and satellite images, study of existing archive and published literature, we have conducted a detailed mapping of selected key test sites:

- Kara-Bura one in a region of the pass with the same name across the Talas Range,
- Sary-Bulak test site in riverhead of the river with the same name - left tributary of the Uzun-Akhmat river and

- Kok-Bel test site in a region of a pass with the same name on the "Bishkek-Osh" highway (Fig. 2).

Fig. 2. Digital map of relief of the western Tian Shan. The locations of the investigated test sites are shown in the map.

On the prospected ranges the TFF line usually goes across the slope of one of river valleys or a ridge (range) slope. Along the line there is usually a fault scarp in the form of a swell. Height of this scarp is usually equal to several dozens centimeters - the first meters. On the investigated ranges the numerous broken forms of a modern relief were found: valleys of temporary waterways and watersheds between them, upper parts of which are shifted in a horizontal direction - to the right to a distance from several dozens to several hundreds meters. The identical width and morphology of the shifted parts of dry valleys above and below the fault line testifies that the shift occurred quickly. It allows linking such shifts with earthquakes (Burtman et al., 1987).

The majority of the shifted valleys of temporary waterways have remained below the fault line, where on the slope at earthquake a fault scarp was formed in the shape of a swell. This scarp has isolated the lower continuation of the broken valley, while seasonal waters found other drain, washing away the scarp in the lowest place. Further the isolated part of the valley could continue to be displaced along the fault.

For definition of time of movements along the TFF indirect method of V.G. Trifonov (1985) was used. At formation of a scarp on the slope along the fault line a depression was also formed. This scarp has impound the waters flowing down the slope and filtering along the fault plane, which created conditions for swamping of the depressions in the vicinity of the fault. The beginning of formation of a peat swamp and a thick soil layer testifies to occurrence of a fault scarp which formed because of horizontal displacement along the fault. For definition of age of these formations samples of organic material for the radiocarbon analysis in bore pits, prospected in impounded parts of valleys, were taken. The received radiocarbon dates, based on the sum of the organic substance which was accumulated during some period of time, are always later dates as compared to the moment of beginning of accumulation. Determination of residual activity of carbon in our samples was carried out on a device QUANTULUS-1220 (Liquid Scintillation Counters) at the Institute of Geology and Mineralogy of the Siberian Branch of the Russian Academy of Science, Novosibirsk. For age calculation the half-life period of ^{14}C was used equal to 5570 years. The age was calculated from 1950. Age determination was done based on fraction of humic acids.

3. Northwestern part of the Talas-Fergana fault

3.1 Investigations in Ara-Beyik – Kara-Kasmak interfluve

In the most west of Kyrgyzstan (the Ara-Beyik river valley) the TFF trace is unclear. Probably, it is divided into several zones parallel to each other and differing in character of the Holocene slip. The fault zone can be identified in outcroppings of rocks along the right slope of the river valley where there is a wide zone of crushed basement rocks.

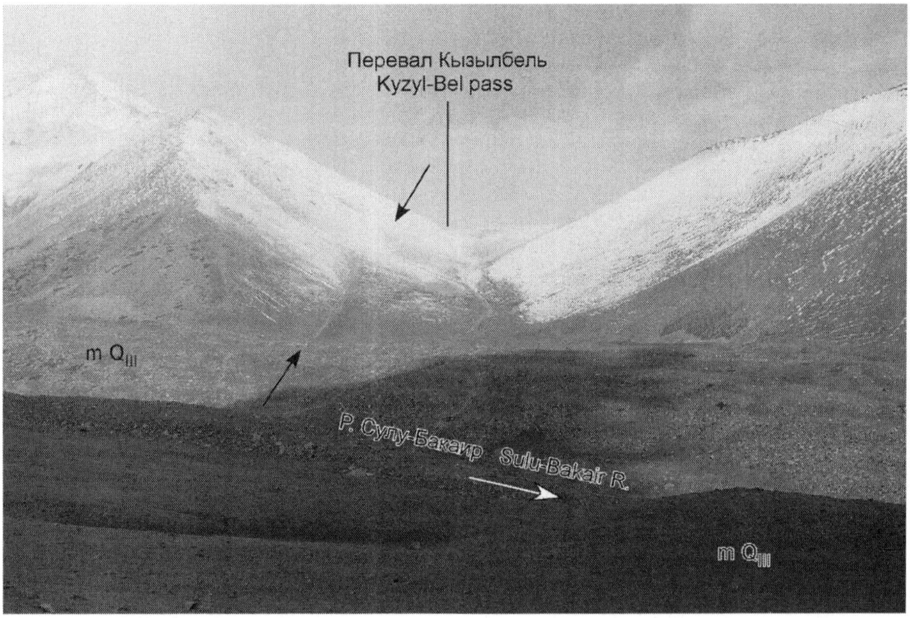

Fig. 3. The right-lateral deviation of dry channels (shown by black arrows) along the Talas-Fergana trace on the left slope of the Sulu-Bakair river valley.

In the Sulu-Bakair River valley the TFF zone cuts a moraine of Middle and Late Pleistocene age and is expressed as a deep and wide upslope facing scarp. Although the moraine is several tens meter thick, the fault zone is marked by springs. Above the TFF line there is a stairs on the slope consisting of upslope facing scarps located one above another and formed in loose colluvial and moraine deposits. Slip displacements are not clearly seen here excluding some cases. Thus, for example, two neighboring scours on the left slope of the Sulu-Bakair river valley within the TFF zone have experienced a right-lateral bending (Fig. 3). On the right bank of the river a small stream channel was beheaded along the fault, and now its upper reaches are at a distance of 128 m from the lower ones (Fig. 4). A neighboring stream channel was shifted to the right on the distance of 57 m, but its valley is smaller than that of previous one.

In a place of crossing of the Korumtor Spring by the Talas-Fergana Fault (Shilbilisay River basin) the fault is expressed by a visible depression in relief. An upslope facing scarp is located along it. Its width - 30 m, depth - down to 3 m (Fig. 5). The upslope facing scarp is also visible in a body of Late Pleistocene (Q_{III}^2) moraine.

Fig. 4. The right-lateral deviation of channels on the distance of 128 m in the Late Quaternary moraine on the right slope of the Sulu-Bakair river valley within the TFF zone.

The Talas-Fergana Fault zone is well exposed in the region of the Kara-Bura pass due to building an automobile route. Here the fault is represented by a zone of hundreds meter wide composed of completely crushed cataclasites and milonites and rocks with traces of initial structural-material features. There are also lenses not modified by tectonic processes. The TFF line is clearly identified by lows in the relief and saddles on watershed spurs.

Probably to the west of the Kara-Bura pass the right-lateral slip turned to a fault plane to the north of the TFF. Thus, in the upper reaches of the Kara-Bura river on the northern slope of the Talas range we observed an upslope facing scarp causing beheading and right-lateral shifting of dry channels and scours (Fig. 6 and 7).

Fig. 5. A photograph of the Talas-Fergana Fault zone. A view NW-ward from Korumtor river basin.

To the south-east of the Kara-Bura pass the Talas-Fergana Fault is well pronounced in three recessional moraines filling the upper valley of the Karakasmak river. Here it is represented by a zone of sagging located at oblique angles to each other and cutting moraine and fluvioglacial sediments. Along the southern line one can observe right-lateral deviation of dry channels (up to 34.7 m) and watersheds between them. We also observed a recessional moraine body deposited at the end of the Late Pleistocene and shifted on 28.3 m. Further to the south there is a stairs consisting of upslope facing scarps on a steep slope.

As we described above, along the whole its length the Talas-Fergana Fault is expressed in relief in a form of a deep depression - upslope facing scarp which cuts both basement rocks and Quaternary deposits (Fig. 8).

Fig. 6. Beheaded dry scours (shown by dashed lines) on the right slope of the upper Kara-Bura river valley. The rupture of upper and lower parts of the scours occurred along the slip fault (show by arrows) located in the north of the main trace of the Talas-Fergana Fault.

Mentioned above upslope facing scarp testifies on active seismic life of NW chain of the Talas-Fergana Fault. A number of features point on the fact that it is a real seismic-rupture form and not a depression of weathering developed along a zone of milonites and cataclazites.

1. Localization to the fault of a number of seismic-gravitation forms (rock- and landslides). For example, a limestone rockslide which locates in 500-600 m east of the Kara-Bura pass (Khodzhaev, 1985).
2. Linear upslope facing scarp is well traced, in spite of it crosses different relief forms, as well as different rock formations. It is particularly well expressed in moraine bodies.
3. The up slope facing scarp cuts channels of temporary springs where the tectonic dams are formed. An existence of the tectonic dams is the direct evidence of impulse

movements along the fault during a moment of its seismic refreshment (Khromovskikh and Nikonov, 1984).

Fig. 7. Upper reaches of the Kara-Bura river, right slope of the valley. Picture of one of beheaded scours (shown by dashed line) along a slip fault (shown by two arrows) in the north of the TFF. The upper portion of the scour is shifted to the right.

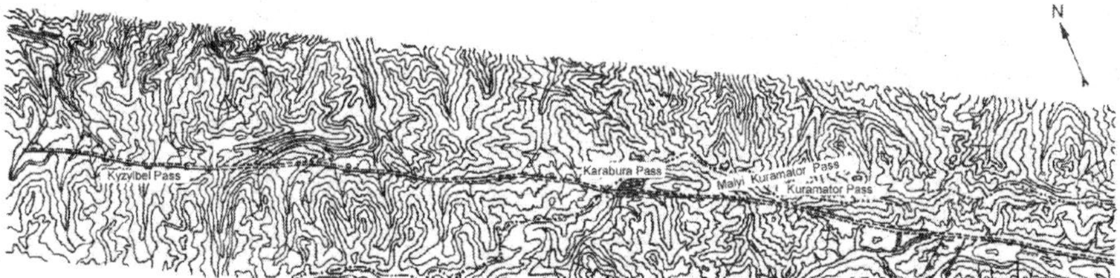

Fig. 8. A map of the Talas seismogenic structure (modified after A.K. Khodzhaev, 1985): 1 - watershed line of the Talas Range; 2 - contour lines in 200 m; 3 - seismogenic ruptures (a - normal faults, b - cracks); 4 - rivers and springs' beds; 5 - river captures; 6 - seismically induced rockslide.

4. An intake of frontal parts of recent taluses by the upslope facing scarp testifies on youth and instantaneity of its formation. This phenomenon one can observe east of the Kara-Bura pass. If not, an ancient upslope facing scarp would be covered by the colluvial material, the talus would gush over the scarp and will continue its movement down the slope.

3.2 Kara-Bura test site

We have mentioned above about our investigations on special geodetic test sites. The first region is located to the south-east of the Kara-Bura pass in the upper reaches of the Kara-Kysmak river (Chatkal region, Djalal-Abad oblast). Here the Talas-Fergana Fault passes downward from the pass and cuts Late Pleistocene recessional moraines. The fault zone is clearly marked by elongated depressions, as well as by right-lateral shifts of stream beds and watersheds between them.

We constructed a digital elevation model of the site (Fig. 9). Using the model and taking into account characteristic elements of the relief we could measure the right-lateral slip along the fault. Thus, for example, a Late Pleistocene recessional moraine (Shubin et al., 1992) was shifted for 28-34 m (B1-B2, Fig. 10). This displacement began, apparently, 5910 ± 130 years ago to what testifies the absolute age of the sample SOAN-6526 selected in detrital deposits which filled the crack, formed in the stretching zone along the fault (Fig. 11). Thus, the calculated rates of displacement in the second half of Holocene on this site of the fault were 4.70 - 4.90 mm/year.

In the body of the phased moraine of the middle of late Pleistocene (Shubin et al., 1992), the stream valley (Fig. 9 and 10) was cut. Its age, hence, is more young - apparently, the end of late Pleistocene. The thalweg of the valley was displaced to the right at 34.78m (A1-A2, Fig. 9). The watershed to the east from this stream was displaced practically on the same distance (31.07 m).

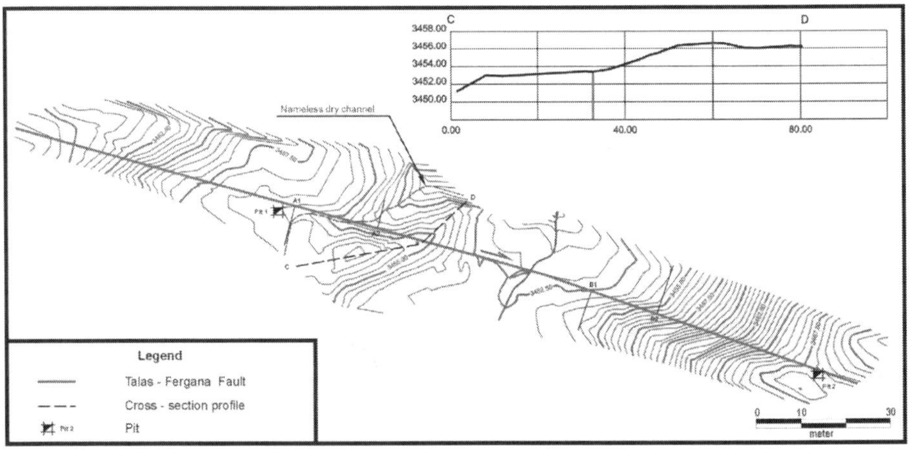

Fig. 9. Digital map of the Kara-Bura test site. Detailed mapping of the Talas-Fergana Fault zone with a use of an electronic tachometer. Pit1, Pit2- pits. Contour lines in 0.5m. Segment A1 - A2 along the fault line – displacement of the dry gully on 34,78 m. Eastern watershed of the dry gully (A – B profile) is displaced on 31,07 m. Segment B1 – B2 – right-lateral displacement of an age of Late Pleistocenemoraine on 28,34 m.

Fig. 10. Photo of the Kara-Bura polygon. View north-westward. The red line is the TFF, the white dashed line is the dry channel, the arrow Pr1 is the profile line in Fig. 9, g2Q_{III} is a recessional moraine of the middle of Late Pleistocene. g3Q_{III} is a recessional moraine of the end of Late Pleistocene.

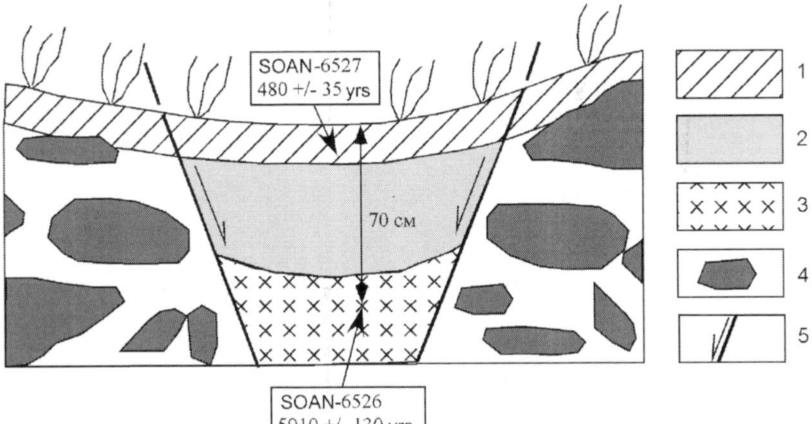

Fig. 11. A schematic sketch of a trench No. 2, which was prospected through graben in a transitive zone of the stretching in the zone of the Talas-Fergana fault. 1 - is characterized by poorly developed mountain soil, 2 - is characterized by loess-like loamy soil with gravel, 3 - sand and detritus deposit/sedimentation, which filled the graben, 4 - moraine deposits/ sedimentations of the end late Pleistocene, 5 - fault planes which have formed the graben.

We examined the bore pit (Fig. 12) in the flood plain of the dry rivulet - in the SW wing of the TFF. Here the tectonic dam and impounded deposits, in connection with right-lateral displacements along the NE wing of the TFF, were formed. In the buried soil formed on a moraine of the middle of late Pleistocene (Shubin et al., 1992), we took a sample with absolute age of 6100 ± 200 years (SOAN-6523). At this particular time, apparently, there was the first earthquake which has displaced for the first time the body of the moraine and fluvial deposits. Thus, the calculated rates of displacement for the last ~ 6 thousand years comprised 4.93 - 5.89 mm/year.

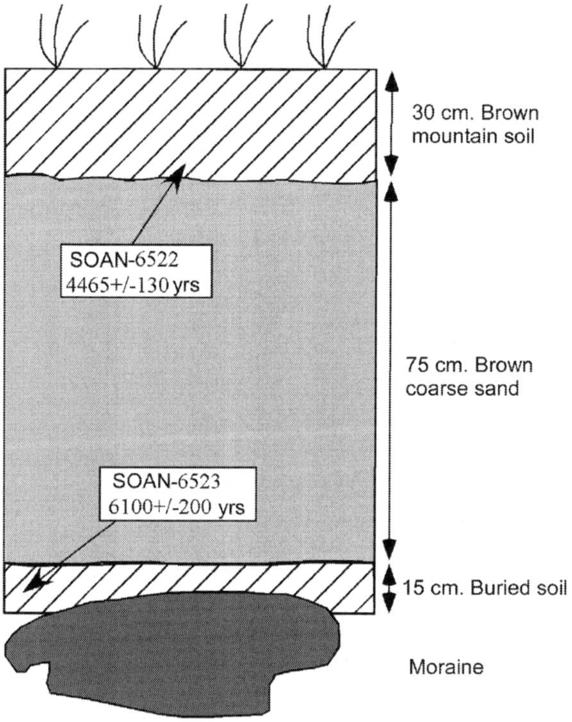

Fig. 12. Impounded deposits/sedimentations in bore pit 1, examined in SW wing of the TFF.

The absolute ages of the deposits formed in the bottom parts of both pits: in bore pit 1 and trench 2, give statistically the same age ~ 6 thousand years which, apparently, is age of the first earthquake in Holocene which we managed to record. However in bore pit 1 (Fig.12) is available one more absolute age determination: in the bottom part of modern soil. Age of this sample by results of the radiocarbon analysis is 4465 ± 130 years old (SOAN-6522). This is the minimum age of the second seismic event, recorded by us: in the middle Holocene. After the 1st earthquake on the examined site of the fault the tectonic dam was formed, which has led to accumulation of impounded deposits on which the soil cover, eventually, was formed.

However the earthquake, which took place 4.5 thousand years ago, was not the last earthquake on given site of the TFF. We have examined two pits in hanging (SW) and foot (NE) wings of the fault scarp formed along TFF (Fig. 9, 10 and 13). In them - samples were taken in the bottom part of the layer of turf, the absolute age was 405±100 years old (SOAN-6525) for a hanging wing of the scarp and 460 ± 40 years old (SOAN-6524) for the foot wing. These two dates indicate in the occurrence of the seismic event 400-500 years ago. These data prove to be true also by the age determination SOAN-6527 (480±35 years old) of the modern soil received from the bottom part in trench 2 (Fig. 11).

Fig. 13. The late Holocenic seismoshoulder/seismoledge along the TFF (it is shown by a red faltering line), which is breaking the flood plan of the unnamed rivulet/say. The SW wing is raised. In both wings pits on depth of 15 cm were selected samples in the bottom part of the turf/peat layer for definition of their absolute age.

Summarizing data on the Kara-Bura test site, we received (calculated) the following rates of horizontal tectonic displacement since the middle Holocene: 4.70-5.90 mm/year. During this time in the named area along the Talas-Fergana fault minimum three strong earthquakes took place: about 6000 years, 4500 years and 400-500 years ago. The data of absolute age, determined on the Kara-Bura test site regarding the latter earthquake, coincide with the archeo-seismologic data on the destroyed caravansary located in the middle part of the Kara-Bura river valley (Korjenkov et al., 2009).

3.3 Investigations in the Kara-Kasmak – Sary-Bulak interfluve

In the interfluve of right inflows of the Karakuldzha-Chatkalskaya river (Dzhashilsai, Dzhosho and others) (Fig. 14A) there are no clear evidences of vertical differential displacements of the TFF walls. However in the south of the region the vertical component can be observed. Thus, above the mouth of the Chiimtash river (right inflow of the Karakuldzha river) and a pass of the same name one can observe a typical upslope facing scarp up to 8 km long half-filled with clastic material that results in only 1.0-1.5 m height difference between the hanging and foot walls of the fault. Nevertheless, it is seen from walls of gullies cutting the seismogenic structure that the depth of the upslope facing scarp is about 4 m (Chediya, 1986). V.S. Burtman et al. (1987) reported a depth of the seismic depression (or the height of the seismic scarp in the shape of swell) up to 5 m (point 23 in Fig. 15). Besides, it is possible to observe a 10 m thick Middle Pleistocene moraine adjoining to the downslope vertically thrusted wall of the fault (the left slope of the Karakuldzha river valley with elevation mark 3296 m), i.e. in this case there is a clearly marked 15-m thrusting of the north-eastern wall in the Late Pleistocene - Holocene. If to prolong mentally toward the fault the Early Pleistocene surface of the fault walls (Fig. 14 B) the stated above will be proved (Chediya, 1986).

V.S. Burtman et al. (1987) noted that the vertical fault plane is clearly marked due to the ragged relief. By the field station 20 (Fig. 15) in a section of a slope of the spring valley one can observe changing of a tilt of the fault plane upward. At the depth 20 m the nearly vertical (80°) fault plane becomes more gentle (up to 45°); near the surface it becomes steeper again (65°). Such phenomenon is probably conditioned by plastic flow of Quaternary alluvial sediments downward the valley slope. The flow was more intense at the depth 5-20 m that resulted in preservation of the steeper fault plane near the surface and tectonic swell near the fault line (Burtman et al., 1987).

At the Karakuldzha-Narynskaya river head (Fig.14 C) the amplitude of the north-eastern wall thrusting can be inferred from the shifted Early Pleistocene surface as 250-300 m (Chediya, 1986). As for horizontal movements along the TFF within the region, V.S. Burtman et al. (1987) found 26 ruptured forms of the modern relief within the site of 16-km long, showing horizontal displacement along the fault line (Table 2). The majority of the forms are small channels. At two points (18 and 24 in Table 2 and Fig. 15) one can observed shifted slopes of channels and at other two points (2 and 20) watersheds of mountain ridges are shifted. Statistically horizontal displacements up to 25-35 m are prevailing (histogram in Fig. 16), but in some cases (marked by a star in Table 2) displacements up to 35-45 m have been measured. In some places the main line of the fault is divided into some branches which either die out (point 11 in Fig. 15), or join the fault line again limiting the tectonic lens (point 23). Moreover, fault-satellites pass parallel to the main fault line at the distance of 70-100 m, along which horizontal displacement of the relief forms from 3 up to 10 m is observed (points 15 and 18). Displacement along the fault-satellites together with plastic deformation of the fault walls compensates changing amplitude of the slip along the main fault plane (Burtman et al., 1987).

For determination of the slip time Burtman et al. (1987) used organic material collected in pits. 5 samples were collected for radiocarbon age determination by Burtman et al. (1987) in pits dug in a drainless sagging (a, Fig. 15 and 16) and depressions (b-d, Fig. 15 and 16) with springs draining along the fault plane. The samples were collected from clays rich in organic material and overlaid by thick soil. Radiocarbon ages (determinations GIN-4300-4304) determined from the total of accumulated organic material is always younger than the time of the accumulation beginning. Therefore, the oldest age (2020±50 years) is starting the

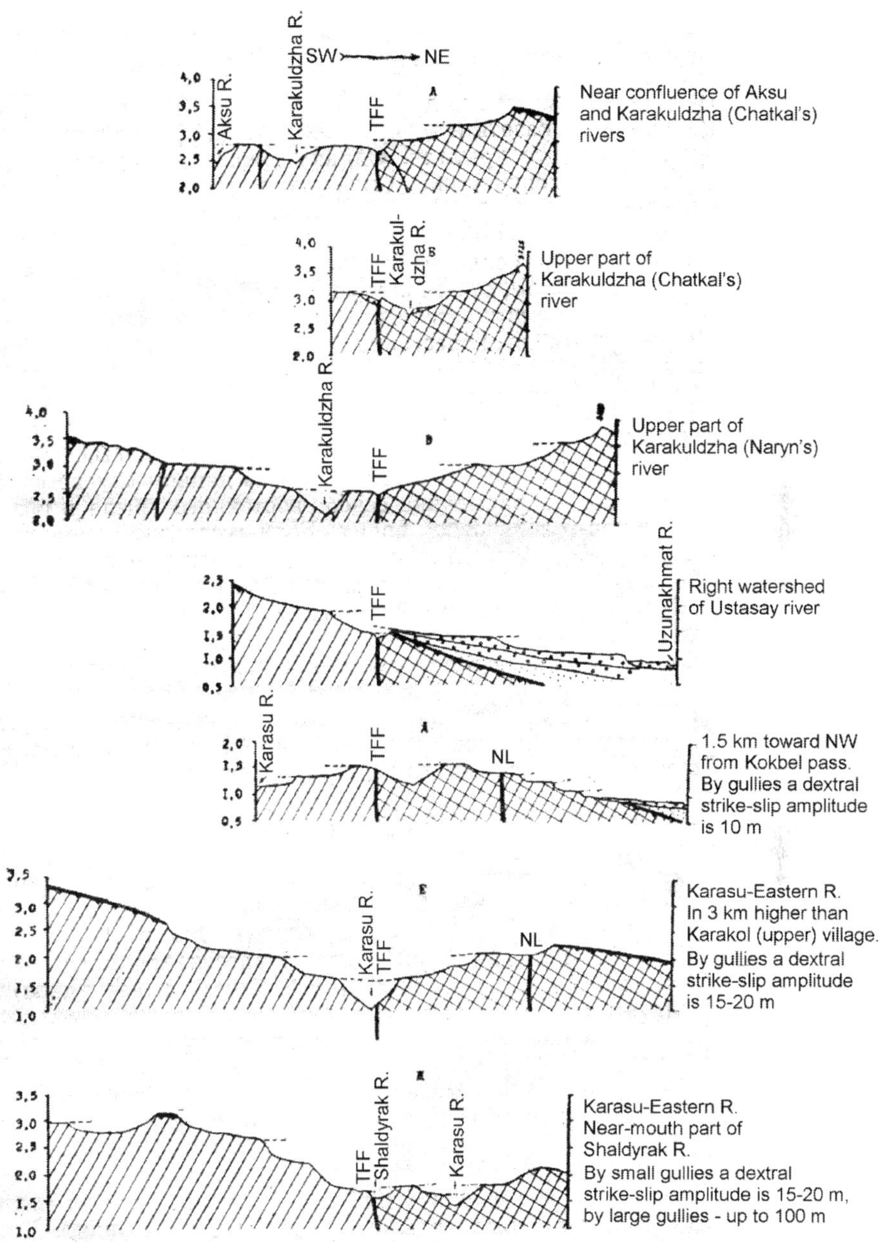

Fig. 14. Geological-geomorphological cross-sections of the Talas-Fergana Fault zone (modified from Chediya, 1986). T-F – Talas-Fergana Fault, L-N – the line of Nikolaev.

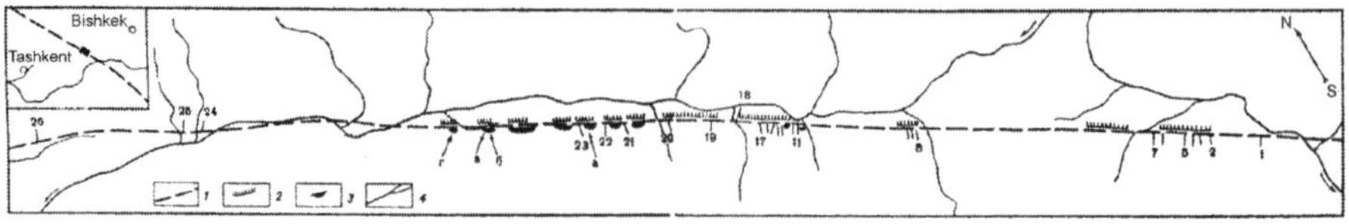

Fig. 15. Talas-Fergana Fault in the Karakuldzha-Narynskaya and Karakuldzha-Chatkalskaya river basins (modified from Burtman et.al., 1987). 1 - line of the modern fault, 2 - tectonic swell, 3 - near-fault valleys, 4 - streams.

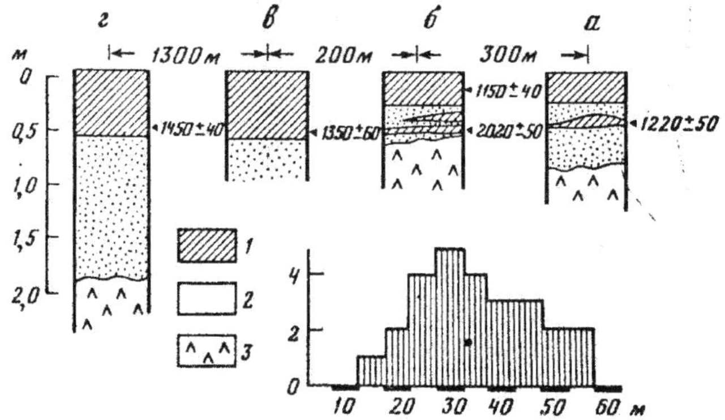

Fig. 16. Position of radiometric samples in sections of near-fault saggings and the histogram of recent movements along the fault (modified from Burtman et.al., 1987): 1 - soil and peat rich in organic material; 2 - clay, loam, clay sand; 3 - basement rocks.

swamping along the line of the Talas-Fergana Fault. V.S. Burtman et al. (1987) also examined a sample (GIN-4299) collected in the upper part of the peat bog at the point 11 (Fig.15). The age of the sample is 250±50 years that does not contradict to the age given above.

The investigation carried out by V.S. Burtman et al. (1987) allows conclusion that a displacement with a horizontal amplitude 30-40 m and small vertical component occurred along the Talas-Fergana Fault at historic time. This displacement took place after an earthquake occurred about 2000 years ago.

Passing downward along the Karakuldzha-Narynskaya river up to the place where the fault divides the Atoinock uplift and Ketmentyube depression one can observe turning of the hanging north-eastern wall of the fault into the foot wall (Fig. 14 D). On the right watershed of the Ustasai river the amplitude of vertical displacement is equal to previous one in the upper reaches of the Karakuldzha-Narynskaya river but with reverse orientation. Near the Naryn river valley the amplitude reaches more than 1000 m (Chediya, 1986).

V.G. Trifonov et al. (1990) found Holocene gullies and small watersheds on the right bank of the Dhzanaryksay river shifted for 6, 8-9 and 14-17 m, respectively. The upper reaches of the river with traces of trough structure and a lateral moraine aging to the end of the Middle Pleistocene are shifted for 1.5 km relatively nowadays cut off lower parts of the valley. In other words the shifting of the Dzhanaryksai river valley started in the beginning of the Late

Pleistocene, therefore the average slip rate is about 1.5 cm/year. On the right bank of the Dzhanaryksai river a terrace aging to the beginning of the Late Pleistocene as well as the river inflows, formed at the same time, are shifted to the right for 550-650 m. Terraces and valleys formed in the end of the Late Pleistocene are shifted for 150 m.

Points	Distance from a previous point	Displacement, m
1	-	35
2	600	35
3	175	25
4	50	30
5	50	30
6	250	24
7	100	45
8	3500	23
9	16	23
10	70	40
11	2200	50
12	10	50
13	120	30
14	33	30
15	30	30
16	100	40-45
17	20	40-45
18	600	50-60
19	110	35
20	600	35
21	700	20
22	300	17
23	400	20
24	3600	40
25	300	40
26	2100	50-60

Table 2. Horizontal displacement of the relief forms along the fault in the Karakuldzha-Narynskaya and Karakuldzha-Chatkalskaya river valleys (modified from Burtman et.al., 1987)

To the north-west the fault is divided into several branches. Thus, there are two parallel branches in the Ustasai river valley. The north-eastern one is younger. One can observed right-lateral displacement of Late Holocene channels along the branch for 10 m. As for the south-western branch one can observe galleys cutting the Late Pleistocene terrace and shifted for 150 m to the right. The last displacement probably characterizes the whole Holocene slip; in this case the slip rates is about 1.5 cm/year. Thus, average rate slips of Holocene and Holocene-Late Pleistocene displacements along the segment of the fault in the north-west of the Toktogul reservoir display the same tendency as those in the Late Holocene: they increase in 1.5-2 times comparing with slip rates on the south-eastern portions of the fault.

We pointed above, that many researchers reported about displacement of channels of the gullies on first tens meters. Thus V.S. Burtman et al. (1987) cited evidences of strike-slip movements in Karakuldzha-Chatkalskaya and Karakuldzha-Narynskaya river basins (Fig. 15, Table 2).

V.S. Burtman et al. (1987) point that the same width and morphology of displaced parts of the dry valley above and below the strike-slip fault testifies on a fact that the displacement occurred fast. This circumstance has allowed V.S. Burtman et al. to tie such displacements with earthquakes. Citing on a fact that every valley has its only one ancient continuation, they came to a conclusion that observed displacements are the result of one strong earthquake.

For us it's seemed impossible that one-act horizontal displacement along the strike-slip fault can reach a value of 60 m during one event. World experience of recent strong and catastrophic earthquakes gives us examples of 10 m, maximum 15 m displacements during one event (Strom and Nikonov, 1997). It is not clear also a value of possible creep displacements, their contribution into total value of observed displacement of valley's thalvegs. Most probably a number of the strong earthquakes, occurred along described segment of the fault, have led to formation of a shift of gullies on few tens meters.

The same authors (Burtman et al., 1987) cite data on radiocarbon dating of the samples collected by them in pits excavated in the fault zone (Fig. 15 and 16).

In Fig 16 one can observed that at investigated segment of the fault the dated soils give us at least 4 events led to accumulation of loose slope material in a near-fault depression where soil was formed later. Minimum ages of these events:

1. 2020±50 years,
2. (1450±40 + 1350±60) / 2 = 1400±50 years,
3. (1150±40 + 1220±50) / 2 = 1185±45 years,
4. 250±50 years.

Stations ## 22 and 23 are most close to places where samples were collected. There 17 m and 20 m displacements correspondently were measured. Most probably for these total (cumulative) values of displacement there were responsible 3 strong earthquakes, that is 6-7 m during one event in average. These values are in agreement with the world data analysied by A.L. Strom and A.A. Nikonov (1997).

3.4 Sary-Bulak test site

The second test site was located in the Sary-Bulak river basin of the Toktogul Region of the Jalal-Abad oblast (Fig. 1 and 2). The mapped area was in the left bank of the upper part of the Sary-Bulak river (the right tributary of the Uzun-Akhmat river). Here we observed the right-lateral displacement of many river beds of small water and dry gullies, as well as watersheds between them.

We performed mapping with electronic tachymeter of the area in the upper part of the valley of a large dry stream along the TFF zone (Fig. 17 and 18). Here systematic displacement to the east (to the right) took place at 65,40 m (section A-Б in Fig. 17); 113,06 m (section A-B in Fig. 17) and 336,49 meters (section A-Г in Fig. 17) of the bottom parts of the dry valley. Age of the displaced dry valley, apparently, is the beginning of late Pleistocene, since it cuts the mid-Quarterly alluvial surface.

Bore pits, dug on the left slope of the «legless» valleys (Fig. 19), give a similar picture of the near-surface deposits developed there. At the top there is the soil developed on the loess-like loam. This layer of loam with scattered detritus fragments, probably, is «a colluvial wedge», which collapsed downwards from the nearby slope during the earthquake. Soil formation processes started on loamy deposits after that collapse about 5 thousand years ago (samples SOAN-6529 and SOAN-6531). The age of the above mentioned samples is the minimum age of the earthquake which had occurred there. The samples, which were taken in the top part of the layer, gave radiocarbon ages of 1130 ± 100 years old (SOAN-6528) and

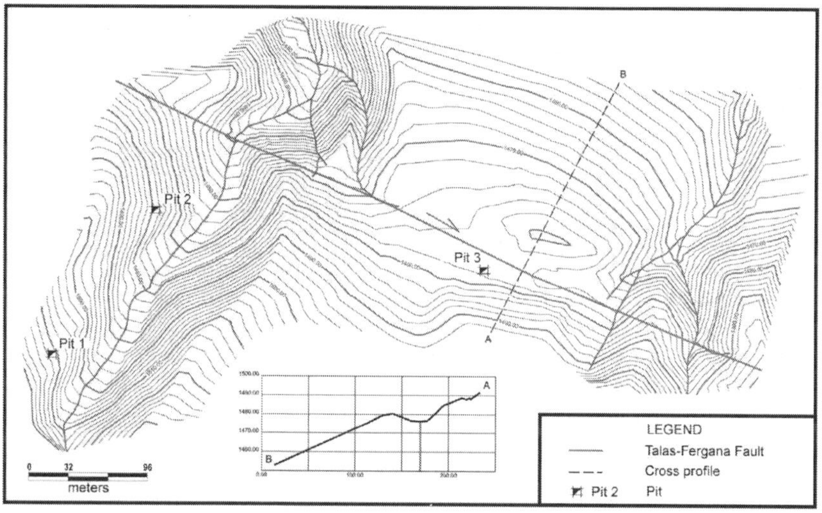

Fig. 17. Sary-Bulak test site and main elements of relief.

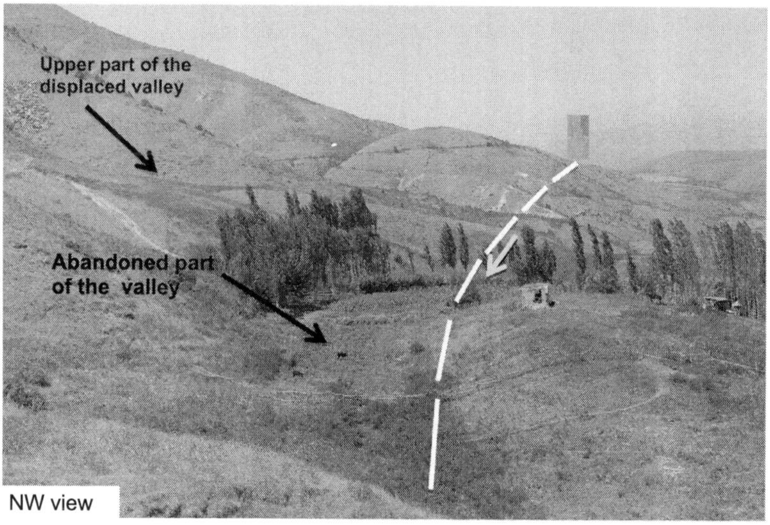

Fig. 18. A fragment of the upper part of the valley of the Sary-Bulak River.

440 ± 45 years old (SOAN-6530). These age determinations indicate the minimum time periods when there were later soil deformations in the investigated area, apparently, caused by the next seismic motions. These deformations have led to burial of soil fragments, where we have taken samples for determination of absolute age, - to switching on in these fragments of "the geological counter of time».

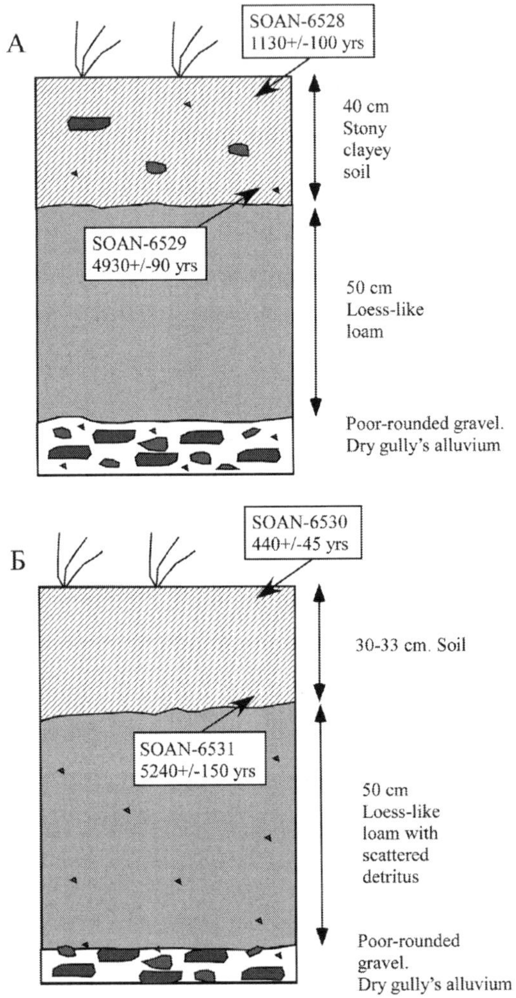

Fig. 19. Bore pits1 (A) and 2 (B) on Sary-Bulak test site.

Bore pit 3 (depth of 155 cm), which was dug directly in the fault zone in the abandoned part of the valley, has shown a more complicated picture (Fig. 20). Here from top we observed a significant layer of reclaimed soil with underlying loess-like loam. Under the loam layer there is a layer of buried soil. Contact of the layer of loam and buried soil is very uneven (see Fig. 20).

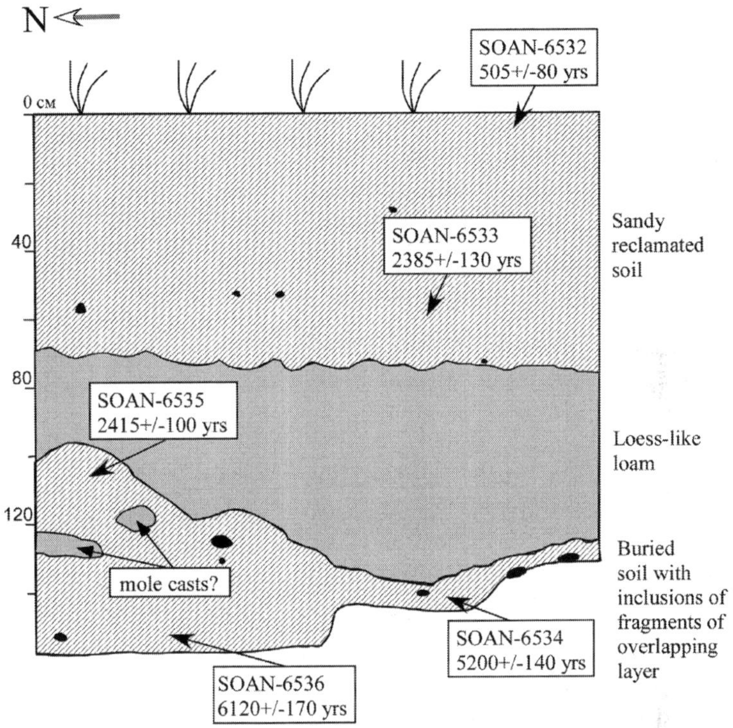

Fig. 20. Bore pit 3 at the Sary-Bulak test site. Recent reclaimed soil is separated from the buried soil by the «colluvial wedge». Black stains in the drawing are pebbles of average roundedness.

The available section gives us sufficient material for sedimentation history reconstruction and tectonic development of the site in mid-late Holocene. The buried soil started to be formed about 6 thousand years ago (sample SOAN-6536), apparently, after the material displaced during the earthquake from the nearby slope was stabilized in the riverbed of the abandoned valley.

The following sample - SOAN-6534 taken above is an evidence of one more earthquake which occurred about 5 thousand years ago, which traces have been found by us in the neighboring bore pits 1 and 2. The next seismic event took place approximately 2 400 years ago (samples SOAN-6533 and SOAN-6535). It has led to destroying of an ancient soil cover, its coverage by the «colluvial wedge» which is in turn covered by modern reclaimed soil.

The latter seismic event known to us occurred about 400-500 years ago (SOAN-6532). Its traces are reflected also in bore pit 2, which was dug nearby.

It is hardly probable, that in past 6 thousand years there were occurred a displacement at 336.49 meters along the fault (Fig. 17, section A-Г). However displacement at 65.40 m (section A-Б) or 113.06 m (section A-B) is plausible, considering the data published in the world literature (see Strom's and Nikonov's executive summary published in 1997). It is necessary to note, that Burtman et al. (1996) informed on regular displacement of valleys of temporary waterways along the Talas-Fergana fault by 110 ± 10 m in the (from the west to the east) Ustasay, Sary-Bulak, Dzhanaryksay river basins.

Thus, average rate of displacement along the TFF at the Sary-Bulak test site, since middle of Holocene, can reach 10.7-18.5 mm/year. Our results are close to the data received by Burtman et al. (1996) on the site located in several kilometers SE from Sary-Bulak. In the Dzhanaryksay river basin the mentioned authors calculated rate of displacement of 9-12 mm/year during last ~1.5 thousand years on 14-metre displacement of a dry gully.

Let's notice, that the closest age determinations of 1150±40 years old (GIN-4302) and 1220±50 (GIN-4304) were received by Burtman et al. (1987) in the bottom part of modern soil along the TFF line in upper area of the Chatkal river, approximately at the distance of 70 km to NW from Sary-Bulak test site. These data assume not shorter length of the seismogenic rapture.

Probably, one more age determination got by Trifonov et al. (1990) concerns this age group: 1240 ± 60 year ago; it was received by them on the right bank of the Keklikbel river. The latter site is also located at a considerable distance of more than 100 km to SE from the Sary-Bulak test site. Then the plane of the seismogene rupture begins with upper reach of the Chatkal river and stretches for the distance of almost 200 km are to the valley of the Keklikbel river. Theoretically it is possible: we can recollect, that the length of the well studied rupture of the Kebin (Kemin) earthquake of 1911 (M=8,2) in Northern Tian-Shan reached this value (Bogdanovich et al., 1914). However in the valley of Keklikbel river probably also that there was occurred an independent earlier event.

Thus, minimum number of the earthquakes which occurred at this site in Holocene is 5. Their minimum ages are ~ 6 thousand years ago, ~ 5 thousand years ago, ~ 2.4 thousand years ago, ~ 1.1 thousand years ago and 400-500 years ago.

The earliest (~ 6 thousand years) and latest age determinations (400-500 years ago) coincide with the absolute data received at the Kara-Bura test site, located at the distance of 90 km to NW from the Sary-Bulak test site. This fact assumes similar length of the seismogenic rupture formed twice in the middle and in the end of the Holocene.

Data of absolute age (400-500 years ago), received on the Kara-Bura and Sary-Bulak test sites coincide also with archeoseismologic data on the destroyed caravanserai located in an middle part of the Kara-Bura river valley (Korjenkov et al., 2009).

4. Middle part of the Talas-Fergana fault

4.1 Investigations in Ustasay-Karasu interfluve

In a distance of 6 km south of the Naryn river the Talas-Fergana fault limits the Ketmentyube depression filled with the Toktogul reservoir. Features of vertical movements are analogous to previous ones. Further, up to the Kokbel pass, the fault plane passes on the left bank of a nameless channel which is drained to the Naryn river. The whole low left slope of the valley is cut with short (up to 500-700 m) small gullies which were probably formed in the Late Pleistocene-Holocene (Chediya, 1986). The whole slope with all gullies is cut by the Talas-Fergana fault. In a profile of the gullies' watersheds one can clearly observe that a straight depression is located along the fault. Mainly it filled by the colluvium-alluvium material of few meters thickness, which is shelved from upper part of the slope, i.e. it is the amplitude of the Holocene thrusting of the Kochkortobe uplift. Comparison of the Early Pleistocene surface (E in Fig. 14) gives the value about 150 m (Chediya, 1986).

Although a trace of the fault in the Ustasay-Zhanaryksay region is less expressed, than in south-east or further north-west, sections of well expressed displacements are distribute along whole zone of the fault, and right-lateral displacement is evident (Fig. 21). Maximum values of displacements of watersheds and spring valleys near station # 14 reach 110 ±10 m (Burtman et al., 1996).

Burtman et al. (1996) have measured a displacement of a small gully - 14±2 m in the field station #14 (Fig. 21). A pit in upper part of the gully near the fault trace reaches hard rocks in a depth of 0.5 m under the soil of 0.45 m thickness. Organic part of the soil - 0.1 m from the whole layer (depth 0.35-0.45 m) gives the radiocarbon age of the seismic event equal 1440±30 years ago.

We have studied segments of the Talas-Fergana Fault zone in 2 km south-east from the Aktaybulak-Korumtokay interfluve – in a region of the Ustasay-Sarybulak pass. Everywhere we measured right-lateral displacement of spring beds and watersheds between them on a value of first ten meters (Fig. 22 a). Eastwards, a displacement of the spring valleys and watersheds between them reaches hundreds meters (as, for example, in the right bank of the Sarybulak River). It is necessary to mention also a vertical uplifting of southwestern limb of the fault on a value of 20 m. In tie with lateral movements along the fault, in some places the consequent spring valleys, flowing down the slope, are blocking by so-called barrier ridges (Fig. 22 b), in front of which there are forming a local depressions, where fine material is accumulated.

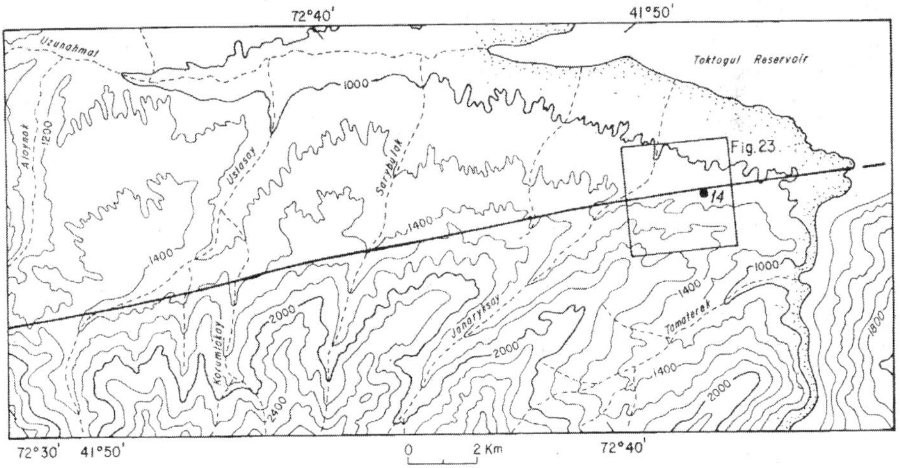

Fig. 21. Topographic map of a fragment of the Talas-Fergana Fault in the Ustasay-Dzhanaryksay interfluve (modified after Burtman et al., 1996). Black line marks a trace of the Talas-Fergana Fault. A location of the field station # 14 is shown. Contour interval in 200 m.

Fig. 22. a. Right-lateral displacement of a channel of a dry gully (white arrows) and watershed ridge along the line of the Talas-Fergana Fault (shown by the dotted line).

Fig. 22. b. Barrier ridge (a shepherd tent is by foot of it) in a zone of the Talas-Fergana Fault (shown by the dotted line) in the Aktaybulak-Korumtokay interfluves.

Fig. 22. c. A structure of the Talas-Fergana Fault zone in the Aktaybulak-Korumtokay interfluves. Fault gouge, filled the fault zone, is overlaid by folded alluvial deposits, in which fragments blue clays laying below are found.

In 500 m south-east of previous field station an irrigation canal exposes the fault zone. In south-western wall of the canal there are exposed (up-down):

- Greyish-black soil;
- Light-brown loess-like loam;
- Folded grey alluvial deposits with inclusions of torn fragments of the blue clays' horizons;
- Folded blue clays with inclusions of deformed loam horizons.

Last two horizons one can observe in Fig. 22 c. A layer, consisting of clays and loams, is representing itself fault gouge, filled the fault zone.

4.2 Kok-Bel test site

The third test site, examined by us, is located approximately at the distance of 30 km to the southeast from Sary-Bulak test site (Fig. 1 and 2). Here during climbing up the pass with the same name from the Ketmen-Tyube depression a systematic right-lateral displacement of small dry gullies and watersheds between them is observed along the TFF zone (Fig. 23). Here we also produced a detailed digital map of the site of the TFF zone by an electronic tachymeter (Fig. 24).

Along one kilometre section of the fault we measured systematic displacement of forms of the modern relief, the upper parts of which are shifted to the right along the fault line in a horizontal direction (Table 3). All measured elements of the relief were small dry valleys of temporary waterways - "says", as well as watersheds between them.

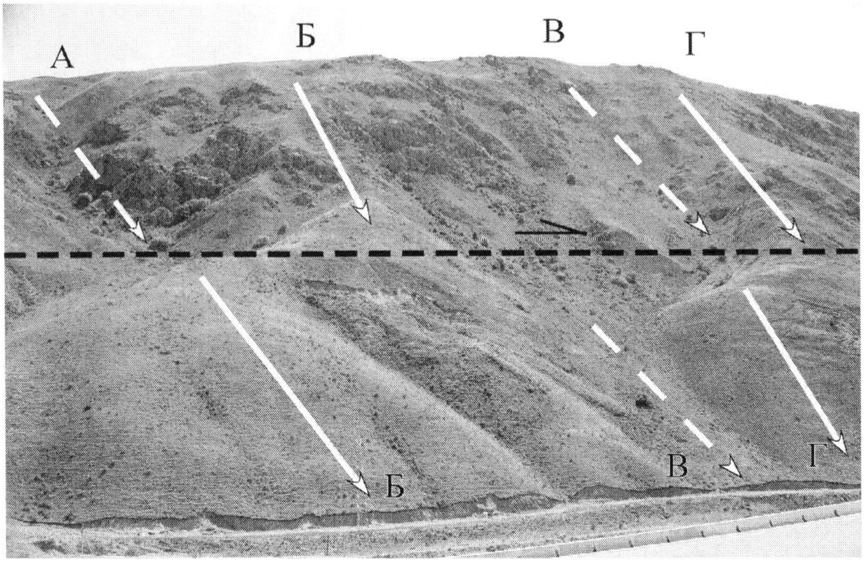

Fig. 23. Displacement to the right of dry rivulets (says) (white faltering lines) and watersheds between them (black faltering lines) along the Talaso-Fergana fault line (a sub horizontal faltering line) to NW from the Cook-Bel pass.

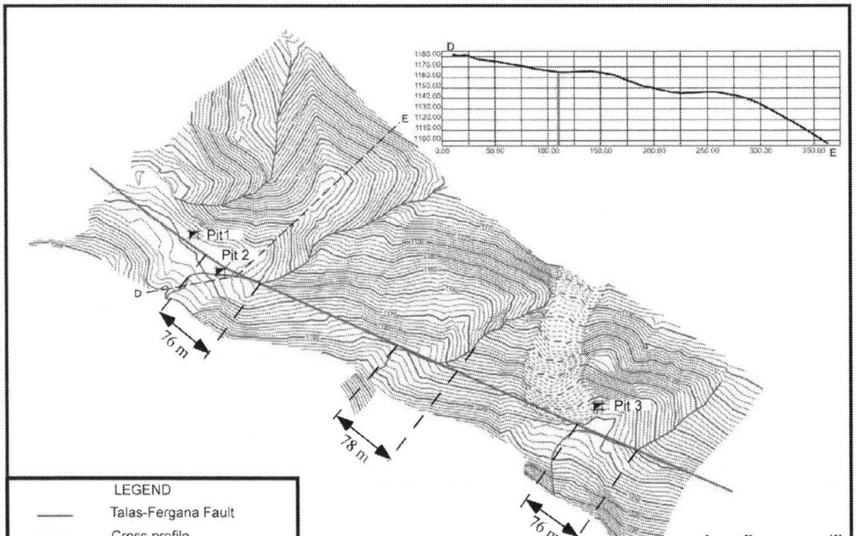

Fig. 24. The Topographic map the Kok-Bel test site, drawn according to shooting by electronic tachometer. Values of displacement of dry waterways beds are shown.

No of points/sites	The displaced element of the relief	Displacement size/value, m
1	Watershed	110,54
2	Dry rivulet (say)	47,62
3	Watershed	97,17
4	Dry rivulet (say)	53,8
5	Watershed	39,62
6	Dry rivulet (say)	15,7
7	Watershed	41.58
8	Dry rivulet (say)	57,63
9	Watershed	76,23
10	Dry rivulet (say)	52,93
11	Watershed	50,14
12	Dry rivulet (say)	63,83
13	Watershed	51,36
14	Dry rivulet (say)	116,3
15	Watershed	68,71
16	Dry rivulet (say)	98,55
17	Watershed	99,03
18	Dry rivulet (say)	13,26
19	Watershed	16,94
20	Dry rivulet (say)	70,11
21	Watershed	77,70
22	Dry rivulet (say)	70,36
23	Watershed	61,96
24	Dry rivulet (say)	64,83
25	Watershed	50,92

Table 3. Horizontal displacement of forms of a relief along the Talas-Fergana fault to NW from the Kok-Bel pass, measured by electronic tachometer

To NW from the Kok-Bel pass we measured 25 valleys of dry gullies (says) and watersheds between them displaced to the right at the distance of up to 120 m. Pay attention at the various representativeness of displacement of different amplitude: some maxima are emphasized whereas intermediate values are absent (Fig. 25). Similar non-uniformity according to Trifonov et al. (1990) indicates the decisive contribution of impulse seismogenic motions into total movement.

To north-east from the the Kok-Bel pass along the Talas-Fergana fault we dug 2 bore pits and cleaned a natural exposure - a scar of a landslide, which was directly in the fault zone (Fig. 26 and 27). Value of horizontal displacement of the dry rivulet and the adjacent watershed, measured by electronic tachymeter, comprised 76 m. We correlated this value with the most ancient date - 4900 ± 230 years. Thus, we receive probable rates of horizontal tectonic movements since mid Holocene equal to 14.81-16.27 mm/year.

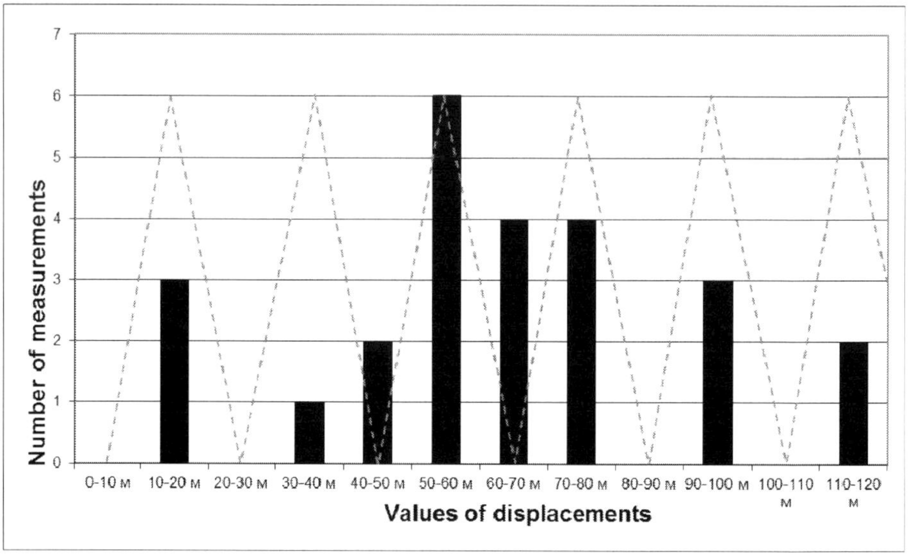

Fig. 25. The histogram of amplitudes distribution of Holocenic right-hand-side displacements of dry water currents and watersheds between them on the Talas-Fergana fault to NW from the Kok-Bel pass (only 25 measurements).

The last strong earthquake on the examined site occurred, apparently, about ~275 years ago. An evidence to that are absolute dates of samples SOAN-7021 (270 ± 85 years), SOAN-7024 (370 ± 90 years) and SOAN-7026 (240 ± 50 years), taken in bore pit No 1 and clearing No 3 in the lower parts of the modern soil developed on the underlying «colluvial wedge». Similar dates were received also by Burtman et al. (1987) along TFF line in upper reach of the Chatkal river - 250 ± 50 years (GIN-4299). Though distance between the Kok-Bel test site and the Karakuldzha-Chatkalskaya river valley is about 100 km, but apparently this was the length of the plane of the seismogenic rupture of this age.

In the same bore pit at the depth of 30 cm we found a fragment of ceramic ware. The form is spherical, the fragment was produced on a potter's wheel from well mixed clay with addition of coarse-grained and fine-grained sand. Firing of the fragment was uniform. The fragment was blackened with smoke. The indicated signs are the most typical for Middle Ages epoch-VIII-XII centuries AD (K.S.Tabaldiev, a written communication). For narrowing down the date determination additional artifacts are necessary. On the other hand, manufacturing of similar thin-walled spherical ceramics in Fergana valley and in its foothills is an ancient tradition. Production of ceramic vessels made on a potter's wheel were introduced in I millennium BC. They are characterized by red angob and painted ceramics which we do not see in this fragment. The described fragment of ceramics had been buried during the collapse of the colluvial wedge which occurred during the last strong earthquake. Subsequently on sediments of this wedge the soil the age of which was determined as 270 ± 85 years old (SOAN-7021) was formed.

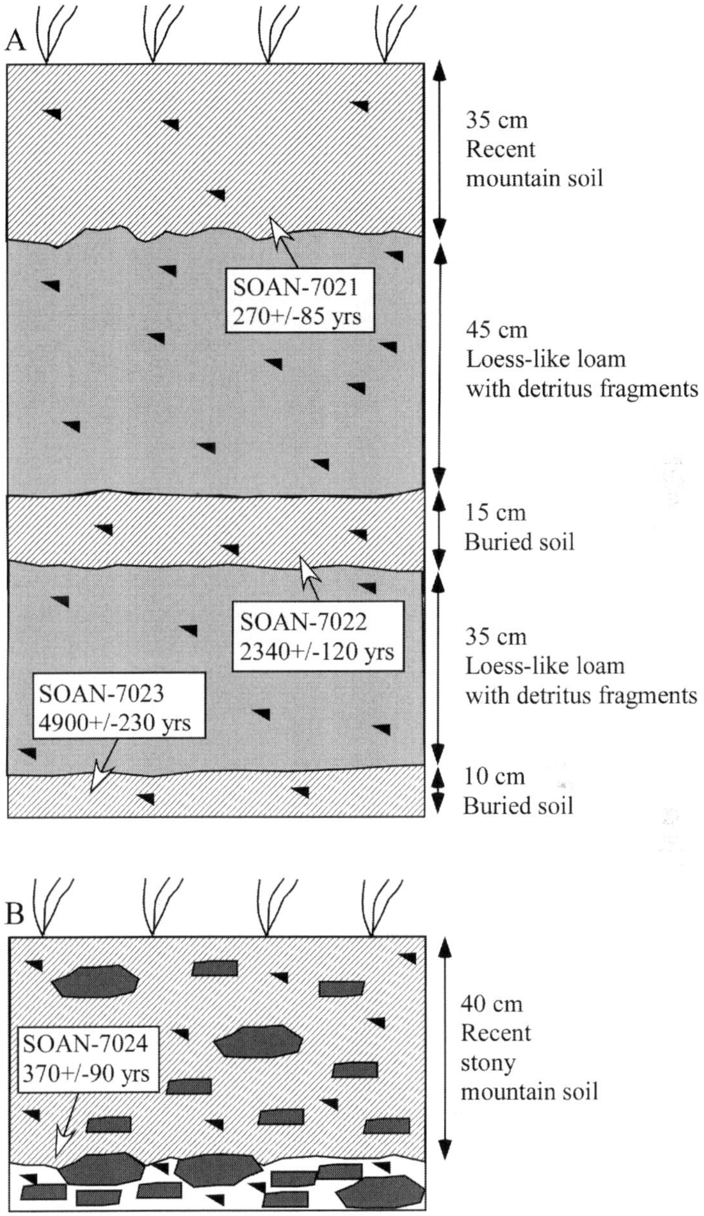

Fig. 26. Schematic drawing of bore pits No 1 (A) and No 2 (B), which were examined in the Talas-Fergana fault zone at the distance of 1 km to NE from the Kok-Bel pass.

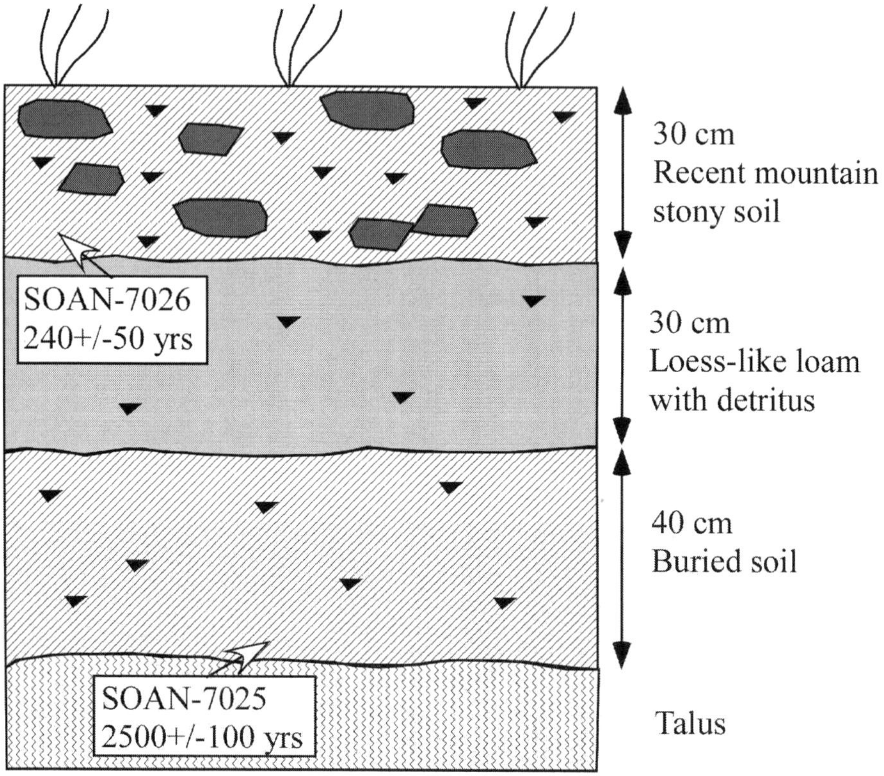

Fig. 27. A schematic drawing of clearing of the natural exposure - a scar, which was formed due to landslide collapse in the zone of the Talas-Fergana fault at the distance of 1 km to NW from the Kok-Bel pass.

The following strong earthquake occurred approximately ~2400 years ago. Absolute date determinations of two samples (SOAN-7022 - 2340 ± 120 years and SOAN-7025 - 2500 ± 100 years) testify to it, the samples were taken in bore pit No 1 and clearing No 3 in the lower part of buried soil. It is not excluded, that this earthquake has left its traces also in the river valleys: Kyldau (dated as 2320 ± 40 years ago according to Trifonov et al. (1990)) and Pchan (dated as 2180 ± 120 years - GIN -7052; 2280 ± 70 years - Beta-47550; 2540 ± 70 years - Beta-47549 according to Burtman et al. (1996)). If it so, the length of this seismogenic rupture reached 100-200 km.

The most ancient earthquake, the traces of which were found at the Kok-Bel test site, took place about 5 thousand years ago to what absolute dating of sample SOAN-7023 testifies.

4.3 Investigations in the Karasu-Kyldau interfluve

Further south-east the zone of the fault crosses the Toktogul water reservoir and stretches along the Karasu river valley (Fig. 1 and 28). However well-expressed displacements (up to tens of meters) along the fault one can observe south-west and north-east from the valley. Watersheds and spring valleys were displaced from ≈40 to 225 m and >2 km. Not far from the Kok-Bel pass, where main highway from the Fargana Valley crosses the Ketmen'-Tyube Depression which is occupied by the Toktogul water reservoir at present time, right-lateral displacements on ≈ 300 m are well expressed. It is especially well visible in south-western slope of a dry valley stretched parallel to the fault trace.

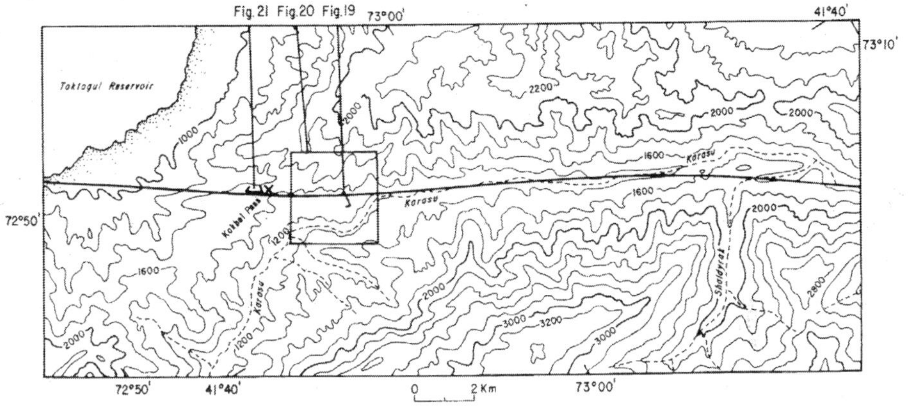

Fig. 28. A topographic map of the Talas-Fergana Fault Fragment along the Karasu River Valley (Burtman et al., 1996). A black line is a fault trace. Isoline intervals - 200 m.

In the south-east of the Kokbel pass the Talas-Fergana fault cuts the lower part of the right slope of the Karasu river valley for the distance of 6 km. The zone is characterized by strongly crushed Devonian deposits and numerous springs, that provokes to wide development of landslides. If to go upward on the Middle Pleistocene terraces of the left slope of the Karasu river one can observe a clearly expressed right-lateral slip along the fault. Here the length of galleys (therefore, time of their development) increases, and the slip amplitude is up to 15-20 m (Chediya, 1986). Analysis of geomorphologic levels does not allow determination of vertical displacement amount.

Further the fault passes on the left slope of the Karasu valley. In two kilometers of the place there is a seismic rockslide (considering its characteristic form) occurred along the fault and dammed the half of the valley. Here on the left slope of the Karasu river there are also gullies large in the length and time of development (since the Middle Pleistocene). One can clearly observe right-lateral displacement of lower parts of the gullies along the fault plane. If the displacement in small galleys reaches 15-20 m, the large ones are characterized by 100 m displacement (Chediya, 1986). If it is so, we can speak of long-term strike-slip fault which was active for the most part of the Quaternary. For example, E.Ya. Rantcman and G.N. Pshenin (1967) reported a 750 m displacement of a Middle Pleistocene moraine in the upper reaches of the Karasu-Eastern river.

A cross zone of the Shaldyrak river by the fault is very interesting. Meeting the fault, the river turns at the angle 90° and flows along the fault for the distance of 2 km, then it turns again and inflows into the Karasu river. The bending of river channel as well as a watershed between its lower parts and the Karasu valley agrees with the scheme of the right-lateral displacement for 2 -2.5 km which could occur during the Pleistocene. However, it would be irresponsibly to infer that basing only on one river bending. Comparison of hypsometrical elevation of the terrace Q_{II}^1 on both sides of the fault testifies to thrusting of north-eastern wall for 150 m (Chediya, 1986).

Above the Shaldyrak river mouth up to the Karasu lake the fault passes on the central part (near-bed) of the valley (Fig. 29); so we cannot speak of any evidences of displacement here.

South-east of the "Karasu" fragment the Talas-Fergana Fault crosses a high-elevated area where a moraine was displaced on ≈30 м (Burtman, 1964).

Further to the south-east along the TFF zone (the Kuroves and Keklikbel rivers' basins) V.G. Trifonov et al. (1990) reported on 32 structures shifted to the right for 36 m (see Fig. 30 c). Displayed maximums of displacement are not so contrasting as those reported by V.G. Trifonov et al.(1990) in the south-east (Fig. 30 a, b), and not correlated to them. V.G. Trifonov et al. (1990) supposed that either other impulses of movements occurred here or creep processes were prevailing. The second supposition is more probable due to abundant elastic clay-shale rocks in the portion of the fault. Movements along the fault and damming caused formation of small depressions with peat bogs on the right slope of the Keklikbel river valley in the surface of a moraine deposited by a small Late Pleistocene glacier. The radiocarbon age of one of samples collected in the bottom of one of the depressions is 1240±60 years (9 in Fig. 31). In the immediate vicinity of the depression one can observe only general displacement of the late Pleistocene moraine occurred during much larger time interval. However, in the north-west there is a number of Late Holocene displacements evidences with prevailing amplitude 10-12 m. If the radiocarbon age of the depression is characterized the displacement, the average slip rate does not exceed 0.8-1 cm/year, i.e. close or a little higher than that on the site to the south-east of the described one (Trifonov et al., 1990).

Fig. 29. Straight-line segment of the Karasu-Eastern River from the Shyldyrak massif to Karasu rockslide. The photograph was made from the rockslide body toward north-west.

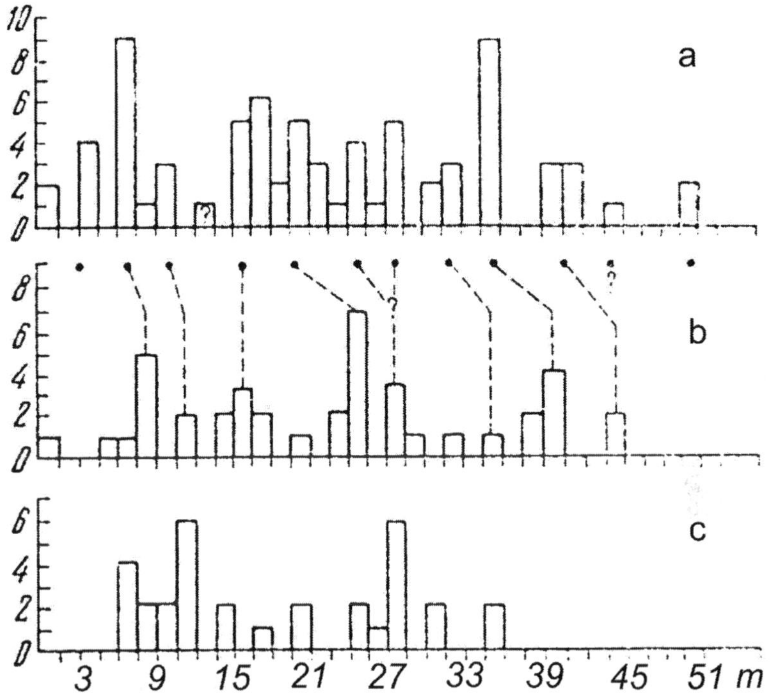

Fig. 30. Histograms of amplitude distribution for the Holocene right-lateral deviation of channels and other small forms of the relief along the Talas-Fergana Fault (modified from V.G.Trifonov et.al., 1990): a - from the Kok-Kiya pass to the Biruza river (points show probable strong earthquakes); б - in the Pchan and Kyldou river valleys; в - in the Kuroves and Keklikbel river valleys. The horizontal axis - amplitude of displacement; the vertical one - number of shifted forms of the relief.

On the right slope of the Keklikbel river valley V.G. Trifonov et al. (1990) found right-lateral displacement of channels for 60, 70-80, 135, 230, 300, 450-500 and 700-800 m. Shifted trough valleys and moraine deposits filling the troughs are of the greatest interest. The Late Pleistocene moraine is shifted for 135 m to the right. It is unclear if the time of its depositing is only the Holocene or the end of the Pleistocene too. More clear evidence is shifting of the Late Pleistocene moraine for 700-800 m relatively its trough valley. This moraine of about 10 m thick covers a depression cut in the surface of the beginning Middle Pleistocene (40-50 m), i.e. it was formed in the end of the Middle Pleistocene. Obviously, the moraine displacement can occur in the Late Pleistocene and Holocene, i.e. for the last 100 000 years, that gives the average slip rate 0.7-0.8 cm/year, which is close to the slip rate of the Late Holocene displacement between the Pchan and Kyldou river valleys. It is necessary to add that the mentioned amplitudes of displacements are accompanied by thrusting of the south-western wall of the fault. For displacement 135 m the thrusting is up to 5-6 m, for 300 m - 8-10 m, for 450-500 m and 700-800 m - several tens meter. Thus, the vertical component is 10-30 times less than the horizontal one (Trifonov et al., 1990).

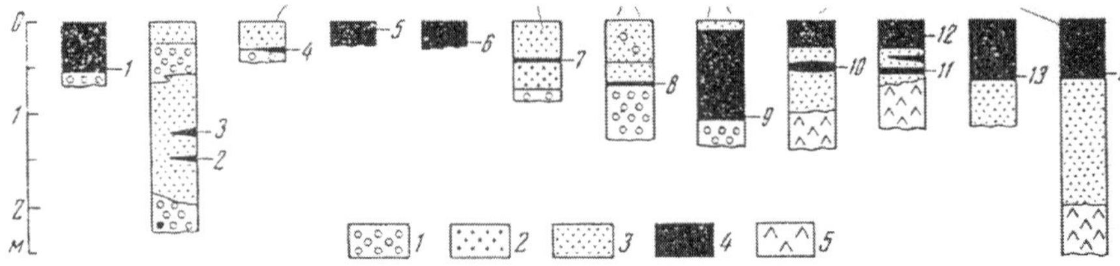

Fig. 31. Sections of near-fault depressions and their radiocarbon ages within the Talas-Fergana Fault (modified from V.G.Trifonov et.al., 1990). 1-3 – the first portion of the fault: 1 - 3970±40 years, left bank of the Bolsun river in the north-west of the Kok-Kiya pass. 2 - 15800±1300 years, 3 - 4590±100 years, the both ages are from a section on the right bank of the Chitta-Severnaya river in the north-west of the Dzhilangach river; 4-7 – the second portion of the fault: 4 - 3740±600 years, interfluve of the Biruza and Pchan rivers in the north-west of the pass between them. 5 - 3150±40 years, to the north-west of the previous point, 6 - 2640±600 years, upper reaches of the Pchan river in the south of the Chityndy pass, 7 - 2320±40 years, upper reaches of the Kyldou river valley; 8 – the third portion of the fault: 1510±60 years, interfluve of the Urumbash and Kuroves rivers; 9 –the firth portion of the fault: 1240±60 years, right bank of the Keklikbel river; 10-14 – the sixth portion of the fault (from V.S.Burtman et.al., 1987): 10 -1220±50 years, upper reaches of the Chatkal river, 11 - 2020±50 years, 12 - 1150±40 years and 13 - 1350±60 years, in 1 km to the north-west of the previous place, 14 - 1450±40 years, in 1,5 km to the north-west of the previous place.
1 – detritus, 2 – sand, 3 – clay, loam, sandy loam, 4 – the same rocks rich in organic material, and peat bogs; 5 – basement rocks.

Right bends of large river valleys (Kugart, Kaldama, Kyzylsu, Urumbash, Kuroves and oth.), described by V.G. Trifonov et al. (1990), allows inferring their right-lateral shifting for 3-4 km. These authors studied some of the bending channels and reconstructed their initial form. Minimal value of the right deviation is: for the Kongurtobe river – 2.8 km, the Kaldama river – 1.8 km, the Molasu river – 1.8 km, the Dzhindisu river – 2 km and the Urumbash river – 1.5 km. Since the mentioned river valleys were formed in the Early Pleistocene (Kostenko et al., 1972; Makarov, 1977), duration of slip processes is hundreds thousand years.

4.4 Investigations in the Kuroves-Malasu interfluve

In mentioned region the Talas-Fergana Fault goes along a foot of north-eastern slope of the Fergana Range. South-western limb of the fault, composed by Paleozoic rocks of the range, is trusted on many hundreds of meters, and north-eastern limb, covered by the Cenozoic deposits, is dropped on the same value. Here, in significant degree there are developed gravitation formations, the scars of which are in north-eastern slope of the Fergana Range, and landslide bodies descend into the Kazarman (Kugart) depression.

5. South-eastern part

5.1 Investigations in the Kyldou-Dzhilangach interfluve

There are a few evidences of Holocene slip along the TFF between the Pchan and Kyldou river valleys, since here the fault passes along river beds. V.G. Trifonov et al. (1990) found right-lateral displacement of small channels for 17-20, 34-37, 60-66 m. The radiocarbon age of a sample collected from loams composing a gulley terrace shifted for 17-20 m (?) is 1510±60 years (section 8 in Fig. 31). This means that the average slip rate is less than 1.1-1.3 cm/year.

On the second section - in the Pchan and Kyldou river valleys (see Fig. 30 b) V.G. Trifonov et al. (1990) marked 41 structures shifted to the right up to 45 m. Measured maximums are comparable with maximum displacements on the first site (from the Kokkiya pass up to the Biruza river), but differ (with 2 exclusions) by 10-20% larger slip amplitudes. This fact reflects increasing intensity of movements from the first section to the second one at a similar regime of development, i.e. prevailing of seismogenic movements.

For calculation of slip rates on the both sites V.G. Trifonov et al. (1990) investigated sections of near-fault depressions formed in result of damming of the channels by shifted gulley slopes and upslope facing scarps occurred simultaneously with the slip events. Sediments at the bottom of the depression correspond to the time of the movements (Fig. 31). They show slip rate about 0.5 cm/year for the first section (the sample age is 3970±40 years at total slip 19 m) and about 0.7 cm/year for the second one (3760±600 years at total slip of the gulley on 27±1 m). The values agree with other age determinations on the first and second section (see Fig. 31). Thus, in section 2 the age of a sample, collected at the bottom of a 1.5 m scarp synchronous to the 40 m slip along the northern branch of the fault, is 15800±1300 years, that gives the rate slip along the branch 0.25 cm/year. In sections 5-7 the age of samples from upper layers is 3150±40, 2640±600 and 2320±40 years at total slip 90±3, 25±1 and 23-24 m, respectively. Moreover, the first measurement was made in a gulley characterizing by a right-lateral displacement of the channel for 7-8 m occurred after filling the near-fault depression with sediments. The displacement gives the probable limit of slip rate 0.24 cm/year (Trifonov et al., 1990).

Between the Kokkiya pass and the Biruza river valley (the right inflow of the Pchan river) V.G. Trifonov et al. (1990) found 75 water channels and other forms of the relief shifted up to 50 m to the right. Different distribution of various amplitudes should be noted: there are 11 or 12 maximums while the intermediate values either are absent or represented by single shifted forms (Fig. 30 a). The irregularity was noted for the first time by R.E. Wallace (1968) at studying of Late Holocene slip along the San Andreas Fault in California. V.G. Trifonov (1985) studied the same phenomenon by the example of the Hangai, Kobdin and Dolinoozersk active faults in Mongolia, where the morphological features of displacement and character of the revealed irregularity are similar to those found on the studied site of the Talas-Fergana Fault. Obviously, here the irregularity testifies to prevailing impulse seismogenic movements in the total slip. The amplitude difference of maximum values on neighboring sites, i.e. slip amount at a single impulse of movement, varies from 3 to 6 m; the average value is 4-4.5 m.

It is interesting that the TFF is not active at present time according to GPS data (Midi and Hugger, 2001), and the maximum right-lateral displacement along the fault is estimated as 2-3 mm/year for the north-western portion and almost zero for the south-eastern one. Apparently, the main slip along the fault was caused by strong movements (several meters) during earthquakes. During the intervals between the earthquakes the fault is blocked and is characterized by strain accumulation, which drops once upon several hundreds-thousands years.

A fragment of the Talas-Fergana Fault "Kyldau" (Burtman et al., 1996) demonstrates a series of right-lateral displacements from 12 m for small gullies to 125 (±25) m for watersheds (Figs. 32 and 33). In a pit section located in north-western slope of the watershed, displaced on 125 (±25) m, in a fault zone a black soil of 0.3 m thickness overlays a brown soil of 0.6 m thickness. The last one overlays a clay and sand strata (Fig. 34). Contact surface of a brown soil and clay-sand layer is tilted on about 20° toward north-east. Radiocarbon dating gives an age 3962–4132 years B.P. (point 11, table 1), which is most ancient from the investigated samples in this part of the Talas-Fergana Fault. Seismotectonic deformations in the Talas-Fergana Fault zone are in association with gravitation formations, probably seismically-induced.

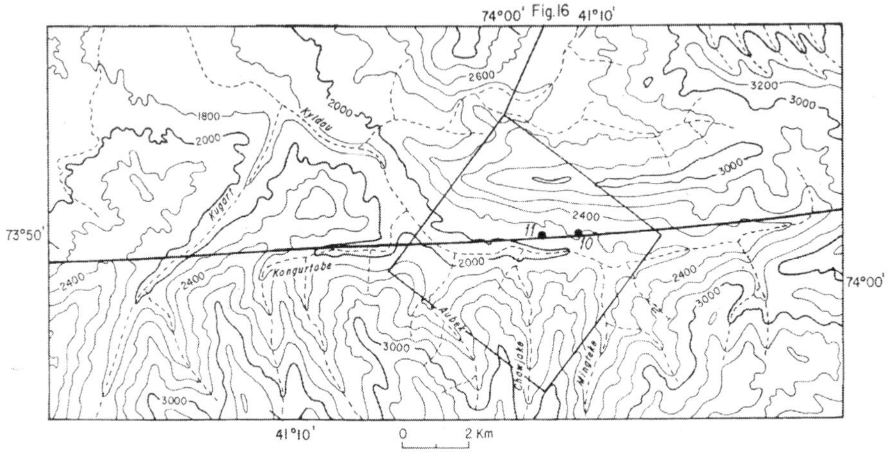

Fig. 32. Topographic map of the Talas-Fergana Fault along the Kyldau River Valley (Burtman et al., 1996). Black line is a fault trace. Figures point location of the field stations 10 and 11.

Fig. 33. The Talas-Fergana Fault goes along the right slope of the Kyldau river valley – south-western slope of the Yangyzkyr ("Lonely") ridge and displaces the watershed right on tens of meters.

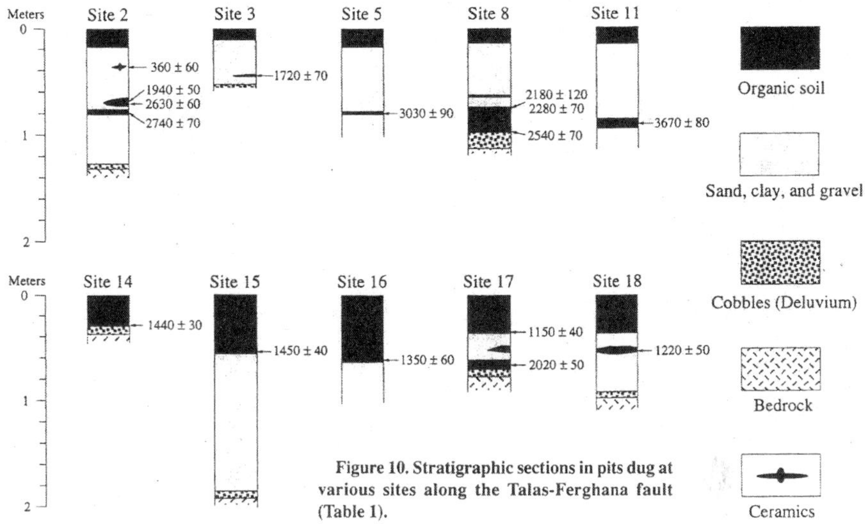

Fig. 34. Stratigraphic sections of pits excavated in different localities along the Talas-Fergana Fault by Burtman et al., 1996 (See also Table 2).

5.2 Investigations in the Dzhilangach-Pchan interfluve

Southern most part of the Talas-Fergana Fault, where there were collected samples for determination of the absolute age (Burtman et al., 1996), includes a segment of the fault located between Dzhilangach and Pchan river basins (Fig. 35). Both in the north and in its middle parts the Talas-Fergana Fault looks as a lengthy up-slope facing scarp of 10-30 m width and down to 5 m depth (Fig. 36). In given locality there are clear evidences of a horizontal component along the fault: displacement of gullies (Fig. 37 and 38) and watersheds (Fig. 39) among them on a value of few tens of meters. Thus along the

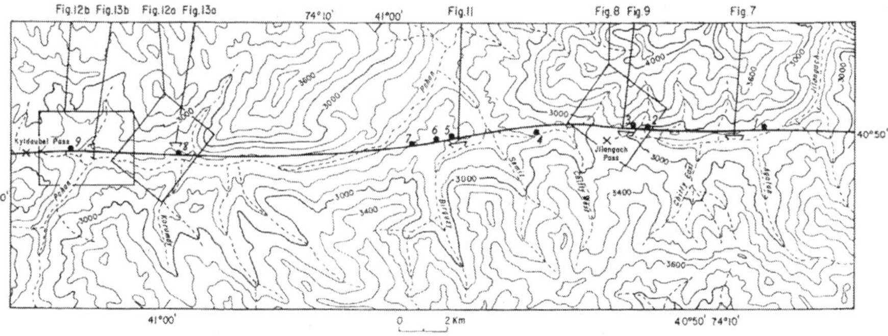

Fig. 35. Topographic map of a fragment of the Talas-Fergana Fault in the Pchan-Dzhilangach interfluves (Burtman et al., 1996). Black line is a fault trace. Numbers are localities of pits where samples for absolute age were collected. Isoline interval - 200 m.

Fig. 36. On whole its length a line of the Talas-Fergana fault is expressed as lengthy up-slope facing scarp (shown by pointers) which cuts any forms of relief. A region of the Dzhilangach pass.

Fig. 37. A region of the Dzhilangach pass. One can observe right-laeral displacement of gullies on 40-50 meters. Arrows placed in left slopes of mentioned gullies.

Dzhilangach river valley a value of horizontal displacement according to geomorphologic data is from 19 to 45 m (See Table 1), in Chitty-Western river basin - 40 m, in the Birguzy river basin - 27-35 m, in the Pchan river basin - from 21 to 90 m.

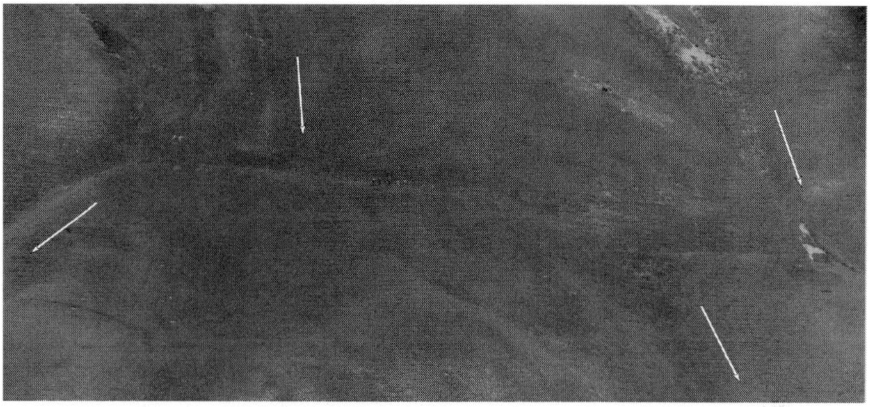

Fig. 38. Close view on a zone of the Talas-Fergana Fault. Left slope of the Dzhilangach river valley. One can observe right-lateral displacement of gullies. For a scale pay attention for a group of people in the center of photograph.

Fig. 39. Displacement of gully channels (solid line) and watersheds along the Talas-Fergana Fault line (dashed line) in a region of the Dzhilangach pass. "B" marks a watershed located higher a slope; "H" point its continuation down the slope. As a scale look on two shepherd tents higher and right of "H" in the center of photograph.

It is clear that such values of displacements cannot be attributed to only one earthquake. Indeed absolute dates (table 1) testifies on minimum eight earthquakes occurred in 15800, 4590, about 3955, about 3095, about 2635, 2230,1940, 1720 years ago.
In tie with mainly strike-slip movements along the fault, often in its zone there are formed specific deformation forms: so-called tectonic swells (Fig. 40), which are like a barrier ridges for consequent drainage and near-fault depressions - places of accumulations of fine material from adjacent slopes. In those depressions one has to conduct pit excavations for samples collection for absolute dating.

Fig. 40. A view toward north-west along the Talas-Fergana Fault. A region of the Semiz river mouth. Here the springs - tributaries o of the river flowing down toward south-west are dammed by the tectonic swells (shown by pointers) serving as barrier or "shatter" ridges for consequent drainage.

6. Discussion of the results - assessment of energy (maximum magnitude) of paleoearthquakes by the length of the seismogenic rupture

We have collected all absolute dates of paleoearthquakes occurred along the Talas-Fergana Fault obtained by us and combined them with the data obtained earlier by V.S. Burtman et al. (1987, 1996) and V.G. Trifonov et al. (1990, 1992) in table 4 and in Fig. 41. Totally there were collected 55 dates in 14 localities appurtenant to Late Pleistocene - Holocene.
An analysis of Fig. 41 has led us to a conclusion that the Talas-Fergana Fault zone can be generally divided onto three chains according to peculiarities of the lateral distribution of earthquakes. The first - north-western chain starts from the Kurkureusu river valley in the far west of Kyrgyzstan and stretches to about Sary-Bulak river valley in the Ketmen-Tyube depression. The second - central chain stretches toward south-east from the Sary-Bulak river valley to the Urumbash River (Kazarman depression) inclusively. Third - south-western

River valley or a name of the fault segment	Obtained radiocarbon ages, years	Average age of the earthquakes, years
Dzhilangach	1720 ±70 1940 ±50 2630 ±70 2740 ±70 3970 ±40	1720 ±70 1940 ±50 2685 ±70 3970 ±40
Chitty-Western	4590 ±100 15800 ±1300	4590 ±100 15800 ±1300
Birguzy	3030 ±90 3740 ±600	3030 ±90 3740 ±600
Pchan	2180 ±120 2280 ±70 2540 ±70 2640 ±600 3150 ±40	2230 ± 120 2590 ± 600 3150 ±40
Kyldau	2320 ±40 3670 ±80 3670 ±80	2320 ±40 3670 ±80
Urumbash	1510 ±60	1510 ±60
Keklikbel	1240 ±60	1240 ±60
Karasu	285 ± 35 975 ± 65 980 ± 55 1015 ± 75	285 ± 35 990 ± 75
Kok-Bel	240 ± 50 270 ± 85 370± 90 2340 ± 120 2500 ± 100 4900 ± 230	295 ± 90 2435 ± 120 4900 ± 230
Dzhanaryksay	1440 ±30	1440 ±30
Sary-Bulak	440 ± 45 505 ± 80 1130 ± 100 2385 ± 130 2415 ± 100 4930 ± 90 5200 ± 140 5240 ± 150 6120 ± 170	475 ± 80 1130 ± 100 2400 ± 130 4930 ± 90 5220 ± 150 6120 ± 170
Chatkal	250 ± 50 1150 ±40 1220 ±50 1220 ±50 1350 ±60	1385 ± 60 2020 ±50 1195 ± 50

	1350 ±60 1450 ±40 2020 ±50 2020 ±50	
Kara-Bura	405 ± 100 460 ± 40 480 ± 35 4465 ± 130 5910 ± 130 6100 ± 200	450 ± 100 4465 ± 130 5910 ± 130 6100 ± 200
Sulu-Bakair	5210 ± 155	5210 ± 155

Table 4. Radiocarbon dates of samples collected from displaced gullies along the Talas-Fergana Fault (by data of Korjenkov et al., 2009, 2010, as well as V.S. Burtman et al., 1987, 1996 and Trifonov et al., 1990, 1992)

chain starts from Kyldau river valley and ends in a region of southern state border of Kyrgyz Republic. We understand whole conventionality of our division, especially without materials on the Talas-Fergana Fault from Kazakhstan and China.

Because of uneven investigations and bareness of existed materials one has to talk very carefully on segmentation of the Talas-Fergana Fault zone. According to existing data we select 13 segments: 3 – in the northwestern chain of the fault, 5 or 6 in the central chain and 4 or 5 in southeastern chain (See Fig. 41).

An analysis and comparison of materials of Table 4 and Fig. 41 have allowed us to reveal 18 paleoseismic events, 17 of which occurred in the second half of Holocene (Table 5 and Fig. 42). We assessed also distances between localities, where there were determined absolute ages of the seismogenic displacements, occurred (supposedly) during one seismic event. We conditionally accepted these distances as minimum lengths of the seismogenic ruptures.

Some of the extreme values of rupture lengths, such as 270 km and 220 km for earthquakes occurred in 4530 and 1980 years ago, provoke a natural doubt. Although such length of the seismogenic ruptures is theoretically possible, however known strong historical earthquakes in the northern Tien Shan demonstrate maximum length of the rupture barely reaching 200 km: for example, Kebin (Kemin) earthquake of 1911 (Bogdanovich et al., 1914). It is possible that in such (and probably in some other) cases it took place an artificial unification of different earthquakes occurred in different parts of the fault, but in close time frames. Nevertheless one cannot exclude a possibility of propagation of many segments along almost whole plane of the Talas-Fergana Fault (during both discussed earthquakes there were united 11 segments in three chains of the disjunctive). An example of such propagation in the Tien Shan is mentioned Kebin earthquake during which there was a propagation of 6 fault segments of the fault zone close in time (Delvaux et al., 2001).

A distribution of paleoearthquakes along the Talas-Fergana Fault in time (for exclusion of individual "jumps") has clear evidence (Fig. 42). In interval 6000-4500 years ago the strong earthquakes occurred in north-western chain of the fault. Then a seismic activity has spread over to the south-eastern chain: 4500-2500 years ago. In the interval 2500-1500 years ago the strong earthquakes occurred in the central and south-eastern chains of the fault. Then the seismic activity is concentrated in north-western and central chains: 1500-250 years ago.

How the fault will behave in future is not clear. Here apparently it can be two variants of events' development: 1) the seismic activity will continue in north-western and central chains of the fault or 2) most probably will spread over to its south-eastern chain, where now there is so-called "a seismic gap": last earthquake occurred here already 1720 year ago!

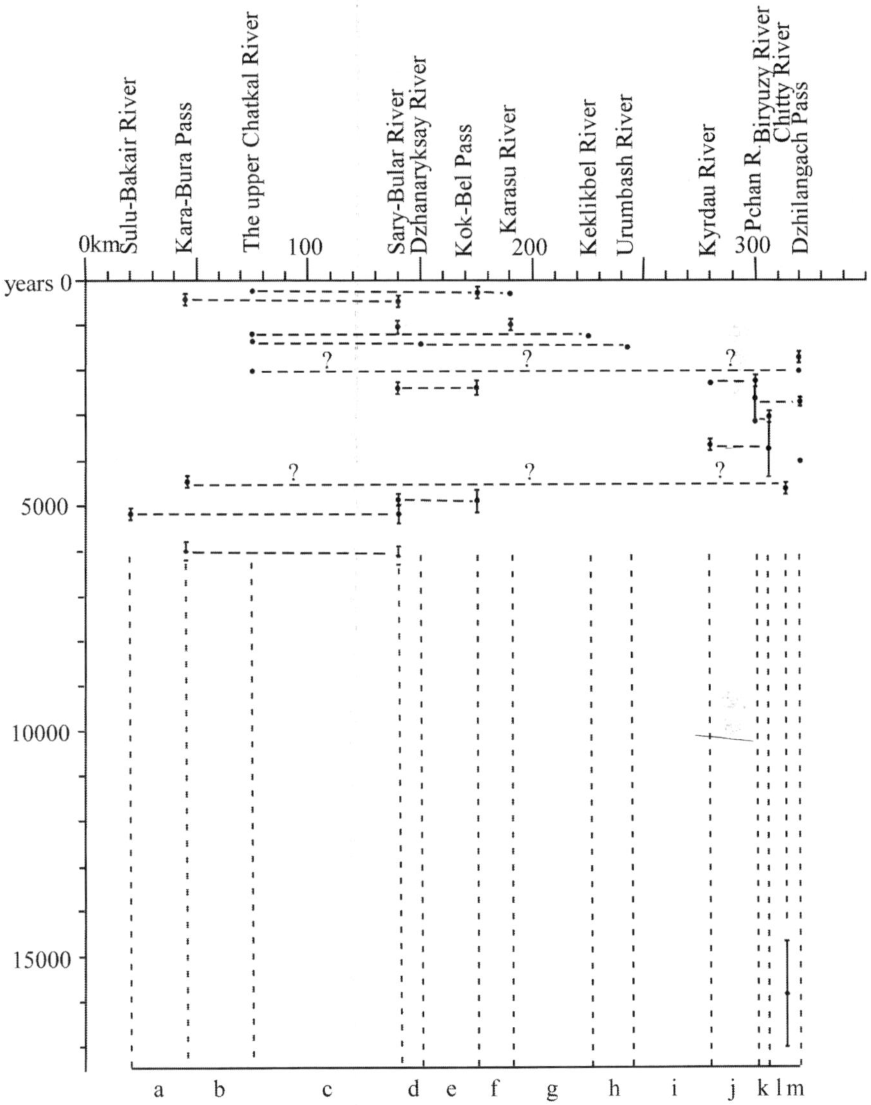

Fig. 41. Distribution of the earthquakes and segmentation along the zone of the Talas-Fergana fault in bounds of the Kyrgyz Republic

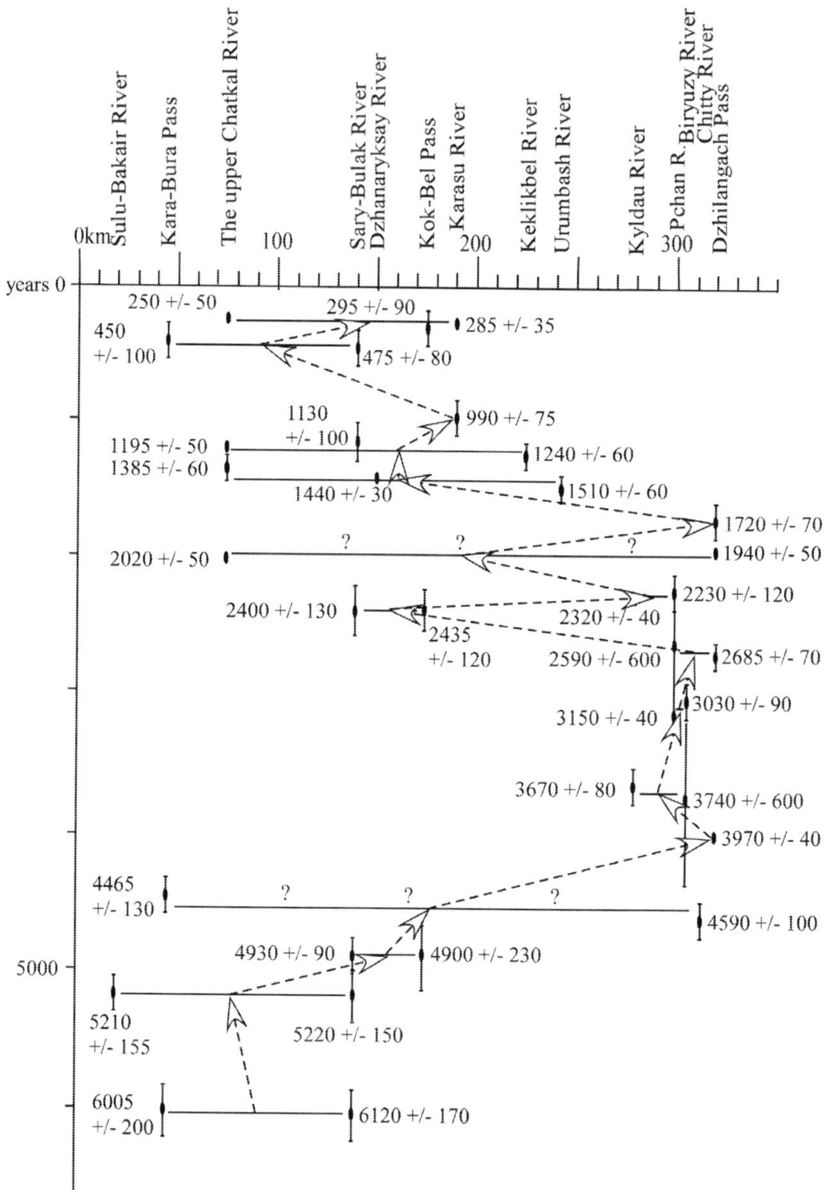

Fig. 42. Migration of the earthquakes along the Talas-Fergana Fault zone in Holocene

##	*River valley or a name of the segment where an earthquake has occurred (from – to)*	*Minimum length of the rupture, km*	*Average (calculated) age of the earthquake, years ago*	*Interval between given and previous earthquakes*
1	Region of the Kara-Bura Pass - the Sary-Bulak river valley	100	6065	
2	The Sulu-Bakair River valley - the Sary-Bulak river valley	120	5215	850
3	The Sary-Bulak River valley - the Kok-Bel Pass	40	4915	300
4	Region of the Kara-Bura Pass - the Chitty-Western River valley	270 ?	4530	385
5	A region of the Dzhilangach Pass	?	3970	560
6	The Kyldau River valley - the Burguzy River valley	30	3705	265
7	The Pchan River valley - the Birguzy river valley	10	3090	615
8	The Pchan River valley - the Dzhilangach Pass	20	2640	450
9	The Sary-Bulak River valley - the Kok-Bel Pass	40	2420	220
10	The Pchan River valley	?	2275	145
11	Head of the Chatkal River valley - a region of the Dzhilangach Pass	220 ?	1980	295
12	A region of the Dzhilangach Pass	?	1720	260
13	Head of the Chatkal River valley - the Urumbash River valley	170	1445	275
14	Head of the Chatkal River valley - the Keklikbel River valley	150	1190	255
15	The Karasu River valley	?	990	200
16	A region of the Kara-Bura Pass - the Sary-Bulak river valley	100	465	525
17	A head of the Chatkal River valley - the Karasu River Valley	120	275	190

Table 5. Average (calculated) ages of strong earthquakes occurred in the Talas-Fergana Fault zone in Holocene, interval between them and minimum length of the seismogenic rupture.

An important question in a concern of the long-term forecast of strong earthquakes is their reoccurrence. We have calculated intervals between strong earthquakes along the Talas-

Fergana Fault during the second half of Holocene (Table 5) where representation of the earthquakes is more complete. They rank from 145 to 850 years. Thus, an average calculated reoccurrence of the earthquakes along the whole zone of the Talas-Fergana Fault is 375 years. However, an arithmetic mean value is not the best characteristics of the natural phenomena. Comparison of number of the strong earthquakes along the fault with interval of their occurrence (Fig. 43) has allowed us to reveal three clear peaks of the earthquake occurrence in the second half of Holocene divided by intervals in 300 years.

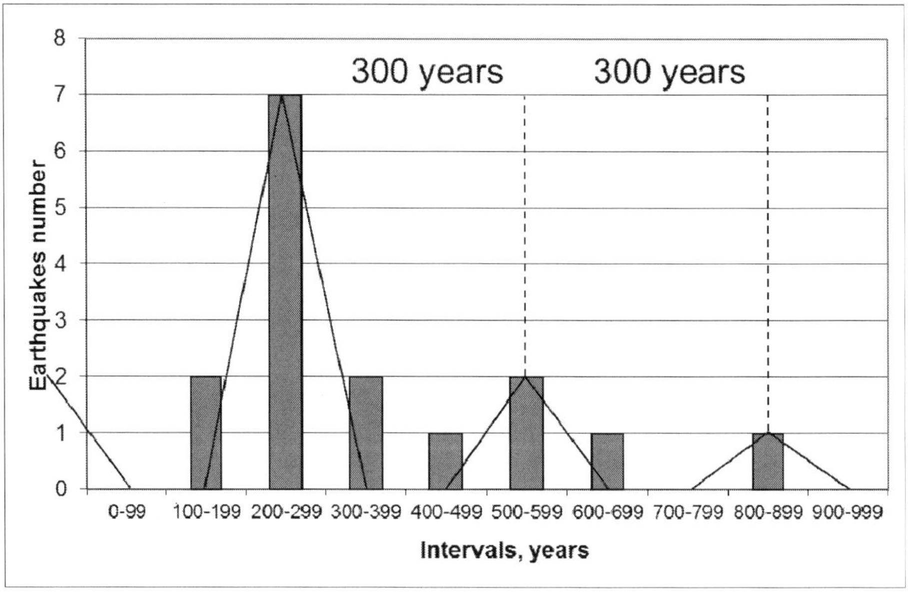

Fig. 43. Comparison of number of the strong earthquakes along the Talas-fergana Fault with intervals of their occurrence

Taking into account all said above, we can supposed that the next strong earthquake (M>7) most probably will occur in approximately 25 years: 300 years minus 275 years (an age of the last strong paleoearthquake) in south-eastern limb of the Talas-Fergana Fault.
We have discussed above that we have conducted a segmentation of the Talas-Fergana Fualt zone by existed data (Fig. 41). In result there were revealed 13 segments: from "a" to "m". We wrote also that by previous studies (Delvaux et al., 2001) there was revealed that during strong Tien Shan earthquakes it takes place a unification of several segments of the fault zone. We pointed above that during Kemin earthquakes of 1911 (M=8.2) it took place a unification of 6 segments of the Chilik-Kemin seismogenic zone of total length to 200 km (Delvaux et al., 2001). The same unification of the segments could take place also during strong earthquakes which occurred along the Talas-Fergana Fault zone (Table 6).

##	*A river valley or fault segment where the earthquake has occurred (from – to)*	*Minimum length of the rupture, km*	*Number of united/ propagated segments (see Fig. 41)*	*Possible maximum magnitudes of the paleoearthquakes*
1	A region of the Karabura Pass - the Sary-Bulak river valley	100	2 (b, c)	7,71
2	The Sulu-Barair River valley - the Sary-Bulak river valley	120	3 (a - c)	7,75
3	The Sary-Bulak River valley - the Kok-Bel Pass	40	2 (d, e)	7,49
4	A region of the Kara-Bura Pass - the Chitty-Western River valley	270	11 (b-l)	7,95
5	A region of the Dzhilangach pass	?	?	?
6	The Kyldau River valley - the Birguzy River valley	30	2 (j, k)	7,42
7	The Pchan River valley - the Birguzy River valley	10	1 (k)	7,16
8	The Pchan River valley - a region of the Dzhilangach pass	20	3 (k-m)	7,33
9	The Sary-Bulak River valley - a region of the Kok-Bel Pass	40	2 (j, k)	7,49
10	The Pchan River valley	?	?	?
11	Upper part of the Chatkal River valley - a region of the Dzhilangach pass	220	11 (c-m)	7,90
12	A region of the Dzhilangach pass	?	?	?
13	An upper part of the Chatkal river valley - the Urumbash River valley	170	6 (c-h)	7,84
14	An upper part of the Chatkal River valley - the Keklikbel River valley	150	5 (b-g)	7,81
15	Karasu River valley	?	?	?
16	A region of the Kara-Bura Pass - the Sary-Bulak River valley	100	2 (b, c)	7,71
17	An upper part of the Chatkal River valley - the Karasu River valley	120	4 (c - f)	7,75

Table 6. Lengths of the seismogenic ruptures along the Talas-Fergana Fault zone and possible magnitudes of earthquakes.

We have analyzed number of formulae for determination of paleoseismic catastrophes' magnitudes according to parameters of seismic rupture, published by different investigators: V.P. Solonenko and V.S. Khromovskikh (1978), A.A. Nikonov (1984) and D.L. Wells and K. J. Coppersmith (1994). Let's investigate their formulae for magnitude assessment by a length of the seismogenic rupture expressed in the surface and check these results on measured parameters of the fault scarps and an instrumental magnitude of the Tien Shan's Suusamyr earthquake of 1992 occurred in the depression with the same name (Bogachkin et al., 1997).

During the earthquake in the surface there were occurred only two short seismogenic ruptures with a total length of 4 km, a distance between then was 26 km (Bogachkin et al., 1997). A magnitude assessed instrumentally was Ms = 7.3. As it was discussed above at a description of the Suusamyr earthquake, in this case we have a deal with so-called "blind" seismogenic rupture, larger part of which did reach the surface. Let's assume that the total length of the rupture (L) was 4 + 26 km = 30 km.

- By formula of V.P. Solonenko and V.S. Khromovskikh (they used the earthquakes of the Baykal Lake and Caucasus regions):

$$M = 0.6 \lg L + 6, \text{ we got } M = 6.89.$$

- By A.A. Nikonov (he used data on Central Asian earthquakes):

$$M = 6.61 + 0.55 \lg L, \text{ we got } M = 7{,}42.$$

- By formula D.L. Wells and K.J. Coppersmith (they use world data):

$$M = 5.08 + 1.16 \lg L, \text{ we got } M = 6.79.$$

Cited above calculations show that the data by V.P. Solonenko and V.S. Khromovskikh (1978) on earthquakes parameters of the Baykal Lake and Caucasus regions, as well as world data by D.L. Wells and K.J. Coppersmith (1994) give underestimated magnitude of the earthquake if compare with the instrumental value. At the same time the formula by A.A. Nikonov (1984), calculated by him for earthquakes of the Central Asia gives a value which only on 0.1 higher than an instrumental value. This is a very good result especially if we are taking into an account that an accuracy of magnitudes determination by such method is in bounds of ±0,5 of the magnitude unit. This is why in our magnitudes assessments we based on formula by A.A. Nikonov (1984), which he specially deduced for a territory of the central Asia.

Our calculation (Table 6) have shown that according to paleoseismological data along the Talas-Fergana Fault zone there are possible earthquakes with magnitude M>7, and during unification of many segments (up to 11) a maximum magnitude can reach M=8. One can not exclude however that along the fault zone there were occurred two or more independent earthquakes divided by short time intervals. This interval we can not reveal because of significant miscalculations of Radiocarbon method of dating. It is possible that there were a clustering of the earthquakes along the seismogenic zone. In a history of strong earthquakes of the Tien Shan such clustering took place in the end of XIX - beginning of XX centuries. Here along so-called Northern Tien Shan Seismic Zone during only 26 years there were occurred 4 strong earthquakes: Belovodsk one of 1885 with M_{LH} = 6,9; Verny earthquake of 1887 with M_{LH} = 7,3; Chilik one of 1889 with $M_{LH} \neq$ 8,3 and Kebin earthquake of 1911 with

M_{LH} = 8,2 (Dzhanuzakov et al., 2003). If paleoseismologists will study consequences of those earthquakes in 3011 using the Radiocarbon method, then because of miscalculations of the method, for them it will be that different segments of the Northern Tien Shan Seismogenic Zone activated simultaneously in 1900 AD plus-minus 50 years...
Thus we understand all conditionality and approximateness of cited above attempts to conduct of the segmentation of the Talas-Fergana Fault zone and calculations of magnitudes of paleoearthquakes with use of so scanty data along the fault zone of 350 km length only in Kyrgyzstan territory. However one has to start from something. The future materials on age of displaced of relief elements, full-fledged paleoseismological trenches, which will cross the whole fault zone, will help to define more precisely cited above numbers.

7. Conclusion

1. 1. The authors' study along the Talas-Fergana Fault zone, as well as analysis of published data have shown that during Neotectonic time the fault developed as a dextral strike-slip fault. It is possible that dextral displacements are spread also on secondary fault planes north and south from the main fault trace.
2. Based on data of absolute dating of authors as well as previous scholars there were determined rates of Holocene and Late Pleistocene dextral movements - 0.2-1.9 cm/ year.
3. The whole zone of the Talas-Fergana Fault is marked by well-developed paleoseismic deformations: up-slope facing scarps and fault scarps, as well as horizontal displacements of the relief forms. In association with them there are revealed numerous seismogravitation forms: rock- and landslides.
4. Collected data on the absolute age determination of mentioned above deformations by the Radiocarbon method point on more than 18 strong earthquake occurred along the fault zone during interval of 275 -15800 years.
5. Reoccurrence of the strong earthquakes along the Talas-Fergana Fault zone during second part of Holocene is about 300 years.
6. The zone of the Talas-Fergana Fault by peculiarities of lateral distribution of the earthquakes can be divided onto 3 chains and 13 segments. First - north-western chain starts from the Kurkureusu River valley in most west of Kyrgyzstan and stretches up to about Sary-Bulak River valley in the Ketmen'-Tyube depression. Second - central chain stretches toward south-east from the Sary-Bulak River valley up to Urumbash River valley (Kazarman depression) inclusively. Third - south-eastern chain starts from the Kyldau River valley and ends in a region of Kyrgyzstan State border.
7. Next strong earthquake along the fault most probably will occur in its south-eastern chain during nearest tens of years.
8. Parameters of the seismotectonic deformations point on $M > 7$ of occurred earthquakes and intensity of the oscillations $I >$ IX. These data have to be taking into account during compiling of a new Map of the Seismic Zoning of Kyrgyz Republic territory.

8. References

Abdrakhmatov K.E. and I.N. Lemzin. 1991. Recent and Quaternary tectonic movements along the Talas-Fergana Fault. - In: (O.K. Chediya - ed.) The Tien Shan in Neotectonic Stage of its Geological Development. Bishkek, Ilim Press. P. 66-74 (in Russian).

Bogachkin B.M., Korzhenkov A.M., Mamyrov E., Nechaev Yu.V., Omuraliev M., Petrosyan A.E., Pletnyov K.G., Rogozhin E.A., and Charimov T.A. The Structure of 1992 Susamyr Earthquake Source Based on Its Geological and Seismological Manifestations // Izvestiya. Physics of the Solid Earth. - Vol. 33. – 1997. - No. 11. - P. 867-882.

Bogdanovich K.I., I.M. Kark, V.Ya. Korol'kov, and I.V. Mushketov. 1914. An Earthquake in Northern Chains of the Tien Shan of December 22, 1910 (January 4, 1911). Transactions of the Geological Committee. Issue 89, 270 p. (in Russian).

Burtman V.S. 1964. The Talas-Fergana Strike-Slip Fault (the Tien Shan). Moscow, Science Press, 143 p. (in Russian).

Burtman V.S., S.F. Skobelev, and L.D. Sulerzhitsky. 1987. The Talas-Fergana Fault: Recent displacements in the Chatkal Region of the Tien Shan // Transactions of USSR Academy of Sciences, V. 296, # 5, P. 1173-1176.

Burtman, V.S., S.F. Skobelev, P. Molnar. Late Cenozoic slip on the Talas-Fergana fault, the Tien Shan, central Asia. 1996. Geol. Soc. Am. Bull. vol. 108, pp. 1004-1021.

Chediya O.K. 1986. Morphostructures and Neotectonic Tectogenesis of the Tien Shan. Frunze: Ilim, 316 p. (in Russian).

Delvaux D., K.E. Abdrakhmatov, I.N. Lemzin, and A.L.Strom. 2001. Landslides and ruptures of the Kemin (Ms 8.2) earthquake of 1911in Kirgizia // Russian Geology and Geophysics. Vol. 42, # 10. P. 1667-1677.

Dzhanuzakov K., M. Omuraliev, A. Omuralieva, B. Iliasov, V.V. Grebennikova. 2003. Strong Earthquakes of the Tien Shan in Bounds of Kyrgyzstan Territory and Adjacent Regions of the Central Asia. Bishkek, Ilim Press, 216 p. (in Russian).

Khodzhaev A. 1985. Pleoseismology of the Chatkal-Kurama Region. Tashkent: Fan Press, 132 p. (in Russian).

Khromovskikh V.S. and Nikonov A.A. 1984. Along Traces of the Strong Earthquakes. Moscow: Nauka Press, 145 p. (in Russian).

Korjenkov A.M. 1993. On Neotectonics of north-western chain of the Talas-Fergana Fault. - In: (T.N. Yankevich - ed.) Actual Questions of Geology, Geophysics and Seismology. Bishkek: Ilim Press. P. 53-64 (in Russian).

Korjenkov A.M. 2006. Seismogeology of the Tien Shan. Bishkek: Ilim Press, 290 p. (in Russian).

Korjenkov A.M., A.V. Bobrovskii, and E.M. Mamyrov. 2010. Evidence for Strong Paleoearthquakes along the Talas-Fergana Fault Near the Kok-Bel Pass, Kyrgyzstan // Geotectonics, vol. 44, No. 3, pp. 262-270.

Korjenkov A.M., A.V. Bobrovskii, E.M. Mamyrov, L.A. Orlova, D. Rust, A.B. Fortuna. 2009.Paleoseismic deformations in the zone of the Talas-Fergana Fault In: (E. Mamyrov - ed.) Geodynamics of the Talas-Fergana Fault of the Tien Shan and Natural Disasters in the Territory of he Central Asia. Bishkek, Arashan Publishers. P. 60-94 (in Russian).

Korjenkov A.M., A.V. Bobrovskii, E.M. Mamyrov, I.E. Povolotskaya, A.B. Fortuna 2006. Paleoseismic deformations in the Talas-Fergana Fault zone in the Kyrgyzstan territory // Science and New Technologies, # 2, P. 81-91 (in Russian).

Korjenkov A.M., A.V. Bobrovskii, E.M. Mamyrov, I.E. Povolotskaya, A.B. Fortuna 2007. Geological and tectonic structure and development of the Talas-Fergana Fault zone // Proceedings of Universities, # 1-2, p. 59-74 (in Russian).

Korjenkov A.M., K.Sh. Tabaldiev, Al.V. Bobrovskii, Ar.V. Bobrovskii, E.M. Mamyrov, L.A. Orlova. 2009. A macroseismic study of the Taldy-Sai caravanserai in the Kara-Bura River valley (Talas basin, Kyrgyzstan) // Russian Geology and Geophysics, vol. 50, # 1, p. 63-69.

Kostenko N.P., Makarov V.I., Soloviyova L.I. 1972. Neotectonics (Kirghiz SSR). Geology of USSR. V. XXV, Part 1, Book 2, Moscow: Nedra Press, p. 249-271 (in Russian).

Makarov V.I. 1977. Neotectonic Structure of the Central Tien Shan. Moscow: Nauka Press. 172 p (in Russian).

Makarov V.I. 1989. On horizontal displacement along the Talas-Fergana Fault during Neotectonic epoch // Transactions of USSR Academy of Sciences, v. 308, # 4, p. 932-937.

Midi B.J. and B.H. Hager 2001. Recent distribution of deformation in the western Tien Shan by block models, based on geodetic data // Russian Geology and Geophysics, v. 42, # 10, p. 1622-1633.

Nikonov A.A. 1984. Development of paleoseismological method for seismic hazard assessment in the central Asia - In: Geological and Geophysical Study of the Seismogenic Zones. Frunze: Ilim Press, p. 192–203 (in Russian).

Rantsman E.Ya., and G.N. Pshenin. 1963. First results of the geomorphologic investigations of horizontal displacements of the earth crust along the Talas-Fergana Fault // Proceedings of USSR Academy of Sciences. Geographical Series, # 5.

Rantsman E.Ya., and G.N. Pshenin. 1967. Neotectonic horizontal movements of the earth crust in the Talas-Fergana Fault zone by data of geomorphologic analysis. – In: Tectonic Movements and Neotectonic Structures of the Earth Crust. Moscow: Nedra Press, p. 155-159 (in Russian).

Rust D., Korjenkov A., Mamyrov E., Bobrovsky A. Evidence for Pre Instrumental Earthquakes on the Talas-Fergana fault, Tien Shan Mountains, Kyrgyzstan // 31st General Assembly of the European Seismological Commission, Crete, 7-12 Sept. 2008 / Programme & Abstracts. P. 301-302.

Shubin V.I. et al. 1992. Open-File Report on Karabura Geological Party on Results of Geological Mapping during 1986-1991. Ivanovka town.

Solonenko V.P. and V.S. Khromovskikh. 1978. Paleoseismological and seismotectonic basis of the seismic zoning of the Great Caucasus – In: Seismotectonics of Southern Regions of USSR. Moscow: Nauka Press, p. 48–63 (in Russian).

Strom A.L. and A.A. Nikonov. 1997. Correlations among parameters of the seismogenic ruptures and earthquakes magnitude // Proceedings of Russian Academy of Sciences. # 12, p. 55-67.

Trifonov V.G. 1985. Peculiarities of development of the active faults // Geotectonics, # 2, p. 16-26.

Trifonov V.G., V.I. Makarov, and Skobelev S.F. 1990. The Talas-Fergana active right-lateral fault. Geotectonics, v. 24, No. 5, p. 435-442.

Trifonov V.G., V.I. Makarov, and Skobelev S.F. 1992. The Talas-Fergana active right-lateral fault. Annales Tectonicae, v. VI, suppl. P. 224-237.

Wallace R.E. Notes on stream channels offset by the San Andreas fault, southern Coast Ranges, California // Proc. of Conf. on Geol. Probl. of San Andreas Fault System. Stanford Univ. Publ., Geol. Sci. 1968. V. 11. P. 6-20.

Wells D.L., Coppersmith K.J. Empirical relationships among magnitude, rupture length, rupture area, and surface displacement // Bull. Seismol. Soc. Am. – 1994. - V. 84. – P. 974–1002.

4

Generation Mechanism of Giant Earthquakes in Subduction Zones with Smaller-Size Interplate Earthquakes During Interseismic Period

Takane Hori[1], Mamoru Hyodo[1] and Shin'ichi Miyazaki[2]
[1]*Japan Agency for Marine-Earth Science and Technology*
[2]*Kyoto University*
Japan

1. Introduction

In the subduction zone along the Japan trench, northeast Japan, only M 7∼8 earthquakes have occurred in the past hundred years (Yamanaka & Kikuchi, 2004). Furthermore, in the surrounding area of such earthquakes, small repeating earthquakes have occurred (Uchida & Matsuzawa, 2011). Because the occurrence of small repeating earthquakes indicates that the plate boundary is creeping (Uchida et al., 2009), it has been considered that only re-rupture of M 7∼8 slip areas (asperities) surrounded by aseismically sliding region can occur in this subduction zone. However, in 2011, a giant earthquake of magnitude (M) 9.0 (we call this as the 2011 Tohoku earthquake hereafter) occurred in the subduction zone and caused destructive tsunami along the pacific coast of Japanese island (Earthquake Research Committee, 2011). The source area extends more than 500km in trench parallel direction and more than 200km in subducting direction including the past M 7∼8 asperities. It should be noted that the occurrence of the M9 earthquake cannot be explained by the combined rupture of the M=7∼8 asperities. The slip amount in the M9 earthquake is one order larger than that in each M=7 ∼8 earthquake. For example, off Miyagi, the central part of the 2011 Tohoku earthquake, the slip amount was 1.8 m for the 1978 off Miyagi earthquake (Yamanaka & Kikuchi, 2004) and more than 10m for the 2011 Tohoku earthquake [e.g., (Iinuma et al., 2011)]. Hence, one of the key questions provoked by this event is: How could an M9 earthquake occur in a subduction zone in which only M = 7∼8 earthquakes have occurred repeatedly in the past 100 years?

This question, however, arises not only for the 2011 Tohoku earthquake, but also for most of other M∼9 earthquakes. In the 2004 Sumatra-Andaman earthquake, for example, an Mw=9.3 event occurred where M=7 ∼ 8 events had occurred separately in space (M=7.7 in 1941, M=7.9 in 1881, and M $\leq$ 7.5 in 1881) (Subarya et al., 2006). In southern Chile, the 1575 and 1960 earthquakes were M $\geq$ 9 giant earthquakes, whose rupture area extended about 1000 km, though two other events in 1737 and 1837 were significantly smaller, with rupture areas limited to about 500 km (Cisternas et al., 2005). Before the 1964 Alaska earthquake, M∼8 earthquakes occurred in 1854, 1855 and 1900 within one of the asperity near Kodiak island (Christensen & Beck, 1994). Such smaller earthquakes occurred not only before M9 earthquakes but also after several decades as follows. For the 1952 Kamchatka earthquake (Mw=8.8∼9.0), M∼7 earthquakes occurred in 1904 also in 1993 within the source area of the 1952 earthquake (Johnson & Satake, 1999). After 29 years of the 1957 Alutian earthquake (Mw=8.6), M=7.7 earthquake occurred within the source area (Johnson et al., 1994). Furethermore, for the 1906 Colombia-Ecuador earthquake (Mw=8.8), three Mw=7.7∼7.9 earthquakes occurred in 1942, 1958 and 1979 (Kanamori & McNally, 1982).

Among the above M9 earthquakes, it is known that some of them have occurred repeatedly as in the southern Chile. For example, in the Kamchatka an M9 earthquake also occurred in 1737 (Johnson & Satake, 1999). For the 2011 Tohoku earthquake, identical tsunami deposite distribution has been found in Sendai area, northeast Japan (Minoura et al., 2001). The estimated recurrence time interval is several hundreds to a thousand years. Hence, we assume that the M9 earthquake occurrence is the fundamental rupture mode in each subduction zone. The key question then becomes: How can M=7 ∼8 earthquakes occur within the source area of an M9 event during the long-term seismic cycle of M9 earthquakes? Based on the concept of a hierarchical asperity model that was applied to the M3 sequence within an M5 asperity off Kamaishi along the Japan trench (Hori and Miyazaki, 2010), we speculate that such events could be explained.

In this chapter, first we will introduce the hierarchical asperity model for M9 earthquakes and explain how to represent them as numerical models. Then, we will describe the simulation results, part of which is identical to our short paper (Hori & Miyazaki, 2011). In discussion, we will introduce the fault strength that is introduced by Nakatani (Nakatani, 2001), and discuss the generation mechanism of M9 recurrence with M=7∼8 earthquakes during the M9 interseismic period. Finally, we will compare our results with observation data, especially for the 2011 Tohoku earthquake.

2. Model and method

2.1 Hierarchical asperity model for M9 earthquakes

2.1.1 Concept

Aseismic sliding occurs during the M9 earthquake cycle if the frictional property in the M9 source area is apparently stable enough, in other words, if the nucleation size is large enough. As a result, the aseismic sliding can load the smaller unstable locked patches within the M9 source area. If the nucleation size and fracture energy of the patches are much smaller than they are in the rest of the source area, unstable slip occurs on the patches but does not propagate to their outside. This is a possible mechanism that can account for M<9 events within the source area of an M9 event. This mechanism is similar to the foreshock model (Matsu'ura et al., 1992) although the space-time scale is much larger here. The events on the unstable patches can occur repeatedly until the accumulated strain energy is sufficient to cause an M9 event in the entire seismogenic zone, as an unstable slip on a patch triggers the rupture of the area including large fracture energy. Note that similar multi-scale heterogeneity in fracture energy is assumed for the 2011 Tohoku earthquake (Aochi & Ide, 2011) although they focus on the dynamic rupture process.

In the following discussion, the unstable patches and the M9 source area, including the unstable patches, are called regular asperities and a hyper asperity, respectively. The source area outside the regular asperities is called a conditional asperity, because in this area both aseismic sliding and seismic slip occur, depending on the stress and fault strength conditions as shown later.

2.1.2 Friction law

In the present study we employ a laboratory-derived rate- and state-dependent friction law (Rice, 1993) that has been succeeded in modeling slip histories over seismic cycles:

$$\mu = \mu_* + a \ln\left(\frac{V}{V_*}\right) + b \ln\left(\frac{V_* \theta}{d_c}\right), \tag{1}$$

where μ is a coefficient of friction, V and θ are slip velocity and state variable on a fault surface, μ_* and V_* are references of frictional coefficients and slip velocity, respectively, a is an increment of frictional coefficients for a velocity step (direct effect), b is a transient decrement of frictional coefficients for the fault to slip over the evolution length d_c.

In the framework of the rate- and state-dependent friction law, nucleation size L_∞ and fracture energy G_c are given by

$$L_\infty = \frac{c}{\pi}\left(\frac{b}{b-a}\right)^2 \frac{G}{b\sigma} d_c, \tag{2}$$

$$G_c = \frac{1}{2}\left(\log\frac{\theta_{in}}{\theta_{out}}\right)^2 b\sigma d_c, \tag{3}$$

respectively (Rubin & Ampuero, 2005), where G, σ, θ_{in} and θ_{out} are shear modulus, effective normal stress, state variables just outside and inside the crack tip, respectively. c is a geometrical factor that is about 2 for a circular crack. Clearly the nucleation size and fracture energy increase with a state evolution length d_c in a similar way as by a slip weakening distance D_c. Bizzarri & Cocco, for example, suggested from fully dynamic rupture simulations that d_c is propotional to D_c (Bizzarri & Cocco, 2003). Furthermore, it is suggested that d_c and D_c depends on observation scales, explaining its discrepancy between laboratory and seismological observations (Perfettini et al., 2003; Shibazaki & Matsu'ura, 1998). A multi-scale patch distribution of scale-dependent D_c is considered to investigate multi-scale heterogeneous dynamic rupture propagation within a continuum media (Ide & Aochi, 2005). They demonstrated that such heterogenities are able to produce realistic features of earthquake size distribution and dynamic rupture patterns on a fault within the continuum class of models.

2.1.3 Representation of a hyper asperity

Based on those previous studies, we assume that a state evolution length d_c is a dominant factor to determine the size of an earthquake. Then we introduce a heterogeneous distribution of d_c in a single asperity (hyper asperity). Areas of smaller d_c in the hyper asperity represent regular asperities and would have small fracture energy G_c. Hence they would produce smaller size of earthquakes. Rest of the same asperity of larger d_c represents conditional asperity and would have larger G_c. This part is persistent to be locked for a longer period and would induce a failure of the entire hyper asperity to generate large size of earthquakes. It should be noted that the size of regular asperities is another factor to determine the style of fault slips. Earthquakes occur if the size is larger than the nucleation length, but otherwise slow slip events occur [e.g., (Kato, 2003)].

On the other hand, we assume that $(a-b)\sigma_{eff}$ $(a-b<0)$ would be uniform over the entire hyper asperity because apparent stress drop is nearly constant over a wide range of earthquake size (Ide & Beroza, 2001). In the following sections we set up a numerical model for a hierarchical asperity model and perform numerical simulations. Then we demonstrate that our model is a candidate to account for the earthquake sequence of M=7~8 within M~9 source area.

2.2 Governing equations

In our model, we devide a flat fault, which represents a subducting plate, into many triangular fault cells and calculate variables for each cell. The governing equations are expressed as following. The first equation is a quasi-dynamic equilibrium of shear stress

$$0 = \sum_j K_{ij}(u_j - V_{pl}t) - \mu_i \sigma_i - \frac{G}{2\beta}\frac{du_i}{dt}, \tag{4}$$

where u_j is a fault slip of j-th fault cell, K_{ij} is a shear stress responce of the i-th cell caused by a unit dislocation of the j-th cell, σ_i is an effective normal stress of the i-th cell, μ_i is a coefficient of friction of the i-th cell, V_{pl} is a plate convergence velocity that is assumed constant over the entire model region, β is a shear wave velocity and G is a rigidity. To calculate the shear stress response at the center of each cell, which is triangular, stress changes calculated with angular dislocations are summed (Comninou & Dundurs, 1975). The last term on the right hand side is a radiation damping term to approximate energy radiation as elastic wave for high velocity slip (Rice, 1993).

The second equation is a laboratory-derived rate- and state-dependent friction law as (1),

$$\mu_i = \mu_* + a_i \ln\left(\frac{V_i}{V_*}\right) + b_i \ln\left(\frac{V_* \theta_i}{d_{ci}}\right), \tag{5}$$

where $V_i = \frac{du_i}{dt}$ is a slip velocity of the i-th cell, θ_i is a state variable for the i-th cell, μ_* is a reference for frictonal coefficients, V_* is a reference velocity and is set to V_{pl} in this study, and a_i, b_i, d_{ci} are friction parameters for the i-th cell. The state variable θ_i follows a state evolution law. Among several versions of evolution laws, we employ a composite law

$$\frac{d\theta_i}{dt} = \exp\left(-\frac{V_i}{V_c}\right) - \left(\frac{\theta_i V_i}{d_{ci}}\right)\ln\left(\frac{\theta_i V_i}{d_{ci}}\right), \tag{6}$$

where V_c is a cut-off velocity and is 1.0×10^{-8} m/s (Kato & Tullis, 2001). This evolution law approximates slip law for $V >> V_c$ and aging law for $V << V_c$.

From the above equations, we derive set of ordinary differential equations for slip velocity V_i and state variable θ_i. The governing equations are solved with adaptive time step 4th-order Runge-Kutta algorithm (Press et al., 1996). For initial condition, slip velocity V_i and θ_i are assumed to be uniform, and set to be $0.9V_{pl}$ and d_{ci}/V_i, respectively.

2.3 Model setup in subduction zone

We show spatial distributions of frictional parameters $A - B$ ($A = a\sigma, B = b\sigma$) and d_c in Figure 1. From rock friction experiments, $A - B$ primarily depends on temperature and hence depends on depths, and d_c increases where temperature is high under wet condition (Blanpied et al., 2001). For simplicity, A is assumed constant on the entire fault. Parameters B and d_c

are, hence, assumed to depend on depths, representing their temperature dependence. B decreases and d_c increases around the shallower and deeper end of seismogenic zone (Hillers et al., 2006). d_c is large enough in conditional asperity so as to become conditionally stable (Boatwright & Cocco, 1996). The large value of d_c may be consistent with the idea that plate boundary is a fairly matured fault composed by thick shear zone and d_c is larger for the thicker shear zone width in direct shear experiments of gouge layer (Marone, 1998). Although existence of thick shear zone is not directly observed in subduction zone, low velocity layer is found on plate boundary in Japan trench area (Miura et al., 2003; Takahashi et al., 2000). So called subduction channel (Vannucchi et al., 2008) is also a candidate of substance for a large d_c fault.

As shown in Figure 1, the entire seismogenic zone is modeled as one large hyper asperity. Of course, this is an over-simplification, and it results in simple recurrence of the entire seismogenic zone rupture for an M9 event, a virtually homogeneous slip distribution in the strike direction, a positive stress drop in the entire M9 source area, no M<9 aftershock occurrence in and just around the source area, and so on. To introduce such realistic complexity associated with an M9 event, heterogeneity both in $A - B$ and d_c should be considered.

In the seismogenic zones, where $A - B$ is negative, we set two regular asperities. d_c is one order smaller in regular asperities than the conditional asperity. The model is a simple and conceptual one and is not intended for any specific subduction zone. The distribution and size of the regular asperities significantly affect the entire rupture patterns and the recurrence time intervals. However, qualitative characteristics, such as M<9 event occurrence within the M9 source area shown below, can be reproduced in many cases that have a different size and distribution of regular asperities. In such cases, d_c in the regular asperity should be one order or more smaller than it is in the conditional asperity. This order difference in d_c on a fault is consistent with the discreteness in distribution function for G_c (Fukao & Furumoto, 1985; Ide & Aochi, 2005).

3. Results

3.1 Slip history and coseismic slip distribution

In order to see total earthquake cycle behavior, slip history at some sampling points is shown first. The points are on the same depth as indicated in Figure 1. Within regular asperities, earthquakes occur repeatedly (Figure 2). The slip amount is several times larger in the entire area rupture (giant earthquakes, EQ0, 5) than in the cases of rupture for each regular asperity (regular earthquakes, EQ1-4). The distribution of coseismic slip, whose velocity is higher than 0.01 m/s, is shown in Figure 3. Each moment magnitude is also shown. The rupture areas of EQ1-4 roughly correspond to the two regular asperities. Including both asperities, EQ5 ruputres almost the entire seismogenic zone.

The recurrence time interval is much longer after the giant earthquake than after the regular earthquakes (Figure 2). This recurrence pattern variation is similar to the one in time-predictable model (Shimazaki & Nakata, 1980). Large and small slips correspond to ruptures of a hyper asperity and a regular asperity, respectively (Figure 3). The hyper asperity ruptures with about 1200 years recurrence interval and its moment magnitude is 9.0. During this hyper-cycle, only in the latter stage of the cycle, two regular asperities within the hyper asperity rupture independently. The recurrence interval for the regular cycles is about 200 $\sim$

300 years. The superposition of the hyper- and regular-cycles results in the time-predictable recurrence pattern in our model.

On the other hand, in the conditional asperity (x=100, 300, 500 in Figure 2), nearly constant aseismic slip and afterslip appear, as well as large coseismic slip and a locked state, depending upon the stage in the M9 earthquake cycle. Such slip pattern variation can also be reproduced in the velocity-strengthening area with constant d_c (Kaneko et al., 2010). Our model and their model are end-members of the heterogeneity in $A - B$ and d_c. We believe that the actual subduction plate boundary has the properties between them.

During a giant earthquake coseismic slip occurs and then locked after long period more than 500 years. After the occurrence of EQ1 or 2, slow sliding appears whose slip velocity is $1/3 \sim 2/3$ of the plate convergence rate. While after EQ3 and/or 4, afterslip occurs and then slow steady slip occurs. Note that interseismic slow sliding speed is almost constant for more than 100 years (Figure 2).

3.2 Space-time evolutions of slip velocity and shear stress

Snap shots of spatio-temporal variation in slip velocity and stress are shown in Figure 4. The entire area of the hyper asperity is locked after a giant earthquake. A snapshot of slip velocity for 47 years after the giant earthquake is shown in Figure 4a, and the timing is shown in Figure 2. During the interseismic period, the locked area shrinks and slow sliding area spreads. As shown in Figure 4b, stress becomes higher along the edge of the locked area. When the slow sliding area reaches the bottom of a regular asperity within the hyper asperity, rupture starts in the regular asperity (Figure 4b, EQ1). The rupture propagates within the regular asperity but decelerates at its outside (Figure 4c). After the event, low velocity slip like afterslip occurs around the asperity, especially in its deeper extent (Figure 4d). Another regular earthquake (Figure 3b, EQ2) occurs at the other regular asperity (Figure 4e). Then both asperities and their surrounding area are locked again (Figure 4f) but the total locked area is smaller than the stage after the giant earthquake (Figure 4a). Similar size of regular earthquakes (Figure 3c, d, EQ3, 4) and following afterslip occurs after stable slip reaches again the regular asperity (Figure 4g, h, i, j). In this case, afterslip more widely spreads than after the former regular earthquakes (see Figure 4d and i). The locked area becomes further shurunk (Figure 4k). When stable slip reaches the regular asperity at the third time, stress level of around the regular asperity is fairly high (Figure 4l). Thus rupture does not decelerate outside the regular asperity and becomes a giant earthquake (Figure 4 m, Figure 3e, EQ5). After the giant earthquake, the shallower part and in and around the regular asperities are locked first, and afterslip occurs in the surrounding area (Figure 4 n). More than a year after EQ5, whole the seismogenic zone is locked and afterslip continues in the deeper extent (Figure 4 o). Then the entire area is locked again like Figure 4a. As shown in Figure 4 b, g, l, rupture initiates near the deeper edge of a regular asperity for all the cases.

4. Discussion

4.1 Mechanism of slip variation

4.1.1 Strength in rate-state friction

To consider the mechanism of space-time variation in slip pattern as shown above, we use the fault strength introduced by Nakatani (Nakatani, 2001). First, we introduce an alternative state variable.

$$\Theta_i = b_i \ln\left(\frac{V_* \theta_i}{d_{ci}}\right). \tag{7}$$

With this state variable, the constitutive equation (5) becomes

$$\frac{\tau_i}{\sigma_i} = (\mu_* + \Theta_i) + a_i \ln\left(\frac{V_i}{V_*}\right) \tag{8}$$

and the state evolution equation (6) becomes

$$\frac{d\Theta_i}{dt} = \frac{b_i}{(d_{ci}/V_*)} \exp\left(-\frac{\Theta_i}{b_i}\right) \exp\left(-\frac{V_i}{V_c}\right) - \frac{V_i}{d_{ci}} \left\{\Theta_i - b_i \ln\left(\frac{V_*}{V_i}\right)\right\}. \tag{9}$$

Here, the equation (8) can be rewritten as follows

$$V_i = V_* \exp\left[\frac{\tau_i - (\mu_* + \Theta_i)\sigma_i}{a_i \sigma_i}\right]. \tag{10}$$

If we define a variable $(\mu_* + \Theta_i)\sigma_i$ as strength of the fault surface, equation (10) gives a slip velocity V_i of a fault surface whose strength is $(\mu_* + \Theta_i)\sigma_i$ when a shear stress τ_i is applied as shown in Figure 5a (Nakatani, 2001). Hence the state evolution equation (9) describes the evolution of the fault strength. The first term of the right hand side represents the healing process of strength which is proportional to $\log(t)$ (Figure 5b), and the second term represents the decrease of strength in terms of slip (i.e., slip-weakening) as shown in Figure 5c. It should be noted that the strength weakening of the rate- and state-friction law is not rate-weakening but slip-weakening. Rate dependence appears in the steady-state strength $\Theta_{SSi} = b_i \ln\left(\frac{V_*}{V_i}\right)$

4.1.2 Mechanism of large variation in coseismic slip amount

The amount of coseismic slip is determined by how the slip is stopped or decelerated. As described in the fault constitutive law (10), slip velocity depends on the difference between stress and strength relation. Here, the stress is the elastic stress $\sum_j K_{ij}(u_j - V_{pl}t)$ minus radiation damping $\mu_i \sigma_i$ (4). We compare the variation in the elastic stress and strength near the edge of the right asperity in Figure 1 between the cases of giant and regular earthquakes (EQ5 and EQ1, 3). As shown in Figure 6a, both the stress and the strength decrease in 8-10 MPa during less than 2 meter slip. And the amount of the stress drop is almost the same for regular earthquakes and the giant earthquake. Thus the difference in slip amount is not due to the difference in the stress drop. Although the strength decreases rapidly during about 2m-slip in both cases, the stress decreases more gently in the giant earthquake than in the regular one. This lower weakening rate of the stress keeps the slip velocity of the giant earthquake higher than that of the regular earthquake. Thus the coseismic slip amount of the former is larger than the latter.

What controls such differences in the weakening behavior? Weakening rate of the strength at a point depends mainly on the own slip rate (9), while the stress weakening rate is affected not only own slip history but also the slip distribution around it (4). Slip weakening rate of the stress should be low when slip occurs in wider area around a monitoring point because in equation (4) $K_{ii} < 0$ and $K_{ij} > 0$ $(i \neq j)$. Thus, the difference in stress weakening rate comes from the extent of the slip area. The slip distribution just before regular earthquakes

(Figure 4b and g) shows that asperities and their surrounding area, especially above 15km in depth, are almost locked. Since the stress is much lower than the strength in such a locked area (equation 10), large stress increase is necessary for rupture propagation. Until such an area starts to slip, the stress weakening rate in the already sliding area should be high. Thus, the slip is decelerated in the asperity before enough slip occurs in the surrounding area. This is the reason why slip amount is small in regular earthquakes.

On the other hand, slip distribution just before the giant earthquake shows that only the small portion of the entire seismogenic zones including the asperities is locked, and slow sliding occurs around the asperities with around half of the plate convergence rate (Figure 4l). In such sliding area, the stress is close to the strength. Thus, slip around the asperities can be easily triggerd with low stress increase and slip can occur in wider area.

4.1.3 Mechanism for variation in slip velocity among earthquake cycles

At the middle point between the regular asperities, not only seismic slip but also slow sliding and afterslip occurred (Figure 2). During the slow sliding, the strength is slightly higher than the stress and the weakening is almost negligible (Figure 6b). After the regular earthquakes (EQ3, 4), the stress increases and afterslip occurs here. During the afterslip, strength weakening occurs with low weakening rate because of the slower slip velocity. Because the strength does not significantly reduce, the slip is decelerated. Hence both the amount of stress drop and slip amount of the afterslip are significantly smaller than those of the giant earthquake. During the giant earthquake, typical slip weakening occurs as similar to the one at the regular asperity (Figure 6a). The amount of the stress drop is also similar. Only the difference is the weakening rate. Because this point has much larger d_c than in the regular asperity, the slip weakening distance is about 12 m, which is 6 times more than the regular asperity. As explained in 4.1.2, because wide area slips in the giant earthquake, weakening rate of the stress is low enough and slip can continue there.

4.1.4 A mechanism of time predictable behavior

As shown in Figure 4, the area around the asperities is locked after the giant earthquakes but slowly sliding after the regular earthquakes. This causes the difference in stressing rate around the hypocenter. After the giant earthquake, stressing rate becomes low because the wide area is locked (Figure 7). On the other hand, the stressing rate is high after the regular earthuqkes because the surrounding area is sliding (Figure 7). Therefore, such inherent stressing rate variation during the cycle of the giant earthquake causes the variation in recurrence interval depending on the earthquake size. This mechanism of time predictable behavior is completely different with the original time predictable model in which the stressing rate and the peak strength are constant and the stress drop varies depending on the earthquake size (Shimazaki & Nakata, 1980). It should be noted that the time predictable behavior here does not necessarily appear in general. If regular asperities exist in the deeper portion of the seismogenic zone, they should be ruptured earlier because of the slow sliding propagation from the bottom of the seismogenic zone.

4.2 Comparison with observation

4.2.1 M<9 earthquakes in M9 source area

As shown in Figure 3, M<9 earthquake occurrence in M9 source area is reproduced. The slip amount of M7 events is nearly one order smaller than that of M9 event at the regular asperites (Figure 3). This is consistent with the slip amount difference between the 1978 off Miyagi earthquake and the 2011 Tohoku earthquake as mentioned in Introduction.

In our simulation, M7 earthquakes occur only in the later half of the hyper cycle. However, they occur decades after the M9 earthquakes in Kamchatka, Alutian and Colombia-Ecuador. For Kamchatka, the time interval between M<9 and M9 earthquakes are 41 years, which is around 20% of the recurrence interval of 215 years. The M<9 earthquake occurrence timing discrepancy after M9 earthquakes between the data and our simulation results can be explained as in 4.1.4. Actually, the above mentioned M<9 earthquake sources located deeper part of the M9 source areas. This indicates that the 1978 off Miyagi type earthquake also can occur in the early stage of M9 cycle because its asperity is located at the deeper edge of the 2011 Tohoku earthquake (Kato & Yoshida, 2011). It should be noted that such M<9 earthquakes are not usual aftershocks just after M9 earthquakes but occurred after more than a few decades.

4.2.2 Interseismic period

As shown in Figure 2, aseismic sliding with the velocity of 30 $\sim$ 60 % of the plate convergence rate occurs before the M9 earthquakes and seismic slip occurs during the M9 earthquakes. This result seems to be consistent with the sliding behavior in the southern half of the 2011 Tohoku earthquake. In the area, aseismic sliding occurred with the velocity of less than 50% of the plate convergence rate (Uchida et al., 2009). In this area, coseismic slip also occurs in the 2011 Tohoku earthquake but the slip amount is relatively small (Iinuma et al., 2011). Furthermore, in the aseismically sliding area, afterslip after M=6$\sim$7 earthquakes extended significantly wider than the coseismic slip area (Suito et al., 2011). Such large area afterslip can be found in the later stage of the M9 earthquake as shown in Figure 4i. Hence, such afterslip occurrence is possibly an indicator of the high stress level before the greater event.

4.2.3 Preseismic slip and rupture initiation

It should be noted here that the wide area afterslip mentioned above decelerates and locked area appears again (Figure 4j). In other words, the afterslip does not accelerate to produce the M9 earthquake as a preslip. As shown in Figure 4l, the initiation of the M9 rupture is triggered by the rupture of a regular asperity, and the nucleation process is similar to the one for the M7 earthquakes in our simulation (Figure 4b and g). This indicates that large preslip is not necessary observed. Actually, preseismic acceleration of the slip near the hypocenter of the 2011 Tohoku earthquake has not been found (Hirose, 2011). On the other hand, it is important to examine whether the initial rupture of the 2011 Tohoku earthquake corresponds to an asperity identified before in this area or not.

4.2.4 Coseismic period

One of the characteristics of the 2011 Tohoku earthquake is the several tens meter slip along the Japan trench axis (Maeda et al., 2011). As shown in Figure 3, the large slip in the shallower part of the fault is reproduced. Since the fault plane cut the free surface, there is no loading source shallower than the fault plane. This causes the accumulation of the slip deficit easier than the deeper portion of the seismogenic zone which is loaded due to the sliding of the deeper extent. Additionally, since the state evolution length d_c is large, slip cannot easily be triggered by M7 earthquakes. Furthermore, when the seismic slip occurs in the M9 earthquake, free surface enhance the sliding. These may explain the observed large slip near the trench axis.

Since our simulation is quasi-dynamic, we cannot estimate rupture propagation velocity quantitatively. However, large d_c in the conditional asperity results in the lower weakening rate and hence slower rupture propagation than the ordinal earthquakes. This is consistent with the slow rupture velocity estimation for the 2011 Tohoku earthquake (Wang & Mori, 2011). Although the weakening rate during the coseismic rupture has not been estimated yet, our model predicts that higher and lower weakening rate in the regular and conditional asperities, respectively (Figure 6). This is another verification of our hierarchical asperity model and friction law used here.

4.2.5 Postseismic period

Furthermore, our model predicts the postseismic transient behavior which can be tested by the observation during the coming years after the 2011 Tohoku earthquake. As shown in Figure 4n, the asperities of the past M=7∼8 events will be locked first, and afterslip occurs in the surrounding area. Then, after a year, the entire seismogenic zone will be locked and afterslip will continue only in the deeper extent. Such spatio-temporal variation in the locked zone after the 2011 Tohoku earthquake can be examined by repeating earthquake distribution and more directly by seafloor deformation observed using ocean bottom pressure gauges off Miyagi. Long term afterslip in the deeper portion will cause uplift along the coast where subsidence occurred seismically. This will be observed by GPS onland for decades.

4.2.6 Other subduction zones

The most important implication from our results and observation is the wide area afterslip observed in the later stage of the M9 earthquake cycle as shown in 4.2.2. Such wide area afterslips comparing with coseismic slip areas are observed in Hyuga-nada, southwest Japan (Yagi et al., 2001). Although no interplate coupling or even forward slip has been obtained in the southern part of this area, a M=7.7 earthquake occurred in 1662, causing strong motion and a tsunami in wide areas of Hyuga (Hatori, 1969). The area off Hyuga may now be in the later stage of a cycle of M∼8 earthquake such as in 1662, while the Japanese Government gives only 10% chance of such an earthquake occurring in this area (Headquarters for Earthquake Research Promotion, 2004). Furthermore, this area may be ruptured with great earthqaukes along the Nankai trough.

Another example of wide area afterslip is observed in Tokachi-Nemuro area, northeast Japan, after the 2003 Tokachi-oki earthquake of Mw∼8 (Miyazaki et al., 2004). In 1600's, significantly larger tsunami than the one caused by 2003 earthquake attacked the coast of Tokachi-Nemuro area (Nanayama et al., 2003). The estimated source area includes the 2003 coseismic slip area, the 1973 Nemuro coseismic slip area (Mw=7.8), and the wide afterslip area. Note that most of the recurrence interval estimated from the tsunami deposits in this area is less than 500 years (Sawai et al., 2009).

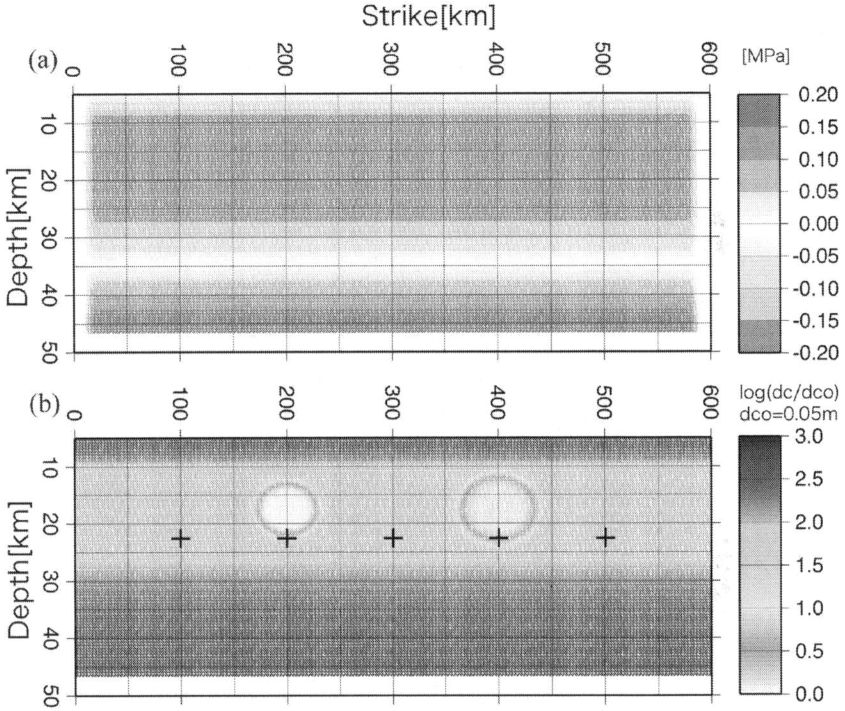

Fig. 1. Spatial distributions of frictional parameters (a) $A - B$ and (b) d_c on the model fault. Crosses indicate the cumulative slip shown in Figure 2.

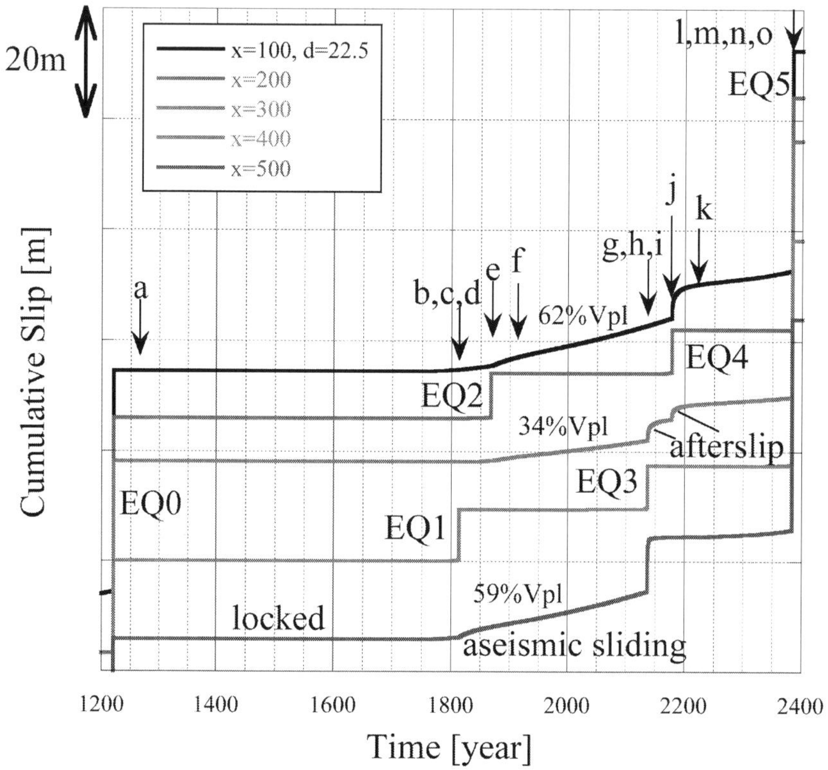

Fig. 2. Temporal variation of slip at the points shown in Figure 1. x and d are strike and depth in Figure 1. Vpl is the plate convergence rate (0.05m/yr). Arrows indicate the timing of images shown in Figure 4.

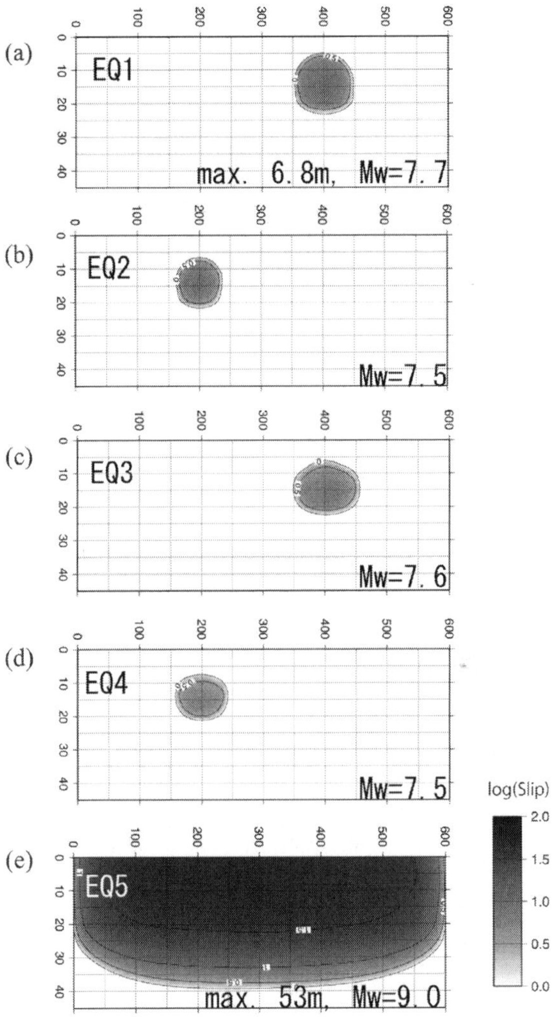

Fig. 3. Coseismic slip distribution and the moment magnitude for each events.

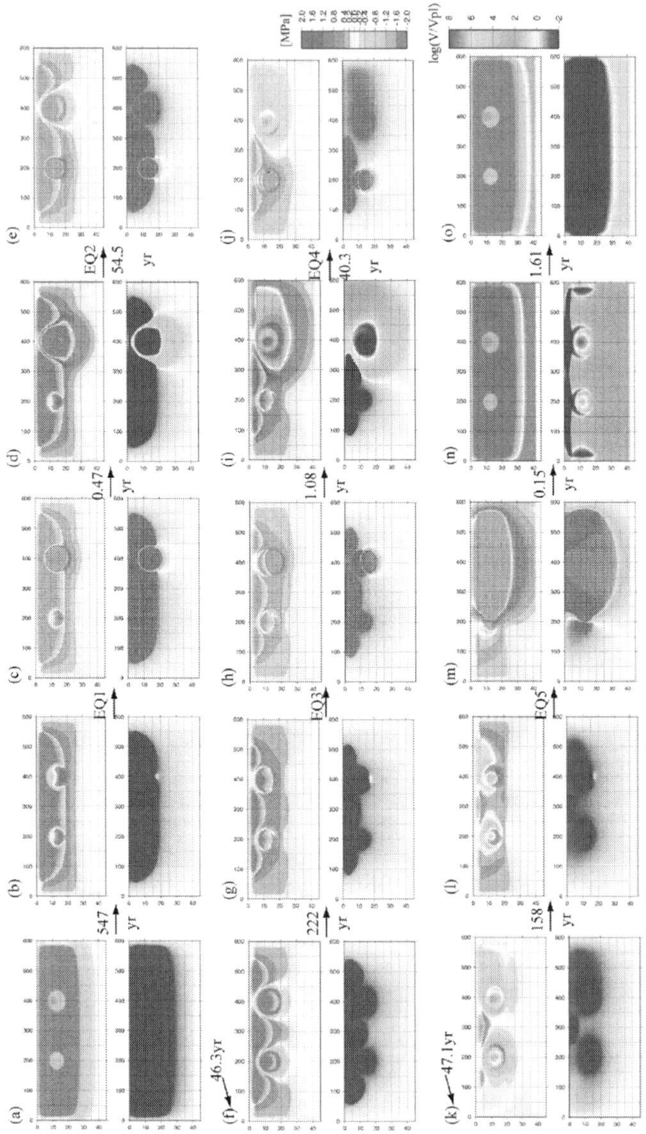

Fig. 4. Spatio-temporal variations in the stress and slip velocity on the fault. Blue, white, yellow-green and red colors for the slip velocity correspond to locking, slip with plate convergence rate, aseismic slip (faster than the plate convergence rate) and seismic slip, respectively. Numerals attached to the arrows indicate the time intervals between two successive snapshots.

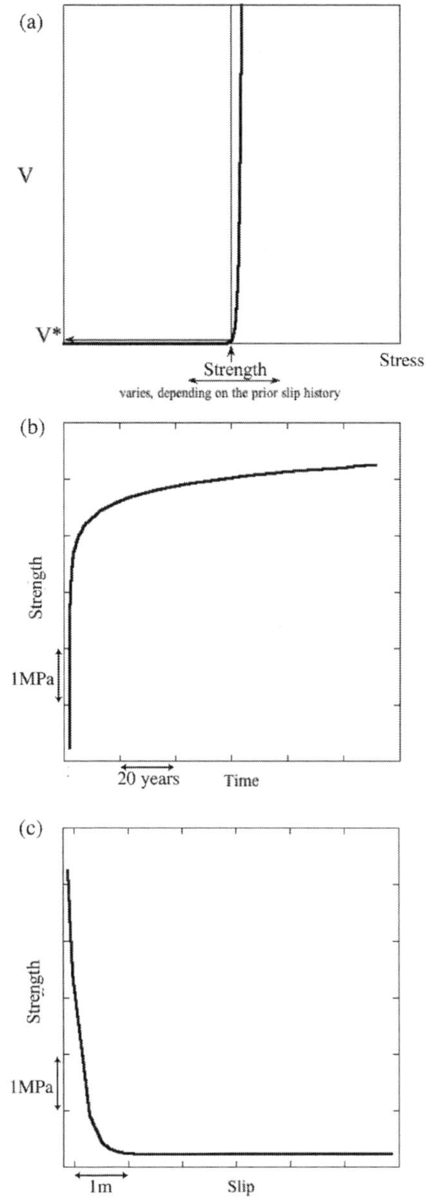

Fig. 5. (a) Relation between stress, strength and slip velocity in the rate- and state-dependent friction law. (b) Time dependent variation in strength. (c) Slip dependent variation in strength.

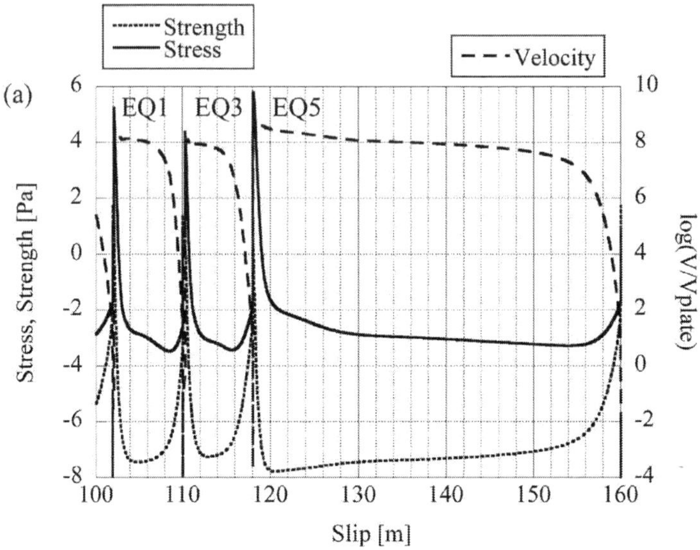

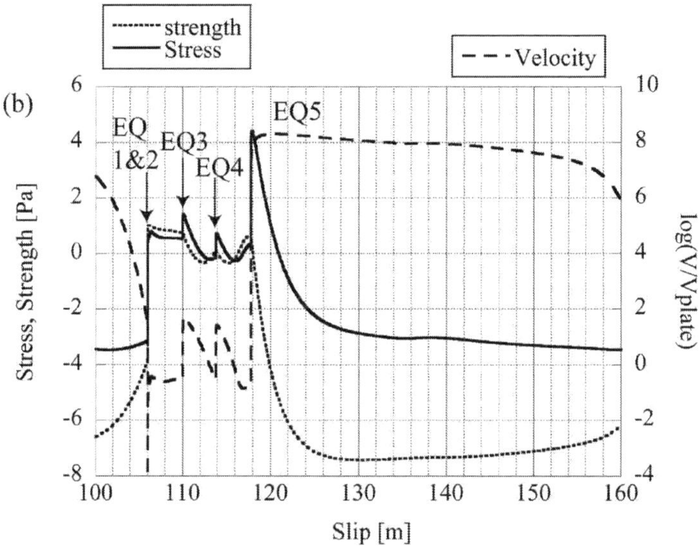

Fig. 6. Solid, dashed and thick dashed lines indicate stress, strength and slip velocity variation, respectively. (a) in a regular asperity. (b) at a point between the two asperities.

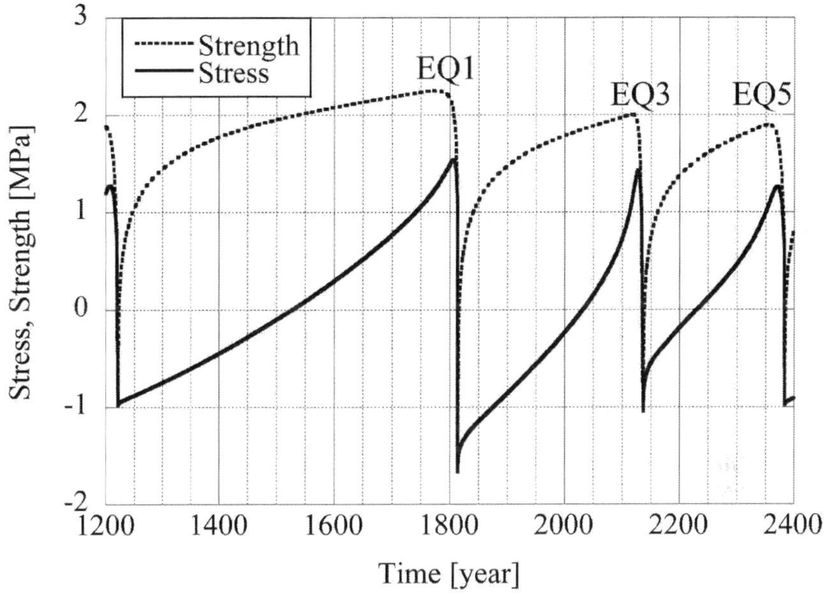

Fig. 7. Stress and strength variation in time at the hypocenter. Solid and dashed lines indicate stress and strength variation, respectively. Coseismic period is excluded.

5. Conclusion

Consequently, the hierarchical asperity model for M9 can explain qualitatively the characteristics of the 2011 Tohoku earthquake and other M∼9 earthquakes with smaller earthquakes in the source area. Some predicted phenomena here will be examined by the analyses of high quality and dense data sets both on the seafloor and on the land. Modeling of forshock, aftershocks or M ∼ 8 occurrence as the combination of M ∼ 7 asperities in off Miyagi area is for the future work. We will evaluate the predictability of our model quantitatively by constructing more realistic model, which includes asperity distribution, slab geometry, and so on for the Japan trench and other areas.

6. Acknowledgement

We thank Dr. William Stuart for providing subroutines for calculating stress fields due to triangular dislocations coded by him and Dr. Robert Simpson. This work is partly supported by the MEXT project named Evaluation and disaster prevention research for the coming Tokai, Tonankai and Nankai earthquakes.

7. References

Aochi, H., and S. Ide (2011) Conceptual multi-scale dynamic rupture model for the 2011 Tohoku earthquake. *Earth Planets Space*, Special Issue: First Results of the 2011 Off the Pacific Coast of Tohoku Earthquake.

Bizzarri, A. & Cocco, M. (2003) Slip-weakening behavior during the propagation of dynamic ruptures obeying rate- and state-dependent friction laws. *J. Geophys. Res.*, 108, B8, 2373, doi:10.1029/2002JB002198.

Blanpied, M.L., Marone, C.J., Lockner, D.A., Byerlee, J.D. & King, D.P. (2001) Quantitative measure of the variation in fault rheology due to fluid-rock interaction. *J. Geophys. Res.*, 103, 9691-9712.

Boatwright, J. & Cocco, M. (1996) Frictional constraints on crustal faulting. *J. Geophys. Res.*, 101, 13895–13909.

Cisternas, M., Atwater, B. F., Torrejon, F., Sawai, Y., Machuca, G., Lagos, M., Eipert, A., Youlton, C., Salgado, I., Kamataki, T., Shishikura, M., Rajendran, C. P., Malik, J. K., Rizal, Y. & Husni, M. (2005) Predecessors of the giant 1960 Chile earthquake. *Nature*, 437, doi:10.1038/nature03943.

Christensen, D.H. & Beck, S.L. (1994) The rupture process and tectonic implications of the great 1964 Prince William Sound earthquake. *Pure Appl. Geophys.*, 142, 29-53.

Comninou, M. A. & Dundurs, J. (1975) The angular dislocation in a half-space. *J. Elasticity*, 5, 203-216.

Earthquake Research Committee (2011) The 2011 off the Pacific coast of Tohoku Earthquake. *Monthly Reports on Evaluation of Seismic Activity in Japan*, http://www.jishin.go.jp/main/index-e.html, March 13, 2011.

Fukao, Y. & Furumoto, M. (1985) Hierarchy in earthquake size distribution. *Phys. Earth Planet. Inter.*, 37, 149-168.

Hatori, T. (1969) A Study of the Wave Sources of the Hiuganada Tsunamis. *Bull. Earthq. Res. Inst.*, 47, 55-63.

Headquarters for Earthquake Research Promotion (2004) Long-term evaluation of seismicity off Hyuga and Southwestern Islands areas along Ryukyu trench.

Hillers, G., Ben-Zion, Y. & Mai, P. M. (2006) Seismicity on a fault controlled by rate- and state-dependent friction with spatial variations of the critical slip distance. *J. Geophys. Res.*, 111, B01403, doi:10.1029/2005JB003859.

G. Hillers, Mai, P. M., Ben-Zion, Y. & Ampuero, J.-P. (2007) Statistical properties of seismicity of fault zones at different evolutionary stages. *Geophys. J. Int.*, 169, 515-533 doi: 10.1111/j.1365-246X.2006.03275.x.

Hirose, H. (2011) Tilt records prior to the 2011 Off the Pacific Coast of Tohoku Earthquake. *Earth Planets Space*, Special Issue: First Results of the 2011 Off the Pacific Coast of Tohoku Earthquake.

Hori, T. & Miyazaki, S. (2011) A possible mechanism of M 9 earthquake generation cycles in the area of repeating M 7 $\sim$ 8 earthquakes surrounded by aseismic sliding. *Earth Planets Space*, Special Issue: First Results of the 2011 Off the Pacific Coast of Tohoku Earthquake.

Ide, S. & Aochi, H. (2005) Earthquakes as multiscale dynamic rupture with heterogeneous fracture surface energy, *J. Geophys. Res.*, 110, B11303, doi:10.1029/2004JB003591.

Ide, S. & Beroza, G. C. (2001) Does apparent stress vary with earthquake size? *Geophys. Res. Lett.*, 28, 17, 3349-3352.

Iinuma, T., Ohzono, M., Ohta, Y. & Miura, S. (2011) Coseismic slip distribution of the 2011 off the Pacific coast of Tohoku Earthquake (M9.0) estimated based on GPS data - Was the asperity in Miyagi-oki ruptured? *Earth Planets Space*, Special Issue: First Results of the 2011 Off the Pacific Coast of Tohoku Earthquake.

Johnson, M. & Satake, K. (1999) Asperity Distribution of the 1952 Great Kamchatka Earthquake and its Relation to Future Earthquake Potential in Kamchatka. *Pure appl. geophys.*, 154, 541-553.

Johnson, M., Satake, K., Holdahl, S. R., Sauber, J. (1996) The 1964 Prince William Sound earthquake: Joint inversion of tsunami and geodetic data. *J. Geophys. Res.*, 101, BI, 523-532.

Johnson, M., Tanioka, Y., Ruff, L. J., Satake, K., Kanamori, H. & Sykes, L. R. (1994) The 1957 Great Aleutian Earthquake. *Pure appl. geophys.*, 142, 1, 3-28.

Kanamori, H. & McNally, K. C. (1982) Variable rupture mode of the subduction zone along the Ecuador-Colombia coast. *Bull. Seism. Soc. Am.*, 72, 4, 1241-1253.

Kaneko, Y., Avouac, J.-P., & Lapusta, N. (2010) Towards inferring earthquake patterns from geodetic observations of interseismic coupling, *Nature Geoscience*, doi:10.1038/ngeo843.

Kato, N. (2003) Repeating Slip Events at a Circular Asperity : Numerical Simulation with a Rate- and State-Dependent Friction Law. *Bull. Earthq. Res. Inst. Univ. Tokyo*, 78, 151-166.

Kato, N. & Tullis, T. E. (2001) A composite rate- and state- dependent law for rock friction. *Geophys. Res. Lett.*, 28, 1103-1106.

Kato, N. & Yoshida, S. (2011) A shallow strong patch model for the 2011 great Tohoku-oki earthquake: A numerical simulation. submitted to

MacInnes, B.T., Weiss, R., Bourgeois, J. & Pinegina, T.K. (2010) Slip Distribution of the 1952 Kamchatka Great Earthquake Based on Near-Field Tsunami Deposits and Historical Records. *Bull. Seism. Soc. Am.*, 100, 4, 1695Ð1709, August 2010, doi: 10.1785/0120090376.

Maeda, T., Furumura, T., Sakai, S. & Shinohara, M. (2011) SigniÞcant tsunami observed at the ocean-bottom pressure gauges at 2011 Off the PaciÞc Coast of Tohoku Earthquake. *Earth Planets Space*, Special Issue: First Results of the 2011 Off the Pacific Coast of Tohoku Earthquake.

Marone, C. (1998) Laboratory-derived friction laws and their application to seismic faulting. *Annu. Rev. Earth Planet. Sci.*, 26, 643-649.

Matsu'ura, M., Kataoka, H. & Shibazaki, B. (1992) Slip-dependent friction law and nucleation processes in earthquake rupture. In: T. Mikumo, K. Aki, M. Ohnaka, L.J. Ruff and P.K.P. Spudich (Editors), Earthquake Source Physics and Earthquake Precursors, *Tectonophysics*, 211, 135-148.

Minoura, K., Imamura, F., Sugawara, D., Kono, Y. & Iwashita, T. (2001) The 869 Jogan tsunami deposit and recurrence interval of large-scale tsunami on the Pacific coast of northeast Japan. *J. Natural Disaster Sci.*, 23, 83-88.

Miura a, S., Kodaira, S., Nakanishi, A., Tsuru, T., Takahashi, N., Hirata, N., & Kaneda, Y. (2003) Structural characteristics controlling the seismicity of southern Japan Trench fore-arc region, revealed by ocean bottom seismographic data. *Tectonophysics*, 363, 79 Ð 102.

Miyazaki, S., Segall, P., Fukuda, J. & Kato, T. (2004) Space time distribution of afterslip following the 2003 Tokachi-oki earthquake: Implications for variations in fault zone frictional properties. *Geophys. Res. Lett.*, 31, L06623, doi:10.1029/2003GL019410.

Nakatani, M. (2001) Conceptual and physical clarification of rate and state friction: Frictional sliding as a thermally activated rheology. *J. Geophys. Res.*, 106, 13347-13380.

Nanayama, F., Satake, K., Furukawa, R., Shimokawa, K., Atwater, B.F., Shigeno, K. & Yamaki, S. (2003) Unusually large earthquakes inferred from tsunami deposits along the Kuril trench. *Nature*, 424, 660-663.

Perfettini, H., Campillo, M. & Ionescu, I. (2003) On the scaling of the slip weakening rate of heterogeneous faults. *J. Geophys. Res.*, 108, B9, 2410, doi:10.1029/2002JB001969.

Press, W.H., Teukolsky, S.A., Vetterling, W.T. & Flannery, B.P. (1996) *Numerical Recipes in Fortran 77: The Art of Scientific Computing, 2nd Edition*, Cambridge University Press, ISBN0-521-43064-X, New York.

Rice, J. (1993) Spatio-temporal complexity of slip on a fault. *J. Geophys. Res.*, 98, 9885-9907.

Rubin, A. M. & Ampuero, J.-P. (2005) Earthquake nucleation on (aging) rate and state faults. *J. Geophys. Res.*, 110, B11312, doi:10.1029/2005JB003686.

Sawai, Y., Kamataki, T., Shishikura M., Nasu, H., Okamura, Y., Satake, K., Thomson, K.H., Matsumoto, D., Fujii, Y., Komatsubara, J. & Aung, T.T. (2009) Aperiodic recurrence of geologically recorded tsunamis during the past 5500 years in eastern Hokkaido, Japan. *J. Geophys. Res.*, 114, B01319, doi:10.1029/2007JB005503.

Shibazaki, B. & Matsu'ura, M. (1998) Transition process from nucleation to high-speed rupture propagation: scaling from stick-slip experiments to natural earthquakes. *Geophys. J. Int.*, 132, 14-30.

Shimazaki, K. & Nakata, T. (1980) Time-predictable recurrence model for large earthquakes. *Geophys. Res. Lett.*, 7, 279-282.

Suito, H., Nishimura, T., Tobita, M., Imakiire, T. & Ozawa, S. (2011) Interplate fault slip along the Japan Trench before the occurrence of the 2011 off the Pacific coast of Tohoku Earthquake as inferred from GPS data. *Earth Planets Space*, Special Issue: First Results of the 2011 Off the Pacific Coast of Tohoku Earthquake.

Subarya, C., Chlieh, M., Prawirodirdjo, L., Avouac, J.P., Bock, Y., Sieh, K., Meltzner,A. J., Natawidjaja, D. H. & McCaffrey, R. (2006) Plate-boundary deformation associated with the great Sumatra-Andaman earthquake, *Nature*, 440, doi:10.1038.

Takahashi, N., Kodaira, S., Tsuru, T., Park, J. O., Kaneda, Y., Kinoshita, H., Abe, S., Nishino, M. & Hino, R. (2000) Detailed plate boundary structure off northeast Japan coast. *Geophys. Res. Lett.*, 27, 1977-1980.

Uchida, N. & Matsuzawa, T. (2011) Coupling coefficient, hierarchical structure, and earthquake cycle for the source area of the 2011 Tohoku earthquake inferred from small repeating earthquake data. *Earth Planets Space*, Special Issue: First Results of the 2011 Off the Pacific Coast of Tohoku Earthquake.

Uchida, N., Nakajima, J., Hasegawa, A., & Matsuzawa, T. (2009) What controls interplate coupling?: Evidence for abrupt change in coupling across a border between two overlying plates in the NE Japan subduction zone. *Earth Planet. Sci. Lett.*, 283, 111-121.

Yagi, Y., Kikuchi, M. & Sagiya, T. (2001) Co-seismic slip, post-seismic slip, and aftershocks associated with two large earthquakes in 1996 in Hyuga-nada, Japan. *Earth, Planets Space*, 53, 793-803.

Yamanaka, Y. & Kikuchi, M. (2004). Asperity map along the subduction zone in northeastern Japan inferred from regional seismic data. *J. Geophys. Res.*, Vol.109, B07307, doi:07310.01029/02003JB002683.

Vannucchi, P., Remitti, F. & Bettelli, G. (2008) Geological record of fluid flow and seismogenesis along an erosive subducting plate boundary. *Nature*, 451 | 7, February 2008 | doi:10.1038/nature06486.

Wang, D. & J. Mori (2011) Rupture Process of the 2011 off the Pacific Coast of Tohoku Earthquake (Mw9.0) as Imaged with Back-Projection of Teleseismic P-waves. *Earth Planets Space*, Special Issue: First Results of the 2011 Off the Pacific Coast of Tohoku Earthquake.

5

Frictional Characteristics in Deeper Part of Seismogenic Transition Zones on a Subduction Plate Boundary

Keisuke Ariyoshi and Yoshiyuki Kaneda
Earthquake and Tsunami Research Project for Disaster Prevention, Japan Agency for Marine-Earth Science and Technology, Yokohama, Japan

1. Introduction

By virtue of dense networks of GPS and highly sensitive seismic stations, occurrences of deep low-frequency tremors/earthquakes have been recognized worldwide in the deep portions of subduction plate boundaries (e.g., Schwartz & Rokosky, 2007). Since these tremors and earthquakes have moment release rate smaller than that of regular earthquakes, the low-frequency tremors (LFTs) /earthquakes are classified into "slow-earthquake" group (Ide et al., 2007). The slow-earthquakes occur sometimes independently and sometimes break into chain-reaction propagating at a speed of about 10 km/day. However, a systematic explanation of the origin of such slow-earthquake migrations in some subduction zones is still lacking.

First of this study, we investigate the mechanisms of slow-earthquake migration by comparing observational results and numerical simulation results. Since some LFTs are found to be modulated by Earth tides (e.g., Nakata et al., 2008) and the moment release rate of the slow earthquake group is much smaller than that of regular earthquakes (Ide et al., 2007), they likely have low-stress drop and are sensitive to shear stress perturbations possibly induced by the preseismic slip of nearby megathrust earthquakes.

Second of this study, we propose a new method to detect precursory change around the large asperity prior to a megathrust earthquake, focusing on the spatio-temporal change of migration speed and moment release rate for nearby slow earthquakes. Recently, low-frequency tremors/earthquakes have been also detected in the shallower portion of frictional transition zone on the subduction plate boundaries (Obana & Kodaira, 2009).

Third of this study, we also perform another numerical simulation of shallow low-frequency tremors/earthquakes, comparing with the characteristics of the deep low-frequency tremors/earthquakes and discussing the strategy of Dense Oceanfloor Network System for Earthquakes and Tsunamis (DONET) toward an anticipated Tonankai Earthquake.

2. What makes low-frequency tremors/earthquakes migrate along strike?

In this section, we investigate the mechanisms of slow-earthquake migration by comparing observational results and numerical simulation results. One of possible mechanism is chain reaction of numerous small asperities (Fig. 1), which is based on the activity of small repeating earthquakes triggered by afterslip of large interplate earthquakes (Matsuzawa et al., 2004).

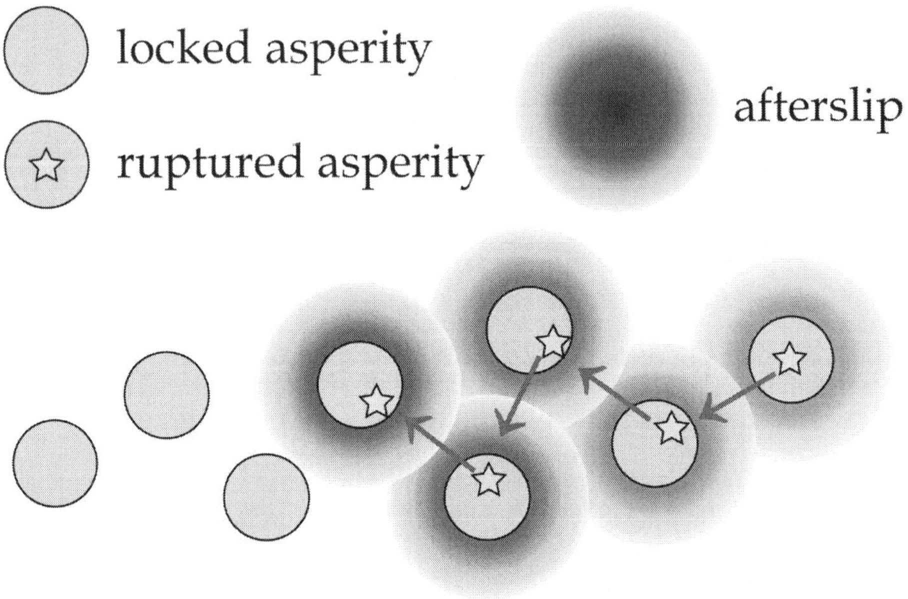

Fig. 1. A schematic model of chain-reaction between asperities after Matsuzawa et al. (2004). Yellow stars represent rupture initiation points of slip events.

On the other hand, some simulation results show slow-earthquake migration is reproduced by models with frictional properties almost uniform along strike and constant effective normal stress independent of depth under initial stress uniform along strike (e.g., Liu & Rice, 2005). Their simulation results suggest that an essential condition for the slow-earthquake migration is neither heterogeneous frictional properties nor heterogeneous initial stress distribution, if effective normal stress is constant at low value and independent of depth.

In the case of effective normal stress proportional to depth for a model with uniform frictional properties, slow-earthquake migration does not occur (e.g., Hirose & Hirahara, 2002). Note that above all simulation studies adopt the same friction law. These previous simulation results raised the following question: For an asperity model with low effective normal stress around 30 km depth, are small asperities essential for the slow-earthquake migration?

In this section, we formulate several test models in order to understand the necessary conditions and characteristics for slow-earthquake migration.

2.1 A test model of chain reaction between numerous small asperities

A test model consists of a planar plate interface dipping at 15 degrees from the free surface in a homogeneous elastic half-space (Fig. 2) with a periodic boundary condition along the strike direction. The plate interface is divided into 1,024 (strike) × 293 (dip) cells.

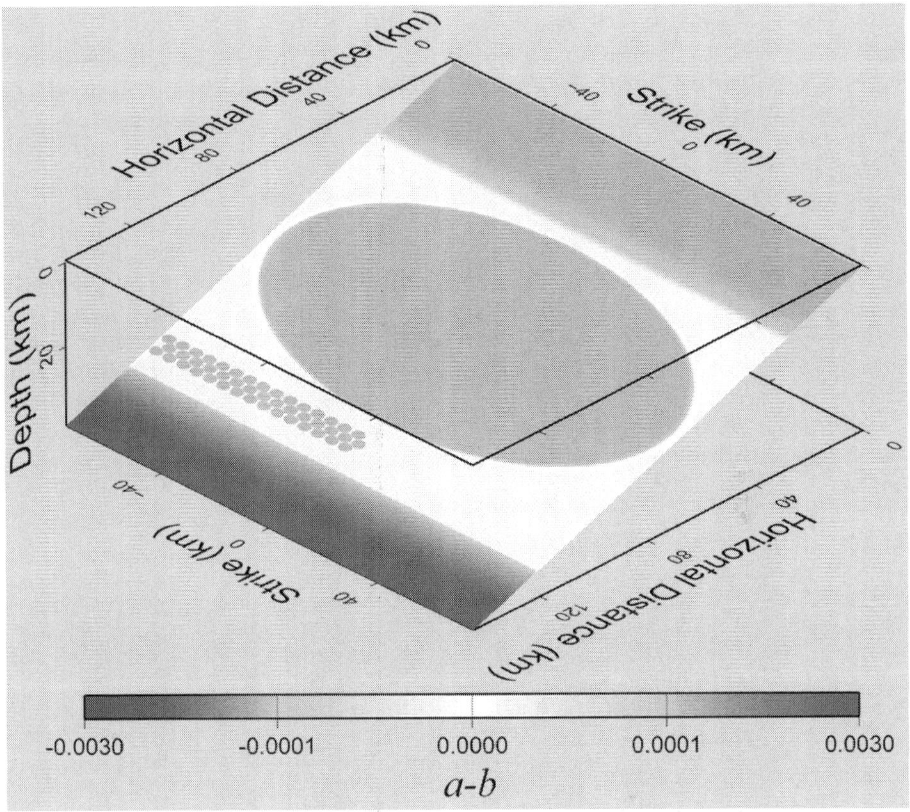

Fig. 2. A 3-Dimensional model of a subduction plate boundary with frictional parameter *a-b* (Eq. 3). Cool color represents asperity.

Slip is assumed to occur in the pure dip direction and to obey the quasi-static equilibrium between shear and frictional stresses by introducing a radiation damping term (Rice, 1993):

$$\mu_i \sigma_i = \sum_{j=1}^{N} K_{ij}(u_j(t) - V_{\rm pl}t) - \frac{G}{2\beta}\frac{{\rm d}u_i}{{\rm d}t} \tag{1}$$

Here, the subscripts i and j denote the location indices of a receiver and a source cell, respectively. The left hand side of equation (1) describes frictional stress, where μ and σ is friction coefficient and effective normal stress, respectively. The right hand side describes the shear stress in the i-th cell caused by dislocations, where K_{ij} is the Green's function for the shear stress (Okada, 1992) on the i-th cell, N is the total number of cells, $V_{\rm pl}$ is the relative speed of the two plates, t denotes time, G is rigidity, β is the shear wave speed. K_{ij} is calculated from the quasi-static solution for uniform pure dip-slip u relative to average slip $V_{\rm pl}t$ (Savage, 1983) over a rectangular dislocation in the j-th cell. Parts of the first term of the right-hand side are written as convolutions, by exploiting the along-strike invariance of the

Green's function, and efficiently computed by the Fast Fourier Transform (e.g., Rice, 1993; Liu & Rice, 2005).

In Eq. (1), the effective normal stress σ is given by

$$\sigma_i(z) = \kappa(z)(\rho_{\text{rock}} - \rho_{\text{w}})gz, \tag{2}$$

where ρ_{rock} and ρ_{w} are the densities of rock and water, respectively, g is the acceleration due to gravity, and z is the depth. The function $\kappa(z)$ is a super-hydrostatic pore pressure factor, as given in Fig. 3. We assume that a high-pore-pressure system locally exists around a depth of 30 km based on the high-V_p/V_s zones in southwestern Japan (e.g., Shelly et al., 2006). The increase in pore pressure is probably due to the dehydration derived from the change in facies in the slab (e.g., Hacker et al., 2003). Ariyoshi et al. (2007) estimated that the value of κ is 0.1 for the deeper part (>30 km depth) based on the post-seismic slip propagation speed. On the basis of the stress field observation in northeastern and southwestern Honshu, Japan, Wang & Suyehiro (1999) suggested that the apparent frictional coefficient is approximately 0.03, which is consistent with $\kappa = 0.1$.

The friction coefficient μ is assumed to obey an RSF law (Dieterich, 1979; Ruina, 1983), as given by

$$\mu = \mu_0 + a \log (V/V_0) + b \log (V_0\theta/d_c), \tag{3}$$

$$d\theta/dt = 1 - V\theta/d_c, \tag{4}$$

where a and b are friction coefficient parameters, d_c is the characteristic slip distance associated with b, θ is a state variable for the plate interface, V is the slip velocity, and μ_0 is a reference friction coefficient defined at a constant reference slip velocity of V_0.

We consider a model with close-set numerous small asperities on the deeper outskirt of a great asperity, as proposed by Dragert et al. (2007). In the present study, an asperity denotes a region with a-b = γ < 0, following Boatwright & Cocco (1996). The plate interface is demarcated into five parts, as shown in Figs. 2 and 3: (i) one large asperity (LA), (ii) 90 small asperities (SAs), (iii) a shallow stable zone, (iv) a deep stable zone, and (v) a transition zone ($\gamma \sim +0$). The values of frictional parameters as described in the caption of Fig. 3 are based on rock laboratory results (e.g., Blanpied et al., 1998), which will be discussed later.

The constant parameters in the present study are V_{pl} = 4.0×10^{-2} m/yr (or 1.3×10^{-9} m/s), G = 30 GPa, β = 3.75 km/s, ρ_{rock} = 2.75×10^{3} kg/m^3, ρ_{w} = 1.0×10^{3} kg/m^3, g = 9.8 m/s^2, V_0 = 1 μm/s, μ_0 = 0.6, and Poisson's ratio ε = 0.25.

2.2 Simulation results of chain reaction effect on low-frequency event migration

Fig. 4 shows the spatial distribution of slip velocity about 37 years before the origin time of a megathrust earthquake in the large asperity. The recurrence interval and magnitude of the megathrust earthquake in our simulation is 116 years and M_w 7.9, respectively, where seismic slip is defined as slip faster than 1 cm/sec. The large asperity (LA) is strongly locked while a slow-earthquake occurs in some of small asperities (SA).

Fig. 5 shows time history of velocity field before and after the megathrust earthquake. Figures 2 and 3 suggest slow-earthquake migration at the migration rate of 0.3~3 km/day, which is driven by the chain reaction of small asperities. On the other hand, these figures also show that slow-earthquake migration does not usually occur in the region without small asperities, where slip velocity is largely stable at values comparable to V_{pl}.

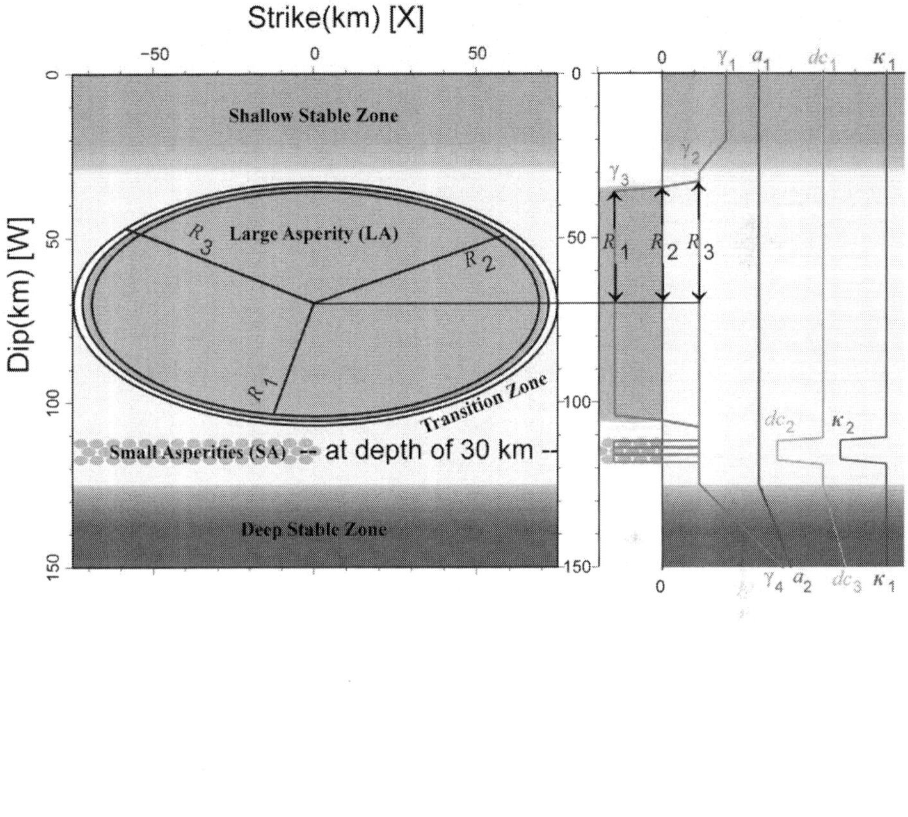

Fig. 3. Frictional parameters (a, γ (= a-b), d_c, κ (See Eq. (2)) as functions of distance along the dip direction from the surface on the plate boundary, where $(a_1, a_2) = (2, 5)$ [$\times 10^{-3}$], $(\gamma_1, \gamma_2, \gamma_3, \gamma_4) = (0.5, 0.01, -0.3, 4.9)$ [$\times 10^{-3}$], $(d_{c1}, d_{c2}, d_{c3}) = (10, 0.43, 100)$ [mm], and $(\kappa_1, \kappa_2) = (1.0, 0.1)$. Half the length of the minor axis (along dip) of the elliptical asperity takes the following values: for LA, $(R_1, R_2, R_3) = (35, 36.25, 37.5)$ [km] and for SA, $(r_1, r_2, r_3) = (1.33, 1.5, 1.67)$ [km], where the aspect ratios for LA and SA are 2.0 and 1.5, respectively. The distance between central points of SA along strike and dip direction is 2 and 2.5 km, respectively. This figure is developed from Ariyoshi et al. (2011a).

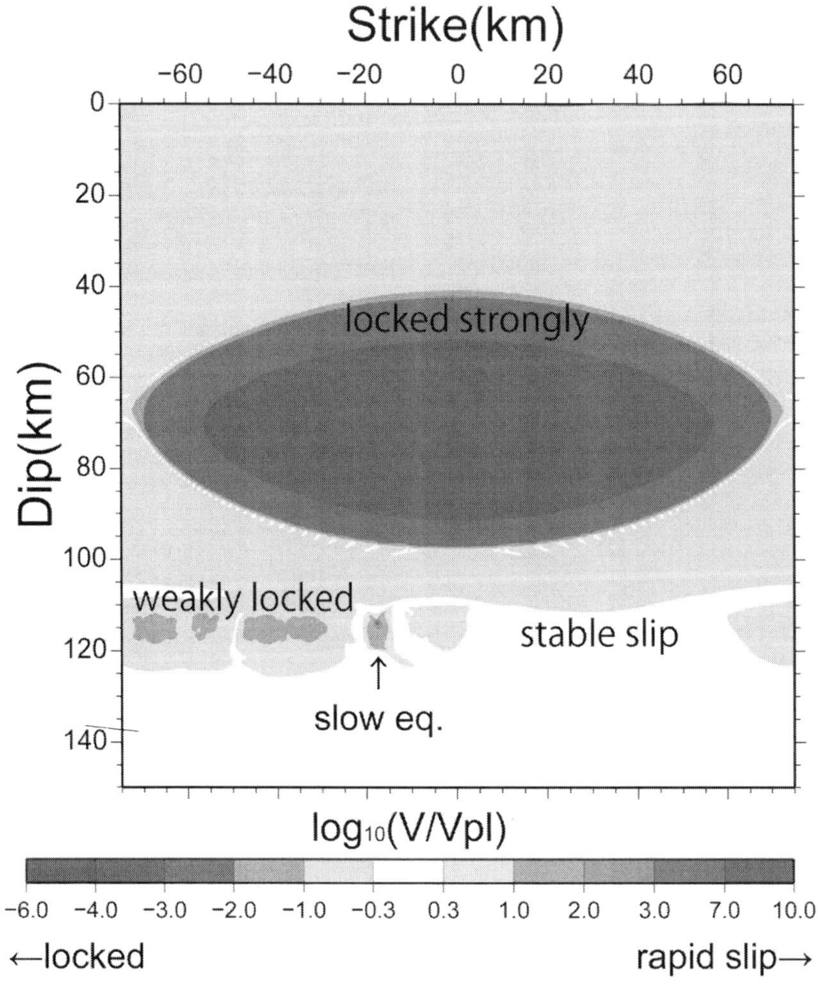

Fig. 4. Velocity field on the subduction plate boundary 36.6 years before the megathrust earthquake. Note that 8 of $\log_{10}(V/V_{pl})$ is about 1 cm/sec. This figure is developed from Ariyoshi et al. (2011a).

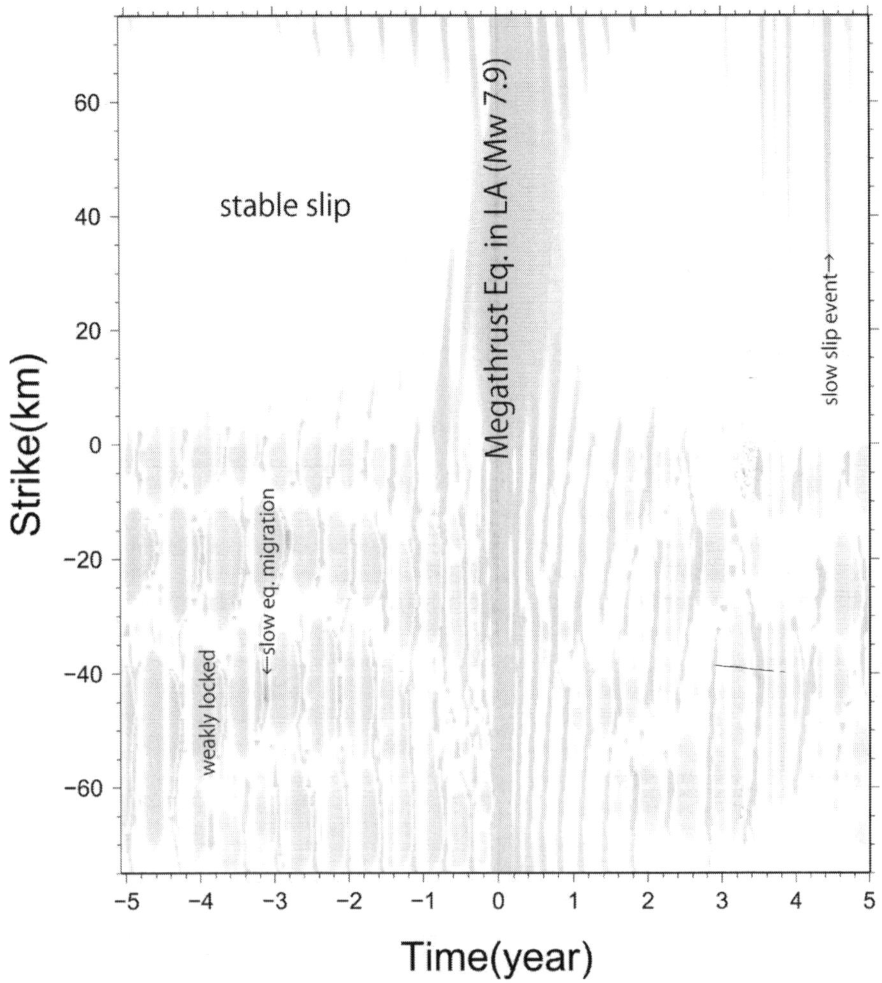

Fig. 5. Spatiotemporal evolution of slip velocities at the "Dip" of 115 km along strike as shown in Fig. 3 before and after the megathrust earthquake in case that numerous small asperities in Fig. 3 are removed for Strike>0. Color scale is the same as Fig. 4. This figure is developed from Ariyoshi et al. (2011a).

After about 4.5 years after the megathrust earthquake in Fig. 5, the largest slow slip event (~ 10 V_{pl} at most; much slower than slow-earthquakes in Fig. 5) occurs in the region without small asperities. Since all of slow slip events in the region without small asperities originate from the transition between SA-belt and no-SA (between Strike = 0 and 75 km due to the cyclic boundary condition along strike direction), these slow slip events are triggered by the chain reaction of SA.

2.3 Size effect of small asperities on chain reaction behaviors

Next, we perform another test model with different size of asperity generating chain reaction as shown in Fig. 6 in order to investigate size effect.

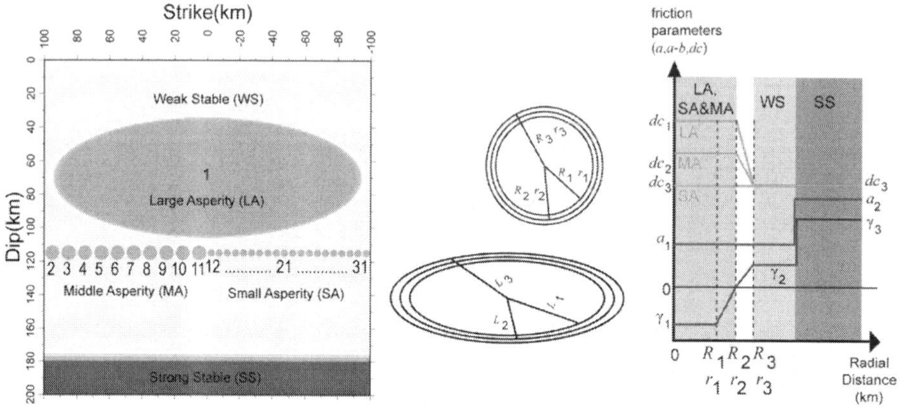

Fig. 6. Spatial distribution of large asperity (LA), middle asperity (MA), small asperity (SA) (a-b < 0) with identification numbers of the asperities, weak stable (WS; a-$b \geq 0$) and strong stable (SS; a-b >> 0). Setting fault geometry, elevated pore pressure, and the constant value of other geophysical parameter is the same as section 2.2. This figure is partly derived from Ariyoshi et al. (2009).

Figs. 7a-7k shows simulation results of slip migration driven by chain reaction of asperities. For MA (left panel of Figs. 7a-7e), unilateral chained propagation process is clearly seen for MA, with propagation speed between asperities No. 7 to No. 5 of about 0.2 km/day (Figs. 7b-d). In the time span indicated by the cyan background in Figs. 7l-m, the patterns of stress drop and averaged slip velocity at asperities No. 3 to 5 appear to be relatively similar but quantitatively different. Especially, the amount of stress drop and averaged slip velocity in asperity No. 3 is greater than in the others. In addition, there is not only leftward chained propagation from the asperity No. 10 but also rightward propagation from the asperity No. 2, where the latter is much slower than the former because of the large area locked in the asperity No. 3. In Fig. 7e, both chained propagations cross at the asperity No. 3, which promotes larger stress drop and higher slip velocity. Since similar behavior seems to be seen in the simulation of Liu & Rice (2005), these results suggests that some phenomena generated by introducing along-strike variations of constitutive parameters or non-uniform initial conditions may be represented by interaction between small asperities.

On the other hand, the propagation process of SA is different from that of MA in some respects. On the right side of Fig. 7 shows leftward propagation from the asperity No. 2 through No. 31 to No. 22, which is due to the periodic boundary condition. In the time span indicated by a green background in Figs. 7n-o, there are two slow earthquakes for each asperity from No. 28 to 31. Propagation speed between asperities No. 31 to 28, No. 28 to 23 and No. 23 to 22 is about 0.2, 0.15 and 0.03 km/day, respectively, which progressively becomes slower than that of MA. Fig. 7j shows that rightward propagation from the asperity No. 27 and 30 is also seen.

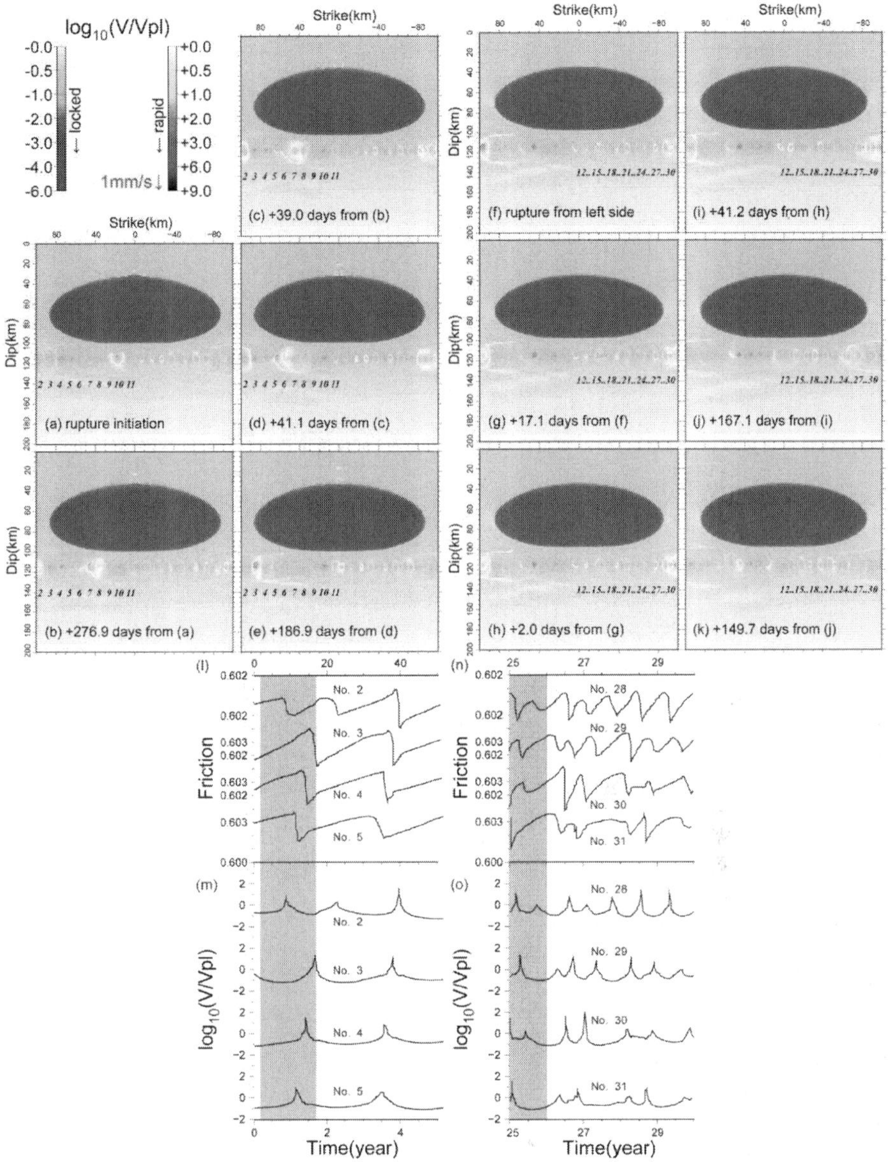

Fig. 7. Examples of chain-reaction between asperities for MA (left panels (a)-(e)) and SA (right panels (f)-(k)) with time history of friction and normalized slip velocity averaged in each asperity (MA; (l)(m), SA; (n)(o)) after Ariyoshi et al. (2009). Italic numbers are identification of asperities. Cyan and green regions correspond to time spans of snapshots for MA and SA, respectively.

These differences from MA are explained as follows. SA has shorter recurrence intervals and smaller moment release because of smaller asperity size with shorter characteristic distance than those of MA. The smaller moment release makes propagation speed slower (0.03 km/day for SA is smaller than 0.2 km/day for LA), which causes that recurrence interval of SA is relatively much shorter than the passage time of aseismic slip across more than twice the SA diameters (10 km). In addition, locking of SA soon after the occurrence of slow earthquakes, due to their short characteristic distance, tends to prevent aseismic slip propagation. Therefore, slow earthquakes occur again soon after the passage of aseismic slip from asperity No. 29 as shown in Figs. 7n-o. This is why propagation process of SA as shown in Figs. 7n-o appears to be bilateral, rather than the unilateral propagation seen in Figs. 7a-7e.

2.4 Discussions of chain reaction effect on slow earthquake migration

Fig. 5 clearly shows that only slow slip events occur at depth of 30 km where frictional property is slightly stable and uniform along strike, while various slow earthquakes including very low-frequency events (orange color in Fig. 5) are generated by chain reaction between small asperities. This result suggests that chain reaction model as shown in Fig. 1 can explain various types of slow earthquakes occur in the same region.

Fig. 7 shows that MA has recurrence interval longer than that of SA, which is also seen for migration distance. These results may suggest that we can estimate asperity size on the basis of migration distance and recurrence interval of slow earthquakes.

For example, migration of low-frequency tremor observed in Kii and Tokai area tends to be unilateral with longer travel distance and loger recurrence interval of slip events, while the tremor in Shikoku tends to be shorter recurrence interval and shorter travel distance (e.g., Obara, 2010). These results suggest that size of asperities generating slow earthquakes in Kii and Tokai is larger than that in Shikoku. Therefore, investigating slow earthquake migration process is important to estimate the characteristics of small asperities.

3. Application to the detection of preseismic slip for megathrust earthquakes

As pre- and post-seismic changes, intense LFT activity began to occur almost directly below the 2004 Parkfield earthquake about three weeks before the earthquake and has continued only apart from the hypocenter over for four years (Nadeau & Guilhem, 2009; Shelly, 2009), which means that the distance from the hypocenter of triggering earthquake may also affect the sensitivity of LFT to pre- and post-seismic slip (Shelly, 2009).

In this section, we try to apply preseismic change of slow earthquake migration to the detection of megathrust earthquakes on the basis of characteristics of slow earthquake migration as described in the section 2.

3.1 A test model of slow earthquake migration for long travel distance

Fig. 8 shows a 3-D model of a subduction plate boundary derived from Fig. 2. Its frictional parameter is the same as Fig. 3 in order to investigate slow earthquake migration for long distance across the center of large asperity along strike direction as observed south-western Japan (Obara, 2010).

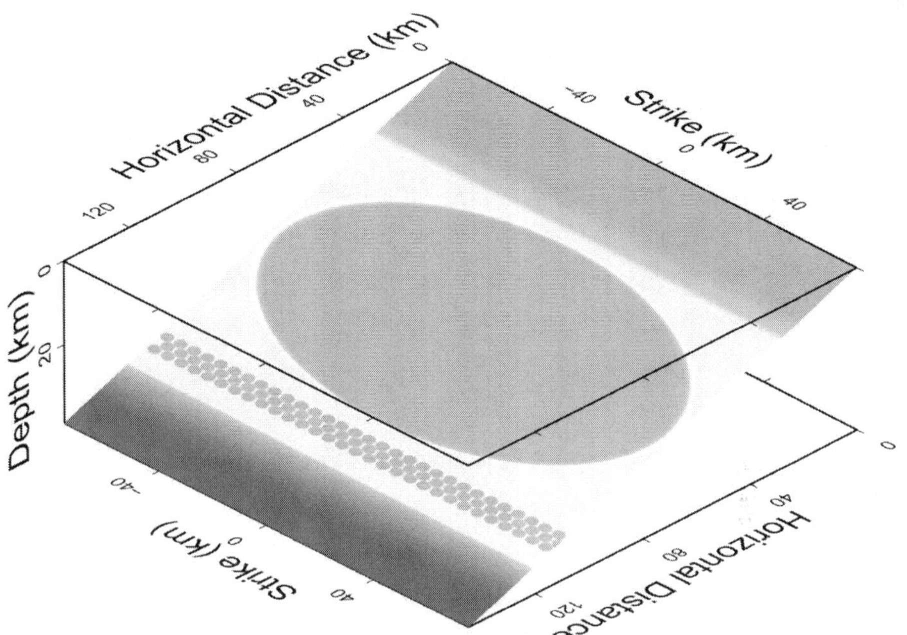

Fig. 8. A 3-Dimensional model of a subduction plate boundary with frictional parameter *a-b* (Eq. 3), where the color scale is the same as Fig. 2.

3.2 Long-term change in the migration speed of slow earthquake swarms

Figs. 9a and 9b show the spatiotemporal evolution of the slip velocity normalized by V_{pl} at 115 km down-dip from the trench (along green line in Figs. 9d and 9e) in the interseismic and preseismic stages, respectively. Close-up of the slip velocity pattern in the rectangle in Fig. 9b is shown in Fig. 9c. Figs. 9d and 9e show the snapshots of the normalized slip velocity 20 years after and 0.86 years before a megathrust earthquake, respectively.

Based on Figs. 9a-9c, we calculate the migration speeds of slow earthquake swarms by tracking transients with slip rate ranging from 2 to 10 V_{pl} (indicated by yellow color). Periods of larger slip rate (from 10 to 100 V_{pl} indicated by orange colors) are difficult to find in Fig. 9b because of their short duration, except for times later than -0.2 years in Fig. 9b.

The dominant migration speed is calculated to be approximately 0.3 to 1 km/day during the interseismic stage (Fig. 9a), while 1 to 3 km/day in the preseismic stage (Fig 9b). Therefore, the simulation results suggest that monitoring of the migration speeds of slow earthquake swarms as well as recurrence intervals are useful to forecast great earthquakes.

Approximately one month before the megathrust earthquake, Fig. 9b shows that the dominant slip velocity for |Strike| < 40 km becomes higher than 10 V_{pl} (orange) and is sustained over a long duration time (more than one month). This implies that the moment release rates of slow earthquake swarms near the locked region of LA just before a megathrust earthquake tend to be significantly higher than that in the interseismic stage.

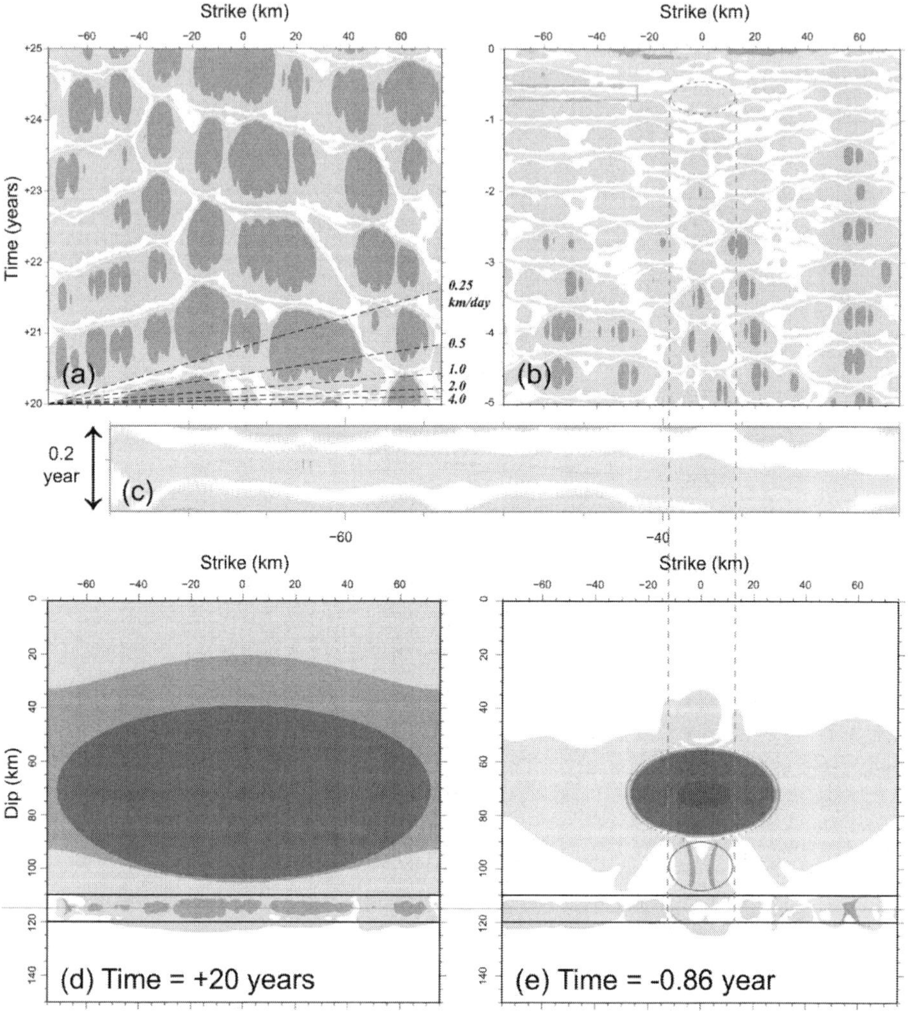

Fig. 9. (a)(b) Spatiotemporal evolution of slip velocities at the "Dip" of 115 km along strike (green line in Figs. 9d and 9e) in the interseismic and preseismic stages, respectively, after Ariyoshi et al. (2011b). Color scale is the same as Fig. 4. The broken lines in (a) denote the migration speed in km/day. (c) Close up of the slip velocity evolution in the spatiotemporal region enclosed by the green rectangle in (b), keeping the aspect ratio of space to time. (d)(e) Snapshots of the slip velocity field (d) 20 years after and (e) 0.86 year before the occurrence time of the megathrust earthquake. The ellipse enclosed by the purple curve in (e) represents a large aseismic slip event activating slow earthquakes as shown by the ellipse in (b).

Fig. 9d suggests that the slip velocity is approximately less than 0.5 V_{pl} (aqua) in the region surrounding the SA (boxed area) and less than 0.1 V_{pl} (blue) dominantly along the center of the SA belt (green line), except for the region where a slow earthquake migration occurs (yellow and orange). Fig. 9e suggests that the area of higher slip velocity (orange) covering SAs in the preseismic stage tends to be larger than in the interseismic stage as shown in Fig. 9d, and there is no region in which the slip velocity is less than 0.1 V_{pl}. Slip velocity in LA becomes higher due to the preseismic slip, especially about one year before the megathrust earthquake. These results mean that preseismic slip of LA promotes higher moment release rates of slow earthquake due to its higher slip velocity.

3.3 Comparison of slow earthquakes between simulation and observational results

Fig. 9e shows that a large aseismic slip event occurs locally between LA and the SA belt (indicated by the ellipse) approximately one year before the megathrust earthquake. The spatiotemporal region enclosed by the ellipse in Fig. 9b shows that the large aseismic slip event triggers slow earthquakes in the SA belt with a shorter recurrence interval compared to other areas ($|\text{Strike}| > 20$ km). The migration distance in the SA belt corresponds to the size of the large aseismic slip region ($|\text{Strike}| < 20$ km). This behavior is similar to the long-term SSE (slow slip event) observed at Bungo Channel in 2003, where nearby LFT (low frequency tremor) migration had occurred either at a shorter recurrence interval or nearly continuously for several months (Obara, 2010). Therefore, the activity of the LFTs may be useful to estimate the duration and the location of local aseismic slip events, such as the long-term SSE and the preseismic slip in the deeper part of LA.

As pre- and post-seismic changes, intense LFT activity began to occur almost directly below the 2004 Parkfield earthquake about three weeks before the earthquake and has continued only apart from the hypocenter over for four years (Nadeau & Guilhem, 2009; Shelly, 2009), which means that the distance from the hypocenter of triggering earthquake may also affect the sensitivity of LFT to pre- and post-seismic slip (Shelly, 2009).

Our simulation shows that the moment release rate of slow earthquakes near the locked region of LA becomes higher about one month before the megathrust earthquake as mentioned in the section 3.2, and Figs. 9ab show that a shorter recurrence interval of slow earthquakes in the SA belt occurs several years after the megathrust earthquake. These simulation results are consistent with those results observed in Parkfield for the pre- and post-seismic stages of nearby large earthquakes.

Since SSE, pre- and post-seismic slips are all transients of interplate-slip faster than V_{pl}, the observations reported by Obara (2010), Nadeau & Guilhem (2009), and Shelly (2009) would support our suggestion that the preseismic slip of megathrust earthquakes can be practically detected by monitoring slow earthquake migrations, even if the actual d_c is less than several centimeters.

4. Monitoring slow earthquakes in shallower part by DONET

Recently, slow earthquakes have been also observed in shallower part of subduction plate boundaries (e.g., Hirose et al., 2010). As seen in Figs. 9d and 9e, stress shadow of LA has more effective on shallower part of transition zone. This means that preseismic change of slow earthquake may be more sensitive in the shallower part. In this section, we develop a model of slow earthquake migration in shallower part of transition zone in addition to deeper part, discussing the detectablity of Dense Oceanfloor Network System for Earthquakes and Tsunamis (DONET) toward an anticipated Tonankai Earthquake (Fig. 10).

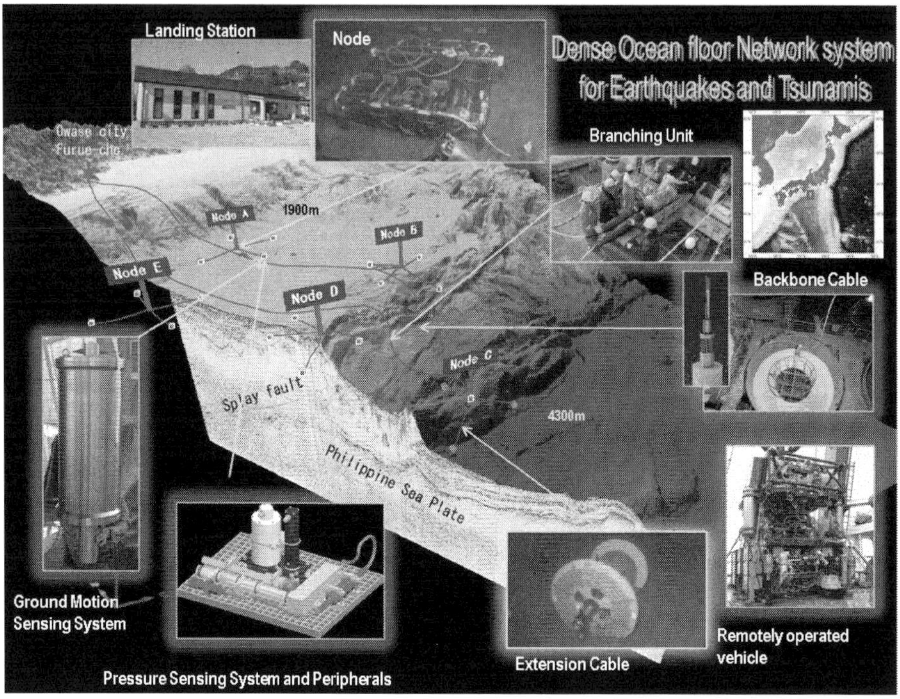

Fig. 10. An overview of Dense Oceanfloor Network System for Earthquakes and Tsunamis (DONET) in Tonankai region.

4.1 Design concept of DONET toward forecasting the Tonankai earthquake

In Fig. 10, DONET has a submarine cabled real-time seafloor observatory network for the precise earthquake and tsunami monitoring. For the purpose of understanding and forecasting the earthquake and related activities underneath the seafloor, the twenty sets of state-of-arts submarine cabled sub-sea measurement instrument will be deployed in seafloor at the interval of 15-20km. All of the twenty sets of preliminary interface have been installed just on July 31, 2011 and are to be prepared in consideration of the improvement of observation capability in the future.

As described by Kawaguchi et al. (2011), operating large-scale subsea infrastructure over a long period of time (20-30 years) is one of a challenge of underwater technology. The increase of measurement instruments has a big influence on the total system reliability, because of the state-of-arts instrument is a bottleneck to maintain long-term reliability. A novel system design concept is necessary for the observatory network development to make two demands such as 'high reliability system design' and 'state-of-arts measurement' united. The observatory network should be able to replace, maintenance and extend while operating, and should be have a redundancy for the internal or external observatory network component failure. To achieve these requirements, DONET adapt a strategy to combining the following three major components with different system reliability: (i) high reliability backbone cable system, (ii) replaceable science node, and (iii) extendable measurement instruments.

4.2 Expectation of preseismic monitoring by DONET

Fig. 11 shows a map projection of 3-D subduction plate boundary model in Tonankai region. In this study, we introduce a plate interface bended by spline curve along dip direction, taking it into account the structural survey published by Nakanishi et al. (2008). hypocenter of shallower part of slow earthquakes is mainly based on the recent studies (e.g., Obara & Shiomi, 2009).

Megathrust earthquake (M_w 8.2) occurring in LA has periodic recurrence time of 113 years. Fig. 12 shows snapshot of slip velocity field in the interseismic and preseismic stage of the megathrust earthquake. Comparing shallower part of slow earthquake activity with deeper part, we found that the shallower part of slow earthquakes is less active than the deeper part in the interseismic stage of the megathrust earthquake (in the top of Fig. 12), because stress shadow from LA has more effective on the shallower part. In the preseismic stage (in the bottom of Fig. 12), the shallower part of slow earthquake comes to be similar to the deeper part and to be more active especially around the center of LA, because locked region is only around the center of LA and slip deficit in the preseismic stage is more than deeper part. These simulation results suggest that monitoring the shallower part of slow earthquakes may be effective on the ground that it is more sensitive to the preseismic change of the megathrust earthquake because of free surface condition.

In order to detect the preseismic slip of the next Tonankai earthquake in the near future, DONET would play an important role in monitoring shallower part of slow earthquake migration from the view of shortening recurrence interval and increasing migration speed as pointed out in the section of 3.2. Considering the location of 20 observation points as shown in Fig. 11 and numerical simulation results in Figs. 11 and 12, we expect DONET to do precise detection of preseismic change of the Tonankai earthquake as listed in Table 1.

	Long	Short	Coverage	Dislocation	Main	Afterslip
Node A	○	◎		○	◎	
Node B	○	◎	◎	◎		
Node C	○	○	◎	◎		○
Node D	◎	○	◎	◎		○
Node E	◎	○		○	◎	

Table 1. DONET's major roles in monitoring seismic & crustal change due to the Tonankai earthquake.

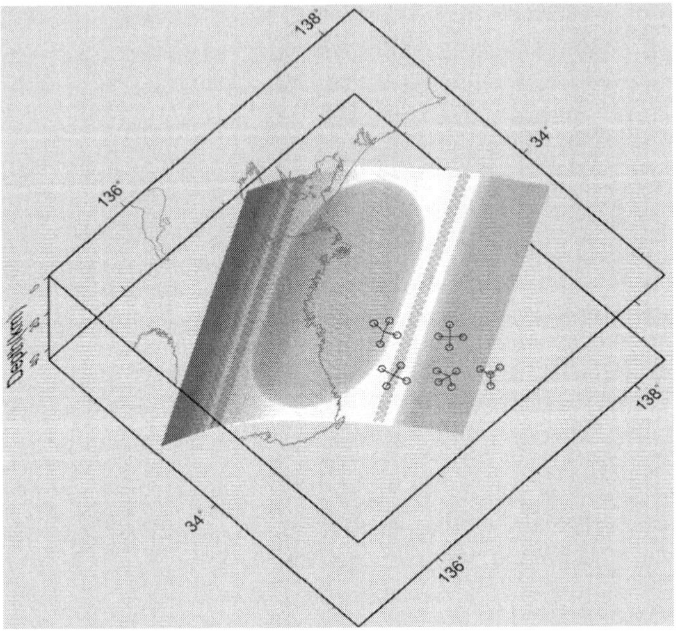

Fig. 11. A 3-D map projection of a subduction plate boundary model for Tonankai earthquake. Frictional parameter for Shallow Stable Zone, Deep Stable Zone, Large & Small Asperity, and Transition Zone is the same as Fig. 3 but (κ_1, κ_2) = (0.3, 0.1) on the basis of afterslip propagation speed investigated by Ariyoshi et al. (2007) and the value of d_{c2} for SA in shallower part is 0.3 mm. Note that we use shade effect on the color scale by applying the command "grdgradient" (Wessel & Smith, 1998) in order to show the bending shape visually. Twenty open circles with five nodes represent observation points of DONET as shown in Fig. 10.

"Long" & "Short" represent long-term & short-term forecast based on recurrence interval and migration speed of slow earthquakes, respectively. "Coverage" represents significant effect on estimating hypocenters of slow earthquakes with high precision due to the coverage from trench side. "Dislocation" represents key observation points for estimating slip amount of continental plate relative to oceanic plate. "Main" represents main part of seismic slip process generated from LA. "Afterslip" represents aseismic slip along trench which may triggers nearby Nankai earthquake.

On the basis of numerical simulation results as shown in Fig. 12 and Table 1, DONET is expected to play important role in monitoring preseismic change of the next Tonankai earthquake from the view of various points including shallower part of slow earthquake activity and in judging the possibility of triggering the nearby Nankai earthquake, which had occurred in 1946 two years after the 1944 Tonankai earthquake.

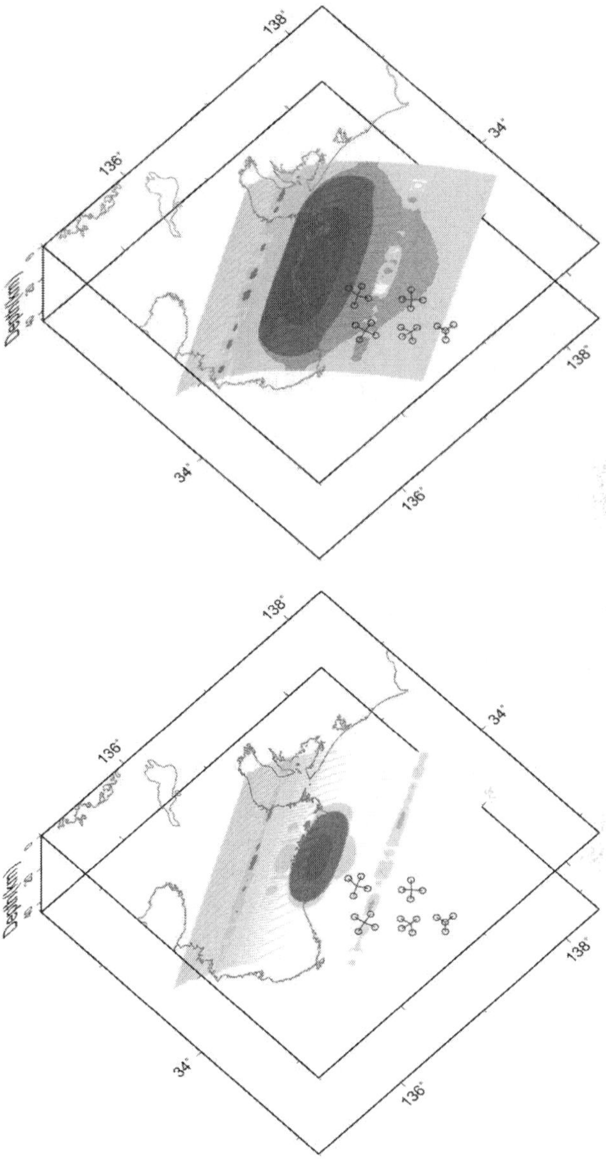

Fig. 12. Snapshots of slip velocity on the plate boundary about 20 years after the megathrust earthquake (top; interseismic period) and 2.5 years before (bottom; preseismic period). Color scale is the same as Fig. 4 but shade effect by applying the command "grdgradient" (Wessel & Smith, 1998) in order to show the bending shape visually. Twenty open circles with five nodes represent observation points of DONET as shown in Fig. 10.

5. Acknowledgment

The authors would like to thank DONET members, Takane Hori and Ryoko Nakata for giving beneficial comments. The authors also would like to thank Tohoku University researchers, Toru Matsuzawa, Ryota Hino and Akira Hasegawa for stimulating our study. J.-P. Ampuero helped us to improve our manuscript with thoughtful discussion. We appreciate publishing managers Igor Babic and Ivana Lorkovic for encouraging our publication, which was partly supported by Grant-in-Aid for Young Scientists (KAKENHI) 23710212.

6. References

Ariyoshi, K.; Matsuzawa, T. & Hasegawa, A. (2007). The key frictional parameters controlling spatial variations in the speed of postseismic-slip propagation on a subduction plate boundary, *Earth Planet. Sci. Lett.*, 256, 136–146, doi:10.1016/j.epsl.2007.01.019.

Ariyoshi, K.; Hori, T.; Ampuero, J.-P.; Kaneda, Y.; Matsuzawa, T.; Hino, R. & Hasegawa, A. (2009). Influence of interaction between small asperities on various types of slow earthquakes in a 3-D simulation for a subduction plate boundary, Gondwana Research, 16(3-4), 534-544, doi: 10.1016/j.gr.2009.03.006.

Ariyoshi, K; Hori, T.; Kaneda, Y.; Ampuero, J.-P.; Matsuzawa, T.; Hino, R. & Hasegawa, A. (2011a). Effects of small asperities on migration of slow-earthquakes along trench direction of subduction plate boundary: investigation through a 3-D numerical simulation, *Adv. in Geosci.*, in press.

Ariyoshi, K; Matsuzawa, T; Ampuero, J.-P.; Nakata, R.; Hori, T.; Kaneda, Y.; Hino, R. & Hasegawa, A. (2011b). Migration process of very low-frequency events based on a chain-reaction model and its application to the detection of preseismic slip for megathrust earthquakes, *Earth Planets Space*, in press.

Blanpied, M. L.; Marone, C. J.; Lockner, D. A.; Byerlee, J. D. & King, D. P. (1998). Quantitative measure of the variation in fault rheology due to fluid-rock interactions, *J. Geophys. Res.*, 103, 9691–9712.

Boatwright, J. & Cocco, M. (1996). Frictional constraints on crustal faulting, *J. Geophys. Res.*, 101, 13895–13909.

Dieterich, J. H. (1979). Modeling of rock friction: 1. Experimental results and constitutive equations, *J. Geophys. Res.*, 84, 2161–2168.

Dragert, H. (2007). Mediating Plate Convergence, *Science*, 315, 471-472, doi: 10.1126/science.1137171.

Hacker, B. R.; Peacock, S. M.; Abers, G. A. & Holloway, S. D. (2003). Subduction factory 2. Are intermediate-depth earthquakes in subducting slabs linked to metamorphic dehydration reactions?, *J. Geophys. Res.*, 108(B1), 2030, doi:10.1029/2001JB001129.

Hirose, H. & Hirahara, K. (2002). A model for complex slip behavior on a large asperity at subduction zones, *Geophysical Research Letters*, 29 (22), 2068.

Hirose, H.; Asano, Y.; Obara, K.; Kimura, T.; Matsuzawa, T.; Tanaka, S. & Maeda, T. (2010). Slow Earthquakes Linked Along Dip in the Nankai Subduction Zone, *Science*, 330(6010), 1502-1502, doi:10.1126/science.1197102.

Ide, S.; Beroza, G. C.; Shelly, D. R. & Uchide, T. (2007). A new scaling law for slow earthquakes, *Nature*, 447, 76–79, doi:10.1038/nature05780.

Kawaguchi, K.; Kaneko, S.; Nishida, T. & Komine, T. (2011). Construction of Real-time Seafloor Observatory for Earthquakes and Tsunami Monitoring, *Praxis book on "Seafloor Observatories"*, in press.

Liu, Y. & Rice, J. R. (2005). Aseismic slip transients emerge spontaneously in three-dimensional rate and state modeling of subduction earthquake sequences, *J. Geophys. Res.*, 110, B08307.

Matsuzawa, T.; Uchida, N.; Igarashi, T.; Okada, T. & Hasegawa, A. (2004). Repeating earthquakes and quasi-static slip on the plate boundary east off northern Honshu, Japan, *Earth Planets Space*, 56, 803–811.

Nakanishi, A.; Kodaira, S.; Miura, S.; Ito, A.; Sato, T.; Park, J.-O.; Kido, Y. & Kaneda, Y. (2008). Detailed structural image around splay-fault branching in the Nankai subduction seismogenic zone: Results from a high-density ocean bottom seismic survey, *J. Geophys. Res.*, 113, B03105.

Nakata, R.; Suda, N. & Tsuruoka, H. (2008). Non-volcanic tremor resulting from the combined effect of Earth tides and slow slip events, *Nature Geosci.*, 1, 676–678, doi:10.1038/ngeo288.

Nadeau, R. M. & Guilhem, A. (2009). Nonvolcanic tremor evolution and the San Simeon and Parkfield, Califorinia, Earthquakes, *Science,* 325, 191-193, doi:10.1126/science.1174155.

Obana, K. & Kodaira, S. (2009), Low-frequency tremors associated with reverse faults in a shallow accretionary prism, *Earth Planet. Sci. Lett.*, 287, 168–174, doi:10.1016/j.epsl.2009.08.005.

Obara, K. (2010). Phenomenology of deep slow earthquake family in southwest Japan: Spatiotemporal characteristics and segmentation, *J. Geophys. Res.*, 115, B00A25, doi:10.1029/2008JB006048.

Obara, K. & Shiomi, K. (2009). Underground Structural Anomalies and Slow Earthquake Activities Around Seismogenic Megathrust Earthquake Zone as Revealed by Inland Seismic Observations, *J. Disaster Res.*, 4(2), 83-93.

Okada, Y. (1992). Internal deformation due to shear and tensile faults in a halfspace, *Bull. Seism. Soc. Am.*, 82, 1018–1040.

Rice, J. R. (1993). Spatio-temporal complexity of slip on a fault, *J. Geophys. Res.*, 98, 9885–9907.

Ruina, A. (1983). Slip instability and state variable friction laws, *J. Geophys. Res.*, 88, 10,359–10,370.

Savage, J. C. (1983). A dislocation model of strain accumulation and release at a subduction zone, *J. Geophys. Res.* 88, 4984–4996.

Schwartz, S. Y. & Rokosky, J. M. (2007). Slow slip events and seismic tremor at circum-Pacific subduction zones, *Rev. Geophys.*, vol.45, RC3004, doi:10.1029/2006RG000208.

Shelly, D. R. (2009). Possible deep fault slip preceding the 2004 Parkfield earthquake, inferred from detailed observations of tectonic tremor, *Geophys. Res. Lett.*, 36, L17318, doi:10.1029/2009GL039589.

Shelly, D. R.; Beroza, G. C.; Ide, S. & Nakamura, S. (2006). Low frequency earthquakes in Shikoku, Japan, and their relationship to episodic tremor and slip, *Nature*, 442, 188–191, doi:10.1038/nature04931.

Wang, K. & Suyehiro, K. (1999). How does plate coupling affect crustal stresses in northeast and southwest Japan, *Geophys. Res. Lett.*, 26, 2307–2310, doi:10.1029/1999GL900528.

Wessel, P. & Smith, W. H. F. (1998). New, improved version of the Generic Mapping Tools released, *Eos Trans. AGU*, 79, 579.

6

Characteristics of Coseismic Thrust-Related Folding from Paleoseismic Investigation Responsible for the 1999 Chi-Chi Earthquake of Central Taiwan

Wen-Shan Chen[1], Nobuhisa Matsuta[2] and Chih-Cheng Yang[3]
[1]*Department of Geosciences, National Taiwan University, Choushan Road, Taipei, Taiwan*
[2]*Research Center for Seismology, Volcanology and Disaster Mitigation Graduate School of Environmental Studies, Nagoya University D2-2(510), Furo-cho, Chikusa-ku, Nagoya, Zip,*
[3]*Taiwan Petroleum Exploration Division, Chinese Petroleum Corporation, Miaoli, Taiwan,*
[1,3]*Republic of China*
[2]*Japan*

1. Introduction

The Western Foothills are a west-verging fold-thrust belt related to the recent arc-continent collision of the Asian continental plate with the Philippine Sea plate (Suppe, 1981). The frontal orogenic belt of Western Foothills consists of well constructed imbricated thrust faults, which have recently been locked and rooted into an aseismic decollement (Chen *et al.*, 2001c; Yue *et al.*, 2005). Over the past tens years, several great intraplate thrust earthquakes (Mw >7.0) located at the orogenic belt occurred in the world (e.g., the 2008 Wenchuan earthquake and the 1999 Chi-Chi earthquake). On September 21, 1999, central Taiwan was hit by an earthquake of magnitude Mw 7.6. Several buildings and infrastructure were severely damaged, and about 2450 people were killed. The Chi-Chi earthquake was caused by a thrust fault located at the boundary between the Western Foothills and the Taichung piggyback basin; the fault produced a spectacular surficial rupture of 100 km in length (Fig. 1; Central Geological Survey, 1999; Chen *et al.*, 2001c). Based on seismic reflection profiles and focal mechanisms of the mainshock, it was inferred that the earthquake occurred on a shallow-dipping (20–30°) thrust ramp of the Chelungpu fault (Chiu, 1971; Kao and Chen, 2000; Wang *et al.*, 2002). The earthquake ruptures show complex structural and stratigraphic relationships that can be divided into two segments: the Shihkang and Chelungpu faults. The Shihkang fault occurs within Pliocene shale and is interpreted as a bed-parallel slip fault. The Chelungpu fault of the southern segment thrust is late Pliocene shale over Quaternary fluvial deposits in the footwall.

In this study, several excavations have been carried out along the Chelungpu fault across the fault trace (Fig. 1); the excavations showed complicated structural features within unconsolidated sediments, including a ductile deformation zone at the fault tip. Because offset measurements within the ductile deformation zone are too variable to estimate the slip, in this study, we attempted other methods to estimate a relatively complete displacement for the thrusting through more detailed observation and discussion of the characteristics of the fault-tip deformation. Here, we also present the results of our paleoseismic studies of the Chelungpu fault to assess the timing, offset, and slip rate of each event. Because active thrust faults commonly show complex geometric patterns of faulting and folding, we present a few examples of excavation that document different styles of surface faulting. We focus on the contribution of the excavation study in providing some information about fault behavior and coseismic fault-tip deformation patterns and in deriving the growth history of the fault-tip fold.

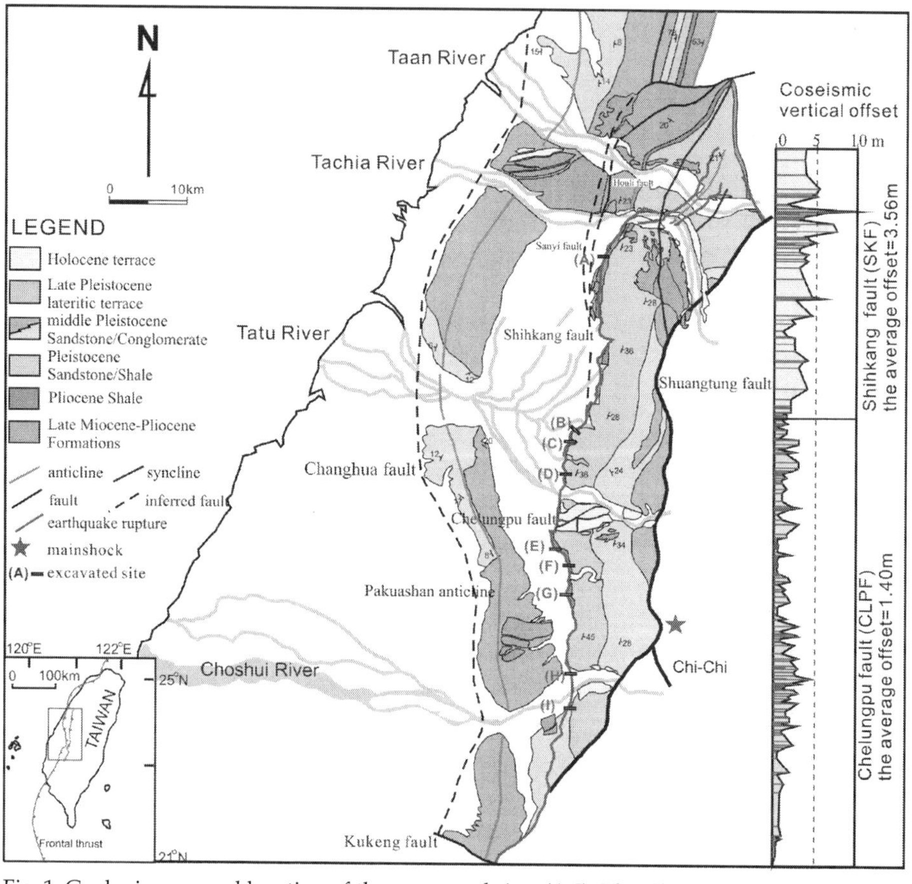

Fig. 1. Geologic map and location of the excavated sites (A-I). The Chi-Chi earthquake rupture is subdivided into the Shihkang and Chelungpu faults along the frontal Western Foothills. (A) the Fengyuan site, (B) the Wenshan farm site, (C) the Pineapple-field site, (D) the Siangong-temple site, (E) the Tsaotun site, (F) the Wanfung site, (G) the Shijia site, (H) the Mingjian site, and (I) the Chushan site. Average vertical displacement of the Chi-Chi earthquake rupture is calculated from the measured vertical displacements measured at 210 locations along the earthquake rupture.

2. The 1999 Chi-Chi earthquake ruptures

The Western Foothills in western Taiwan consist of a series of west-verging subparallel thrusts and fault-related fold structures, which have been a zone of active folding and thrusting throughout the late Quaternary (Suppe, 1981). The Chi-Chi earthquake exhibits a fault-bend fold geometry, forming a flat-ramp structure on the frontal upthrown block of the Western Foothills (Yue *et al.*, 2005; Lai *et al.*, 2006); the concealed Changhua fault of a fault propagation fold that forms the Paukashan anticline in the western side of the earthquake rupture (Fig. 1). The earthquake ruptures have complex structural and stratigraphic characteristics and can be split into two principal segments, namely, the Shihkang and Chelungpu faults. Surface slip was greatest along the northern portion of the Shihkang fault, with a maximum observed vertical slip of 9.5 m. The Shihkang fault, which has a length of about 45 km and is confined within a Pliocene shale, is interpreted as a bed-parallel slip fault, based on extrapolation from the Taiwan Core Drilling Project deep borehole. The Chelungpu fault is about 55 km in length and differs from the Shihkang fault in that the Pliocene shale overthrusts a thick sequence of Quaternary deposit. The Shihkang fault consists of a N30-40°W verging oblique thrust with vertical offsets ranging from 3-9.5 m, while the Chelungpu fault is made up of a N70-90°W verging pure thrust with vertical offsets ranging from 0.2-4 m. According to the seismicity, GPS, and SPOT data, the coseismic deformation also defines two different structural domains (Dominguez *et al.*, 2003; Pathier *et al.*, 2003; Cattin *et al.*, 2004).

3. Paleoseismologic analysis

Although the Taiwan seismic catalogs provide a historical record of strong earthquakes over the past three centuries, there is no record of any large earthquake associated with the Shihkang-Chelungpu fault (Chen *et al.*, 2004). Paleoseismic study is one of the best methods for characterizing the earthquake behavior along the Shihkang-Chelungpu fault. Over the past ten years, we have excavated numerous sites and continuously cored borings on the Chelungpu fault, where the surface rupture shows surface deformation within the Holocene deposits; our results have provided quantitative data for late Holocene slip rates (Fig. 1; Chen *et al.*, 2001a, b, 2004, 2007a, b; Ota *et al.*, 2001, 2005; Streig *et al.*, 2007). Bedrock exposed along the Shihkang fault has not been characterized in previous paleoseismic studies. Here, we integrate paleoseismologic data derived from four excavations of the Shijia, Siangong-temple, Pineapple-field, and Chushan sites to understand the characteristics of coseismic deformation of the Chelungpu fault.

3.1 Shijia site

The Shijia site is located along the frontal foothills; the surface features of the site before the earthquake show a gently alluvial fan slope of approximately 3°, which was an undeveloped terrace scarp. The 1999 fault runs along the fan slope with a hangingwall uplift of gently dipping coseismic fold scarps that are about 1 to 2 m higher than those of the footwall. We excavated a 7-m-deep, 27-m-long trench across the coseismic fold scarp formed during the Chi-Chi earthquake. The excavation showed clear exposures of well-sorted silty sand interbedded with mud and humic soil; these exposures represent overbank deposits. The exposures show three depositional units of silty sand layers (cw1, cw2, and cw3) which are defined by the onlap of H1 and H2 humic soils (Fig. 2). The upper sequence includes two wedge-shaped deposits of cw1 and cw2 units associated with deposition across the scarp; these deposits are defined by the onlap of H1 and H2 humic soils across the forelimb. The onlapped relation indicates the occurrence of two folding events after the deposition of the H1 and H2 humic soils (Fig. 2). Detrital charcoal that was collected from the alluvial

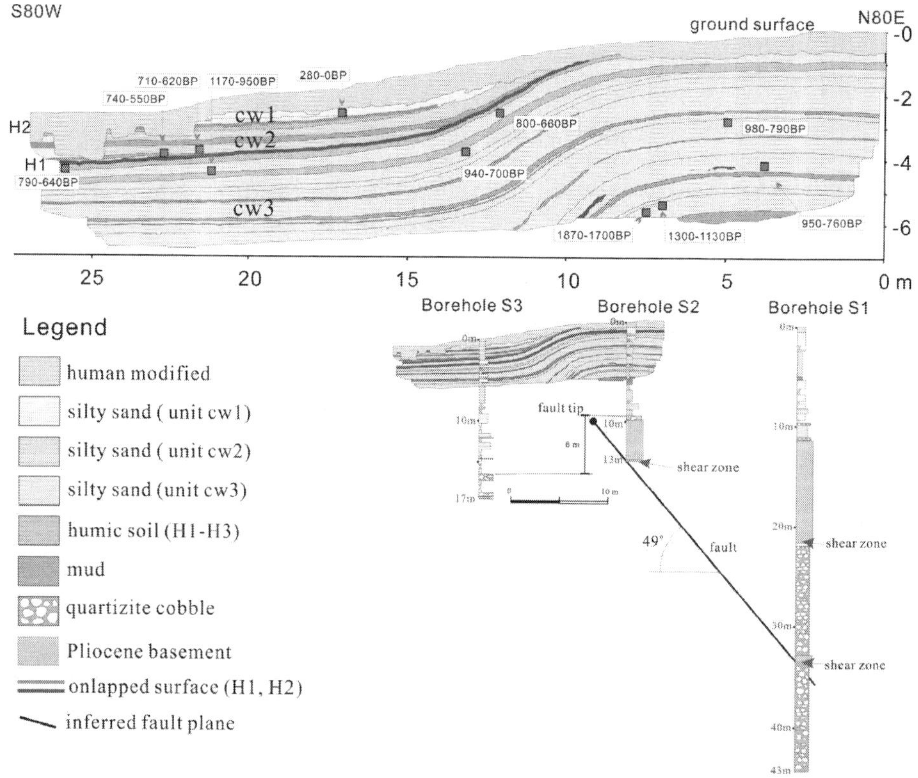

Fig. 2. The contact of alluvial deposits and humic soil (H1 and H2) occurs a distinct onlapped structure interpreted to represent a paleoearthquake event. Radiocarbon dates are used to constrain the timing of S1 and S2 paleoseismic events. Fault dip of fault depends on correlation between the borehole S1 and S2 of 49° dips. The Pliocene formation shows inclined bedding plane and is weakly sheared. The fluvial cobble deposits between boreholes S2 and S3 offset about 6-m high in both sides of the fold scarp.

deposits yields eleven radiocarbon ages; the upper part of cw3 unit: 800–660 yr BP and 940–700 yr BP, thus indicating the depositional age of 940–660 yr BP; cw2 unit: 790–640 yr BP, 740–550 yr BP, and 710–620 yr BP, indicating the depositional age of 790–550 yr BP; cw1 unit: 280–0 yr BP (Chen *et al.*, 2007b).

Borehole S1 and S2, which were drilled on the hangingwall, showed the location of the fault zone at a depth of 20.7 m and 30 m in borehole S1 and 13 m in borehole S2, where Pliocene shale is displaced over gravel deposits (Fig. 2). Based on the depth of the shear zones in both the boreholes and location of fold scarp in the excavated exposures, the fault must dip about 49° along the shallowest portion of the thrust ramp (Fig. 2). The S2 and S3 borehole logging of the top of gravel bed on both sides of the main thrust reveal a total vertical offset of 6 m, indicating that repeated large earthquakes have occurred. However, based on onlapping feature of H1 and H2 humic soils and wedge-shaped deposits of cw1 and cw2 units, we can identify two paleoearthquake events which occurred at >280 yr BP (S1 event) and 790–680 yr BP (S2 event).

3.2 Siangong-temple site

The earthquake ruptures cut through the foot of an alluvial fan and produced a gentle 1-m-high monoclinal scarp of approximately 11°. The pre-existing ground surface exhibits a gentle westward-dipping slope of approximately 2°. We excavated a 4- to 6-m-deep, 38-m-long trench across the earthquake scarp; the excavation exposed six wedge-shaped alluvial units: aw1, aw2, aw3, aw4, aw5, and aw6 (Fig. 3). A distinctive humic soil overprints the top of each colluvial unit, which consists of a well-sorted sand layer with a channelized gravel bed. The wedge-shaped colluvial deposits were well-defined by Os1, Os2, Os3, Os4, and Os5 humic soils, which were unconformably overlain by colluvial deposits after a large earthquake. Unit aw2, for example, pinched out over the Os3 soil of the forelimb or was eroded on the uplifted side of the fold crest. However, based on onlapping relations, it was deduced that the colluvium was deposited across the forelimb of a coseismic fold scarp and that at times, the colluvium onlapped and overlapped against the forelimb. The dips of soil in the forelimb progressively decrease toward the ground surface from 38° (Os5) to 11° for the Chi-Chi earthquake-induced slope (Fig. 3). Upward change in dips, hindward thinning, and angular unconformites between soils and overlying colluvium are interpreted as indicating the occurrence of repeated large earthquakes on the Chelungpu fault. The radiocarbon ages obtained from detrital charcoal that was collected from units aw1, upper aw4, lower aw4, and aw5 are <300 yr BP, 1960–1810 yr BP, 3000–2840 yr BP, and 3100–2920 yr BP, respectively.

Boreholes T1 and T2 were drilled to a depth of 40 m on the hangingwall and to a depth of 50 m on the footwall. Borehole T1 constrains the location of the fault zone at a depth of 22.8 m between the Pliocene shale and Holocene colluvial deposits. We link the fault zones in borehole T1 and the synclinal fold axis in the excavated exposures; the results suggest a 30° east-dipping blind fault (Fig. 3). A corresponding gravel bed between boreholes T1 and T2 is offset to a height of 40 m on both sides of the fold scarp, indicating that several large earthquakes have occurred. Based on the syntectonic sedimentary structure, we infer that five paleoearthquake events including the Chi-Chi earthquake occurred near 1960–1810 yr BP (G3 event) and 3160–2840 yr BP (G4 event); however, we have not been able to obtain age control for the G1 and G2 events (Chen *et al.*, 2007b).

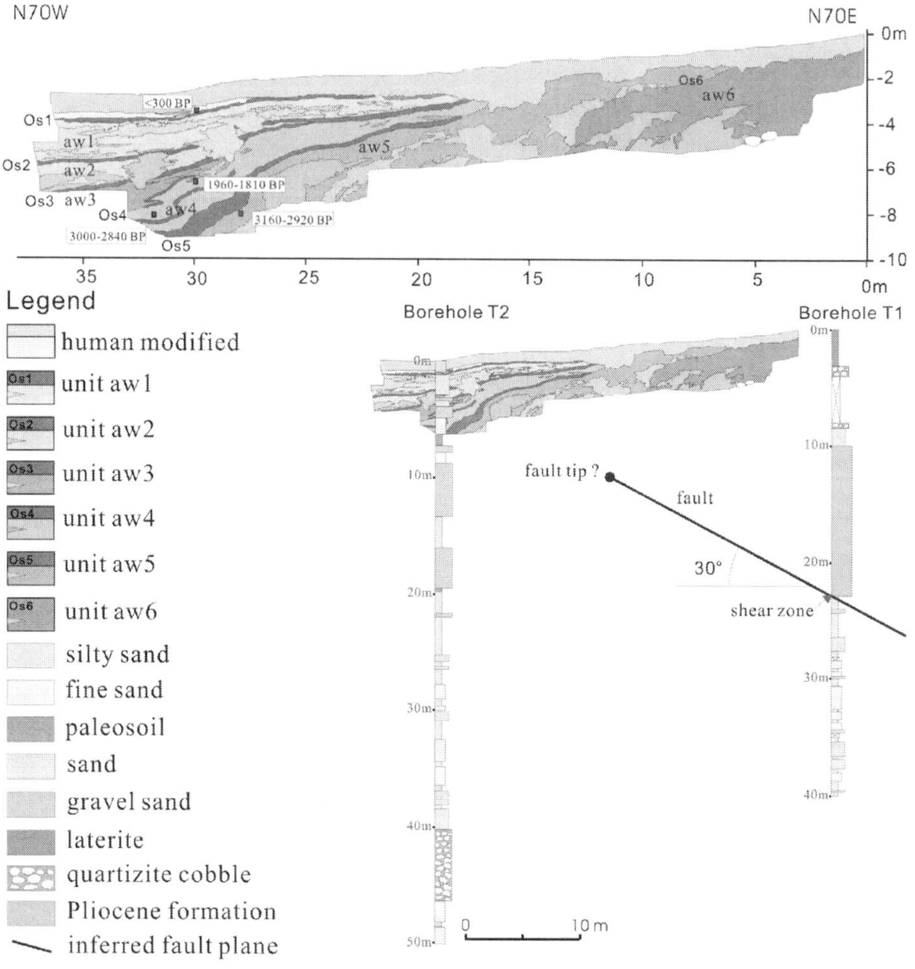

Fig. 3. Sketch of trench-wall emphasizing the top of beds of soil (Os) and gravel depositional wedges (aw). Alluvial gravels onlap onto each soil bed forming a wedge-shaped depositional unit. Borehole T1 reveals a shear zone at the bottom of Pliocene formation which shows inclined bed and slightly shearing at the excavated site. Correlation with a shear zone at depth in the borehole T1 and the synclinal fold axis in the excavated exposures identifies a fault-plane dipping of 30°.

3.3 Pineapple-field site

Repeated coseismic displacements show clear morphological expressions on both sides of the fault zone; the expressions commonly occur as a lineament scarp between the foothills and the basin plain (Chen *et al.*, 2003). The site was dug across a pre-existing terrace scarp, where the earthquake rupture clearly uplifted the hangingwall by about 1.6–2.1 m. We

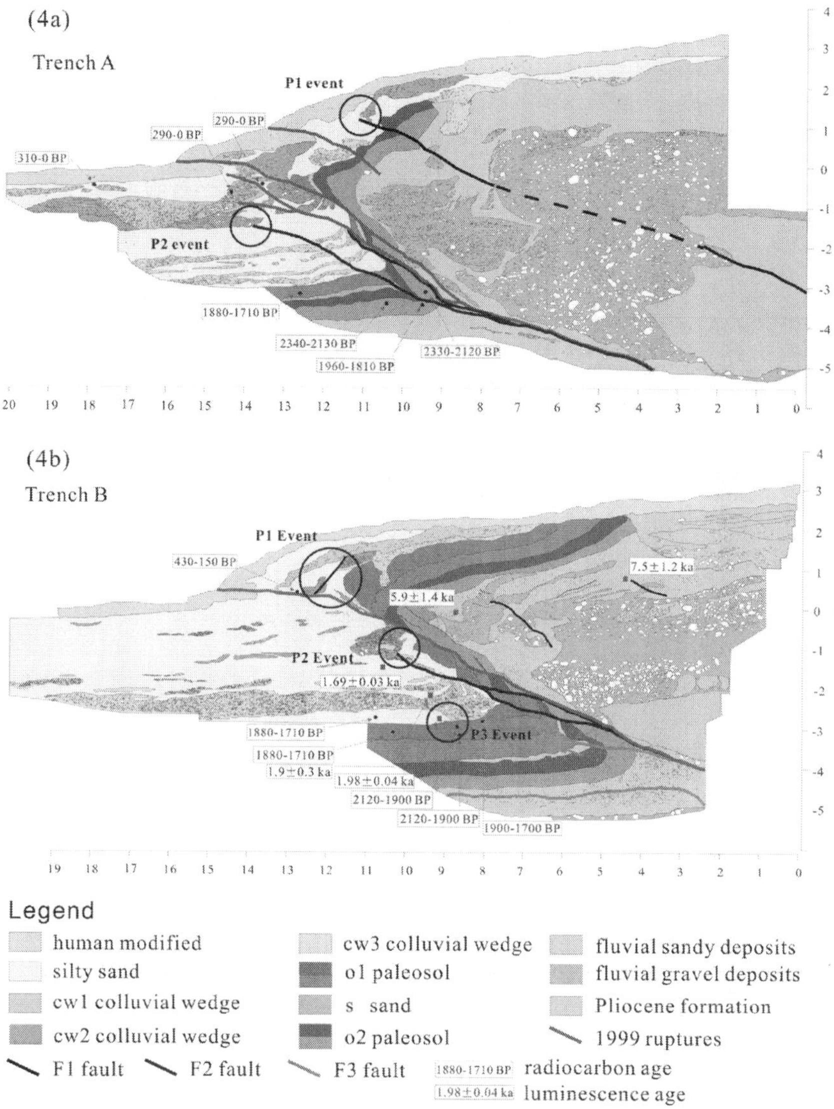

Fig. 4. (a, b) Five types of deposits, the 1999 earthquake ruptures, and three paleo-coseismic faults (F1, F2, and F3) are observed in the excavation of the Pineapple-field sites. The Pliocene basement is overthrust on the hangingwall. The older sediments of the fluvial gravel deposits reached at 2 m depth on the hangingwall and 5 m depth on the footwall. The fluvial gravel deposits have 6 m vertical offset across the main fault. The top soil is offset 1.6 m at trench A (Fig. 4a) and 2.1 m at trench B (Fig. 4b) in both sides of the 1999 earthquake ruptures. Radiocarbon dates are used to constrain the timing of P1, P2, and P2 paleoseismic events.

excavated a 25-m-long, 8-m-deep trench, which showed clear exposures of the Pliocene rock, Holocene fluvial, and colluvial deposits. The Pliocene basement is preserved on the upthrown block, which is unconformably covered by Holocene fluvial gravel and boulder beds (Figs. 4a and 4b). The downthrown block is deposited by Holocene fluvial deposits of soils (s1 and s2 units), and sand layers are interbedded with colluvial wedge-shaped gravel beds (cw1, cw2, and cw3 units). Wedge-shaped colluvium overlies the paleoearthquake ruptures, indicating that the colluvium evidently represents the earthquake-induced deposits.

The excavation at the Pineapple-field site indicates several shear planes including the Chi-Chi earthquake and other paleoearthquake ruptures. The character deformation at the fault front indicates ductile deformation comprising a hangingwall anticline and a footwall syncline with an overturned limb and increasing thickness in the anticlinal hinge. The Chi-Chi earthquake rupture here indicates a fault-bend fold geometry, forming a flat-ramp structure cutting through the ground surface. The ruptures branch into several reverse faults, which form about N42°E trending and a 26° southeast-dipping fault. The excavation exposes three paleoearthquake thrust faults, namely, F1, F2, and F3, which are defined by a wide thrust zone with fully developed thin shear bands (Figs. 4a and 4b).

Luminescence dating constrains the age of the top layer of alluvial deposits (f unit) to about 4 ± 0.3 ka (Chen *et al.*, 2009). Except for luminescence date, all the other twelve detrital charcoal samples collected from fluvial deposits yield radiocarbon ages as follows: s2 unit: from 2340–2130 yr BP to 1910–1810 yr BP; s1 unit: from 2120–1900 yr BP to 1880–1710 yr BP; cw3 unit: 1880–1690 yr BP; cw2 unit: < 1690 yr BP; cw1 unit: 430–150 yr BP. However, based on structural and depositional relations, it is inferred four paleoearthquake events including the Chi-Chi earthquake occurred at the Pineapple-field site. Based on the radiocarbon age constraints of the paleoearthquakes, it is inferred that the four events occurred at 430–150 yr BP (P1 event), <1690 yr BP (P2), 1900–1710 yr BP (P3 event), and the 1999 Chi-Chi earthquake (Chen *et al.*, 2004).

3.4 Chushan site

The Chushan site is located on an alluvial fan; the Chi-Chi earthquake rupture was located between this fan and the Western Foothills. We excavated an approximately southeast-trending, 35-m-long, 14-m-wide, 8-m-deep trench, which exposed two depositional sequences; these sequences represent an abrupt change in the local depositional environment from fluvial to alluvial fans (Figs. 5a and 5b). The lower sequence of thick-bedded boulder beds represents fluvial deposition. The upper sequence is divided into six units (units A–F) comprising sand with lenticular gravel beds and silt with thin mud based on their distinctive lithologies, which the alluvial deposits derived from the hangingwall strata of the Western Foothills.

The Chi-Chi earthquake rupture produced a 1.7-m-high vertical offset and about 3.5 m of horizontal shortening on an east-dipping fault with thrust displacement that is oriented 32° in the south wall and 20–24° in the north wall. The excavation shows that fault-tip deformation can be subdivided into breakthrough fault-propagation and fault-bend folds. Fault-tip deformation in the south wall, which is the main fault that cuts through the ground surface, is exposed as a fault-bend fold with an asymmetric anticlinal fold and a ramp-flat thrust (Fig. 5a). On the north wall of the excavation, the folding style is similar to a fault-propagation fold, and the tip of the main fault cuts through the ground surface during the Chi-Chi earthquake (Fig. 5b). A ductile shear zone near the ramp thrust forms rollover folds that show an overturned forelimb on the hangingwall.

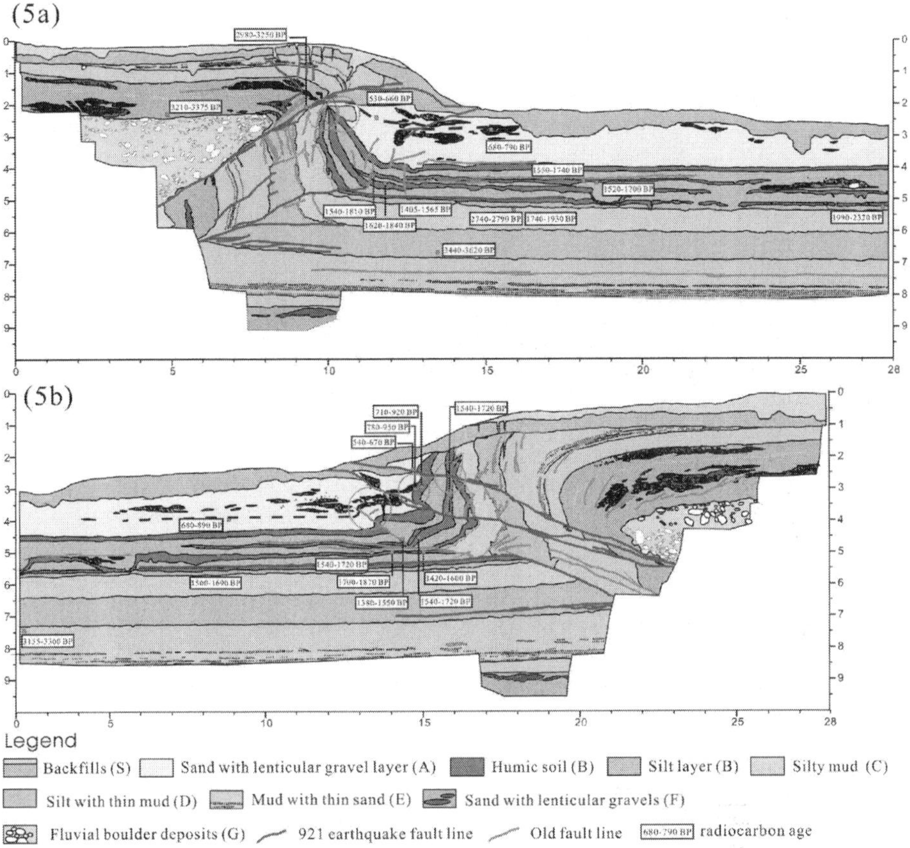

Fig. 5. (a) The south wall of the Chushan excavation shows a breakthrough thrust fault, where the tip of the thrust ramp follows the anticlinal axial surface. The fault tip deformation causes a fault-bend fold, forming a broadly open anticline and a ramp-flat thrust. (b) The north wall of the Chushan excavation shows a breakthrough fault-propagation folding geometry, which contains a recumbent anticline with an overturned forelimb. Radiocarbon dates are used to constrain the timing of stratigraphic units.

In borehole C1 on the hangingwall, the fault zone is located at a depth of 54.3 m, where Pliocene shale is displaced over Holocene deposits. We link the fault zones in boreholes C1 and the synclinal fold axis in the excavated exposures; the results suggest a fault with a dip of 24°, which is consistent with the measurement of the fault plane of about 20-32° in the excavation. A corresponding gravel bed between borehole C2 and the excavation on the hangingwall is offset to a height of 7 m on both sides of the main fault, indicating that repeated large earthquakes have occurred.

One wood sample that was collected from fluvial terrace deposits near the excavated site and interpreted to be correlative with unit G yielded a radiocarbon age of 3470–3820 yr BP (Figs. 5a and 5b). All of the 26 detrital charcoal samples collected from the upper sequence of fluvial deposits and divided into six units (units A–F) yield radiocarbon ages as follows: F unit: 3380–2980 yr BP; E unit: 3360–3155 yr BP; C unit: 2790–2740 yr BP; B4 subunit: 2320–1740 yr BP; B3 subunit: 1710–1540 yr BP; B2 subunit: 1700–1550 yr BP; B1 subunit: 1550–710 yr BP (the top of soil: 950–710 yr BP); lower A unit: 890–680 yr BP; upper A unit: 670–530 yr BP. The timing of the paleoearthquake events can be deduced from the observed earthquake-induced wedge-shaped colluvium, progressive fault displacements on the stratigraphic units, and fault intersections and terminations. However, based on structural and depositional relations, it is inferred that five paleoearthquake events including the Chi-Chi earthquake occurred at the Chushan site at 790–680 yr BP (C1 event), 950–790 yr BP (C2 event), 1550–1380 yr BP (C3 event), and 1930–1710 yr BP (C4 event) (Chen *et al.*, 2007a).

4. Discussion

Repeated coseismic displacements commonly show tectonic-geomorphic features related to displace Holocene sediments, which form a fault or a fold scarp, but it was difficult to determine the related fault or fold. Active thrust faults commonly show complex geometric patterns of faulting and folding in response to different sedimentary facies, thickness, and lithologic characters of Holocene deposits. Previous studies along the Chi-Chi earthquake rupture have shown that fault-tip deformation can be subdivided into breakthrough thrust and blind-thrust components (Chen *et al.*, 2001a, b, 2004, 2007a, b; Ota *et al.*, 2001, 2005; Streig *et al.*, 2007). We will integrate these excavated profiles across the earthquake rupture to delineate the characteristics of thrust-related fold and to reconstruct the growth history of the fault tip fold. Moreover, evidences for repeated coseismic deformation can be identified by relationships between the structural and sedimentological features, such as colluvial wedges, unconformity, and onlap and overlap geometries. We will exploit these types of relationships to characterize the surface deformation that is associated with the sedimentary features. Therefore, in the following, we present the observation of deformation geometries that document the fault-propagation fold of the Shijia and Siangong-temple sites, breakthrough fault-propagation fold of the Pineapple-field site and the north wall of the Chushan site, and fault-bend fold of the south wall of the Chushan site.

4.1 Fault-propagation fold

The Shijia and Siangong-temple excavations show a monoclinal fold that produced a gentle fold scarp, where the fault tip did not propagate through the surface deposits. Coseismic uplift during the Chi-Chi earthquake produced a vertical displacement of 0.8 m at both sites, which is less than the average vertical offset of about 1.4 m along the Chelungpu fault. The excavated exposures reveal a structural complexity resulting from ductile deformation, which makes it difficult to account for the complete deformation field associated with each coseismic event. The trishear model provides an important alternative to restore deformation for a fault-propagation fold that exhibits fault-tip ductile deformation within folded strata (Erslev, 1991; Hardy and Ford, 1997).

The Shijia and Siangong-temple excavations indicate that the cohesive strength of sediments is low, which apparently allows trishear deformation within the folded strata in the shallow subsurface. The main focus of this study is to understand coseismic growth folding characters across a monoclinal fold above a trishear deformation zone. The excavation shows several alluvial units that onlap soil layers on the forelimb of a monoclinal fold. The thickness of each alluvial deposit gradually decreases across the forelimb, forming wedge-shaped deposits, which were laid to onlap and overlap the forelimb. The wedge-shaped bed, which represents material reworked from the hangingwall immediately after a surface-rupturing earthquake, commonly occurred at a location above an unconformity on the hangingwall. This indicates that folding produces an uplift of the fold crest and a forelimb with a large deposition on the footwall. The depositional architecture of an alluvial unit shows an aggradational forward depositional sequence on the forelimb (Figs. 6 and 7). The wedge-shaped deposits bounded by the soil layers represent the postseismic sediments, and the soils form during periods with low depositional rates, which encourage soil development (Machette *et al.*, 1992). Therefore, the wedge-shaped depositional unit brackets coseismic uplift and interseismic deposition, which provide evidence for a surface-rupturing earthquake event.

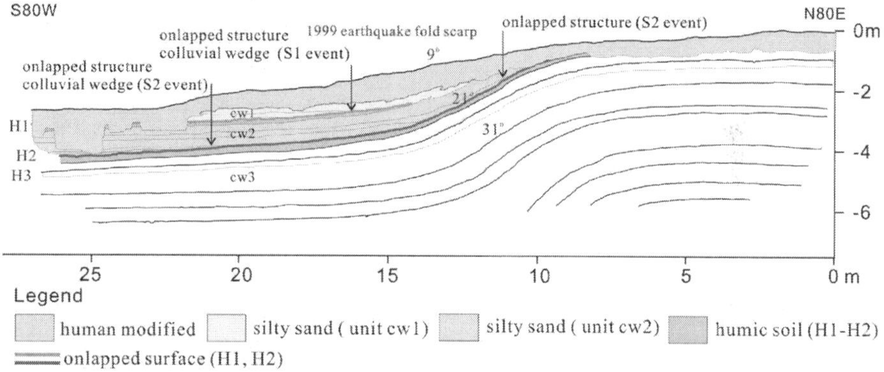

Fig. 6. The Chi-Chi earthquake formed a westward-dipping scarp dipping approximately 9°on the Shijia site. Sketch of trench-wall highlights the top of the ground surface, humic soil, and mud layers. Soil dip within the forelimb increases from 9° to 31°. The silty and sandy sediments onlap onto the H1 and H2 humic soils form a wedge-shaped alluvium.

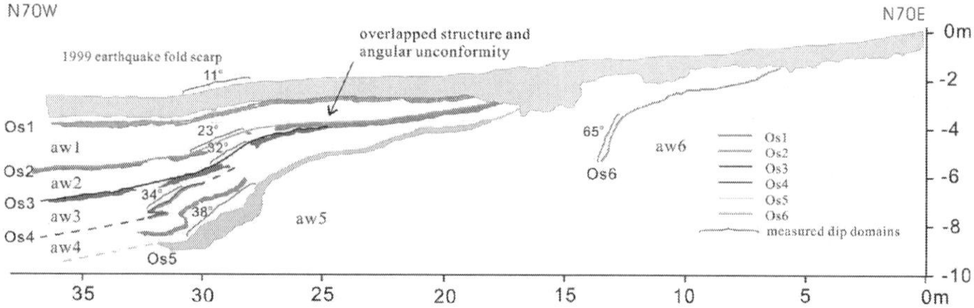

Fig. 7. Siangong-temple excavated profile shows a monoclinal fold with a gently dipping west-facing forelimb of approximately 11° that formed during the Chi-Chi earthquake. The dips of soil within the forelimb increase from 11° to 38° and show a fanning of bed dips. Soils (Os1, Os2, Os3, Os4, and Os5) can provide useful controls on the interpretation of trench stratigraphy which bound five alluvial wedges (aw1, aw2, aw3, aw4, and aw5) representing several episodic seismic events.

Based on structural and depositional relations in excavated exposures, it is inferred that the Shijia and Siangong-temple sites show a progressive fanning and an increase in dips for soils of low stratigraphic units, which is consistent with growth strata geometries of the trishear deformation model (Erslev, 1991; Hardy and Ford, 1997; Allmendinger, 1998). Growth strata show the progressive tilted forelimb of the monoclinal fold that is caused by changes in the dip during the earthquake event. In addition, fanning of the soil dip within the forelimb indicates that the monoclinal fold grew by repeated coseismic deformation. At the Shijia site, strata of backfills and H1 and H2 soils above the forelimb show an increase in the dip of 9°, 21°, and 31°, respectively, which represent three paleoearthquake events including the Chi-Chi earthquake (Fig. 6). The Siangong-temple site also shows fanning of the soil dip within the forelimb, this indicates an increase in the dip of 11°, 23°, 32°, 34°, and 38°, which represent four paleoearthquake events including the Chi-Chi earthquake (Fig. 7). The growth fault-propagation fold in the excavation showed growth folding characters that revealed at onlap and overlap features, unconformity, fanning forelimb dip, and wedge-shaped deposits, that each wedged depositional unit in the Siangong-temple and Shijia sites respectively represents a surface-rupturing paleoearthquake event.

4.2 Breakthrough fault-propagation fold

The geomorphologic features of the Pineapple-field and Chushan sites show a visible fold scarp where the Chi-Chi earthquake rupture followed the pre-existing Holocene scarp. The geomorphologic evidence indicates that the observable vertical separation of terrace scarps may have been produced by several paleoearthquakes along the toe of scarp. The structural characteristics obtained during excavation at the two sites indicate breakthrough faulting, which occurs when the fault-tip deformation causes a fault-bend fold and a breakthrough fault-propagation fold. The Pineapple-field excavation (Figs. 8a and 8b) and the north wall of the Chushan excavation (Fig. 9a) show a breakthrough fault-propagation folding geometry. The two excavations contain a recumbent anticline with an overturned forelimb, which yields a tightly anticlinal hinge with bed thickening, and a steeply narrow forelimb with pronounced bed thinning. The forelimb is dominated by flexural slip within the deformation zone owing to fold-forward rotation. We interpret the deformation within the forelimb as being the product of hangingwall strata rolling into the deformation zone, which resulted in the growth of numerous layer-parallel slip shear planes. The forelimb may increase in length owing to hinge migration and limb rotation (Mitra, 2002). Hence, the bedding slip indicates that the overall deformations may have been quite large in the forelimb. Though the main fault tip breaks through the ground surface at the Pineapple-field excavation and the north wall of the Chushan excavation during the Chi-Chi earthquake, the fault tip of the main thrust faults occurred slip displacement of 1.2–1.8 m and 1.5 m less than that toward the lower reaches displacement of 4.2 m and 3.8 m, respectively. The displacement along the fault reaches found here seem to be consistent with that obtained using the trishear model of displacement variations, in which the amount of slip along the fault near the surface is much less than that toward the lower reaches (Erslev, 1991). Therefore, it is difficult to obtain the actual displacement at the fault tip within the trishear zone. However, the structural relief on the hangingwall is equal to the vertical displacement along the lower reaches of the fault beyond the trishear zone. Outside of the trishear zone, the actual slip on the upthrown block can be determined from the structural relief and dip of the fault. For these reasons, the deformation of the breakthrough fault-propagation fold is studies by focusing on the measurement of the vertical displacement on the hangingwall away from the trishear zone, which produces reliable measurements for the estimation of the actual displacement.

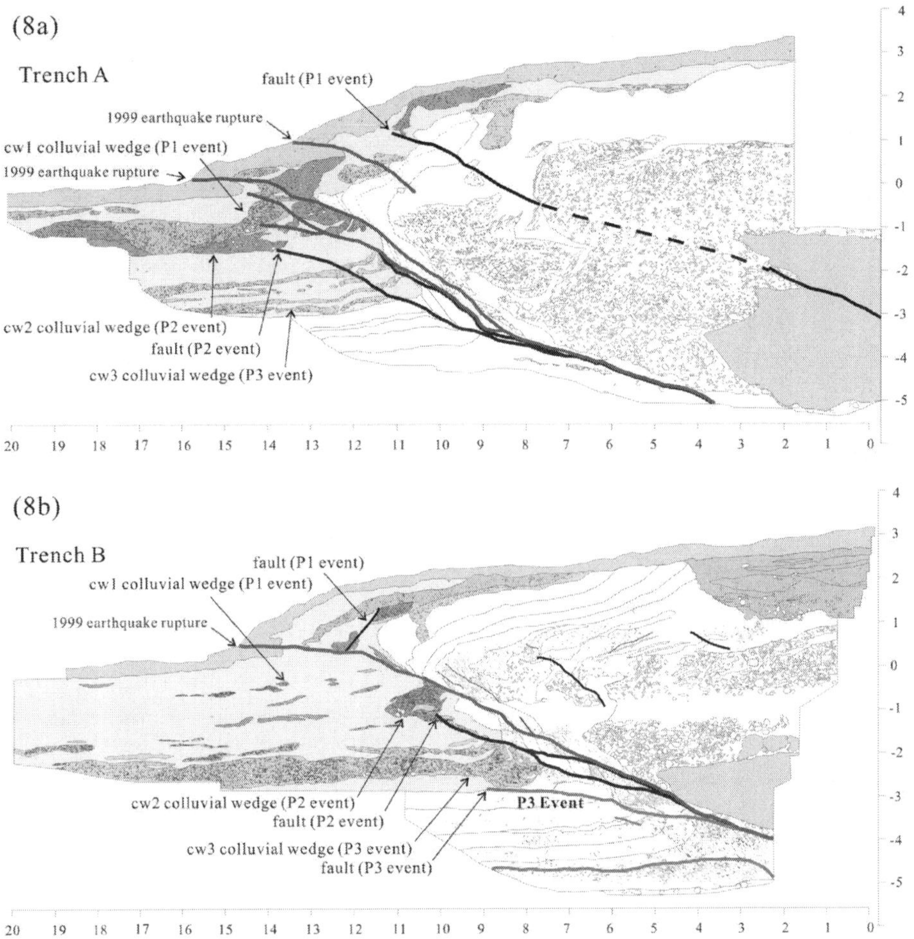

Fig. 8. The structural and stratigraphic features of the earthquake rupture on the A profile (8a) and B profile (8b) of the Pineapple-field site. The colored pattern of stratigraphic units shows the wedge-shaped colluvial deposits. The structural and stratigraphic features also show onlapped features and colluvial wedges that inferred the P1, P2 and P3 paleoearthquake events.

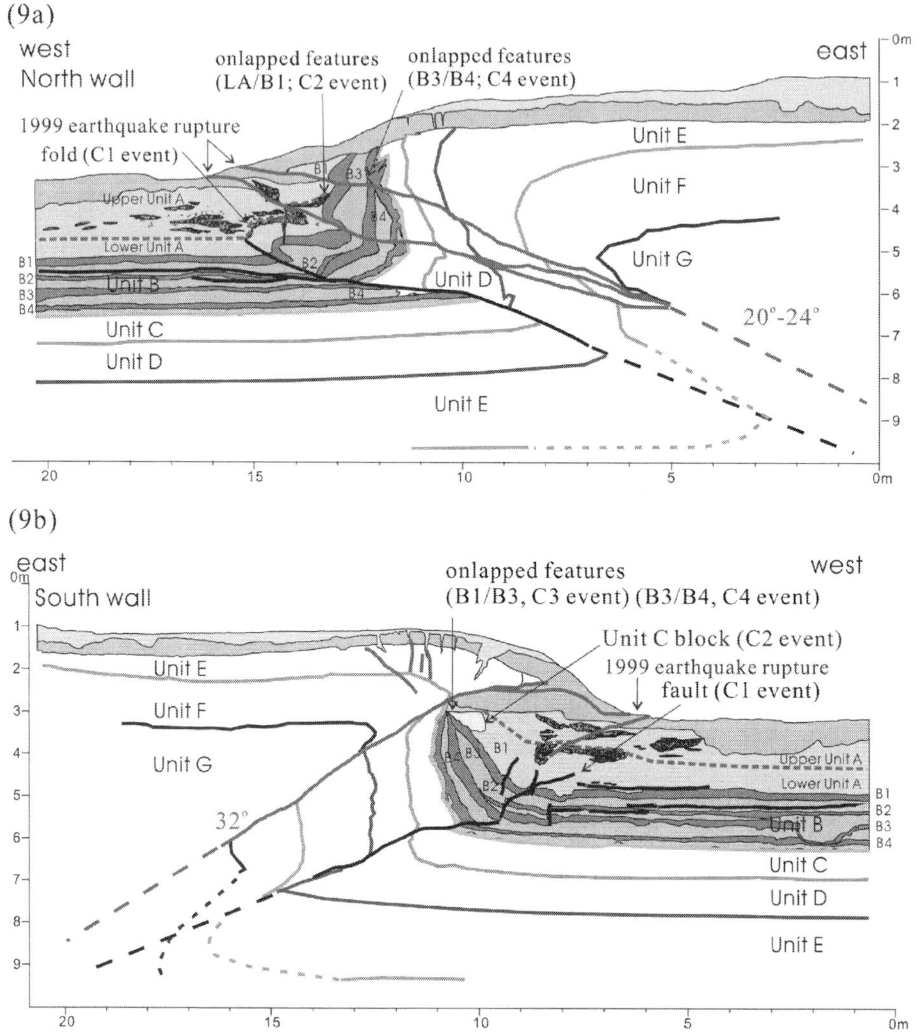

Fig. 9. The structural and stratigraphic features of the earthquake rupture on the north wall (9a) and south wall (9b) of the Chushan site. The red lines show the Chi-Chi earthquake fault strands. The black lines show the paleoearthquake fault strands. The colored pattern of stratigraphic units shows the wedge-shaped colluvial deposits. The structural and stratigraphic features also show folded and onlapped features that inferred the C1, C2, C3, and C4 paleoearthquake events.

4.3 Fault-bend fold

The south wall of the Chushan excavation is found to be a breakthrough thrust fault, where the tip of the thrust ramp follows the anticlinal axial surface, showing a clear displacement along the fault ramp. The fault-tip deformation causes a fault-bend fold, forming a broadly open anticline and a ramp-flat thrust. The anticline on the frontal upthrown block produced by the Chi-Chi earthquake is associated with a minor west-dipping backthrust, tensile cracks, and normal faults at the anticline crest along a main east-dipping 32° breakthrough thrust (Fig. 9b). A footwall syncline is found to have produced a thinning and oversteepened forelimb due to the shearing and dragging produced by the repeated paleoearthquakes. The synclinal forelimb below the fault ramp is defined by a 2-m-wide deformation zone with numerous thin shear planes. The unfolded strata on the hangingwall suggest that the fault tip broke through the ground surface at an earlier stage, even as the fault continued to propagate and the footwall continued to fold; this suggests a hangingwall-fixed fault-bend folding mechanism. However, the fault-bend fold associated with a steep ramp in the underlying thrust propagated through the unfolded strata on the upthrown block. The slip along the thrust may be transferred with an equal displacement during the coseismic deformation. For this calculation, the actual slip on the upthrown block can be determined from the structural relief and dip of the fault.

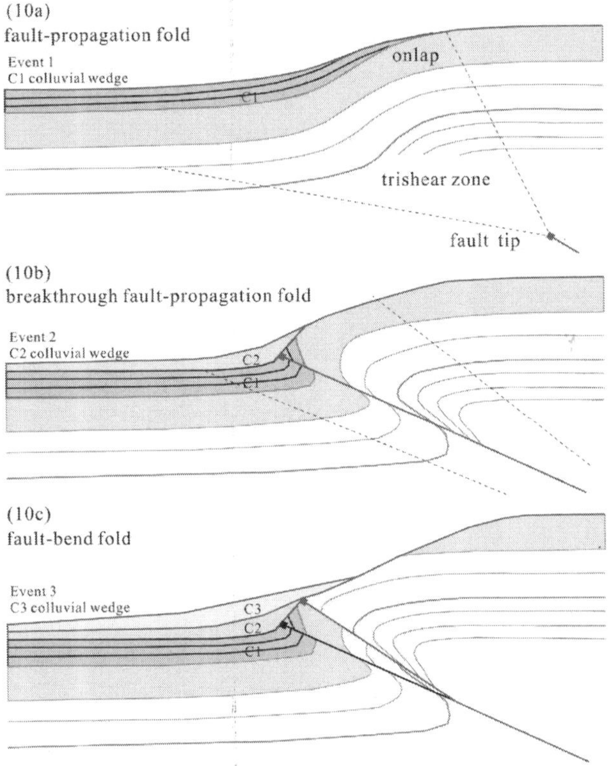

Fig. 10. Sketch diagrams of the structural and stratigraphic features show that the evolutionary path of thrust-related folds in the excavation develops from fault-propagation folding to breakthrough fault-propagation folding and finally to fault-bend folding.

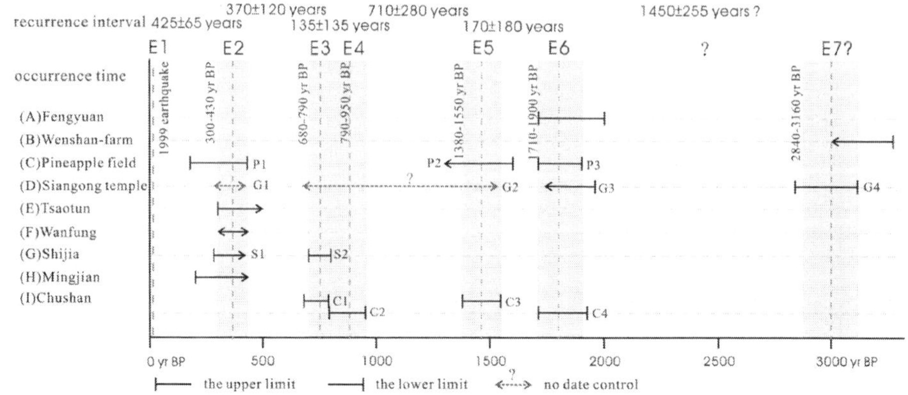

Fig. 11. Correlation between earthquake events at nine excavated sites on the Shihkang and Chelungpu faults, showing occurrence time and recurrence interval. The occurrence time estimates corresponds to the color rectangles based on radiocarbon dating. The horizontal black lines indicate the age range for each event at each site. Lines terminated by vertical bars have defined upper or lower age limits; arrowheads indicate undetermined limits. The locations of excavation sites (A-I) are shown in Figure 1.

Our analysis of the configuration of paleoseismic excavations along the 1999 Chi-Chi earthquake rupture identified three structural features of the fault tip deformation associated with thrust-related folding, suggesting that the evolutionary path of thrust-related folds in the excavation develops from fault-propagation folding to breakthrough fault-propagation folding and finally to fault-bend folding (Fig. 10). Stage I. The structural configuration of the Shijia and Siangong-temple excavations is interpreted to be the product of a monoclinal fold above the blind thrust fault during the earliest part of thrust-related fold development; this indicates the formation of a growth fault-propagation fold based on the trishear deformation model (Fig. 10a). The folding characters reveal onlap and overlap geometries, unconformity, fanning forelimb dip, and wedge-shaped deposits; each wedged depositional unit represents a surface-rupturing paleoearthquake event. Stage II. The structural configuration of the Pineapple-field excavation and the north wall of the Chushan excavation is interpreted to be a breakthrough fault-propagation fold showing a recumbent anticline with an overturned forelimb. The advancing thrust fault is found to have broken through the ground surface, and the continuing fault propagation rotates the forelimb in both the hangingwall and footwall. The folding characters yield a tightly folded hinge, a steep and narrow folded forelimb with pronounced bed thinning, and an anticlinal hinge with bed thickening (Fig. 10b). Stage III. The structural configuration of the south wall of the Chushan excavation is interpreted to be a fault-bend fold. The tip of the main thrust fault breaks through the anticlinal hinge, forming a hangingwall-fixed fold (Fig. 10c).

Over the past tens years, we have excavated numerous trenches on the Chelungpu fault, where the surface rupture shows surface deformation within the Holocene deposits (Fig. 1). Here, we integrate paleoseismologic data with the data derived from excavations to understand the behavior of the Chelungpu fault. By excavating numerous sites along the strike of the Chelungpu fault, we document six surface-rupturing events that have occurred over the last 2000 years. These are the events that occurred in 1999 A.D. (E1), 430–300 yr BP (E2), 790–680 yr BP (E3), 950–790 yr BP (E4), 1550–1380 yr BP (E5), 1900–1710 yr BP (E6), and

3160–2840 yr BP (E7?), respectively (Fig. 11). The E1, E2, E3, and E6 events were well exposed in most excavations and are extremely well-resolved events. The E4 and E5 events have been documented only at the Chushan site in the southernmost part of the Chelungpu fault. The paleoseismic data allow us to define recurrence intervals of 425±65, 370±120, 135±135, 710±280, and 170±180 years along the Chelungpu fault. The recurrence intervals have not been uniform over the past two millennia (Fig. 11). The data from individual paleoseismic sites indicate that the dip slip rate ranges from 4.2 to 9.6 mm/yr.

5. Conclusions

The deformed configuration at the fault tip, resulting from the four excavated sites along the 1999 Chi-Chi earthquake rupture, showed three geometric patterns: fault-propagation, breakthrough fault-propagation, and fault-bend folds. The development deformation is modified by subsequent translation on the propagating fault tip. During the initial deformation, the thrust fault propagates upward through the undeformed units, producing a fault-propagation fold. Then, the fault tip breaks through the ground surface, resulting in a transition from breakthrough fault-propagation folding to fault-bend folding.

The paleoearthquake events are determined by radiocarbon dating of wedge-shaped colluvial deposits exposed in trenches. Evidence obtained from excavation across the Chi-Chi earthquake rupture indicates that there have been six large earthquake events during the past two millennia, which occurred in 1999 A.D. (E1), 430–300 yr BP (E2), 790–680 yr BP (E3), 950–790 yr BP (E4), 1550–1380 yr BP (E5), and 1900–1710 yr BP (E6), suggesting relatively short recurrence intervals in the order of 150–700 years.

6. References

Allmendinger, R.W. (1998) Inverse and forward numerical modeling of trishear fault-propagation folds, *Tectonics* 17, 640–656.

Cattin, R., Loevenbruck, A., Le Pichon, X. (2004) Why does the co-seismic slip of the 1999 Chi-Chi (Taiwan) earthquake increase progressively northwestward on the plane of rupture? *Tectonophysics* 386, 67–80.

Central Geology Survey (1999) Report of the geological survey of the 1999 Chi-Chi earthquake (in Chinese), Central Geological Survey, Taipei.

Chen, W.S., Chen, Y.G., Chang, H.C. (2001a) Paleoseismic study of the Chelungpu fault in the Mingjian area, *West. Pacific Earth Sci.* 1, no. 3, 351–358.

Chen, W.S., Chen, Y.G., Chang, H.C., Lee, Y.H., Lee, C.C. (2001b) Paleoseismic study of the Chelungpu fault in the Wanfung area, *West. Pacific Earth Sci.* 1, no. 4, 43–72.

Chen, W.S., Huang, B.S., Chen, Y.G., Lee, Y.H., Yang, C.N., Lo, C.H., Chang, H.C., Sung, Q.C., Huang, N.W., Lin, C.C., Sung, S.H., Lee, K.J. (2001c) Chi-Chi earthquake, 1999 September 21: a case study on the role of thrust-ramp structures for generating earthquakes, *Bull. Seism. Soc. Am.* 91, no. 5, 986–994.

Chen, W.S., Chen, Y.G., Shih, R.C., Liu, T.K., Huang, N.W., Lin, C.C., Sung, S.H., Lee, K.J. (2003) A modern analog of the out-of sequence thrust system in relation with the Chi-Chi earthquake ruptures in the Western Foothills, central Taiwan, *J. Asian Earth Sci.* 21, 473–480.

Chen, W.S., Lee, K.J., Lee, L.S., Ponti, D.J., Prentice, C., Chen, Y.G., Chang, H.C., Lee, Y.H. (2004) Slip rate and recurrence interval of the Chelungpu fault during the past 1900 years, *Quat. Int.* 115–116, 167–176.

Chen, W.S., Yang, C.C., Yen, Y,C., Lee, L.S., Lee, K.J., Yang, H.C., Chang, H.C., Ota Y., Lin, C.W., Lin, W.H., Shih, T.S., Lu, S.T. (2007a) Late Holocene paleoseismicity of the southern portion of the Chelungpu fault, central Taiwan: Evidence from the Chushan excavation site, *Bull. Seism. Soc. Am.*, 97(1B), 1-13.

Chen, W.S., Lee, K.J., Lee, L.S., Streig, A.R., Chang, H.C., Lin, C.W. (2007b) Paleoseismic evidence for coseismic growth-fold in the 1999 Chichi earthquake and earlier earthquakes, central Taiwan, *J. Asian Earth Sci.* 31, 204-213.

Chen, Y.G., Chen, Y.M., Chen W.S., Lee, K.J., Lee, L.S., Lu, S.T., Lee, Y.H., Watanuki, T., Lin, Y.N. (2009) Optical dating of a sedimentary sequence in a trenching site on the source fault of the 1999 Chi-Chi earthquake, Taiwan, *Quat. Int.* 199, 25-33.

Chiu, H.T. (1971) Folds in the Northern Half of Western Taiwan, *Petrol. Geol. Taiwan* 8, 7-19.

Dominguez, S., Avouac, J.P., Michel, R. (2003) Horizontal coseismic deformation of the 1999 Chi-Chi earthquake measured from SPOT satellite images: Implications for the seismic cycle along the western foothills of central Taiwan, *J. Geophy. Res.* 108, no. B2, 2083, doi:10.1029/2001JB000951

Erslev, E.A. (1991) Trishear fault-propagation folding, *Geology* 19, 617-620.

Hardy, S., Ford, M. (1997) Numerical modeling of trishear fault-propagation folding, *Tectonics* 16, 841-854.

Kao, H., and W.P. Chen (2000) The Chi-Chi Earthquake sequence: active, out-of-sequence thrust faulting in Taiwan, *Science* 288, 2346-2349.

Lai, K.Y., Chen, Y.G., Hung, J.H., Suppe, J., Yue, L.F., Chen, Y.W. (2006) Surface deformation related to kink-folding above an active fault: Evidence from geomorphic features and co-seismic slips, *Quat. Int.*147, 44-54.

Machette, M.N., Personius, S.F., Nelson, A.R. (1992) Paleoseismicity of the Wasatch Fault zones: a summary of recent investigation, interpretations and conclusions, In Gori, P.L. (Ed.), Assessment of regional earthquake hazards and risk along the Wasatch Front, Utah, *U.S. Geol. Surv. Professional Paper* 1500, p. A1-A30.

Mitra, S. (2002) Fold-accommodation faults, *Am. Asso. Petrol. Geol. Bull.* 86, 4, 671-693.

Ota, Y., Huang, C.Y., Yuan, P.B., Sugiyama, Y., Lee, Y.H., Watanabe, M., Sawa, H., Yanagida, M., Sasake, S., Tanifuchi, K. (2001). Trenching study at the Tsaotun site in the central part of the Chelungpu Fault, Taiwan, *West. Pacific Earth Sci.* 1, no. 4, 487-498.

Ota. Y., Chen, Y.G., Chen, W.S. (2005) A review on paleoseismological and active fault study in Taiwan, *Tectonophysics* 408, 63-77.

Pathier, E., Fruneau, B., Deffontaines, B., Angelier, J., Chang, C.P., Yue, S.B., Lee, C.T. (2003) Coseismic displacements of the footwall of the Chelungpu fault caused by the 1999, Taiwan, Chi-Chi earthquake from InSAR and GPS data, *Earth Planet. Sci. Lett.* 212, 73-88.

Streig, A.R., Rubin, C.M., Chen, W.S., Chen, Y.G., Lee, L.S., Thompson, S., Madden, C., Lu, S.T. (2007) Evidence for prehistoric coseismic folding along the Tsaotun segment of the Chelungpu fault near Nan-Tou, Taiwan, *J. Geophy. Res.* 112, B03S06, dori:10.1029/2006JB004493.

Suppe, J. (1981) Mechanics of mountain building and metamorphism in Taiwan, *Geol. Soc. China Mem.* 4, 67-89.

Wang, C.Y., Li, C.L., Su, F.C., Leu, M.T., Wu, M.S., Lai, S.H., Chern, C.C. (2002) Structural mapping of the 1999 Chi-chi earthquake fault, Taiwan by seismic reflection methods, *Terr. Atmo. Ocea. Sci.* 13, 211-226.

Yue, L.F., Suppe, J., Hung, J.H. (2005) Structural geology of a classic thrust belt earthquake: The 1999 Chi-Chi earthquake Taiwan (Mw = 7.6), *J. Struct. Geol.* 27, 2058-2083.

7

Focal Depth Determination for Moderate and Small Earthquakes by Modeling Regional Depth Phases *sPg*, *sPmP*, and *sPn**

Shutian Ma
Carleton University
Canada

1. Introduction

Earthquake focal depth is a critical parameter for seismological research, seismotectonic study, seismic hazard assessment, and event discrimination. For most earthquakes with M_W ≥4.5, the focal depth can be estimated from the arrival times of the teleseismic depth phase *sP* (or *pP*) and its reference phase *P*. Many seismologists have studied how to detect and use teleseismic depth phases to estimate focal depth (e.g., Goldstein and Dodge, 1999). For smaller earthquakes, focal depths can be estimated jointly while being located with the arrival times of the *Pg* and *Sg* phases recorded at close stations. Because stations in a regional network are generally not dense enough to control focal depth, operators often use default focal depths for regional events.

If regional depth phases can be identified, an alternative solution for moderate and small earthquakes is to use regional depth phases to estimate focal depth. The *P* portion of regional waveform records contains three major parts: (1) the *P*-wave travels directly to the station; (2) the *P*- or *S*-wave travels upward to the surface in the source region, is reflected or converted at the surface and then travels downward to the Moho (or interfaces), is reflected or refracted there, and then travels upward to the station; and (3) the *P*-wave travels downward to the Moho (or interfaces), is reflected there and then travels upward to the station. One feature of *P*- and *S*-waves is that the amplitude of the *S*-wave radiated from the source is generally stronger than that of the *P*-wave by about five times (Aki and Richards, 1980) and the period of the *S*-wave is longer than that of the *P*-wave on the same record.

From this analysis we know that there are regional depth phases in the *P* portion of the record and the usable regional depth phases are (1) *sPg* (the *S*-wave travels upward to the surface, is converted to a *P*-wave at the critical angle, then the *P*-wave travels along or close the surface to the station), (2) *sPmP* (the *S*-wave travels upward to the surface, is converted to a *P*-wave, then the *P*-wave travels downward to the Moho, is reflected there and travels upward to the station; Langston *et al.*, 2003), and (3) *sPn* (the *S*-wave travels upward to the surface, is converted to a *P*-wave, then the *P*-wave travels along the *Pn* path to the station;

---

* This chapter is adapted from the paper "Focal Depth Determination for Moderate and Small Earthquakes by Modeling Regional Depth Phases *sPg, sPmP*, and *sPn*", *Bull. Seism. Soc. Am.* **100,** 1073-1088

Zonno and Kind, 1984). Fig. 1 (Ma and Eaton, 2011) shows the sketch paths of these regional depth phases. Many scientists have studied regional depth phases to some extent (e.g., King, 1979; Helmberger and Engen, 1980; Langston, 1987, 1996; Mulder and Lamontagne, 1990; Zhao and Helmberger, 1991, 1993; Bock, 1993; Ebel, 1995; Bock *et al.*, 1996; Zhu and Helmberger, 1997; Saikia, 2000; Saikia *et al.*, 2001; Bent and Perry, 2002; Savage *et al.*, 2003; Uski *et al.*, 2003).

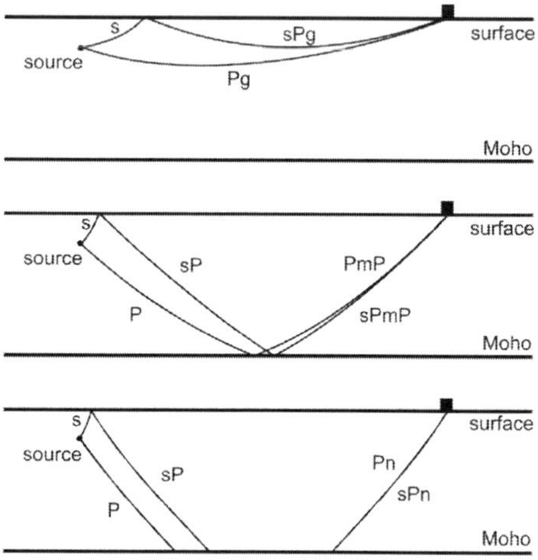

Fig. 1. Sketch figures for regional depth phase *sPg* (upper panel), *sPmP* (middle), and *sPn* (bottom).

Regional depth phases (*sPg, sPmP,* and *sPn*) can be used to estimate focal depth if they and their reference phases (*Pg, PmP,* and *Pn*) can be correctly identified. Following Langston (1987) and Bock *et al.* (1996), we developed a method to use the regional depth phases to determine focal depths. The principle is: (1) calculate synthetics with the reflectivity method (Randall, 1994) at a station with a reasonable range of depths; (2) compare the synthetics with the observed values at the same station; and (3) take as the focal depth of the earthquake the depth at which the synthetic and the observation have similar time differentials (regional depth phase to its reference phase).

We previously reported some aspects of the regional depth-phase modeling (RDPM) method (e.g., Ma *et al.*, 2003; Ma and Atkinson, 2006). Here we introduce the RDPM method more systematically and describe the principles and features of the three depth phases in detail. We have proved that the assumptions of depth phase *sPmP* and its reference phase *PmP* are correct, and by conducting several tests, found in which regions the regional depth phases are developed and in which they are not, and which factors contribute to errors in the modeled focal depths. We also found that the contents of *PmP* and *sPmP* come from different interfaces beneath the source. These findings are useful for researchers who want to use the RDPM method, and especially for identifying the regional depth phases and their reference phases.

Because we use regional synthetics as a "ruler" to measure focal depth from observed waveforms, we first describe how to generate synthetics that are suitable to be used as the "ruler" and discuss some features of the regional depth phases.

2. Synthetic regional depth phases *sPg*, *sPmP*, and *sPn*

To generate synthetics we need a crustal model, earthquake location, focal mechanism, and focal depth. To generate synthetics for smaller earthquakes, the source time function is not important. We use a triangle as the source time function. Because the focal mechanism does not determine the arrival times of the seismic phases—the crustal structures determine the arrival times—the crustal model is a key factor in generating synthetic regional depth phases. Western Quebec is one of the more active seismic zones in eastern North America and the crustal structures are relatively well known. Mereu *et al.* (1986) "conducted a major long-range seismic refraction and wide-angle reflection experiment across the Grenville province of Canadian Shield," and obtained some crustal models. After studying these crustal models and modifying them slightly, we obtained one crustal model and put it in our program package as the default crustal model (Fig. 2; model 1 in Table 1). Because the focal mechanisms in western Quebec are predominantly thrust type (e.g., Adams *et al.*, 1989; Bent and Perry, 1999; Ma and Eaton, 2007), we used a thrust type focal mechanism as the default (Fig. 2, bottom left).

Because regional depth phases are easier to discern on displacement records than on velocity records, we used displacement records in the RDPM. All the synthetic and observed waveforms in this chapter are the vertical component.

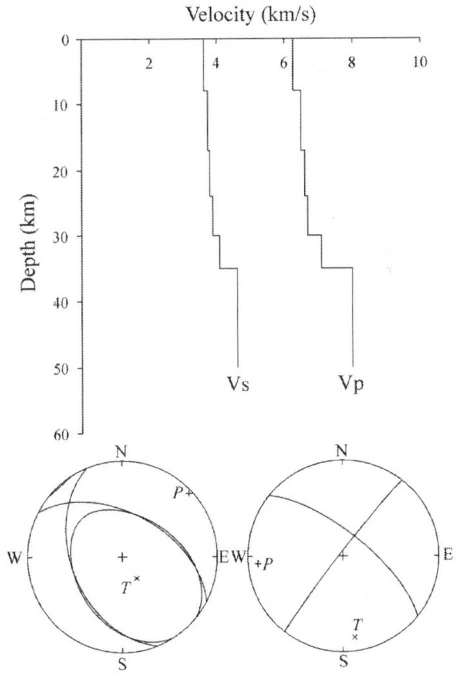

Fig. 2. The default crustal model and default focal mechanism (left; right focal mechanism is for comparison) in the RDPM method. Except if specified, all synthetic waveforms were generated with the default crustal model and default focal mechanism.

Model 1 (6 layers) h Vp Vs ρ	Model 2 (5 layers)	Model 3 (4 layers)	Model 4 (3 layers)	Model 5 (2 layers)
8 6.25 3.61 2.53	8 6.25 3.61 2.53	8 6.25 3.61 2.53	8 6.25 3.61 2.53	8 6.25 3.61 2.53
9 6.50 3.75 2.63	9 6.50 3.75 2.63	9 6.50 3.75 2.63	9 6.50 3.75 2.63	0 6.50 3.75 2.63
7 6.60 3.81 2.67	7 6.60 3.81 2.67	7 6.60 3.81 2.67	0 6.60 3.81 2.67	
6 6.70 3.87 2.71	6 6.70 3.87 2.71	0 6.70 3.87 2.71		
5 7.10 4.10 2.87	0 7.10 4.10 2.87			
0 8.00 4.62 3.23				

Table 1. The crustal models. Model 1 is the default in the RDPM program package. Crustal models 2, 3, 4, and 5 were formed by deleting the last layer successively from model 1. *h* = layer thickness (km); *Vp* = velocity of the *P*-wave (km/sec); *Vs* = velocity of the *S*-wave (km/sec); *ρ* = crustal density (g/cm^3).

2.1 Synthetics generated at different distances with a fixed focal depth

To observe features of the regional depth phases that are displayed when the distance changes, we generated synthetic waveforms at distances ranging from 0.3° to 4.8° and plotted them (Figs. 3, 4, and 5). Fig. 3 shows that the *sPg* phase is well developed at distances ranging from 0.7° to 0.9° (trace 070 to 090). The distance range within which *sPg* is well

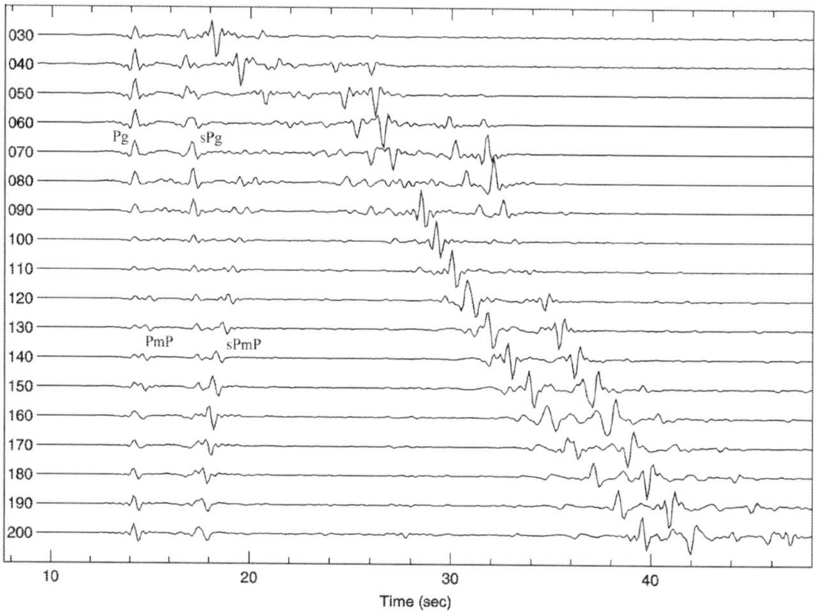

Fig. 3. Synthetic waveforms generated with depth 12 km, azimuth 236°, at distances 0.3° to 2.0° (also used to generate Figs. 4, 5, 6, 7, and 8). Trace number = distance in degrees × 100. Trace 030 was generated at distance 0.3°. Traces are aligned on the first phase. On trace 070 phases *Pg* and *sPg* and on trace 140 phases *PmP* and *sPmP* are labeled.

developed changes with focal depth: the range shifts farther as the focal depth increases. The time difference *sPg–Pg* changes very slightly with distance. For the distance range of about 1.0° to 1.7° (trace 100 to 170), *Pg, PmP, sPg,* and *sPmP* co-exist. Fig. 4 shows that *Pg* disappears at 1.6° (or *Pg* and *PmP* merge there; trace 160); *sPg* disappears at 1.9° (trace 190). For the distance window of about 1.8° to 2.8° (trace 180 to 280), the waveforms are quite simple. The first phase is *Pn* (generally weak); the second phase is *PmP* and the third phase is *sPmP*. Fig. 5 shows that *sPn* stands out at about 3.0° (trace 300). The time difference *sPn–Pn* is independent of distance. For distances larger than 2.9° (trace 290), waveforms become complex. At about 4.1° (trace 410) there is another distance window in which waveforms are relatively simple.

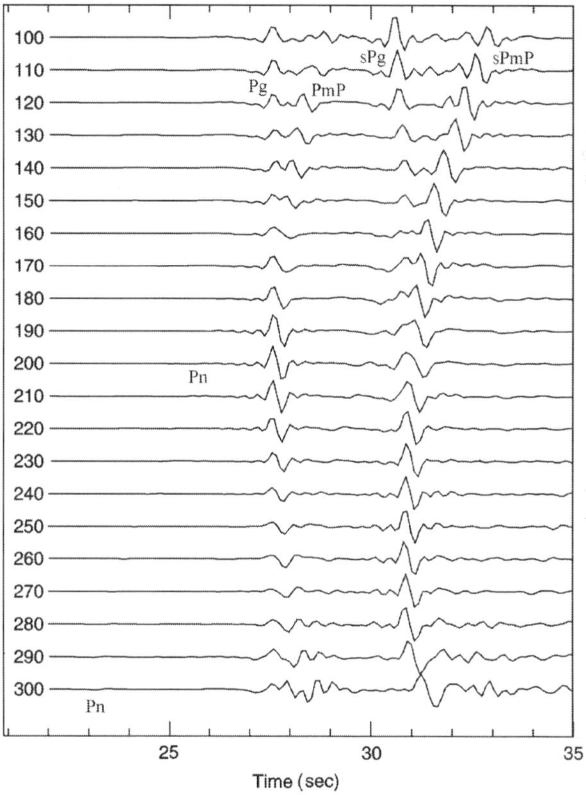

Fig. 4. Synthetic waveforms generated with depth 12 km, azimuth 236°, at distances 1.0° to 3.0°. Trace 100 was generated at distance 1.0°. Traces are aligned on *Pg* or *PmP*. The distance window in which waveforms are simple is from about 200 to 300 km (trace 180 to 280). The *Pn* phase is weak. Traces 100 to 200 correspond to the early parts of those traces with the same labels in Fig. 3, but with amplitude enlarged and timescale expanded.

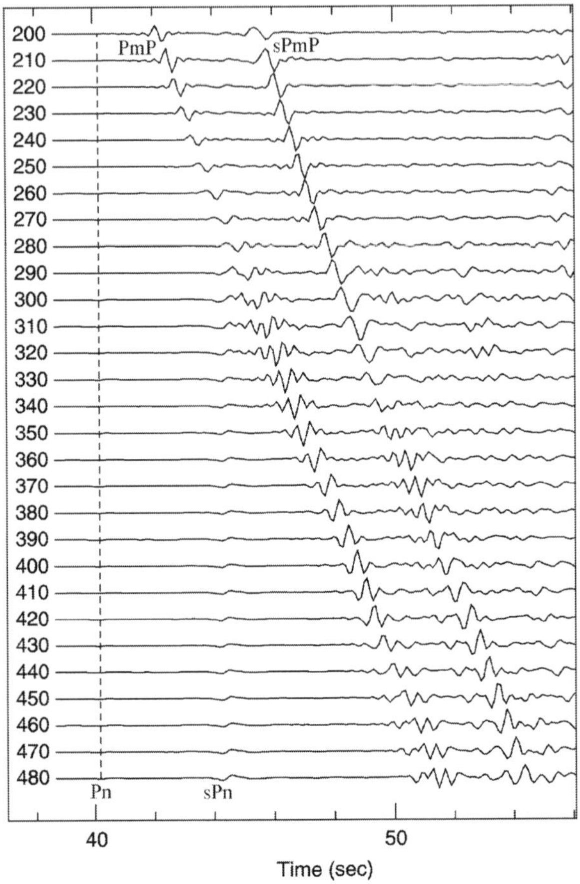

Fig. 5. Synthetic waveforms generated with depth 12 km, azimuth 236°, at distances 2.0° to 4.8°. Top trace 200 was generated at distance 2.0°. Traces are aligned on *Pn*. The *sPn* phase stands out at 3.0° (trace 300), but is buried at closer distances. After 2.8° (trace 280) waveforms become complex. Around trace 410 (4.1°) waveforms are simple again.

2.2 Synthetics generated with a range of focal depths at fixed distances

To observe how regional depth phases change with focal depth, we generated synthetic *sPg*, *sPmP*, and *sPn* with a range of depths at fixed distances 0.9°, 2.1°, and 4.1°. Fig. 6 shows that the time difference *sPg*-*Pg* becomes progressively larger with focal depth. The position of *sPg* shifts by about half a cycle when the depth changes by 1 km. This means that the time difference *sPg*-*Pg* is very sensitive to focal depth. At distance 0.9°, *sPmP* is not well developed. Fig. 7 shows that the time difference *sPmP*-*PmP* becomes larger as depth increases. The position of *sPmP* also shifts by about half a cycle when focal depth changes by 1 km. The *Pn* phase is also a depth phase, but it is not as sensitive as *sPmP* to focal depth. For example, on trace 210, the time difference between *Pn* and *PmP* is about half that between *sPmP* and *PmP*. The time difference *Pn*–*PmP* changes obviously with distance (Fig. 5). These features of *Pn* can be used to identify *sPmP* in its distance window (200 to 300 km). Fig. 8 shows how the time difference *sPn*–*Pn* changes with focal depth. Because the *sPn* phase is stronger than *Pn*, it is possible that some of the observed "*Pn*" phase beyond 300 km is *sPn*.

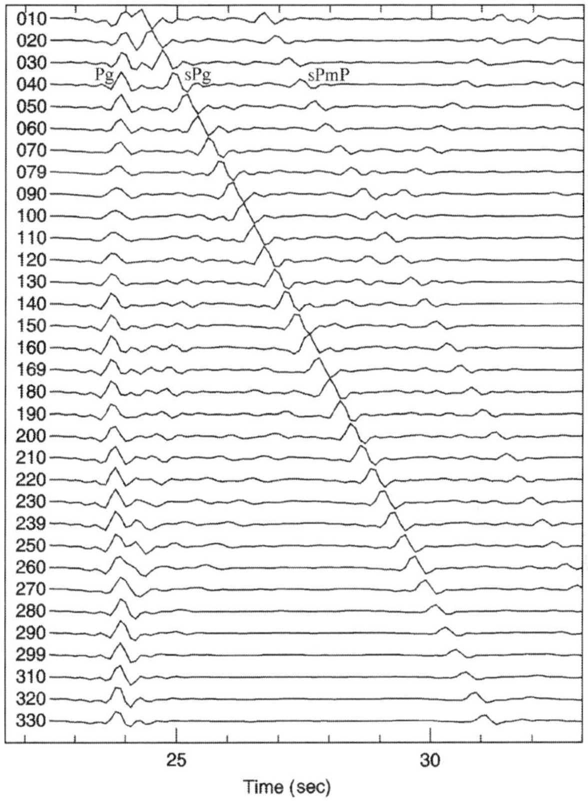

Fig. 6. Synthetic waveforms generated at distance 0.9° with depths from 1 to 33 km. Trace number = depth in km × 10. Trace 010 was generated with depth 1 km at distance 0.9°.

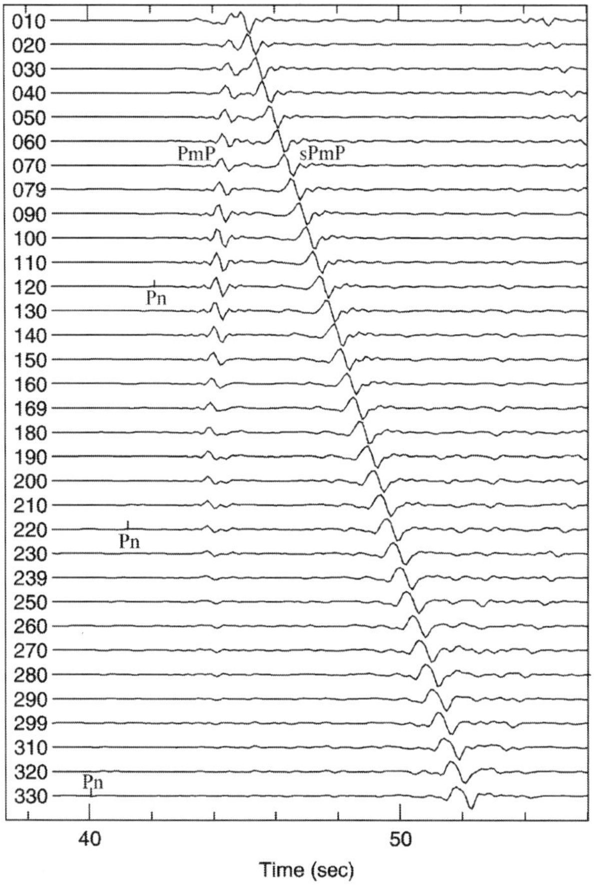

Fig. 7. Synthetic waveforms generated at distance 2.1° with depths from 1 to 33 km. Trace 010 was generated with depth 1 km at distance 2.1°. On trace 070 phases *PmP* and *sPmP* are labeled. The *Pn* phase is weak, and is labeled on traces 120, 220, and 330.

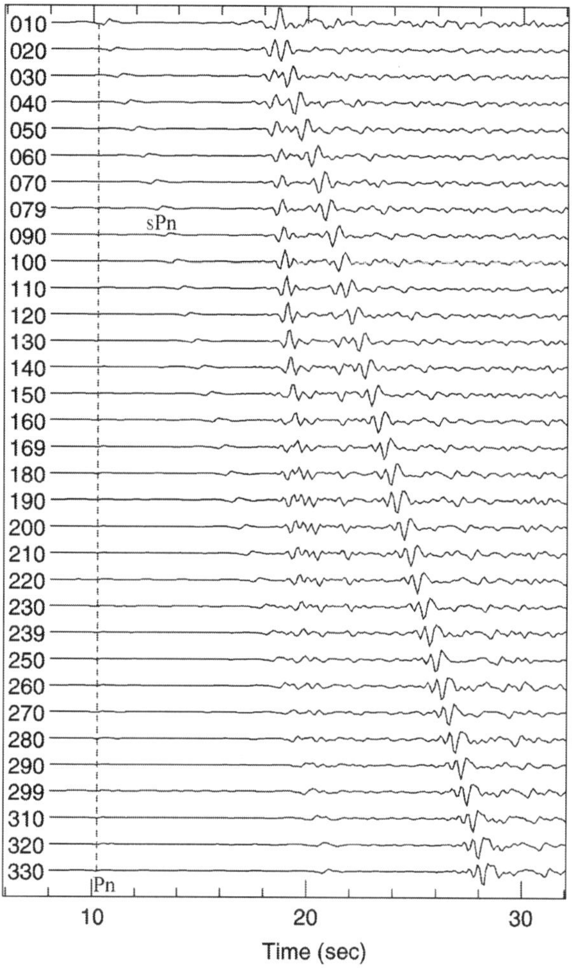

Fig. 8. Synthetic waveforms generated with depths of 1 to 33 km at distance 4.1°. Trace 010 was generated with depth 1 km at distance 4.1°. The *Pn* phase is weak; *sPn* is stronger than *Pn*. From trace 260 (26 km) *sPn* merges with other phases. Traces are aligned on the *Pn* phase.

2.3 Synthetics generated with different focal mechanisms

To generate synthetics we need a focal mechanism. Generally, no focal mechanism solutions are available for small earthquakes. To solve this problem, we used a default focal mechanism for all earthquakes to generate synthetics to measure focal depths. Because focal mechanisms do not determine the arrival times of seismic phases, we used a default focal mechanism to generate synthetic phases for their arrival times. Because we do not use waveform shapes to estimate focal depth, we do not have to use a strike-slip focal mechanism to generate synthetics for earthquakes with strike-slip focal mechanisms.

3. Demonstration for the assumed *sPmP* and *PmP* phases

In eastern North America many *P* portions of waveform records are similar to trace CRLO/EHZ in Fig. 9. On this trace, the first weak phase is *Pn*. We assumed that the second phase is *PmP* and the third phase (the largest) is *sPmP*. To demonstrate that the assumptions are correct, we prepared Fig. 9 using explosive and earthquake source models. On trace EXPL/140, no strong assumed depth phase appears at the position corresponding to that on trace CRLO/140 which was generated with the same depth and crustal model. The reason is that an explosive source does not directly generate *S*-waves. According to the definition of *sPmP*, the phase should arrive at a station progressively later as the focal depth increase because the total path length becomes longer. The third phase on traces CRLO/130, 135, 140, 145, and 150 arrives progressively later as the depth increases. This feature and the absence of the strong phase on EXPL/140 indicate that the "assumed *sPmP*" is *sPmP*.

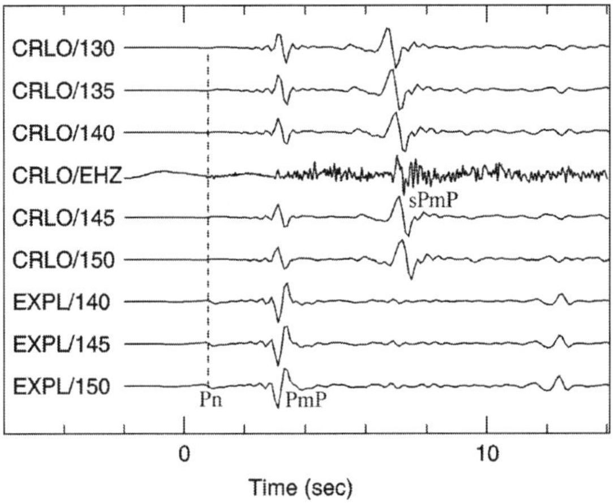

Fig. 9. Synthetic waveforms generated with an earthquake source model (Fig. 2, bottom-left) (traces CRLO/130 to 150), an explosive source model (traces EXPL/140 to 150), and the *P* portion recorded at station CRLO generated by a small earthquake (1995/09/12, m_N 3.7, west Quebec; trace CRLO/EHZ). CRLO/130 was generated with depth 13 km at distance 2.10° and azimuth 283°. EXPL/140 was generated with depths 14.0 km. Trace CRLO/EHZ is the *P* portion generated by the small earthquake. The modeled focal depth for this earthquake is 14.5 km.

To demonstrate that the phase has experienced reflection from the Moho, we prepared Fig. 10 by using different crustal models and depths. In group 2 (2/130, 2/140, 2/150), a weak *sPmP* appears. This might be because the interface at depth 8 km is above the sources. When we put the sources at the same depths in a half space, the weak *sPmP* disappears. In groups 3 (3/130, 3/140, 3/150), 4 (4/130, 4/140, 4/150), 5 (5/130, 5/140, 5/150), and 6 (6/130, 6/140, 6/150), the assumed *PmP* and *sPmP* appear and are clear. The time differences *sPmP*–*PmP* are almost the same on traces 3/130, 4/130, 5/130, and 6/130, etc. "*PmP*" and "*sPmP*" from different interfaces can pile up at similar positions. Fig. 10 demonstrates that the assumed depth phase *sPmP* experienced reflection at the Moho, because its shapes, for example, on traces 3/150, 4/150, 5/150, and 6/150, are different.

The waveform contents of the *sPmP* on trace 6/150, for example, contain contributions from all interfaces beneath the source.

From the definition of *PmP* we know that as focal depth increases, *PmP* should arrive earlier, because the total travel path becomes shorter. Generally, we use the phase recorded in the distance window of about 200 to 300 km. At these distances, the feature that *PmP* arrives earlier as focal depth increases is not easy to examine without *Pg* as the reference phase. So we traced the assumed *PmP* to close distances where *Pg* exists. Fig. 11 is the synthetic waveforms generated with an explosive source model. Trace 210 in Fig. 11 is similar to trace EXPL/140 in Fig. 9, which was generated with depth 14 km at distance 2.10°. Let us trace the assumed *PmP* in Fig. 11 from trace 080 to trace 150 where both *Pg* and the assumed *PmP* exist. At such a close distance, the first phase is *Pg*. We then generated synthetics at fixed distance 0.8° with depths of 1.0 to 35.0 km (Moho depth). Fig. 12 shows that the two phases become closer as depth increases, and they merge at the Moho. This test shows that the assumed *PmP* is *PmP*. Based on these tests, the assumed phases *PmP* and *sPmP* on trace CRLO/EHZ are *PmP* and *sPmP*, because these two phases have the same features as those on traces CRLO/130 to CRLO/150.

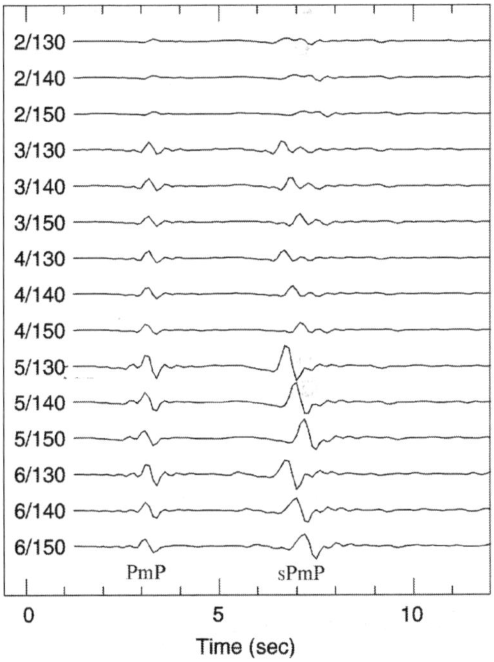

Fig. 10. Synthetic waveforms generated with the five crustal models listed in Table 1 at distance 2.10° and with depths 13, 14, and 15 km. Trace Group 2 (2/130, 2/140, 2/150) was generated with a two-layer crustal model (Model 5 in Table 1); Group 3 with a three-layer crustal model (Model 4 in Table 1); Group 4 with a four-layer crustal model (Model 3 in Table 1); Group 5 with a five-layer crustal model (Model 2 in Table 1); Group 6 with a six-layer crustal model (Model 1 in Table 1).

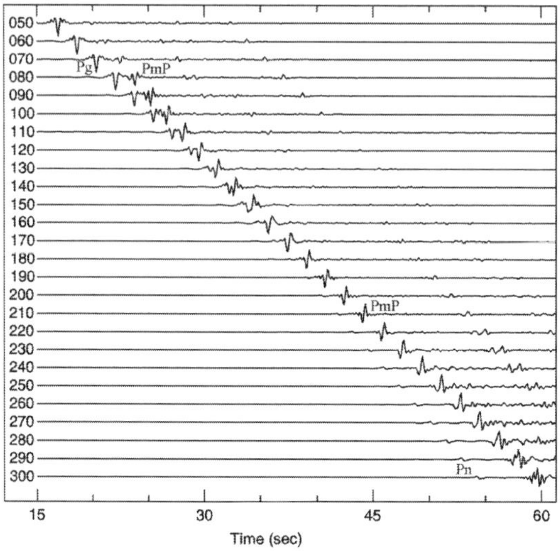

Fig. 11. Synthetic waveforms generated with depth 13 km, an explosive source model, and crustal model 1 in Table 1, at distances of 0.5° to 3.0°. Trace 210 was generated at 2.10°. The *Pg* phase disappears at about 1.5° (trace 150). The *PmP* phase is clear from 1.7° to 2.8° (traces 170 to 280).

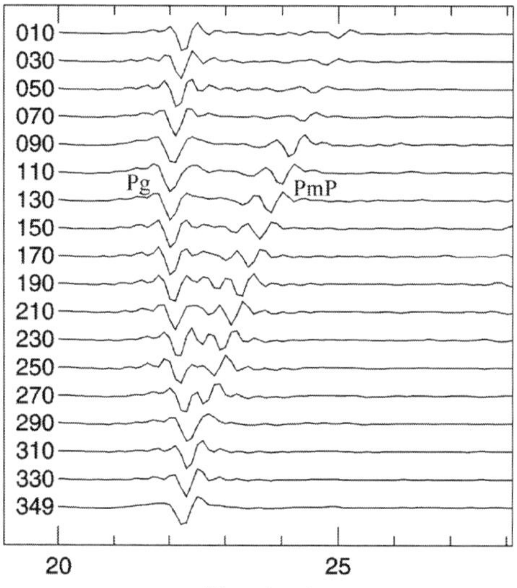

Fig. 12. Synthetic waveforms generated at distance 0.8° with depths of 1.0 to 34.9 km, an explosive source model, and crustal model 1 in Table 1. Trace 130 in this figure is the same as trace 080 in Fig. 11. Because the total path length becomes shorter with focal depth, the *PmP* phase arrives progressively earlier with depth.

4. Comparison of RDPM focal depths with those from other methods

In eastern Canada and the New York State region, some moderate and strong earthquakes have reliable focal depth solutions. For the same earthquakes we compared focal depth solutions obtained by RDPM with those obtained by other methods. We found that the consistency is good.

4.1 *sPn* modeling for the M_W 5.0 2002/04/20 Au Sable Forks, New York earthquake

We analyzed the waveform records for this earthquake and found many *sPn* records. We modeled the *sPn* and *Pn* phases recorded at a POLARIS station, Canada (Fig. 13). From *sPn* and *Pn* paths in a one-layered crustal model of Poisson medium, the following equation calculates focal depth using the time difference *sPn*–*Pn*:

$$t(h;sPn - Pn) = \frac{h}{V_{p1}}\left[\sqrt{3-k^2} + \sqrt{1-k^2}\right]; \qquad (1)$$

where $k = V_{p1} / V_{p2}$; V_{p1} and V_{p2} are *P*-wave velocities in and beneath the crust, respectively; h is the focal depth; and t is the differential time. The differential time is independent of station distance and crustal thickness. Even if we have only one pair of reliable *sPn* and *Pn* observations, we can obtain a focal depth solution with a small error without considering the earthquake location error. The time difference *sPn*–*Pn* on trace ACTO/HHZ is 4.06 sec. When we take V_{p1} = 6.25 km/sec and V_{p2} = 8.0 km/sec, the focal depth from equation (1) is 11.7 km; this depth is close to the focal depth (about 11.5 km) determined by RDPM.

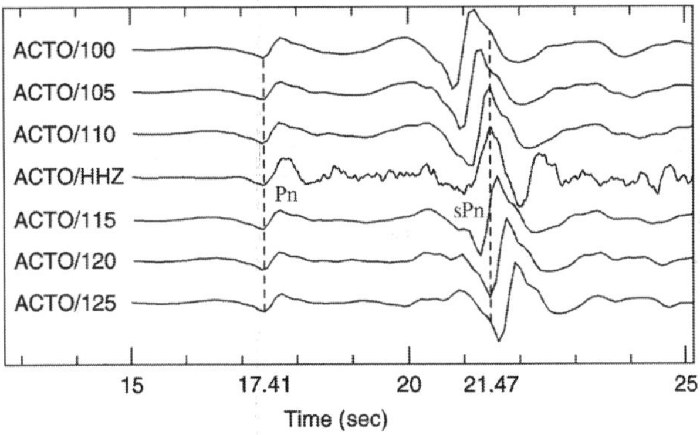

Fig. 13. Phase *sPn* modeling for 2002/04/20 Au Sable Forks, New York, M_W 5.0 earthquake. Trace ACTO/100 was generated with depth 10 km at station ACTO (517 km). Trace ACTO/HHZ is the earthquake observation; it matches a trace somewhere between ACTO/110 and ACTO/115. We can take 11 km or 11.5 km as the modeled focal depth. The time difference *sPn*–*Pn* on trace ACTO/HHZ is 4.06 sec. Using this number in equation (1) gives a focal depth of 11.7 km.

To compare the RDPM depth, we analyzed the teleseismic depth phase *sP* and its reference phase *P* at YKA (Yellowknife Array, Canada). Fig. 14 shows some of the records. The time difference *sP*–*P* is 4.85 sec. The following equation calculates focal depth from differential time:

$$t(h,\alpha;sP-P)=\frac{h}{V_P\sqrt{1-\frac{1}{3}\sin^2\alpha}}\left(\sqrt{3}+1-\frac{4}{3}\sin^2\alpha\right); \qquad (2)$$

where V_P is the P-wave velocity in a one-layered crustal model of Poisson medium, a is the P-wave take-off angle, t is the differential time, and h is the focal depth. If we take V_P = 6.25 km/sec and a = 29.68° the focal depth from equation (2) is 12 km.

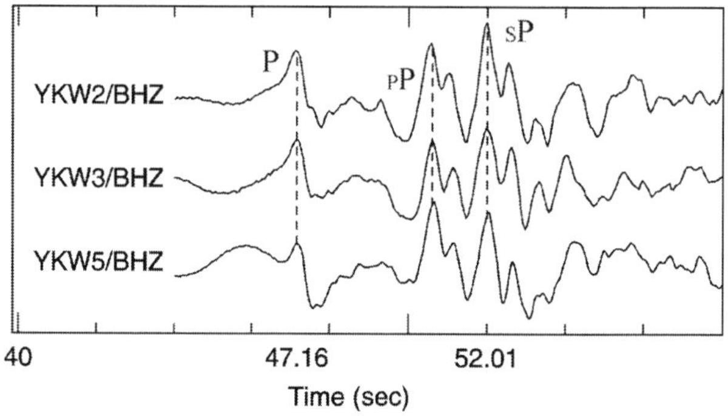

Fig. 14. Teleseismic depth phases records of 2002/04/20 Au Sable Forks, New York, M_W 5.0 earthquake at YKA (Yellowknife Array, Canada). Based on the time difference *sP–P* (4.85 sec), the focal depth from equation (2) is about 12 km.

4.2 *sPg* modeling for the m_N 4.3 1993/11/16 Montreal south earthquake and its aftershocks

Many *sPg* (and *sPmP*) records are available in eastern Canada and New York region. As an example we selected the records from station MNT (37 km from a main shock and its

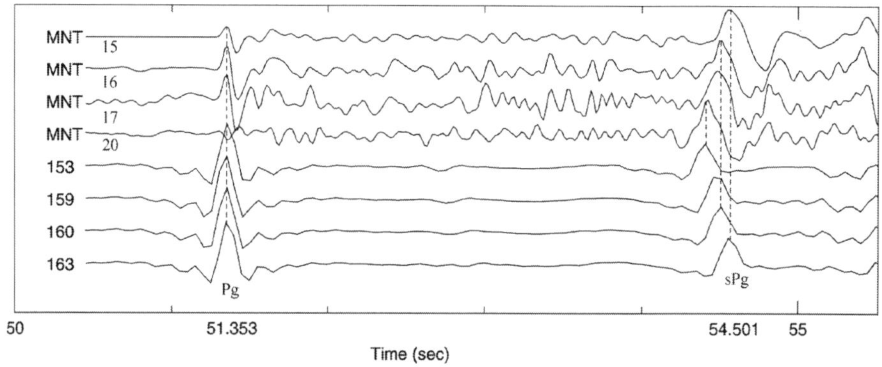

Fig. 15. *sPg* and *Pg* modeling for earthquakes No. 15, 16, 17, and 20 in Table 2 of Ma, 2010. Traces labeled 15, 16, 17, and 20 are *P* portions recorded at station MNT (37 km). Traces 153, 159, 160, and 163 are synthetics generated with depths 15.3, 15.9, 16.0, and 16.3 km at station MNT. The subtle differential times *sPg–Pg* show that these earthquakes occurred on a fault that ruptured about 1 km at depth about 15 km.

aftershocks; No. 15, 16, 17, and 20 in Table 2 of Ma, 2010). Fig. 15 shows that the main shock is the deepest and the last aftershock is the shallowest, and the depth difference between, for example, No. 15 and 16, is about 0.3 km.

4.3 *sPmP* modeling for the M_L 5.1 1983/10/07 New York region earthquake and its aftershock (No. 5 in Table 2 of Ma 2010)

We selected the records from station SBQ (244 km) as an example. In Fig. 16, trace SBQ/SHZ 10:18:12.0 is the record of the main shock at station SBQ, and trace SBQ/SHZ 10:39:06.0 is the record of its aftershock. From the time differences *sPmP*–*PmP* on the synthetics SBQ/079 and SBQ/085, we found that the aftershock is shallower than the main shock by about 0.5 km. The modeled focal depth for the main shock is 8.5 km. The focal depth obtained by the Geological Survey of Canada is 10.0 km. From the first record we see the *Pn* phase clearly, but from the second record we cannot. This means that the *Pn* phase disappeared or was too weak to be measured.

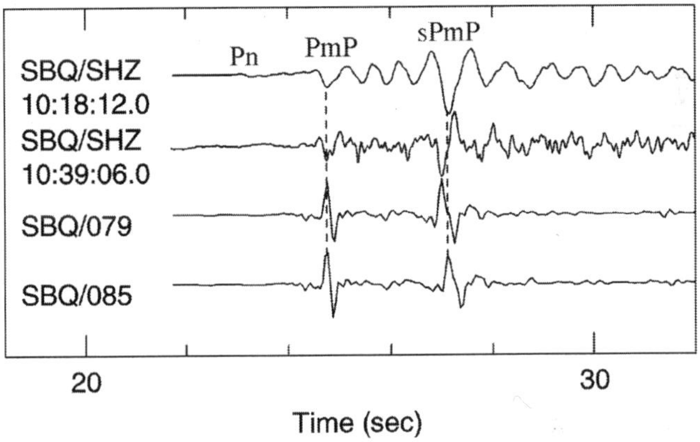

Fig. 16. *sPmP* and *PmP* modeling for earthquake No. 5 in Table 2 of Ma, 2010 and one of its aftershocks (1983/10/07 10:39:39.0, M_L3.5). The top trace is the record of the earthquake at station SBQ; the second trace is the record of the aftershock. Traces SBQ/079 and SBQ/085 are synthetic waveforms generated with depths 7.98 and 8.5 km at station SBQ. The subtle differential times *sPmP*–*PmP* show that the two earthquakes occurred at different depths. On the second trace the *Pn* phase is not measurable.

4.4 *sPmP* modeling for two aftershocks of the M_W 6.5 2003/12/22 California earthquake

Many regional records are available for the main shock and its aftershocks. We retrieved the records of some aftershocks at stations within 5° from IRIS for analysis. We found that two aftershocks have clear *sPmP* phase records at stations VCS (2.76°), CHF (2.89°), PAS (2.87°), and MWC (2.92°). We modeled the *sPmP* and *PmP* at station PAS, and obtained focal depths 6.5 km for the 05:30 event and 6 km for the 18:17 event (Fig. 17). The preliminary focal depths obtained by the local network for the two aftershocks are 5.9 km and 6.9 km, respectively.

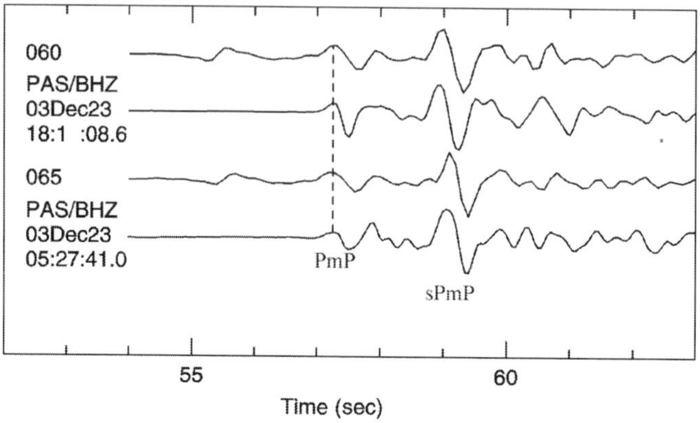

Fig. 17. *sPmP* and *PmP* modeling for two aftershocks of the 2003/12/22 Central California M_W 6.5 earthquake. Trace PAS/BHZ/03Dec23 18:15:08.6 is the record of the aftershock 2003/12/23/ 18:17:11.0 *M* 4.9 at station PAS. Trace PAS/BHZ/03Dec23 05:27:41.0 is the record of the aftershock 2003/12/23/ 05:30:19.0 *M* 4.5. Traces 060 and 065 are the synthetic waveforms generated with depths 6 and 6.5 km at the same station (320 km). The focal depth solution for the 18:17 event is 6 km and for the 05:30 event is 6.5 km.

5. Possible errors in the modeled focal depth

We used the differences in arrival times between synthetic regional depth phases and their reference phases to measure focal depth from the observations. The *P*- and *S*-wave velocities in the crustal model determine the arrival times of these phases. When we generate synthetics we also need the focal mechanism and the earthquake location, but the errors generated by these two factors are negligible.

5.1 The error in the modeled focal depth caused by the crustal model

Travel times of regional depth phases and their reference phases are determined by the crustal structures through which the phases propagate. As such, most of the error in modeled focal depths comes from the crustal velocity model used.

(A) The error caused by the velocity model

To evaluate errors arising from velocity uncertainty in the crustal model, we generated synthetic seismograms (at an epicentral distance of 2.16°) using our default crustal model given in Table 1 (model 1), with focal depths from 2 to 23 km. We then reduced the *P*- and *S*-wave velocities of the crustal model by 10% and generated another set of synthetics with the same distance and depths. Fig. 18 (Ma and Eaton, 2011) shows the differential times between *sPmP* and *PmP* phases for these two models. The time delay obtained by subtracting these differential times is approximately linear with focal depth. Focal depths, estimated using the RDPM method by treating one set of synthetic traces as observed seismograms, differ by 9.5 - 12% (Fig. 18, bottom). The differences are on the order of 11%, slightly greater than the 10% change in velocity. These numerical tests indicate that the level of uncertainty in the velocity model propagates, at approximately the same order of magnitude, into focal-depth uncertainty.

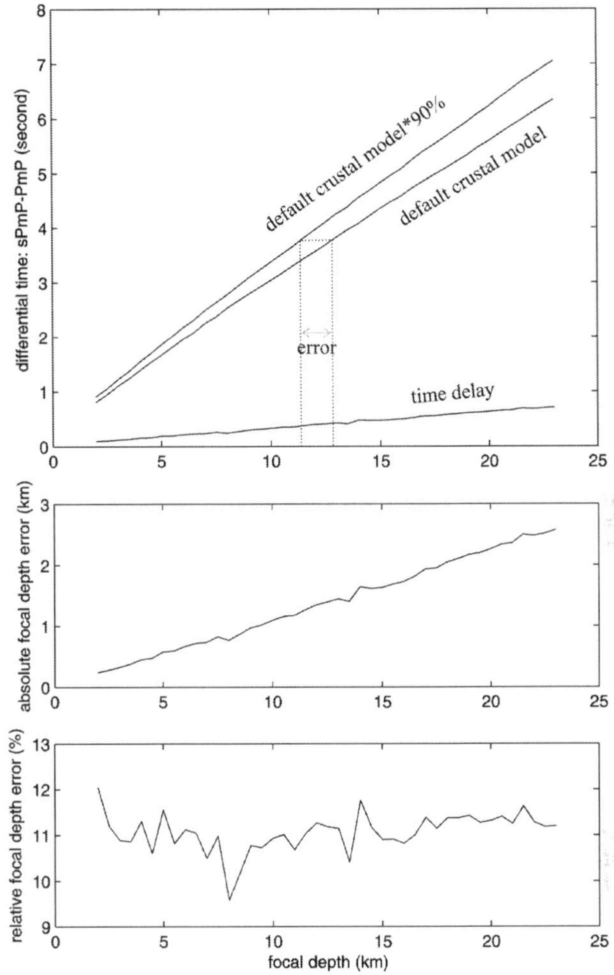

Fig. 18. Errors in the modeled focal depth caused by errors in the crustal model. The upper panel shows the differential times between *sPmP* and *PmP* generated by the default crustal model and by a low-velocity model (90% of the default crustal model). The intersection points between the vertical faint lines and the depth axis are focal depth solutions obtained by the two tiled lines from the same differential time *sPmP-PmP* (the height of the faint line bar). The difference between the two solutions is the absolute error. The middle panel shows how the absolute errors change with focal depth. The bottom panel shows the relative errors.

(B) The error caused by the Vp/Vs ratio

We assumed that the crustal media are Poisson type in which Vp/Vs is 1.732. To examine the possible error in our modeled focal depth caused by the Poisson assumption, we made the following tests: (1) We made one new ratio by adding 5% to 1.732 and used the ratio and the Vp values in our default crustal model to create one crustal model *M*1. (2) We subtracted 5% from 1.732 to form a second new ratio and used this ratio to create crustal model *M*2. We compared the synthetics generated using these two crustal models and the default model and found that the time differences *sPmP*–*PmP* on traces generated with *M*1 and depth 11.2 km, generated with *M*2 and depth 12.9 km, and generated with the default crustal model and depth 12 km are approximately equal. This shows that when the crustal medium differs from the Poisson medium by 5%, the relative error in modeled focal depth is less than 8%.

(C) The error caused by strong interfaces in the crust

Our default crustal model assumes five layers. The thickness of the fourth layer is 6 km (Model 1 in Table 1). We divided the layer into two parts of equal thickness, keeping the original velocities in the upper layer, but changing the velocities in the lower layer to those of the third layer. We generated synthetics with this new crustal model and the default focal mechanism at distance 2.13° (Fig. 19). The time difference *sPmP*–*PmP* does not change noticeably (on traces 120 and STD), but the shape of "*sPmP*" broadened. This change shows that the *sPmP* phase is not a simple phase; it has the "*sPmP*" from the interface where Vp = 6.6 and 7.1 km/sec; Vs = 3.81 and 4.1 km/sec in the new crustal model. This change demonstrates that if there are strong interfaces above the Moho, the *sPmP* phase can be complex, and can cause time-reading errors.

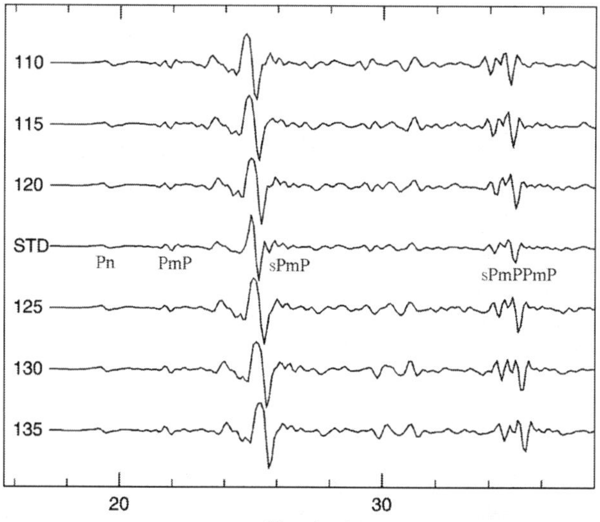

Fig. 19. Synthetic waveforms generated with the default crustal model and a new crustal model that contains a weak lower velocity layer. Trace 110, 115, 120, 125, 130, and 135 were generated with the new model and depth 11, 11.5, 12, 12.5, 13, and 13.5 km. Trace STD was generated with the default crustal model and depth 12 km.

Model 1	Model 2	Model 3	Model 4	Model 5	Model 6
h Vp Vs ρ					
1 1.25 0.72 0.51	1 2.25 1.30 0.91	1 3.25 1.88 1.31	1 4.25 2.45 1.72	1 5.25 3.03 2.12	1 6.25 3.61 2.53
7 6.25 3.61 2.53	7 6.25 3.61 2.53	7 6.25 3.61 2.53	7 6.25 3.61 2.53	7 6.25 3.61 2.53	7 6.25 3.61 2.53
9 6.50 3.75 2.63	9 6.50 3.75 2.63	9 6.50 3.75 2.63	9 6.50 3.75 2.63	9 6.50 3.75 2.63	9 6.50 3.75 2.63
7 6.60 3.81 2.67	7 6.60 3.81 2.67	7 6.60 3.81 2.67	7 6.60 3.81 2.67	7 6.60 3.81 2.67	7 6.60 3.81 2.67
6 6.70 3.87 2.71	6 6.70 3.87 2.71	6 6.70 3.87 2.71	6 6.70 3.87 2.71	6 6.70 3.87 2.71	6 6.70 3.87 2.71
5 7.10 4.10 2.87	5 7.10 4.10 2.87	5 7.10 4.10 2.87	5 7.10 4.10 2.87	5 7.10 4.10 2.87	5 7.10 4.10 2.87
0 8.00 4.62 3.23	0 8.00 4.62 3.23	0 8.00 4.62 3.23	0 8.00 4.62 3.23	0 8.00 4.62 3.23	0 8.00 4.62 3.23

Table 2. New crustal models generated by dividing the first layer in the default model (Table 1) into two parts, making the first part 1-km thick, and changing the *P*- and *S*-wave velocities in that layer in steps. h = layer thickness (km); Vp = velocity of the *P*-wave (km/sec); Vs = velocity of the *S*-wave (km/sec); ρ = crustal density (g/cm^3).

5.2 The error caused by an error in earthquake location

To estimate the error in the modeled focal depth caused by the error in earthquake location we can observe Fig. 4 or 5 (or Uski *et al.*, 2003; their Fig. 2). In the distance window of 1.8° to 3.0°, when the distance changes, for example, 0.1° (~11 km) at distance 2.2°, the time difference *sPmP–PmP* is almost constant. This means that when the earthquake location has an 11-km error in the above distance window, the error in the modeled focal depth caused by the error in earthquake location is negligible.

5.3 The error caused by the focal mechanism

Focal mechanism determines the radiation pattern of seismic waves but it does not determine arrival times of seismic phases (e.g., Ma and Atkinson, 2006; their Fig. 11). However, the focal mechanism partially determines the shapes and amplitudes of waveforms and thus may cause an error in time readings. If the observed *PmP* and *sPmP* are clear, but the synthetic *PmP* or *sPmP* are not, we can change the station azimuth, generate new synthetics, and then make comparisons again. So, the arrival time reading error caused by the focal mechanism is negligible. To reduce the reading error, we use the arrival times of peaks of *sPmP* and *PmP*.

6. Possible factors that determine the development of the regional depth phases

Many factors determine the development of regional depth phases. The key factors are the crustal structures at the free surface and at the Moho. We tested the effect of crustal structure at the free surface on the development of the regional depth phases. We divided the first layer of the default crustal model into two parts, making the first layer 1-km thick. We changed the velocities in the new first layer gradually to obtain new crustal models (Table 2). For each new crustal model, we generated synthetics with depth 12 km at distance 2.20°. Fig. 20 shows that on traces 020, 030, 040, 050, 125, and 225, the depth phase *sPmP* is not discernible. On traces 010, 325, and 425, *sPmP* is not prominent. On traces 001, 005, 525, and 625, *sPmP* is clear. Based on this test, we may say that if there is a sedimentary layer in the source region (reflecting sites) and the layer is sufficiently thick, the *sPmP* phase is not developed.

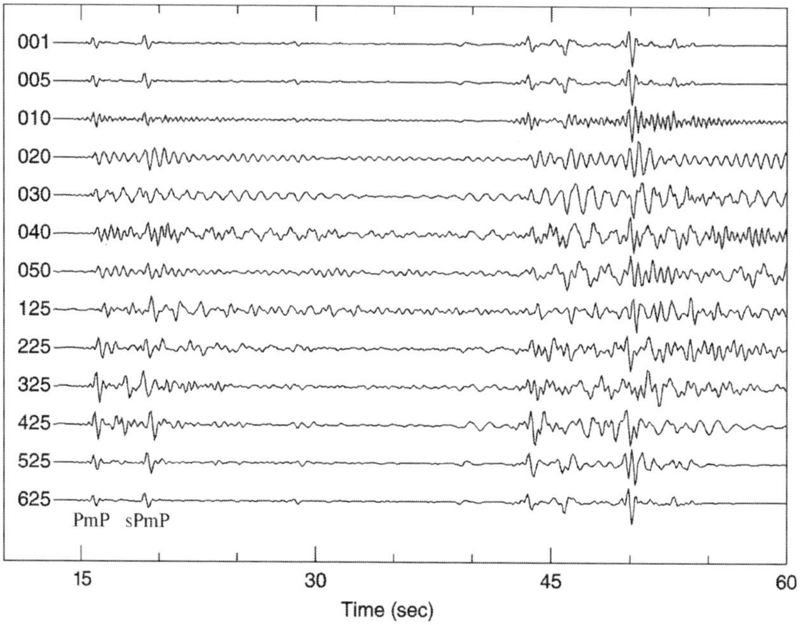

Fig. 20. Synthetic waveforms generated with the new crustal models listed in Table 2 at distance 2.20° and depth 12 km. Trace 125 was generated with Model 1 in Table 2; trace 225 with Model 2; trace 325 Model 3; trace 425 Model 4; trace 525 Model 5; trace 625 Model 6. Trace 001, 005, 010, 020, 030, 040 and 050 were generated with Model 1, but the thickness of the first layer was changed in steps to 0.01, 0.05, 0.1, 0.2, 0.3 0.4, and 0.5 km, successively.

7. Outline of the procedure to identify the regional depth phases

There are distance windows in which regional depth phases are developed. Phase *sPg* is developed well within 100 km, *sPmP* is developed well within about 200 to 300 km, and *sPn* is developed at more than 300 km. Figs. 3, 4, and 5 show these distance windows. Because time differences *sPg-Pg*, *sPmP-PmP*, and *sPn-Pn* are not sensitive to station distance, we can align all records generated by an earthquake by station distance to identify these regional depth phases.

The procedure has the following steps:

1. Retrieve the catalogue from the official website of GSC (Geological Survey of Canada);
2. Select an earthquake to determine its focal depth using RDPM;
3. Retrieve the pick file for the earthquake;
4. Retrieve all the available waveform records from the same website;
5. Based on the station distances in the pick file, arrange/display the waveform records;
6. In the above distance windows, align the records first and then search the regional depth phase and its reference phase pairs.

If two pairs (one pair = regional depth phase and its reference phase) on two records at two stations have similar differential times, the pairs may be treated as candidates for modeling. We can model one pair or two pairs using RDPM to obtain the focal depth. The modeling procedure is shown in Fig. 9. If only one record is available at a distance window, the period feature (*sPg*, *sPmP*, and *sPn* have longer periods than their respective reference phases) may be used for the identification. The feature that *Pn*, *PmP*, and *sPmP* have relative positions on the record (see Fig. 7) can also be used to identify *sPmP* and *PmP*. Dineva *et al.* (2007; their Fig. 9) provided an excellent example for regional depth phase identification.

If more than one station has regional depth phase records for the same earthquake, the focal depth solutions obtained at different stations should be similar (see Ma and Atkinson, 2006; their Table 1).

8. Summary and discussion

There are many small earthquakes in eastern North America. These earthquakes do not have measurable teleseismic depth phases, and generally do not have close (<40 km) waveform records. No focal depths can be reliably estimated with either teleseismic depth phases or close seismic signals for most of these earthquakes, but the depths are crucial information for many topics, both theoretical and applied.

On regional waveform records, one or more phases are well developed between the first arrivals (*Pg* or *Pn*) and the *S*-wave train, and one or two of them are regional depth phases. Within about 100 km, the *sPg* phase is well developed on some records of earthquakes as small as m_N 1.5. In the distance window from about 200 to 300 km, the *sPmP* phase is well developed on some records of earthquakes as small as m_N 2.0. Beyond 300 km, the *sPn* phase is developed on some records of moderate and sub-moderate earthquakes. All these regional depth phases can be used to estimate focal depth.

When we generate synthetic depth phases, we use a default crustal model and default focal mechanism, and even a default station azimuth and default instrument response. In this way, we can conveniently generate synthetics for any small earthquake, and use the synthetic depth phases to estimate focal depth for that earthquake, if the arrival time difference between one observed depth phase and its reference phase is available.

We compared our modeled focal depths with those that were reliably obtained by other methods and found that the consistency in the comparisons is good.

The errors in our modeled focal depths are caused mainly by the crustal model used. The relative errors due to crustal model are estimated as within 15% when the error in the crustal velocity model is 10%, or when the thicknesses of layers in the crustal model have some errors, or the crustal medium differs from the assumed Poisson medium.

We have analyzed the regional depth phases for many earthquakes (Ma and Atkinson, 2006; Kim *et al.*, 2006; Dineva *et al.*, 2007; Ma and Eaton, 2007; Atkinson *et al.*, 2008). For this chapter we selected some special cases to show that our modeling method is simple, reliable, and suitable for all regions where regional depth phases are developed.

9. Acknowledgments

This chapter was adapted with permission from the paper "Focal Depth Determination for Moderate and Small Earthquakes by Modeling Regional Depth Phases *sPg*, *sPmP*, and *sPn*" published by *Bull. Seism. Soc. Am.* It covers the author's long-time research results. Most of the research is from time spent as a Visiting Fellow at the Geological Survey of Canada, Ottawa, where colleagues Dr John Adams, David McCormack, and Veronika Peci provided key references and made useful suggestions. Part of the research is from post-doctoral work at Carleton University, Ottawa. Two anonymous reviewers provided constructive recommendations that improved that paper dramatically. BSSA Associate Editor Lorraine W. Wolf proved helpful suggestions for publication of this superficially simple but important paper. The efforts from the staff at the InTech - Open Access Publisher make the publication of this chapter possible. The author is grateful for all assistance listed above.

10. References

Adams, J., A. Vonk, D. Pittman, and H. Vatcher (1989). *New focal mechanisms for south eastern Canada earthquakes*, Vol. II, Geol. Surv. Canada, Open File 1995.

Aki, K., and P. G. Richards (1980). *Quantitative seismology, theory and methods*, Vol. II., W.H. Freeman and Co., San Francisco.

Atkinson, G., S. Kaka, D. Eaton, A. Bent, V. Peci, and S. Halchuck (2008). A very close look at a moderate earthquake near Sudbury, Ontario, *Seism. Res. Lett.*, 79, 119–131.

Bent, A. L., and C. Perry (1999). *Focal mechanisms for eastern Canada earthquakes: 1 January 1996–30 June 1998*, Geol. Surv. Canada, Open File 3698.

Bent, A. L., and C. Perry (2002). Depths of eastern Canadian earthquakes from regional data, *Seism. Res. Lett.* 73, 273–284. Bock, G. (1993). Depth phases from local earthquakes, *BMR J. Austr. Geol. Geophy.*, 13, 275–279.

Bock, G., G. Grunthal, and K. Wylegalla (1996). The 1985/86 Western Bohemia earthquakes: Modeling source parameters with synthetic seismograms, *Tectonophysics* 261, 139–146.

Dineva, S., D. Eaton, S. Ma, and R. Mereu (2007). The October 2005 Georgian Bay (Canada) earthquake sequence: Mafic dykes and their role in the mechanical heterogeneity of Precambrian crust, *Bull. Seism. Soc. Am.* 97, 457–473.

Ebel, J. (1995). Analysis of digital waveforms in the northeastern U.S. for source depth and strong ground motion information, report to USGS, Weston Observatory, Dep. Geol. and Geophys., Boston College.

Goldstein, P., and D. Dodge (1999). Fast and accurate depth and source mechanism estimation using *P*-waveform modeling: A tool for special event analysis, event screening, and regional calibration, *Geophys. Res. Lett.* 26, 2569–2572.

Helmberger, D., and G. Engen (1980). Modeling the long-period body waves from shallow earthquakes at regional ranges, *Bull. Seism. Soc. Am.* 70, 1699–1714.

Kim, Won-Young, S. Dineva, S. Ma, and D. Eaton (2006). The 4 August 2004, Lake Ontario, earthquake, *Seism. Res. Lett.* 77, 65–73. King, R. (1979). Observations of *sPn* from Swabian Alb earthquakes at the GRF array, *J. Geophys.* 45, 337–340.

Langston, C. (1987). Depth of faulting during the 1968 Meckering, Australia, earthquake sequence determined from waveform analysis of local seismograms, *J. Geophys. Res.* 92, 11561–11574.

Langston, C. (1996). The S*sPmP* phase in regional wave propagation, *Bull. Seism. Soc. Am.*, 86, 133–143.

Langston, C., A. Nyblade, and T. Owens (2003). Regional wave propagation in Tanzania, East Africa, *J. Geophys. Res.* 107, ESE 1-1 to 1-18.

Ma, S., and G. Atkinson (2006). Focal depth distribution for earthquakes with $m_N \geq 2.8$ in western Quebec, southern Ontario and northern New York, *Bull. Seism. Soc. Am.* 96, 609–623.

Ma, S., and D. Eaton (2007). The western Quebec seismic zone (Canada): Clustered, mid-crustal seismicity on a Mesozoic hotspot track, *J. Geophys. Res.* 112, B06305, doi:10.1029/2006JB004827.

Ma, S., V. Peci, J. Adams, and D. McCormack (2003). Routine estimate of focal depths for moderate and small earthquakes by modeling regional depth phase *sPmP* in eastern Canada, 2003 EGS-AGU-EUG Joint Assembly, Nice, France, April 2003; contribution EAE03-A-06176.

Ma, S. (2010). Focal Depth Determination for Moderate and Small Earthquakes by Modeling Regional Depth Phases *sPg*, *sPmP*, and *sPn*, *Bull. Seism. Soc. Am.* 100, 1073-1088.

Ma, S., and D. Eaton (2011). Combining double-difference relocation with regional depth-phase modelling to improve hypocentre accuracy, *Geophysical Journal International*; DOI: 10.1111/j.1365-246X.2011.04972.

Mereu, R., D. Wang, O. Kuhn, D. Forsyth, A. Green, P. Morel, G. Buchbinder, D. Crossley, E. Schwarz, R. DuBerger, C. Brooks, and R. Clowes (1986). The 1982 COCRUST seismic experiment across the Ottawa–Bonnechere Graben and Grenville Front in Ontario and Quebec, *Geophy. J. R. Astr. Soc.* 84, 491–514.

Mulder, T., and M. Lamontagne (1990). Analysis of CLTN traces from events in the Saguenay region, internal report, Geol. Surv. Canada (Chalevoix Local Telemetered Network).

North, R., R. Wetmiller, J. Adams, F. Anglin, H. Hasegawa, M. Lamontagne, Du Berger, L. Seeber, and J. Armbruster (1989). Preliminary results from the November 25, 1988 Saguenay (Quebec) earthquake, *Seism. Res. Lett.* 60, 89–93.

Randall, G. (1994). Efficient calculation of complete differential seismograms for laterally homogeneous earth models, *Geophys. J. Int.* 118, 245–254.

Saikia, C. (2000). A method for path calibration using regional and teleseismic broadband seismograms: Application to the 21 May 1997 Jabalpur, India earthquake (M_W 5.8), *Curr. Sci.* 79, 1301–1315.

Saikia, C., B. Woods, and H. Thio (2001). Calibration of the regional crustal waveguide and the retrieval of source parameters using waveform modeling, *Pure Appl. Geophys.* 158, 1301–1338.

Savage, B., J. Chen, and D. Helmberger (2003). Velocity variation in the upper-most mantle beneath the southern Sierra Nevada and Walker Lane, *J. Geophys. Res.* 108, B7.

Somerville, P., J. McLaren, C. Saikia, and D. Helmberger (1990). The 25 November 1988 Saguenay, Quebec, earthquake: Source parameters and the attenuation of strong ground motion, *Bull. Seism. Soc. Am.* 80, 1118–1143.

Uski, M., T. Hyvonen, A. Korja, and M. Airo (2003). Focal mechanisms of three earthquakes in Finland and their relation to surface faults, *Tectonophysics* 363, 141–157.

Zhao, L., and D. Helmberger (1991). Broadband modelling on a regional shield path, Harvard recording of the Saguenay earthquake, *Geophys. J. Int.* 105, 301–312

Zhao, L., and D. Helmberger (1993). Source retrieval from broadband regional seismograms: Hindu Kush region, *Phys. Earth and Planet. Inter.* 78, 69–95.

Zhu, L., and D. Helmberger (1997). Regional waveform calibration in the Pamir-Hindu Kush region, *J. Geophys. Res.* 102, 22799–22813.

Zonno, G., and R. Kind, (1984). Depth Determination of north Italian earthquakes using Grafenberg data, *Bull. Seism. Soc. Am.* 74, 1645–1659.

8

Relating Crustal Structure and Stress Indicators in the Azores Islands

Nuno Dias[1,2] and Luís Matias[2,3]
[1]*Instituto D. Luiz,*
[2]*Instituto Superior Engenharia de Lisboa,*
[3]*Faculty of Sciences, University of Lisbon*
Portugal

1. Introduction

The Azores Archipelago is a region of intense geodynamical activity, with particular morphology and tectonics, which results in an intense seismic and volcanic activity (cf. Fig. 1). The seismic activity recorded in the region is very high, predominating low magnitude events organized in space and time in seismic crisis, sometimes triggered by stronger

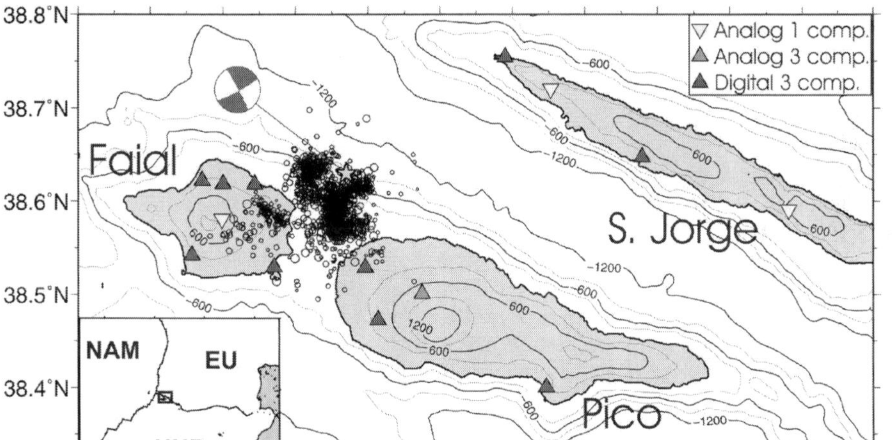

Fig. 1. Epicentral distribution of the aftershocks recorded between 9 and 31 of July 1998, located with MAC model (Senos et al., 1980). The Harvard-CMT focal mechanism is showed, pointing to the main shock location. Location of SIVISA permanent seismic stations and the temporary portable stations of IM and IGIDL/CGUL as triangles. Isobaths interval of 300m (Lourenço et al., 1998). Inset: Azores Triple Junction between North-America, Eurasia and Nubia plates.

earthquakes such as the 1980 (Ml~7.2) Terceira earthquake or the 1998 (Ml~5.8) Faial earthquake. Besides the interest of such a natural laboratory for Geosciences, the study and understanding of such phenomena is also critical in terms of civil protection and associated risk assessment, because such seismic activity has often a significant impact in economic and human terms. However, the available knowledge on the mechanisms responsible for the observed seismicity pattern and its relationship with tectonic and volcanic features of the region is still unsatisfactory.

The occurrence of the July 9th 1998 earthquake near the island of Faial, together with the fact that the instrumental coverage of the seismic sequence generated by this earthquake applied for the first time digital seismic stations, opened a window of opportunity for a comprehensive analysis on the mechanisms involved in the generation of such seismicity. The main event triggered a seismic crisis of thousands of aftershocks, strongly monitored in the weeks following the event, whose analysis allowed defining the characteristics of such seismicity and its relationship with volcanic and tectonic phenomena associated with the Tectonic Triple Junction environment.

The analysis of the aftershocks aimed at one hand the spatiotemporal characterization, by reviewing and refining the hypocentral locations, the determination of the seismic generation mechanisms and identification of the main active structures (Dias, 2005; Matias et al., 2007); on the other hand, to determine the crustal structure, with the refinement of the existing 1-D models, determining the local three-dimensional structure, analysis of seismic anisotropy and correlation with the state of crustal stress (Dias, 2005; Dias et al., 2007).

2. Temporary seismic network and data selection

At the time of the July 9th 1998 earthquake, the existing permanent seismic network of SIVISA (Seismological Surveillance System of Azores) was composed by 17 analog single-component (vertical) stations and 8 three-component analog stations, all with A/D converters, plus 4 digital stations. All stations were short-period, the analog stations - equipped with MARK L4C seismometers of 0.5 s (2 Hz) and the digitals with Lennartz LE-3D seismometers, 1 Hz. The only exception was the broad-band station of CMLA in São Miguel island, belonging to the worldwide network IRIS/IDA, equipped with a very broad-band triaxial Streckeisen STS-2 seismometer. This network was installed in 8 islands of the Azores Archipelago.

In the days following the main earthquake, the Meteorology Institute (IM), the Infante D. Luíz Geophysical Institute (IGIDL) and the Geophysical Centre of the University of Lisbon (CGUL)[1], in order to improve the instrumental coverage of the aftershocks by the SIVISA permanent network, deployed several portable digital seismic stations: IM placed 5 stations on Faial and Pico islands, while 2 additional stations were placed by IGIDL-CGUL on S. Jorge island. The seismometers used on all temporary stations were Lennartz LE-3D sensors, analogue to those used in the permanent digital stations. This increased the seismic monitoring network to a total of 14 stations located on the three islands surrounding the epicentral area (cf. Fig. 1). The stations sampling rates were 100 Hz for the IGIDL-CGUL stations, 125 Hz for IM stations; due to limited storage capacity, the data recorded by the 5 IM temporary stations were latter decimated to half, being stored with 62.5 Hz.

The temporary stations were operational until the end of July 1998, allowing a significant increase in the network registering capability, lowering the magnitude detection threshold

[1] IGIDL and CGUL were latter merged into Institute D. Luiz (IDL).

to events between 0.9 and 1.1 ML (Dias, 2005). During this period thousands of events were detected, with 4627 located aftershocks (Matias et al., 2007).

In order to maximize the quality and reliability of the results, for the one-dimensional modelling a set of 692 events was selected with the following criteria (Dias, 2005; Matias et al., 2007): minimum number of 6 stations recording an event, with one station at least located on S. Jorge island and azimuthal gap (GAP) smaller than 180°. For the 3-D modelling and seismic anisotropy analysis the criteria were more severe, with the used data set reduced to 688 and 438 events, respectively.

3. Tomographic inversion and crustal seismic structure

The inversion of a 1-D or 3-D seismic model of the crustal structure is based on the source-station travel times of seismic waves, generally by minimizing the time residual (difference between observed and theoretical travel times) due to disturbances introduced in the hypocentral coordinates and model properties. The dependence or coupling between the hypocentral parameters and the velocity structure being modelled is usually resolved by way of damped least-squares (Menke, 1984); a more detailed discussion on these methodologies can be found at Thurber (1993), Kissling et al. (1994), Thurber et al. (2000) or Thurber (2003). The resolution of this coupling problem requires the use of seismic events with stable hypocentral solutions, a criterion which must be balanced with the best spatial coverage possible (Thurber, 1993; Kissling et al., 1994; Thurber et al., 1999).

The methods used for the inversion of the velocity structure are implemented in programs like VELEST (Kissling et al., 1994; Kissling, 1995) for 1-D structure and SIMUL2000/SIMULPS (Thurber, 1983; Thurber et al., 1999) for 3-D structure. In the first case the model is parameterized by plane horizontal layers where seismic properties like P-wave velocity, Vp, or the Vp/Vs ratio remain constant, while in the former the model parameterization is done using a three-dimensional grid where on each node the value of these properties is set, with the values in other regions obtained by linear interpolation. Given the lower quality usually associated with S-waves readings compared to P-waves, both methods invert directly to the Vp and Vp/Vs structures, the Vs structure being latter derived from them.

Figure 2 displays the best (minimum) 1-D model obtained for the area between the islands of Faial, Pico and São Jorge, labelled FAIAL98 (Matias et al., 2007). Considering the P and S velocities and the Vp/Vs ratio of the model, and comparing with values associated with typical oceanic crusts (Tanimoto, 1995; Karson, 1998) it is possible to make a generic interpretation in petrological terms of the crustal structure of the area under study. The results indicate a basaltic composition for the layer located between 1 and 3 km in depth (Layer 2), with the velocities values suggesting a succession from lava flows to pillow lava and basaltic dikes. Between 3 and 12.5 km, the values of Vp and Vp/Vs indicate the presence of a Layer 3 with increased thickness, whose composition corresponds to Gabbro type mafic rocks. The location of the Moho at 12.5 km is an effect arising from the numerical modelling, the crust-mantle transition being more gradual and located between 12 and 14 km depth. The Moho associated Vp values will be intermediate values between 7.3 and 7.7 km/s, a possible evidence of a gradual increase of ultra-mafic rocks intrusion in the lower crust or the effect of variations in the depth of the Moho.

In order to determine the 3-D crustal structure in the tomographic inversion process (Dias et al. 2007) the above 1-D model was used as initial model, parameterized in the 3-D grid of Figure 3. After relocation of the earthquakes used in the 1-D modelling, the stability criteria for the hypocentral solutions were strengthened, with only 688 events used (cf. Fig. 3).

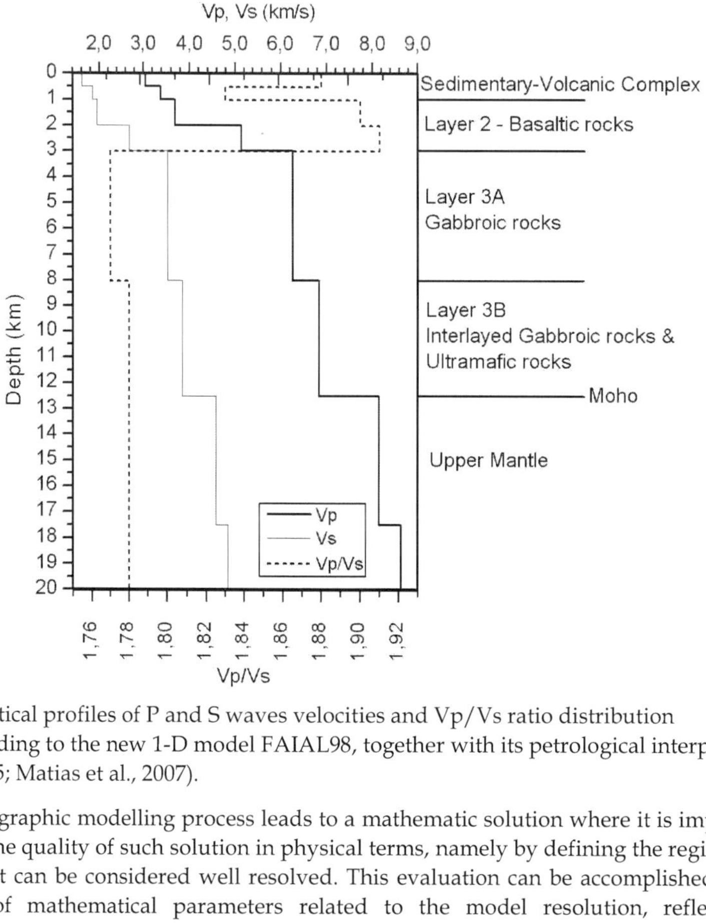

Fig. 2. Vertical profiles of P and S waves velocities and Vp/Vs ratio distribution corresponding to the new 1-D model FAIAL98, together with its petrological interpretation (Dias, 2005; Matias et al., 2007).

Any tomographic modelling process leads to a mathematic solution where it is important to estimate the quality of such solution in physical terms, namely by defining the regions of the model that can be considered well resolved. This evaluation can be accomplished through analysis of mathematical parameters related to the model resolution, reflecting the distribution of information and associated resolution, and also by performing sensitivity tests that use synthetic models or data. Only after such tests were performed, and together with joint analysis of the resolution parameters, is possible to define the resolved areas that will later be interpreted in physical terms. For discussion on the tomographic resolution of tomographic models see for example Menke (1984), Eberhart-Phillips (1993), Kissling et al. (1994), or Kissling et al. (2001); for the quality assessment of the model presented here see Dias (2005) and Dias et al. (2007).

Figure 4 shows the tomographic model obtained, in horizontal planes of Vp and Vp/Vs structures at depth, with resolved areas outlined. Given the information distribution (source-station paths), the resolved areas are located around the NNW area of the Faial -Pico channel and essentially between depths of 4 to 10 km. At the shallower level (1.0 km) there is good agreement between the anomalies of the tomographic model and the surface geological structure, particularly with the several volcanic units, but the most relevant feature of the tomographic model is the presence of a high-velocity body (Vp>6.3 km/s) located under the NW area of the island of Faial and extending from 6-7 km to a depth of approximately 13 km, which is enclosed in the northeast and east by deeper seismicity.

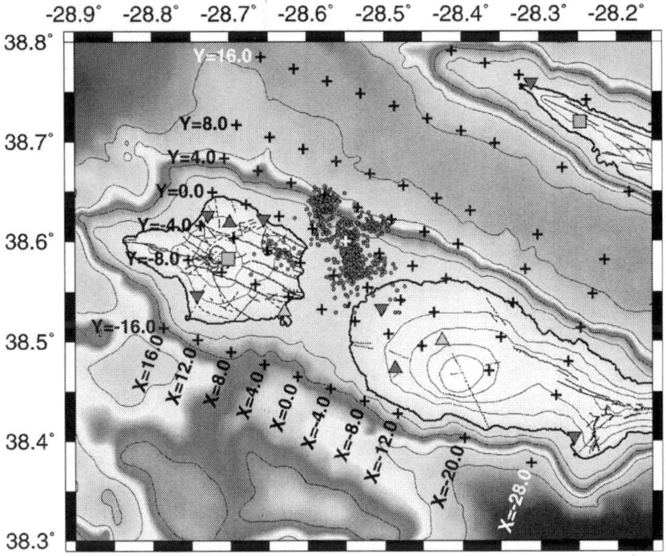

Fig. 3. Horizontal grid nodes positions (crosses) of the 3-D grid, seismic network (cf. Fig. 1) and epicenters (red dots) of the 688 selected events. A white cross marks the center of the grid, and the X and Y coordinates of Figure 4 are represented laterally to the grid. The horizontal distance between nodes was of 4-8 km, and vertically the nodes were placed in layers located at depths Z = 1, 4, 7, 10, 13, 17 km.

In the deepest levels (Z = 7, 10 km, see Figure 4) there is a strong lateral gradient to NNE in the direction of the S. Jorge channel, mainly of the Vp/Vs ratio that drops from 1.82 to less than 1.72; since this transition is partly coincident with the NNE limit of the high-velocity body and simultaneously runs parallel to the Faial-Pico alignment, it probably reflects the presence of an important tectonic feature. Under the central area of Faial, there is the presence of a low velocity zone (Vp <6.0 km/s), with a 5-10% lateral gradient on Vp and extending between 4 and 7 km in depth; under Pico there is the suggestion of the presence of a similar anomaly. The location of these anomalies is consistent with the estimated position for the magmatic chambers of the main volcanic edifices of Faial and Pico (Machado et al., 1994).

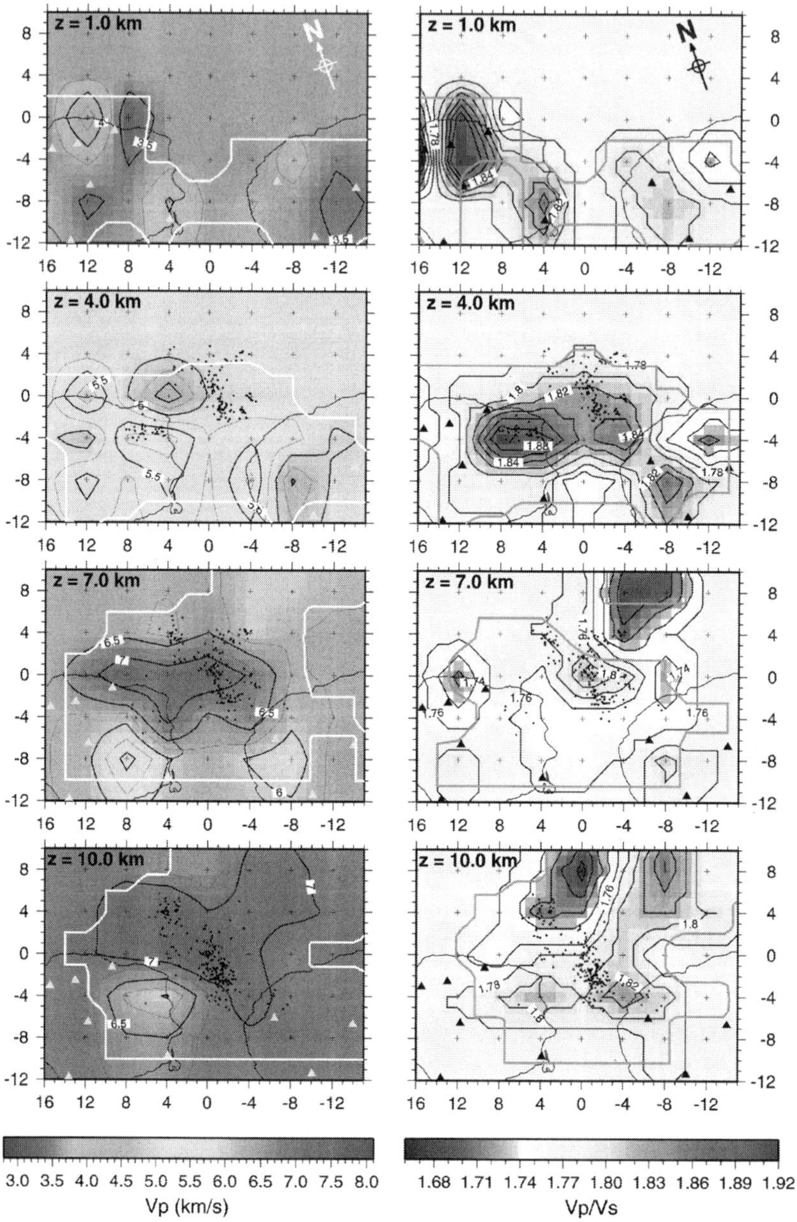

Fig. 4. Horizontal depth planes of the 3-D Vp (left) and Vp/Vs (right) models, at selected depths. Triangles: seismic stations, crosses: grid nodes, dots: epicenters within 1 km vertical range of the plane. Lateral XY coordinates according to the 3-D grid of Fig. 3.

Overall, the distribution of the observed seismicity is linked to areas of high value of the Vp/Vs ratio (> 1.84), also linked in some places to the high values of Vp, which should reflect changes in the mechanical properties of the rocky surroundings in fault areas.

4. Hypocentral relocation

The accurate determination of an earthquake hypocenter parameters, that is geographical coordinates, focal depth and time of occurrence, requires some knowledge on the crustal structure between the seismic focus and the receiving station. With the knowledge of such structure, usually parameterized in a seismic velocities model, the space-time coordinates of an event can be determined from the observed arrival moments of the seismic waves recorded at several stations (inverse problem). Depending on the specific structure of the used velocity model, the hypocentral location and especially the determination of the focal depth can undergo major changes.

Regarding the aftershocks of the July 9th 1998 earthquake, the preliminary location made by SIVISA (cf. Fig. 1) was carried out using model MAC (Hirn et al., 1980; Senos et al., 1980). MAC is a 1-D model developed with the purpose of locating all offshore seismicity in the Azores region, and reflects the average crustal structure of the plateau. The determination of the new 1-D model of the Faial-Pico-São Jorge area and of the tomographic 3-D model automatically entails a revision of the seismic hypocentral parameters used in the modelling.

In mathematical terms, the relocations by the new 1-D and 3-D models leads to substantial reduction of the data root mean square (RMS): relatively to the MAC model the new 1-D model allows for a reduction of 63% of the RMS, from 0,286 s to 0,103 s, while the use of the 3-D model leads to a further reduction of 32% of this parameter reaching 0,070 s (75.5% compared to the MAC model). In spatial terms, the changes entailed by the 1-D modelling generally do not exceed 2 km shifts of the epicentral position (cf. Fig. 5), with an additional 1 km in the tomographic inversion. The changes in focal depths are more relevant because they reduce the thickness of the fragile layer to the upper 15 km (22 km with the MAC model).

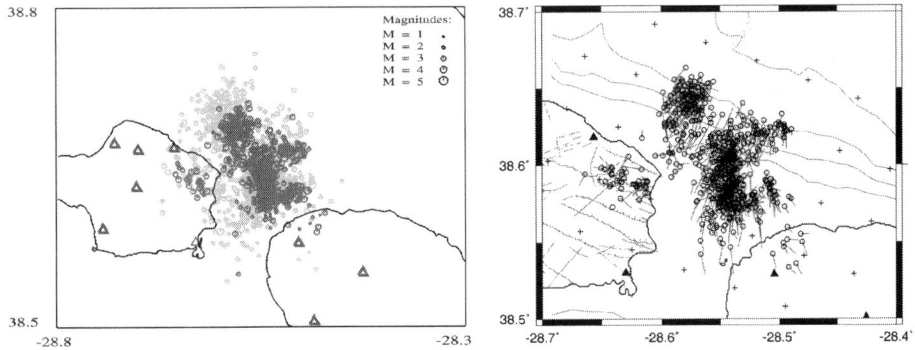

Fig. 5. Left: Comparison between the original epicenter locations (blue) of the 692 events with MAC model and the revised epicenter locations (red) with the new 1-D model FAIAL98. Right: Relocated events following the tomographic inversion; circles represent the final positions obtained with the 3-D model, line segments the position shift.

In terms of the seismicity spatial organization following the relocation (cf. Fig. 5), both models revealed two main alignments compatible with the two nodal planes of the focal mechanism of the main earthquake (cf. Fig. 1): one striking NNW-SSE, containing the majority of the aftershocks, and a secondary one with ENE-WSW direction. Most of the seismicity occurs in the Faial-Pico channel area, the inland events being less relevant both in terms of occurrence rates and magnitudes (Dias, 2005). This inland seismicity presents a NW-SE alignment, especially in Faial.

5. Seismic anisotropy and crustal stress

Seismic anisotropy is a three-dimensional phenomenon that takes different forms, corresponding to the variation in the wave's propagation velocity with azimuth and eventual splitting of an S-wave into several pulses (birefringence). The interpretation of the crustal seismic anisotropy is sometimes carried out according to the Extensive Dilatancy Anisotropy model (Crampin et al. 1984), which states that as result of a crustal stress field, a initially isotropic homogeneous medium undergoes micro-fracturing (or previously existing micro-fractures are reoriented), the assumed plane fractures adjusting to directions roughly parallel to the direction of maximum horizontal stress (and perpendicular to the direction of minimum horizontal stress). The presence of fluids in the crust (H_2O, CO_2) usually leads to the filling up of these micro-fractures, changing the propagation properties according to direction. An S-wave with sub-vertical incidence in such medium will split into two orthogonally polarized quasi-S-waves, propagating with different velocities due to the variations of the mechanical properties in the parallel and perpendicular directions to the micro-fractures. As a result seismic anisotropy will be observed, its level being evaluated from the measurement of the time difference between the arrivals of these two S-waves.

Following the method of Bouin et al. (1996) 438 events were selected, the analysis of their records suggesting signs of anisotropy. Although the 3D revised hypocenters solutions are more accurate, the error involved in the measurement of the splitting directions allows using the simpler 1D approach: the already referred sampling rate of 62.5 Hz, for the data stored for the majority of the digital temporary stations, coupled with the difficulties in the north alignment of the seismic sensors in basaltic oceanic islands, outmanoeuvre the accuracy of using 3D locations instead of 1D. On the other hand, most of the programs available for crustal stress modelling are 1D based (Robinson & McGinty, 2000).

The detection and quantification of such anisotropy was possible only in some of the digital stations located on the islands of Faial and Pico (cf. Fig. 6). In each station, the presence of anisotropy in the S-wave window records was shown by a systematic polarization of the first pulse of this kind of wave, sometimes followed by a second pulse showing a roughly orthogonal polarization. Figure 6 represents the results obtained for each station, with the rose diagrams representing the statistical direction of polarization of the first (blue) and second (red) S-waves. In a general, the observed polarization is very stable and independent of the epicentral distribution, ranging from approximately NW-SE direction in the northern Faial and north-western Pico to a significantly orthogonal WSW-ENE direction in the eastern area of Faial island.

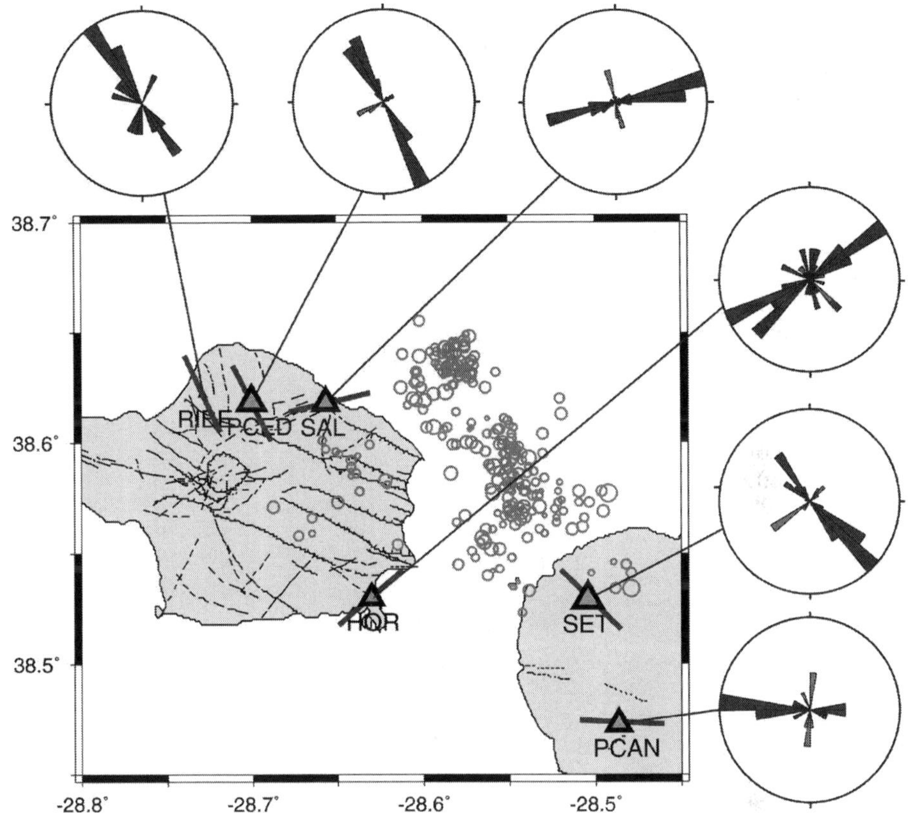

Fig. 6. Epicenters of the 438 selected events together with rose diagrams (10° intervals) of the polarizations directions, as measured in each station, of the fast (blue) and slow (red) S-waves. The size of each station symbol is proportional to the respective quantified anisotropy level, with the mean polarization direction of the fast S-wave projected over each station.

To relate the crustal stress state in the area affected by the seismic crisis, there are two seismologic indicators to determine the direction of maximum horizontal stress, SHmax or σ1: the analysis of the polarization of the S-wave and the computation of focal mechanisms. The markers associated with focal mechanisms (usually the direction of the P axis) reflect the state of stress in the source area, while the direction of polarization of the first wave associated with the birefringence of S-waves reflects the direction of maximum horizontal stress in the shallower structure located beneath the station. In case of focal mechanisms, single or composed, the estimate of SHmax is made according to the criteria defined by Zoback (1992), which relates the orientation of tension T or pressure P axis with the direction of SHmax.

To calculate single focal mechanisms, 18 events were selected with a minimum of 11 polarities for the P-wave, the hypocentral location used in calculating the focal parameters obtained from the 1-D model FAIAL98 (Matias et al., 2007). In the case of composite focal mechanisms, a similarity analysis of the recorded waveforms was performed by cross-correlation, which established a classification of "similar" earthquake clusters; subsequently, the joint focal mechanism was calculated for the 16 more numerous and stable clusters (Dias, 2005). The compilation of these results is represented in Figure 7 together with the estimated SHmax direction for both indicators.

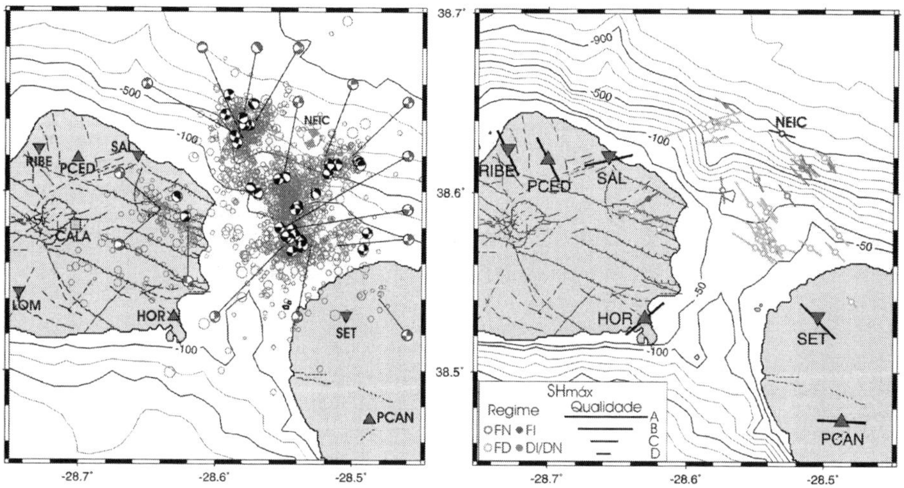

Fig. 7. Left: Aftershocks distribution of 9-7-1998 main shock, for events recorded by at least - 4 stations readings and relocated with the 1-D model FAIAL98; single (black) and composite (grey) focal mechanisms, together with the Harvard-CMT solution for the main shock. Right: maximum horizontal stress (SHmax) directions obtained from focal mechanisms and S-waves polarization analysis. The segments length is proportional to the quality of the stress measurement following Zoback (1992). The regime indicates the focal mechanism type: FN - normal fault, FD - stryke-slip fault, FI – inverse fault, DI/DN –oblique fault with inverse/normal component. Batimetry of Lourenço et al. (1998).

Figure 7 shows two dominant almost orthogonal general orientations for the maximum horizontal stress, N220°E -N260°E and N90°E-N130°E, limited to two distinct areas, with an apparent sharp transition of SHmax between them. As this indicator corresponds to the horizontal projection (i.e. two-dimensional) of a three-dimensional crustal stress vector, this sharp transition may be apparent, since the horizontal projection of T and P axis of the focal mechanisms suggests a continuous rotation (albeit fast) in the orientation of these axis (Dias, 2005).

The orientations obtained for the dominant polarization directions show some correlation with the tectonic alignments of Faial and Pico, which could suggest a tectonic control of the crustal fracturing near the stations. The observed anisotropy is consistent with the presence beneath the stations of anisotropy of the type envisaged by the EDA model, with the crack planes parallel to the direction of maximum horizontal stress. The major uncertainty is related with the depth extension of the anisotropic structures. The estimated directions for the maximum horizontal stress (SHmax), obtained from single and composed focal mechanisms solutions, is around 50°-80° in the NE zone of the Faial island, and a SHmax direction of around 130°-140° for the NW area of Pico island. The different orientations in the polarization direction obtained for the stations located in the north of Faial appears to be also related with the effect observed in the macroseismic effects (substantial mitigation of intensities) and the blocking of the progression of seismic activity to NW (Matias et al., 2007).

6. Coulomb stress variation

The aftershock distribution (Fig. 5) presents an odd distribution, with events aligned along roughly perpendicular directions, and the area surrounding the main shock seems to be devoided of significant aftershock activity. Furthermore, the focal mechanisms obtained from the best-constrained aftershocks indicate both strike-slip and normal faulting, suggesting that different faults have ruptured after the main shock.

Following the revision by Das & Henry (2003) on the relationship between the main earthquake slip and its aftershock distribution, the occurrence of low magnitude events in the high-slip regions of the fault that ruptures is rare. Instead, the authors have found that, in most cases, the aftershocks occur in nearby faults that are reactivated by a favourably oriented post-seismic static stress.

To investigate this possibility we have used the method implemented in the program GNStress (Robinson & McGinty, 2000), that calculates the crustal strains and their derivatives in a homogeneous half-space due to slip in a rectangular fault. To convert strains to stresses in the half-space a constant rigidity of 2.68×10^{10} Pa was used. A regional stress field with $\sigma 1$ (maximum compressive stress) oriented N135E, was assumed according to the transtensional tectonic regime deduced for this area of the Azores archipelago from neotectonic investigation (Madeira & Ribeiro, 1990). For the main shock source description we used the parameters derived from GPS data (Fernandes et al., 2002), namely the solution of a rectangular fault 9 km long x 4.5 km wide, oriented N165E, with the top of the fault plane lying 2 km below the topographic zero. A detailed discussion of the procedure can be found in Matias et al. (2007).

The calculated distribution of the Coulomb failure stress is presented in Fig. 8. Looking at the depth distribution of the variation in the Coulomb failure stress, and its relationship with aftershock distribution, we conclude that the main shock ruptured a very shallow fault, from a depth of 2 to 7 km, and then most of the aftershocks occurred below this depth along faults that were favourably orientated in relation to the post-seismic stress field. This very simple model doesn't explain the shallowest activity, above 7 km depth. It may be a consequence of the simple half-space model assumed for the stress computations.

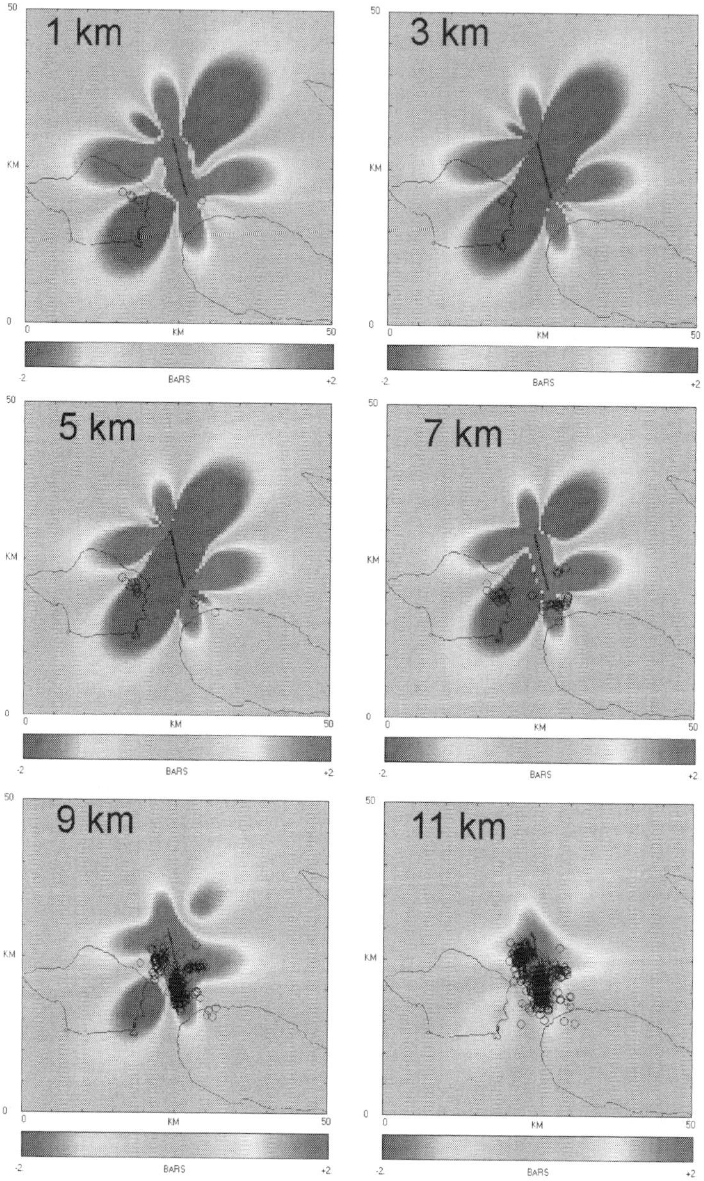

Fig. 8. Induced changes in Coulomb failure stress (δCFS) due to the main shock and its comparison to aftershock distribution. The δCFS was computed for 6 reference depths: 1, 3, 5, 7, 9 and 11 km. The aftershocks were plotted according to the depth ranges. The assumed main shock source plane is indicated in each figure.

7. Conclusion

The results obtained from the 1-D and 3-D inversions allow an insight of the crustal structure of the Azorean plateau region, primarily in the Central Group area between the islands of Faial and Pico. In terms of crustal thickness, the 1-D model points to a Moho located at 12.5 km depth while the 3-D model, though not allowing an evident definition in the velocities transitions, corroborates this suggestion with the Vp = 7.8 km/s isoline fluctuating around 14 km depth. Although this crustal thickness is higher than those determined in other studies, pointing to values around 8-11 km (Luis et al., 1998; Miranda et al., 1998; Luis & Neves, 2006), these works were performed at the plateau scale or in offshore areas while the models presented here necessarily include the structure of the islands (where the stations are located), thus justifying a further thickness increase of about 2-3 km. Additionally, a crustal thickness of 14 km fits better the estimated parameters for the largest instrumental earthquakes, both in terms of the sources estimated geometry and the correlation with the recorded magnitudes (Madeira et al., 2003).

The resulting distribution of seismicity indicates that for the main observable tectonic structures, the most active corresponds to the one of NNW-SSE orientation with the focal mechanisms indicating a dominant left strike-slip movement, with a secondary WSW-ENE direction of dominant right strike-slip. The NW-SE structures have a minor role, mainly confining seismicity.

The presence of seismic anisotropy was detected beneath the seismic stations (cf. Fig. 6), compatible with the EDA model, with a correspondence between the fracturing index and the positioning of each station. The estimated orientation for the direction of maximum horizontal stress from the S-wave polarization analysis, combined with single and composite focal mechanisms solutions, indicates a complex pattern in the crustal stress paths rotating from a general NW-SE direction to NE-SW in the eastern section of the island of Faial and extending into the sea. The comparison with the crustal stress distribution at the shallow layer of 1 km depth (cf. Fig. 8), shows that the measured direction of polarization of the first (fast) S-wave appear to follow the lobules shape, at least in northern Faial (RIBE, PCED, SAL stations) and NW Pico island (SET station). As for HOR station, in SE Faial island, is inside a low stress lobule and the polarization direction appears not to be correlated with the stress distribution although as it can be seen from Fig. 6, this station shows a bigger fluctuation in the measurements of the first polarization.

Comparing the distribution of seismicity and the rotation of SHmax direction (Fig. 5 and 7) with the tomographic model (Fig. 4), there is an apparent tectonic control by the high-velocity intrusion lying NE of Faial. It is essentially a crustal asseismic area enclosed by more active seismogenic areas, its form "conditioned" by the distribution of seismicity; in addition, the rotations in the directions of SHmax appeared to be sub-paralleled to the limits of this anomaly. Comparing with the Coulomb failure stress (cf. Fig. 8) planes coincident with the location of the main seismicity (layers between km depth) the SHmax directions show good correlation with the lobules shape, namely the rotation observed between NNW-SSE and WSW-ENE directions. The main exceptions appear to be the shallow seismicity observed inland in NE Faial island, in the area of Faial Graben.

The results shown here suggest a crustal thickening due to an increase in thickness of layer 3, the 1-D model presenting a total thickness of 9.5 km. This indicates that the accretion of the crust in the Azores plateau was essentially due to a relatively high magmatic supply rate, which cooled and crystallized at depth, with the rate of extruded

volcanism though important playing a small role in the crustal build-up. The source of such high magmatism is still a matter under debate, maybe the result of the actual presence of a "hot-spot" in the Azores or, alternatively, be only the result of the condition of the plateau as a tectonic triple junction.
The mentioned presence of a high seismic waves velocity volume, interpreted as a crystalline intrusion of mafic (Gabbro) and ultra-mafic rocks, apparently constraining the observed seismicity and stress state, appears to corroborate the last assumption. Under the main volcanic edifices of Faial and Pico there is no clear signal of the presence of magmatic chambers, but this maybe result of to the poor coverage of seismic paths; the presence of a low-velocity anomaly is revealed, but this may be due only to the effect of a volcanic feeding system and not the presence of a magmatic chamber, something that concurs with the suggestion of Nunes et al. (2006) of a lack of magmatic chamber under Pico's volcano.

8. Acknowledgments

We would like to acknowledge the effort of all the people involved in the seismic monitoring campaign that took place in the Azores during July 1998, namely, C. Corela, A. Carvalho, and J. Simões. The figures were produced using the GMT software (Wessel and Smith 1991) and SEISAN software (Havskov and Ottemöller 1999).This work was supported by the Portuguese Fundação para a Ciência e Tecnologia, under projects MASHA (POCTI/CTA/39158/2001) and STAMINA (PDCTM1999MAR15255). Nuno A. Dias had a fellowship under the PRODEP III program, Action 5.3. This is a contribution to PEST-OE/CTE/LA0019/2011 - IDL.

9. References

Bouin, M.P., Tellez, J. & Bernard, P. (1996). Seismic anisotropy around the Gulf of Corinth, Greece, deduced from three-component seismograms of local earthquakes and its relationship with crustal strain. *Journal of Geophysical Research,* 101(B3), pp. 5797-5811.

Crampin, S., R. Evans & B.K. Atkinson (1984). Earthquake Prediction - a New Physical Basis. *Geophysical Journal of the Royal astronomical Society,* 76(1), pp. 147-156.

Das, S. & Henry, C. (2003). Spatial relation between main earthquake slip and its aftershock distribution. *Reviews of Geophysics,* 41(3), pp. 1013-1036.

Dias, N.A. (2005). Estudo da sequência sísmica gerada pelo sismo do Faial de 9 de Julho de 1998: anisotropia crustal, inversão tomográfica e caracterização sismo-tectónica. *PhD Thesis* (in Portuguese), Lisbon University, Lisboa, 540 pp.

Dias, N.A., L. Matias, N. Lourenco, J. Madeira, F. Carrilho & J.L. Gaspar (2007). Crustal seismic velocity structure near Faial and Pico Islands (AZORES), from local earthquake tomography. *Tectonophysics,* 445(3-4), pp. 301-317.

Eberhart-Phillips, D. (1993). Local earthquake tomography: earthquake source regions. In: H.M. Iyer and K. Hirahara (Editors), Seismic Tomography: Theory and practice. Chapman & Hall, London, pp. 613-643.

Fernandes, R.M.S., Miranda, J.M., Catalao, J., Luis, J.F., Bastos, L. & Ambrosius, B.A.C. (2002). Coseismic displacements of the M-W=6.1, July 9, 1998, Faial earthquake (Azores, North Atlantic). *Geophysical Research Letters,* 29(16), pp. 21 1-4.

Havskov, J. & Ottemöller, L. (1999). Seisan earthquake analysis software. *Seismological Research Letters*, 70, pp. 532-534.

Hirn, A., H. Haessler, P. Hoangtrong, G. Wittlinger & L.A. Mendes-Victor, 1980. Aftershock Sequence of the January 1st, 1980, Earthquake and Present-Day Tectonics in the Azores. *Geophysical Research Letters*, 7(7), pp. 501-504.

Karson, J.A. (1998). Internal Structure of Oceanic Lithosphere: A Perspective from Tectonic Windows. In: W.R. Buck, P.T. Delaney, J.A. Jeffery A. Karson and Y. Lagabrielle (Editors), Faulting and Magmatism at Mid-Ocean Ridges. Geophysical Monograph. American Geophysical Union, Washington, D.C., pp. 177-218.

Kissling, E. (1995). Program VELEST user's guide - Short introduction, ETH Zuerich, pp. 26.

Kissling, E., W.L. Ellsworth, D. Eberhart-Phillips & U. Kradolfer (1994). Initial Reference Models in Local Earthquake Tomography. *Journal of Geophysical Research*, 99(B10), pp. 19635-19646.

Kissling, E., S. Husen & F. Haslinger (2001). Model parametrization in seismic tomography: a choice of consequence for the solution quality. *Physics of the Earth and Planetary Interiors.*, 123(2-4), pp. 89-101.

Lourenço, N., J.M. Miranda, J.F. Luis, A. Ribeiro, L.A.M. Victor, J. Madeira & H.D. Needham (1998). Morpho-tectonic analysis of the Azores Volcanic Plateau from a new bathymetric compilation of the area. *Marine Geophysical Researches*, 20(3), pp. 141-156.

Luis, J.F., J.M. Miranda, A. Galdeano & P. Patriat (1998). Constraints on the structure of the Azores spreading center from gravity data. *Marine Geophysical Researches*, 20(3), pp. 157-170.

Luis, J.F. & Neves, M.C. (2006). The isostatic compensation of the Azores Plateau: A 3D admittance and coherence analysis. *Journal of Volcanology and Geothermal Research*, 156(1-2), pp. 10-22.

Machado, F., Matias, L. & Lemos, A. (1994). Vibration of magma under the volcano of Pico (Azores Islands). Acta Vulcanologica, 5, pp. 147-151.

Madeira, J. & Ribeiro, A. (1990). Geodynamic Models for the Azores Triple Junction - a Contribution from Tectonics. *Tectonophysics*, 184(3-4), pp. 405-415.

Madeira, J. & Silveira, A.B. (2003). Active tectonics and first paleoseismological results in Faial, Pico and S. Jorge Islands (Azores, Portugal). *Annals of Geophysics*, 46(5), pp. 733-761.

Matias, L., Dias, N., Morais, I., Vales, D., Carrilho, F., Madeira, J., Gaspar, J., Senos, L. & Silveira, A. (2007). The 9th of July 1998 Faial Island (Azores, North Atlantic) seismic sequence. *Journal of Seismology*, 11(3), pp. 275-298.

Menke, W., 1984. Geophysical Data Analysis: Discrete Inverse Theory. Academic Press, 260 pp.

Miranda, J.M., Victor, L.A.M., Simoes, J.Z., Luis, J.F., Matias, L., Shimamura, H., Shiobara, H., Nemoto, H., Mochizuki, H., Hirn, A. & Lepine, J.C. (1998). Tectonic setting of the Azores Plateau deduced from a OBS survey. *Marine Geophysical Researches*, 20(3), pp. 171-182.

Nunes, J.C., A. Camacho, Z. Franca, F.G. Montesinos, M. Alves, R. Vieira, E. Velez & E. Ortiz (2006). Gravity anomalies and crustal signature of volcano-tectonic structures of Pico Island (Azores). *Journal of Volcanology and Geothermal Research*, 156(1-2), pp. 55-70.

Robinson, R. & McGinty, P.J. (2000). The enigma of the Arthur's Pass, New Zealand, earthquake 2. The aftershock distribution and its relation to regional and induced stress fields. *Journal of Geophysical Research*, 105(B7), pp. 16139-16150.

Senos, M.L., J.C. Nunes & V.S. Moreira (1980). Estudos da estrutura da crosta e manto superior nos Açores, INMG, Lisboa.

Tanimoto, T. (1995). Crustal structure of the Earth. In: T. Ahrens (Editor), A Handbook of Physical Constants: Global Earth Physics. AGU Reference Shelf Series. AGU, pp. 214-224.

Thurber, C. & D. Eberhart-Phillips (1999). Local earthquake tomography with flexible gridding. *Computers & Geosciences*, 25(7), pp. 809-818.

Thurber, C.H. (1983). Earthquake Locations and 3-Dimensional Crustal Structure in the Coyote Lake Area, Central California. *Journal of Geophysical Research*, 88(B10), pp. 8226-8236.

Thurber, C.H. (1993). Local earthquake tomography: velocities and Vp/Vs - theory. In: H.M. Iyer and K. Hirahara (Editors), Seismic Tomography: Theory and practice. Chapman & Hall, London, pp. 563-583.

Thurber, C.H. (2003). Seismic tomography of the lithosphere with body waves. *Pure and Applied Geophysics.*, 160(3-4), pp. 717-737.

Thurber, C.H. & Kissling, E. (2000). Advances in travel-time calculations for three-dimensional structures. In: C.H. Thurber and N. Rabinowitz (Editors), Advances in Seismic Event Location. Modern Approaches in Geophysics. Kluwer, Dordrecht, pp. 71-100.

Wessel, P. & Smith, W.H.F. (1991). Free software helps map and display data. EOS Trans. Amer. Geophys. U., 72(41), pp. 441 e 445-446.

Zoback, M.L. (1992). 1st-Order and 2nd-Order Patterns of Stress in the Lithosphere - the World Stress Map Project. *Journal of Geophysical Research*, 97(B8), pp. 11703-11728.

9

Electromagnetic View of the Seismogenic Zones Beneath Island Arcs

Hiroaki Toh and Takuto Minami
Division of Earth and Planetary Sciences, Graduate School of Science, Kyoto University
Japan

1. Introduction

The ring of fire is well-known by its intense seismic as well as volcanic activities. Most of those activities in the world are concentrated in this narrow circle around the Pacific rim. The Japanese Islands are one of the greatest island arcs in the circum-Pacific area and thus are located along the so-called seismogenic zone which is globally very unique in the sense that it consists of not only a very cold (i.e., old) but also a warm (young) subduction zone (Fig. 1). Namely, the very thick Pacific plate is subducting beneath northeast Japan while the relatively thin and young Philippine Sea plate is sliding under southwest Japan. The oldest parts of each subducting plate are approximately ~140 (Hirano et al., 2006) and ~30 Ma (Wang et al., 1995) respectively for the Pacific plate and the Philippine Sea plate at each trench's outer rise, although the latter age is more variable along the Nankai trough to the Ryukyu trench than the former along the Japan trench. The two subducting plates with different thermal states affect differently the seismic as well as the volcanic activity on the Japanese Islands, which makes the island arc an ideal place to conduct a comparative study on seismogenic zones with a contrasting tectonic setting. We, therefore, will pursue this theme as the main topic of this chapter.

From a geophysical point of view, there are a lot of physical properties that can give constraints on the current dynamics taking place beneath island arcs. Electrical conductivity is a physical property that can be determined independently of elastic properties such as seismic P- and S-wave velocities. Furthermore, it is known to be a strong function of subsurface temperature and very sensitive to presence of melts and/or fluids such as water. The electrical conductivity of the Earth's mantle as well as the seismic velocities changes discontinuously when phase changes dominate in the ongoing physical process at critical depths such as 410km and 660km. These features are very useful in interpreting the characteristics of the geophysical structure beneath an island arc because a joint interpretation of the electrical and elastic structures can provide further constraints on the island arc's thermal and physical states. For instance, fraction of fluids and/or melts can be estimated more precisely if the electrical properties are determined with known seismic geometries/boundaries as a priori information.

In addition, subduction-driven injection of surface water into the Earth's crust and mantle, and its circulation in the Earth have been a recent matter of hot debate in various disciplines in the geoscience community since Karato (1990) first pointed out a possible strong effect of water on the upper mantle properties (especially on electrical conductivity). A series of laboratory measurements has been conducted so far. However, those experimental results seem to differ severely in a quantitative sense. For example, Huang et al. (2005) reported a huge effect of water on electrical conductivity of both wadsleyite and ringwoodite, the two major minerals in the mantle transition zone. On the contrary, Yoshino et al. (2008) claimed that there is no need to include hydrous minerals in the mantle transition zone in order to explain the electrical conductivity profile determined by electromagnetic (EM) field works. The same contradiction is applicable for the asthenosphere, viz., there exist contrasting experimental results of strong (Wang et al., 2006) and weak (Yoshino et al., 2006) dependence of olivine conductivity on water content. Abundance of water in the mantle transition zone is important in the sense that it can form a filter just above the 410km discontinuity (Sakamaki et al., 2006) to segregate geochemical components in the lower mantle from those in the upper mantle (Bercovici & Karato, 2003). Coexistence of the geochemically enriched lower mantle and the very depleted upper mantle has been a long-lasting enigma in the geosciences community, and thus we need to identify a reasonable differentiation mechanism. Because there is no doubt that the mantle transition zone has high potential as a water reservoir (Inoue et al., 1995), this issue definitely requires further research. On the other hand, water in the crust is important in the sense that it can be the sources for shallow earthquakes, deep low-frequency tremors and volcanic activities in the seismogenic zones of the island arcs (e.g., Obara & Hirose, 2006). It seems to have become a consensus that the subduction-driven water injection is strongly dependent on thermal states of each subducting plate beneath the island arcs. Namely, cold and warm subduction zones behave quite differently in terms of the amount and depth of water release from the slabs. This means that a report on electrical images beneath different parts of the Japanese Islands is nothing but the aforementioned comparative study itself.

Northeast Japan can be classified into the cold subduction regime and thus thought to have high potential, in turn, for water supply to the deep mantle (Iwamori, 2004). Injection of water into the deep mantle seems to produce electrical conductivity anomalies of regional to semi-global scale beneath back-arc regions. Furthermore, those electrical anomalies are present irrespective to whether the subducting slab is stagnant at the 660 km seismic discontinuity (Ichiki et al., 2006) or plunging into the lower mantle (Booker et al., 2004)., although their surface manifestations look, naturally, quite different (Worzewski et al., 2010). Arc volcanism in northeast Japan is known to be three-dimensional (3-D) as typically depicted by Tamura et al's (2002) hot-finger model. A two-dimensional (2-D) east-west slice of a 3-D P-wave seismic tomography (Mishra et al., 2003) at 39.5N showed a nearly uniform distribution of moderately fast velocity above the subducting Pacific plate within the slice. It can be attributed to the fact that the slice covers a non-volcanic part, viz., a region between the hot fingers, of the well-developed island arc. An electrical section (Toh et al., 2006), which covers the non-volcanic part of northeast Japan, reveals a resistive shallow mantle and a conductive anomaly beneath the back-arc region at depths 150-200 km. The electrical conductivity anomaly can be interpreted as a direct manifestation of slab dehydration associated with collapse of the high-temperature type serpentine such as antigorite. An EM 2-D section of northeast Japan at crustal depths (Ogawa et al., 2001) revealed several high conductive anomalies in the lower crust that are considered to bear geofluid. The source of the lower crustal geofluid is attributed to the convection in the wedge mantle beneath northeast Japan, which is compatible with the distribution of the Quarternary volcanoes on the volcanic front as well.

Although the arc volcanism relevant to northeast Japan looks 3-D in terms of its structure, the magma source can be unique and simple. It stems from the deep mantle behind the mature island arc. On the contrary, that of southwest Japan cannot be presumed as simple as northeast Japan, if one studies basalt samples of this area (Iwamori, 1991; Kimura et al., 2003). The alkaline, sub-alkaline and adakite basalt magmas of southwest Japan imply multiple sources for its magma production in the mantle including slab-melting of the hot and young Philippine Sea plate. The presence of the adakite magma is a signature of fluid originating from the slab. Toh and Honma (2008) reported a possible mantle plume in the west of the Kyushu Island of southwest Japan, which can be another candidate of the multiple magma sources. On the other hand, seismic and EM observations on land have revealed coincidence of lower crustal conductors and epicenters of both deep low-frequency events and earthquakes in the upper crust, which suggests presence of crustal fluid (e.g., Kawanishi et al., 2009). 2-D slices of Nakajima & Hasegawa's (2007) 3-D seismic tomography results beneath southwest Japan imply the presence of a deep mantle plume released not from the Philippine Sea plate but from the Pacific plate that is located well below the younger plate. However, the link between the two kinds of fluid is still missing and needs further research, especially based on marine geophysical data or a combination of land and marine data.

In the following, we will first describe the principle and methods of electrical conductivity determination by EM field works. The principle and methods section will be followed by an EM case study on northeast Japan to illustrate usefulness of the principle as well as the methods. Thirdly, the EM image beneath southwest Japan will be presented and discussed in contrast to that of northeast Japan. Finally, results of the whole comparative study will be summarized and concluded.

2. Principles and methods in EM field works

There exist lots of methods for delineating subsurface electrical conductivity structures by field works. They are classified broadly into two categories: one to make use of transient response of the conducting Earth in time domain, and the other is to derive the Earth's stationary response as a function of location in frequency domain. The latter category is often adopted irrespective to observation locations (viz., on land or at sea) and hence readers are advised to refer to standard textbooks for the former category (e.g., Kaufman & Keller, 1983). We will describe briefly the principles and typical methods in the frequency domain here in this section.

The principle of the EM methods in frequency domain is to use amplitude ratios and phase differences between different time-varying EM components observed on the Earth's surface including the seafloor rather than to model observed time-series themselves. Amplitudes of the raw time-series are dependent on each event, i.e., they differ from time to time. However, their ratios and phase differences are constant for a particular frequency and a fixed site, provided that one is really looking at induced parts of temporal variations by external

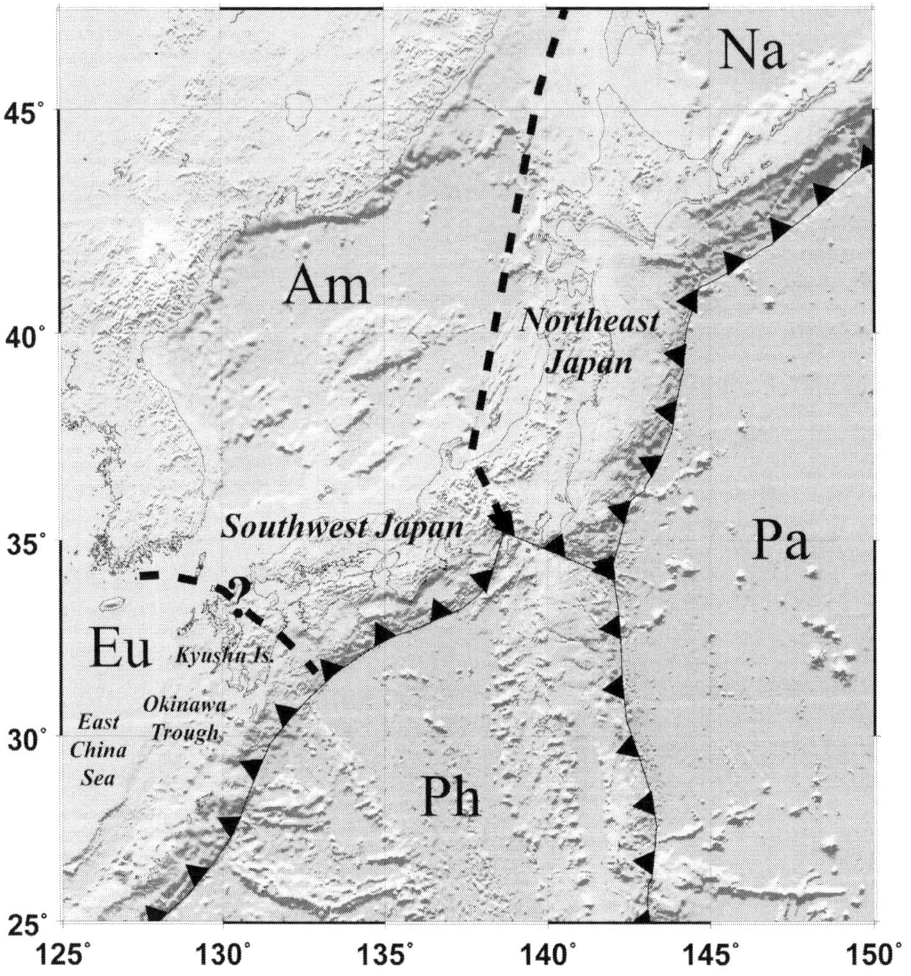

Fig. 1. Tectonic plates around the Japanese Islands. Pa: Pacific Plate, Ph: Philippine Sea Plate, Na: North American Plate, Eu: Eurasia Plate, Am: Amurian Plate.

geomagnetic disturbances. Another fundamental assumption of the frequency-domain EM methods is that the external source fields are either stationary plane waves or at least waves with known simple geometries such as dipole fields. If this assumption is applicable, the EM responses, p and q, of the conducting Earth can be estimated by the following linear regression formula in frequency domain:

$$W(f) = p(f) \cdot U(f) + q(f) \cdot V(f) \tag{1}$$

where U, V and W are the observed EM components and f is the frequency in concern. U, V and W are given by Fourier transforms of the observed time-series for respective EM components. If the source field is stationary enough, stable estimates of the Earth's EM responses (p and q) can then be yielded by standard stacking methods that divide the whole time-series into a number of segments of a suitable length.

There are several variants of the frequency-domain EM method represented by Eq. (1) because many combinations of EM components in the regression equation are possible. Of those, the geomagnetic depth sounding (GDS) method, the magnetotelluric (MT) method and the horizontal geomagnetic transfer function (HGTF) method are often favoured in actual field works since each method has its own distinct physical meaning and advantages. We will give succinct summaries of those methods in the following three subsections.

2.1 Geomagnetic depth sounding method

This method is a case of $U=B_x$, $V=B_y$ and $W=B_z$, where B_x, B_y and B_z denote the northward, eastward and downward geomagnetic components, respectively. The GDS method is usually applied when lateral contrast of the subsurface electrical structure is expected to be strong. This is because the anomalous B_z is most likely to be induced by the external inducing field of plain wave form than any other EM components. Vertically propagationg plane waves have B_x and B_y components alone. As for its detailed physical meaning and the graphical representation of the method, refer to Section 2 of Toh & Honma (2008).

In cases where the inducing source field can be approximated by a global-scale axial dipole, it is known that the Earth's scalar impedance Z is given by the following formula (e.g., Schultz & Larsen, 1987);

$$Z(f) = -\pi i f R_E \tan(\theta) \frac{B_r(f)}{B_\theta(f)}, \tag{2}$$

where R_E, θ, B_r and B_θ are the mean radius of the Earth, co-latitude, radial and southward geomagnetic components, respectively. The ratio of the two geomagnetic components is equivalent to the Earth's EM response function $p(f)$ if $W=-B_z$, $U=-B_x$ and $V=0$. Eq. (2) is often invoked to estimate the Earth's one-dimensional (1-D) impedance at long periods (typically $T > 4$ days) for EM forcing by the magnetospheric ring currents. The global-scale ring currents can produce temporal variations of axially symmetric magnetic dipole fields that are the premises of the valid application of Eq. (2). In order to determine tensor impedances, however, it is required to measure not only the vector geomagnetic field but also the vector (or rather 'horizontal') geoelectric field, which will be described in the next subsection.

2.2 Magnetotelluric method

The MT method needs two linear regression equations in which two sets of (U, V, W) are substituted: (B_x, B_y, E_x) and (B_x, B_y, E_y). E_x and E_y are the horizontal geoelectric components. The resultant EM response functions are neither scalar nor vector but tensor, which are elements of the so-called 'MT impedance tensor'. Namely, the MT impedance tensor is defined by the following matrix formula:

$$\begin{pmatrix} E_x \\ E_y \end{pmatrix} = \begin{pmatrix} Z_{xx} & Z_{xy} \\ Z_{yx} & Z_{yy} \end{pmatrix} \begin{pmatrix} B_x \\ B_y \end{pmatrix}. \tag{3}$$

Z_{ij} $(i, j = x, y)$ denotes each tensor element. The frequency f is intentionally dropped off from each variable in Eq. (3) for simplicity.

The MT impedance tensor is originated from the MT scalar impedance as in Eq. (2), which is a complex ratio of mutually orthogonal geomagnetic and geoelectric components. If the Earth's electrical conductivity varies in the vertical direction only, an external geomagnetic field variations polarized in a particular horizontal direction induces toroidal telluric currents in the Earth perpendicular to the magnetic field. It is the 1-D complex ratio of the MT scalar impedance $Z(=E/B)$, which has already appeared in Eq. (2). In this case, it follows that $Z_{xx}=Z_{yy}=0$ *and* $Z_{xy}=-Z_{yx}=Z$. If we substitute the magnetic field $(H_x, H_y)^T$ into Eq. (3) instead of the magnetic induction $(B_x, B_y)^T$, it is straightforward to show that the physical dimension of each impedance tensor element becomes ohm. Thus, the primary physical meaning of the MT impedance is the resistance of the Earth. However, if we use 'magnetic induction' in place of 'magnetic field', the MT impedance has a physical dimension of 'velocity'.

When the subsurface electrical structure elongates in a specific direction and x-axis is aligned to the structural strike, the diagonal elements of the MT impedance tensor vanish again while the absolute values of off-diagonal elements are not necessarily equal to each other. It is well-known that the Maxwell equations decouple into two independent modes in 2-D cases: one mode involves E_x, B_y, B_z and Z_{xy} alone and the other E_y, E_z, B_x and Z_{yx}. The former combination is called 'TE-mode' or 'E-polarization' while the latter is called 'TM-mode' or 'B-polarization' borrowing the nomenclatures of the EM wave-guide theory. It is evident that the GDS responses appear only in TE-mode for 2-D cases. The MT impedance tensors can be defined even for 3-D cases and remain being powerful tools in estimation of electrical structures. However, we need to work with full tensors rather than more simpler antisymmetric tensors after pertinent coordinate rotations in the horizontal plane.

2.3 Horizontal geomagnetic transfer function method

If one substitutes two sets of (U, V, W), (B_x^0, B_y^0, B_x) and (B_x^0, B_y^0, B_y), into the linear regression formula (Eq. (1)), you will end up with the following matrix equation with a 2 x 2 matrix:

$$\begin{pmatrix} B_x \\ B_y \end{pmatrix} = \begin{pmatrix} K_{xx} & K_{xy} \\ K_{yx} & K_{yy} \end{pmatrix} \begin{pmatrix} B_x^{\ 0} \\ B_y^{\ 0} \end{pmatrix}. \tag{4}$$

The vector $(B_x, B_y)^T$ is horizontal geomagnetic variations at the observation site in concern while $(B_x^0, B_y^0)^T$ is that of a reference site. Both the observation site and the reference site can be either on land or at the seafloor. In any land-sea combinations, each element, K_{ij} $(i, j = x, y)$, of the horizontal transfer function matrix in Eq. (4) constitues another set of EM response functions representative of the electrical properties of the Earth. Among the various combinations, an interesting pair is a seafloor observation site and a near-by reference site on land. This combination involves vertical shears of each horizontal geomagnetic component, which are measures of the net electric currents induced in the ocean. The pair, therefore, is called as the vertical gradient sounding (VGS) method, which is a good alternative of the seafloor MT method when geoelectric measurements at the sealoor are missing. As for details of the VGS method, refer to Ferguson et al. (1990) and references therein.

The goal of the EM methods described above is to derive electrical conductivity structures that can explain the spatial distribution as well as the frequency dependence of the Earth's

EM response functions thus derived. We will illustrate how to estimate electrical structures that are compatible with EM field works by introducing a few case studies in and around the Japanese Islands using the two subsequent sections.

3. EM view of the seismogenic zone beneath northeast Japan

Northeast Japan is classified into the cold subduction regime and thus said to be the very spot of on-going water supply into the deep mantle (Iwamori, 2004). Injection of water into the deep mantle can produce electrical conductivity anomalies beneath back-arc regions. In order to image such kind of anomalies, we constructed a seafloor MT array in the Japan Sea (Toh et al., 2006).

The seafloor array consisted of six ocean bottom electromagnetometers (OBEMs) that are capable of measuring both vector geomagnetic and horizontal geoelectric fields in addition to horizontal tilt variations. The attitude data were used to rotate each measuring frame at the seafloor back to a common reference frame in the horizontal plane. Directions of the geomagnetic north at each site were estimated using the averages of the horizontal geomagnetic components in order to carry out azimuthal corrections. The thus corrected EM time-series were further processed by the robust remote reference MT response estimator in frequency domain (Chave et al., 1987) to yield MT impedance tensors in Eq. (3). The seafloor MT response functions, together with those at four sites on land, were used in the subsequent 2-D inversion to explain their spatial distribution as well as the frequency dependence.

To construct a 2-D electrical model of northeast Japan, a Reduced Basis OCCam's (REBOCC) inversion method (Siripunvaraporn & Egbert, 2000) was applied to the land and sea MT impedances observed at the latitude of 39.5N. The REBOCC inversion is a variant of Occam inversion (Constable et al., 1987), which works with sensitivity matrices in data space instead of the conventional model space. This significantly reduces the size of the sensitivity matrices required in the course of the inversion procedure. Because the REBOCC inversion prefers high correlation of the final model with a priori model, it was possible to build in the basic tectonic model of northeast Japan such as the presence of the thick and resistive subducting Pacific plate. The known 2-D section for crustal depths (Ogawa et al., 2001) was also included in the REBOCC inversion as another a priori information. The inversion converged at an rms of 3.55 using both TE and TM mode responses.

The derived electrical 2-D section (Toh et al., 2006), which is an EW slice of the non-volcanic part of northeast Japan, reveals a resistive shallow mantle and a conductive anomaly beneath the back-arc region at depths 150-200 km (Fig. 2). The electrical conductivity anomaly can be interpreted as a direct manifestation of slab dehydration associated with collapse of the high-temperature type serpentine (Iwamori, 1998) rather than that of a group of minor hydrous phases such as phlogopite (Tatsumi, 1989). To test the robustness of the anomaly, we examined changes in the rms when the anomaly was replaced by a normal and uniform mantle conductivity of $3.3\text{x}10^{-2}$ S/m. It turned out that the rms increase is too large to explain the observed MT data by the normal conductivity if we mask the anomaly surrounded by the black-dashed lines in Fig. 2. However, the increase was marginal if we do the same thing for the anomaly surrounded by the white-dashed lines. We, therefore, concluded that the anomaly surrounded by the black-dashed lines is required by the MT data.

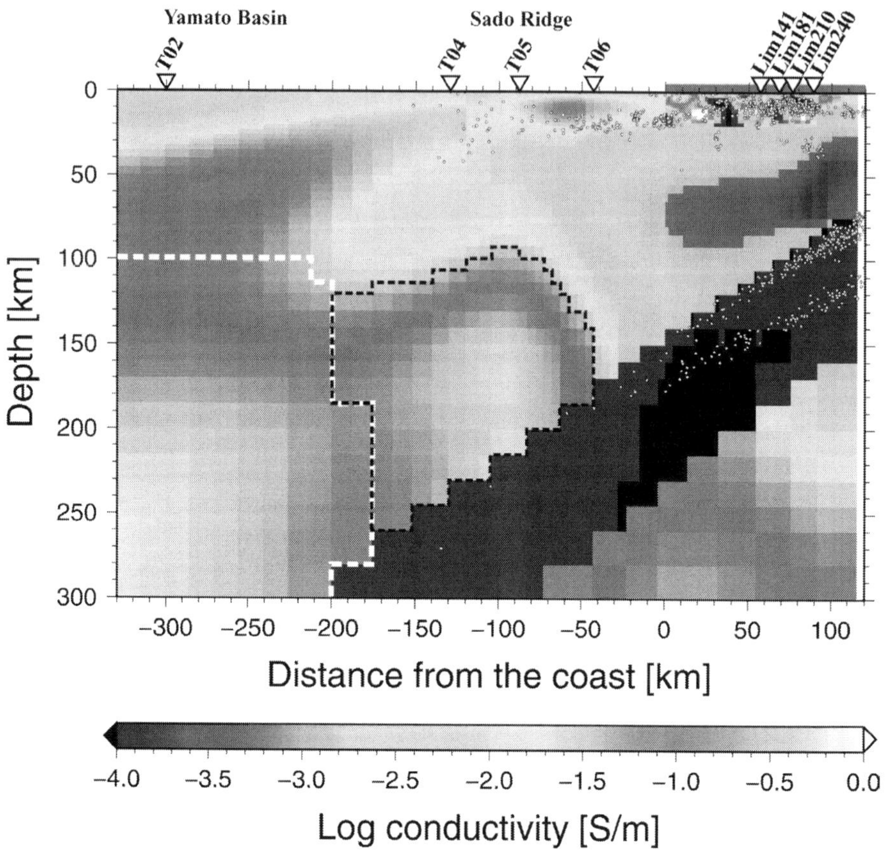

Fig. 2. The 2-D electrical conductivity model of northeast Japan at the latitude of 39.5N. The inverted triangles with labels indicate the locations and names of the observation sites. The horizontal gray bar represents the island arc of northeast Japan, while dots denote hypocenters in this region. The areas surrounded by black- and white-dashed lines are those masked in the F-tests (See text for details). Reproduced from Toh et al. (2006).

The Pacific plate is subducting beneath New Zealand as well. Wannamaker et al. (2009) derived a 2-D electrical cross-section beneath the South Island of New Zealand using a wide-band MT dataset on a densely distributed profile perpendicular to the island arc strike. They found three conductivity anomalies beneath the fore-arc region, the volcanic front and the back-arc region. The fore-arc conductor can be regarded as a natural result of dehydration from a younger and thus relatively hot subducting plate. The age of the Pacific plate there is approximately twice as young as northeast Japan (70-75 Myr). The conductor at the volcanic front is no wonder if the MT transect traverses a volcanic part of the island arc. However, there is a significant difference in the depth of dehydration beneath the back-arc region. They concluded the back-arc dehydration to occur at depths ranging from 75 to 100 km and attributed the process to breakdown of amphibole-zoisite. It is natural that the depth of the back-arc dehydration as well as the collapsing minerals at that depth differs in the case of northeast Japan and the South Island of New Zealand, since the different ages mean different thermal effects of the subducting plates on the wedge mantle. In the case of the South Island of New Zealand, the temperature may be too high for the hydrous minerals to penetrate deep into the mantle beneath its back-arc region.

4. EM view of the seismogenic zone beneath southwest Japan

Southwest Japan makes a good contrast to northeast Japan in terms of Volcanology, Seismology and Geomagnetism in the sense that:

1. The subducting slab is much younger (and thus warmer) than that of northeast Japan (e.g., Iwamori, 1998).
2. Not only the structure but also the source of volcanism seems 3-D with its peculiar distribution of volcanoes and multiple sources for the magma production (e.g., Iwamori, 1991; Kimura et al., 2003).
3. The region shows a more distinct relation of its seismic and volcanic activity to 'geofluid' typically appearing to the very linear distribution of the upper crustal hypocenters along the Japan Sea coast (Kawanishi et al., 2009), which coincide well with the volcanic front and the lower crustal conductors. It is also noteworthy that deep low frequency events, which are also signatures of the presence of 'geofluid', occur in this region as well (e.g., Obara & Hirose, 2006).

In the following, we will try to illustrate how the magma source can be multiple in southwest Japan and how the subducting young slab influences the seismogenic zone beneath southwest Japan by reviewing an EM study in and around the Kyushu Island and the ongoing field work in the back-arc region of southwest Japan.

4.1 Mantle plume in the west of Kyushu Island

It has been long known that if one calculates GDS responses using short-period vector geomagnetic variations observed in southwest Japan, they end up with $|p(f)| << |q(f)|$ and $q(f) > 0$ where $p(f)$ and $q(f)$ are the transfer functions (i.e., the GDS responses) appeared in Eq. (1). This implies that there exists a prominent electrical conductivity anomaly in the west of southwest Japan. In order to identify the intensity of the conductivity anomaly and its spatial extent, a genetic algorithm inversion (e.g., Sambridge & Drijkoningen, 1992) of the observed GDS responses using non-uniform thin sheet approximation (McKirdy et al., 1985) was applied.

The genetic algorithm inversion converged at an rms of 3.20 after 112 iterations. Because we worked with 50 models per iteration, the best model in Fig. 3 is the result of more than 5000 forward calculations using the non-uniform thin-sheet approximation. It is evident that the model can give no constraints on the spatial extent both in westward and southward directions. This is due to the spatial distribution of the original GDS dataset that were mainly collected on land in southwest Japan. Direct EM measurements at the seafloor both in the south and west of the model in Fig. 3 will be indispensable for any further improvement in spatial resolution of the derived model.

The surface conductance (a product of the layer thickness and its electrical conductivity) anomaly model in Fig. 3 has two intriguing features in terms of arc volcanism if the distribution can be interpreted as that of a subsurface mantle plume:

1. The plume can be one candidate of the multiple magma sources suggested by Iwamori (1991).
2. There seems to exist a short volcanic chain that branches out from the main volcanic front on the Kyushu Island.

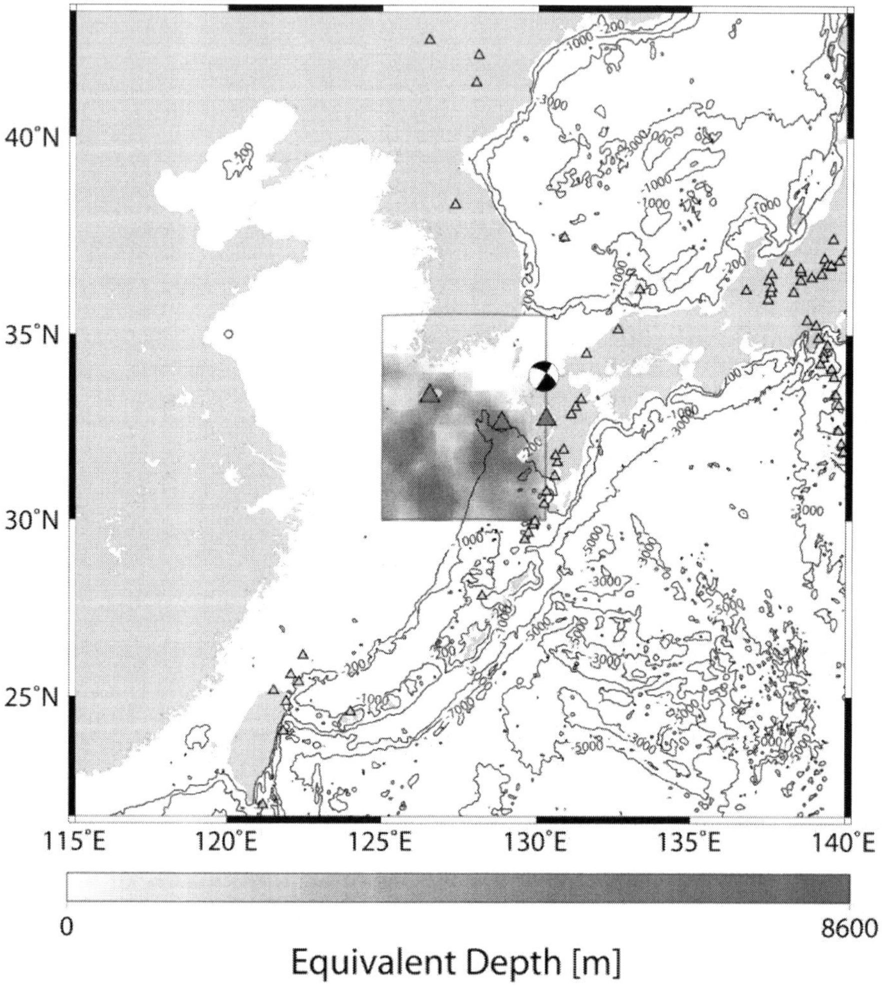

Fig. 3. Result of the genetic algorism (GA) inversion in search for the optimized conductance distribution within the surface thin sheet. Additional conductance [S] required by the GA inversion is shown by the white-to-red scale. Bathymetric contours are also shown. Triangles denote Quarternary volcanoes in the East Asia (yellow) and on the small volcanic branch (magenta) such as the Unzen Volcano and the Cheju Island. The focal mechanism of the main shock of the west off Fukuoka Prefecture earthquake (Mw 6.7) is also shown. Reproduced from Toh & Honma (2008).

Presence of a regional-scale mantle plume in the middle of the East China Sea has been favoured by many researchers (e.g., Ichiki et al., 2006) since Miyashiro (1986) first claimed its existence. It is noteworthy that the geological strike of the volcanic branch is nearly parallel to the northeastern boundary of the partly discovered anomaly. A vertical slice of a seismic tomography result (Zhao et al., 2000) cutting through the northern Kyushu Island from the northwest to southeast direction has also imaged a low velocity branch toward the back-arc region. Another interesting evidence is the occurrence of the west off Fukuoka Prefecture earthquake near the edge of the electrical anomaly. Even though the direct cause of the earthquake is probably due to the extensional regional stress field in the back-arc region of the Kyushu arc induced by the mantle upwelling (Wei & Seno, 1998), the focal mechanism implies a lateral motion between the Eurasian plate and the Amurian plate.

4.2 2-D electrical section of southwest Japan

One major characteristic of the seismicity in southwest Japan is that the epicenters tend to concentrate within a belt of about 4-9 km wide parallel to the coast line of the Japan Sea (Kawanishi et al., 2009). Most of the focal depths there are shallower than approximately 10km and thus the earthquakes are occurring in the upper crust. In the seismic belt, several large earthquakes of M6.2-7.4 also occurred in 1943, 1983 and 2000, respectively. Furthermore, quaternary volcanoes, such as the Daisen and Oginosen Volcanoes are located in the seismic belt while the basalt that formed the Oki Islands in the back-arc region is much older (> 5 Myr) and of different composition (Kimura et al., 2003).

Wide-band MT observations have been made along a number of north-south profiles on land since 1998 so as to reveal high conductivity regions beneath the seismic belt on each MT profile. It is noteworthy that the earthquakes seem to occur on the boundary between the upper resistive crust and the highly conductive body in the lower crust. The high conductivity regions found beneath each wide-band MT profile may form a connected conductive zone extending in an almost east-west direction. Coincidence of the hypocenter distribution with the upper surface of the conductive zone as well as the presence of deep low-frequency events suggests that crustal fluid must involve the focal mechanism in the seismogenic zone.

In order to clarify the relation among the mantle dynamics in the back-arc region, the lower crustal conductors found on land and the complicated volcanism, seafloor EM observations were conducted off southwest Japan together with MT measurements on land. We laid out two seafloor MT arrays, one traversing the non-volcanic region in the eastern part of southwest Japan and the other running through a volcanic ridge including the Oki Islands. These seafloor arrays are indispensable to image the subducting Philippine Sea plate, a possible source of the crustal fluid and seismicity of the region.

Figure 4 shows the result of 2-D finite element forward modeling of the eastern profile, which ended up with an rms of 3.3. Note that the electrical section is a product of not inversion but forward modeling, although the finite element code was improved to give high precision even at locations very close to the coastline as well as those on the rugged bathymetry/topography. It was not until a superior differential scheme (Li et al., 2008) and triangular elements suitable for describing complicated bathymetry/topography had been adopted that the good precision in forward calculation was achieved.

The 2-D model shows that the lower crustal conductor has seaward extension at least more than 30 km further north of the coastline. Because we were unable to identify the tip of the subducted Philippine Sea plate beneath the volcanic front of the profile, it is unlikely that the lower crustal conductor beneath the volcanic front stems from the slab melting at least for this particular portion of the island arc. Another major feature of the model is that it has a deep (> 100 km) conductor in the back-arc region. This conductor may be attributed to the magma source for the volcanism that made the Oki Islands. However, it is difficult to relate the conductor to the slab melting or the dehydration from the Philippine Sea plate, since the plate initiated its subduction too recently to allow itself enough penetration toward the back-arc region. It is more appropriate to regard it as a result of mantle upwelling from the deeper slab, i.e., the Pacific plate, whose subduction beneath the back-arc region of southwest Japan has been clearly imaged by recent seismic tomography studies (e.g., Nakajima & Hasegawa, 2007).

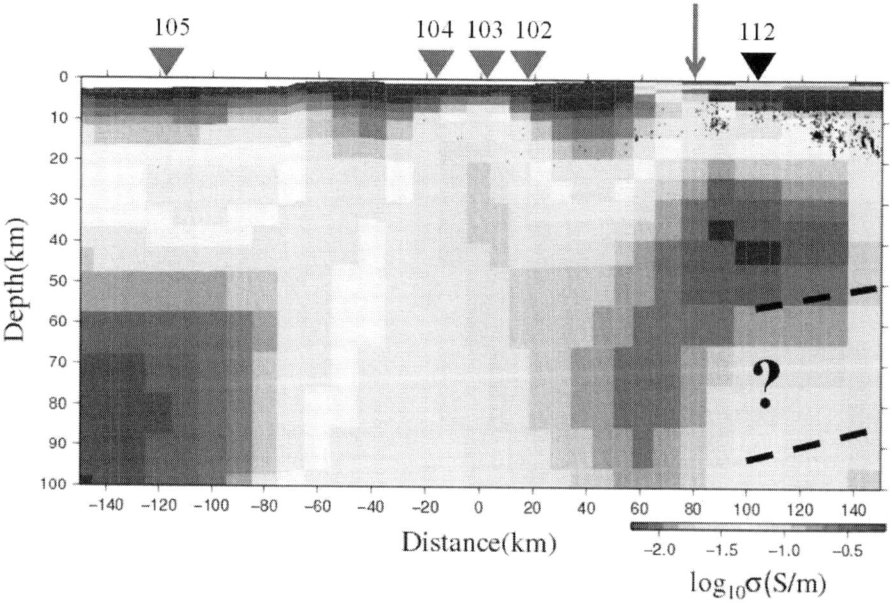

Fig. 4. The 2-D electrical section around the land-sea boundary of southwest Japan. The red vertical arrow indicates the location of the coastline of the Japan Sea. The red and black inverse triangles are the seafloor and land EM observation site, respectively. Small black dots show the distribution of hypocenters. Estimated location of the edge of the subducted Philippine Sea plate is also shown by thick dashed lines.

5. Conclusion

The very cold and thick subducting plate beneath northeast Japan can supply water deep into the back-arc region (Iwamori, 1998), which forms 3-D counter flows to generate arc volcanism of that part of the Japanese Islands (Tamura et al., 2002). Because of this scenario, the magmatic source of northeast Japan can be simple enough to be approximated by 'uni-source' magmatism even though the magmatic structure itself can remain 3-D. The seismogenic zone beneath northeast Japan is governed by its regional stress field rather than by the presence of 'geofluid'. Major earthquakes of this region are mostly related to the processes involved with the plate boundary. However, the geofluid in northwest Japan is important for generation of hazardous inland earthquakes as well as water circulation in the wedge mantle.

The scenario by the warm and thin subducting plate beneath southwest Japan is more complicated than that of northeast Japan. In terms of Volcanology, the structure as well as the magma source is 3-D in the sense that there are several firm evidences for the presence of multiple magma sources (Iwamori, 1991; Toh and Honma, 2008) and 3-D seismic structures (Nakajima & Hasegawa, 2007). Because the subducting plate is too warm to carry the surface water deep into the mantle, there occurs major dehydration from the slab beneath the fore-arc region that causes much more geofluid-related seismic activities in southwest Japan (e.g., Obara & Hirose, 2006) than in northeast Japan. The young slab commenced its subduction several million years ago to have the penetration of its edge as close as just beneath the Japan Sea coast. The shallow penetration resulted in production of adakite magmas, which is a signature of slab melting and gives the volcanism in southwest Japan further complexity. However, it also turned out that the mantle upwelling in the back-arc region of southwest Japan is governed by not the warm slab but the cold slab further below the warm slab.

6. Acknowledgment

We are indebted to captains and crew members of R/V Tansei-Maru, Ocean Reserach Institute, University of Tokyo for their skilful help at the time of the sea experiments in both back-arc regions of northeast and southwest Japan. The geomagnetic data presented in the subsection 4.1 were kindly provided by Geographical Survey Institute, Japan. Earthquake Research Institute, University of Tokyo is acknowledged for its continuous support throughout this work. The land and seafloor EM data necessary for obtaining the 2-D electrical section of southwest Japan were kindly provided by Disaster Prevention Research Institute at Kyoto University, Japan Agency for Marine-Earth Science and Technology and Kyushu Polytechnic College.

7. References

Bercovici, D. & Karato, S. (2003). Whole-mantle convection and the transition-zone water filter. *Nature*, Vol.425, (September 2003), pp. 39–44, doi:10.1038/nature01918

Booker, J.R.; Favetto, A. & Pomposiello, M.C. (2004). Low electrical resistivity associated with plunging of the Nazca flat slab beneath Argentina. *Nature*, Vol.429, (May 2004), pp. 399-403, doi:10.1038/nature02565

Chave, A.D.; Thomson, D.J. & Ander, M.E. (1987). On the robust estimation of power spectra, coherences, and transfer functions. *Journal of Geophysical Research*, Vol.92, (January 1987), pp. 633– 648, doi:10.1029/JB092iB01p00633

Constable, C.S., Parker, R.L. & Constable, C.G. (1987). Occam's inversion: A practical algorithm for generating smooth models from electromagnetic sounding data. *Geophysics*, Vol.52, (March 1987), pp. 289–300

Ferguson, I.J.; Lilley, F.E.M. & Filloux, J.H. (1990). Geomagnetic induction in the Tasman Sea and electrical conductivity structure beneath the Tasman Seafloor. *Geophysical*

Journal International, Vol.102, (August 1990), pp. 299-312, doi: 10.1111/j.1365-246X.1990.tb04468.x

Hirano, N.; Takahashi, E., Yamamoto, J., Abe, N., Ingle, S.P., Kaneoka, I., Hirata, T., Kimura, J., Ishii, T., Ogawa, Y., Machida, S., Suyehiro, K. (2006). Volcanism in response to plate flexure. *Science*, Vol.313, (September 2006), pp. 1426-1428, doi: 10.1126/science.1128235

Huang, X.; Xu, Y. & Karato, S. (2005). Water content in the transition zone from electrical conductivity of wadsleyite and ringwoodite. *Nature*, Vol.434, (April 2005), pp. 746–749, doi:10.1038/nature03426

Ichiki, M.; Baba, K., Obayashi, M. & Utada, H. (2006). Water content and geotherm in the upper mantle above the stagnant slab: Interpretation of electrical conductivity and seismic P-wave velocity models. *Physics of the Earth and Planetary Interiors*, Vol.155, (April 2006), pp. 1 - 15, doi:10.1016/j.pepi.2005.09.010

Inoue, T.; Yurimoto, H. & Kudoh, Y. (1995). Hydrous modified spinel, $Mg_{1.75}SiH_{0.5}O_4$: a new water reservoir in the mantle transition region. *Geophysical Research Letters*, Vol.22, (January 1995), pp. 117–120

Iwamori, H. (1991). Zonal structure of Cenozoic basalts related to mantle upwelling in southwest Japan. *Journal of Geophysical Research*, Vol.96, No.B4, (April 1991), pp. 6157-6170, doi:10.1029/90JB02399

Iwamori, H. (1998). Transportation of H_2O and melting in subduction zones. *Earth and Planetary Science Letters*, Vol.160, (July 1998), pp. 65- 80

Iwamori, H. (2004). Phase relations of peridotites under H_2O-saturated conditions and ability of subducting plates for transportation of H_2O. *Earth and Planetary Science Letters*, Vol. 227, (September 2004), pp. 57- 71, doi:10.1016/j.epsl.2004.08.013

Karato, S. (1990). The role of hydrogen in the electrical conductivity of the upper mantle. *Nature*, Vol.347, (September 1990), pp. 272-273, doi:10.1038/347272a0

Kaufman, A.A. & Keller, G.V. (1983). *Frequency and transient soundings*, Elsevier, ISBN 0-444-42032-0, Amsterdam, Netherlands

Kawanishi, R., Iio, Y., Yukutake, Y., Shibutani, T. & Katao, H. (2009). Local stress concentration in the seismic belt along the Japan Sea coast inferred from precise focal mechanisms: Implications for the stress accumulation process on intraplate earthquake faults. *Journal of Geophysical Research*, Vol.114, (January 2009), B01309, doi:10.1029/2008JB005765

Kimura, J.; Kunikiyo, T., Osaka, I., Nagao, T., Yamauchi, S., Kakubuchi, S., Okada, S., Fujibayashi, N., Okada, R., Murakami, H., Kusano, T., Umeda, K., Hayashi, S., Ishimaru, T., Ninomiya, A. & Tanase, A. (2003). Late Cenozoic volcanic activity in the Chugoku area, southwest Japan arc during back-arc basin opening and reinitiation of subduction. *The Island Arc*, Vol.12, (March 2003), pp. 22–45, doi: 10.1046/j.1440-1738.2003.00377.x

Li, S.; Booker, J.R. & Aprea, C. (2008). Inversion of magnetotelluric data in the presence of strong bathymetry/topography. *Geophysical Prospecting*, Vol.56, (January 2008), pp. 259-268, doi: 10.1111/j.1365-2478.2007.00677.x

McKirdy, D.M.A.; Weaver, J.T. & Dawson, T.W. (1985). Induction in a thin sheet of variable conductance at the surface of a stratified Earth-II. Three-dimensional theory. *Geophysical Journal of the Royal Astronomical Society*, Vol.80, (January 1985), pp. 177–194, doi: 10.1111/j.1365-246X.1985.tb05084.x

Mishra, O.P.; Zhao, D., Umino, N. & Hasegawa, A. (2003). Tomography of northeast Japan forearc and its implications for interplate seismic coupling. *Geophysical Research Letters*, Vol.30, (August 2003), doi:10.1029/2003GL017736

Miyashiro, A. (1986). Hot regions and the origin of marginal basins in the western Pacific. *Tectonophysics*, Vol.122, (February 1986), pp. 195–216, doi:10.1016/0040-1951(86)90145-9

Nakajima, J. & Hasegawa, A. (2007). Tomographic evidence for the mantle upwelling beneath southwestern Japan and its implications for arc magmatism. *Earth and Planetary Science Letters*, Vol. 254, (January 2007), pp. 90 - 105, doi:10.1016/j.epsl.2006.11.024

Obara, K. & Hirose, H. (2006). Non-volcanic deep low-frequency tremors accompanying slow slips in the southwest Japan subduction zone. *Tectonophysics*, Vol.417, (January 2006), pp. 33-51, doi:10.1016/j.tecto.2005.04.013

Ogawa Y.; Mishina, M., Goto, T., Satoh, H., Oshiman, N., Kasaya, T., Takahashi, Y., Nisitani, T., Sakanaka, S., Uyeshima, M., Takahashi, Y., Honkura, Y. & Matsushima, M. (2001). Magnetotelluric imaging of fluids in intraplate earthquakes zones, NE Japan back arc. *Geophysical Research Letters*, Vol.28, (October 2001), pp. 3741-3744, doi:10.1029/2001GL013269

Sakamaki, T.; Suzuki, A. & Ohtani, E. (2006). Stability of hydrous melt at the base of the Earth's upper mantle. *Nature*, Vol.439, (January 2006), pp. 192–194, doi:10.1038/nature04352

Sambridge, M.S. & Drijkoningen, G.G. (1992). Genetic algorithms in seismic waveform inversion. *Geophysical Journal International*, Vol.109, (May 1992), pp. 323 -342, doi:10.1111/j.1365-246X.1992.tb00100.x.

Schultz, A. & Larsen, J. (1987). On the electrical conductivity of the mid mantle - I. Calculation of equivalent scalar magnetotelluric response functions. *Geophysical Journal of the Royal Astronomical Society*, Vol.88, (March 1987), pp. 733-761, doi: 10.1111/j.1365-246X.1987.tb01654.x

Siripunvaraporn, W. & Egbert, G. (2000). An efficient data-subspace inversion method for 2–D magnetotelluric data. *Geophysics*, Vol.65, (May 2000), pp. 791–803

Tamura, Y.; Tatsumi, Y., Zhao, D., Kido, Y. & Shukuno, H. (2002). Hot fingers in the mantle wedge: new insights into magma genesis in subduction zones. *Earth and Planetary Science Letters*, Vol.197, (March 2002), pp. 105-116, doi:10.1016/S0012-821X(02)00465-X

Tatsumi, Y. (1989). Migration of fluid phases and genesis of basalt magmas in subduction zones. *Journal of Geophysical Research*, Vol.94, (April 1989), pp. 4697-4707, doi:10.1029/JB094iB04p04697

Toh, H.; Baba, K., Ichiki, M., Motobayashi, T., Ogawa, Y., Mishina, M. & Takahashi, I. (2006). Two-dimensional electrical section beneath the eastern margin of Japan Sea. *Geophysical Research Letters*, Vol.33, (November 2006), L22309, doi:10.1029/2006GL027435

Toh, H. & Honma, S. (2008). Mantle upwelling revealed by genetic algorithm inversion of the magnetovariational anomaly around Kyushu island, Japan. *Journal of Geophysical Research*, Vol.113, (October 2008), B10103, doi:10.1029/2006JB004891

Wang, K.; Hyndman, R.D. & Yamano, M. (1995). Thermal regime of the Southwest Japan subduction zone: effects of age history of the subducting plate. *Tectonophysics*, Vol.248, (August 1995), pp. 53-69, doi:10.1016/0040-1951(95)00028-L

Wang, D.; Mookherjee, M., Xu, Y. & Karato, S. (2006). The effect of water on the electrical conductivity of olivine. *Nature*, Vol.443, (October 2006), pp. 977-980, doi:10.1038/nature05256

Wannamaker, P.E.; Caldwell, T.G., Jiracek, G.R., Maris, V., Hill, G.J., Ogawa, Y., Bibby, H.M., Bennie, S.B. & Heise, W. (2009). The fluid and deformation regime of an advancing subduction system; Marlborough, New Zealand. *Nature*, Vol.460, (August 2009), pp. 733-739, doi:10.1038/nature08204

Wei, D. & Seno, T. (1998). Determination of the Amurian plate motion, In *Mantle Dynamics and Plate Interactions in East Asia*, M. Flower et al., (Ed.), 337-346, AGU, ISBN 0875905293 9780875905297, Washington, D.C., USA

Worzewski, T.; Jegen, M., Kopp, H., Brasse, H. & Castillo, W.T. (2010). Magnetotelluric image of the fluid cycle in the Costa Rican subduction zone. *Nature Geoscience*, (December 2010), doi: 10.1038/ngeo1041

Yoshino, T.; Matsuzaki, T., Yamashita, S. & Katsura, T. (2006). Hydrous olivine unable to account for conductivity anomaly at the top of the asthenosphere. *Nature*, Vol.443, (October 2006), pp. 973-976, doi:10.1038/nature05223

Yoshino, T.; Manthilake, G., Matsuzaki, T. & Katsura, T. (2008). Dry mantle transition zone inferred from the conductivity of wadsleyite and ringwoodite. *Nature*, Vol.451, (January 2008), pp. 326-329, doi:10.1038/nature06427

Zhao, D.; Asamori, K. & Iwamori, H. (2000). Seismic structure and magmatism of the young Kyushu subduction zone. *Geophysical Research Letters*, Vol.27, (July 2000), pp. 2057-2060, doi:10.1029/2000GL011512

10

S Wave Velocity Structure in Non-Tectonic SE Asia by Seismogram Analysis of the Earthquakes in Sumatra-Java at TATO Station, Taiwan

Bagus Jaya Santosa
Physics Dept., Institut Teknologi Sepuluh Nopember, Kampus ITS, Surabaya Indonesia

1. Introduction

The Earth's solid mantle is about 2900-km thick and divided into upper and lower mantle (Dziewonski and Anderson, 1981; Kennett, 1991). The lower mantle comprises the region located between the seismically defined discontinuity at 670-km depth and the core–mantle boundary (CMB) at 2900-km depth. It represents 70% of the volume of the mantle and is still being a controversy in Solid Earth Geophysics, whether the convection is the driving force for the plate tectonics movement. A result of plate tectonic movement is the collision between the earth plates, for example the small India and Asia continents, following the closure of the Thetys Ocean in the Mesozoic time. The collision created the Himalaya and Plato Tibet mountain ranges, also induced an extensive slacking strain in South East Asia and China. Due to the slacking strain, Replumaz et al. (2004), Engdahl et al. (1998) and Grand et al. (1997) have interpreted that the earth structures below the area of South China Sea and South East Asia have a negative anomaly of the P wave velocity. Romanowicz (2003) has also interpreted the S wave velocity beneath South East Asia with negative S anomaly in the upper 410 km earth mantle and without anomaly in the mantle layers below, refer to fig. 1.

The western part of Indonesia, especially Java and Sumatra Islands, is an area with a complex tectonic-condition. In this area, the Indian Ocean plate collides with the continental plate. The ridge in the Antarctic sea produces new rocks in lateral direction toward north and south, as a part of the mantle rotation. Since the new produced rock expands laterally, the new rocks push the older plates of ocean in the lateral direction of the ridge field. This pushes the Indian Ocean and Australia Continent to the north, until they collided with the shelf of the South East Asia, and subducted ocean plate under the Eurasia continent. The region in this research lies also in the front area of subduction zone due to collision between the Pacific plate and the Asian plate. The ocean plate moves in an average velocity of 11 cm/year. The movement is still ongoing, and yield big tectonic earthquakes. It is characterized by the occurrence of earthquakes in this area.

In recent years, tomography is routinely used to process the travel time data from seismic waves, such as International Seismological Center (ISC). This routine especially has become very successful in the mapping on the submergence of the cold lithosphere material into the mantle, along and under the active trench from the edge of the big plates, in three dimensions.

According to Replumaz et al. (2004), the front area of the subduction field in the continental shelf (Java and Sumatra) experiences a compression and the P wave structure on the upper mantle layers has a positive anomaly. Such velocity structure can be found by inverting the travel-time data of direct P wave amounting to $\propto 8 \times 10^6$, the reflected wave phase Pp amounting to $\propto 0.6 \times 10^6$, and biased in the earth core PKP amounting to almost 1×10^6. This amount of data are collected from 300.000 earthquakes in the time interval from 1/1/1964 to 31/12/2000 (Li et al., 2006); in the same vein, a small amount of the absolute difference of travel times PP-P, PKP-P data $_{diff}$, is accurately measured by the cross correlation of waveform from the broadband digital waveform data (Grand et al., 1997). Fig. 1 presents the interpreted earth structure beneath SE Asia using the P travel time (Romanowicz, 2003).

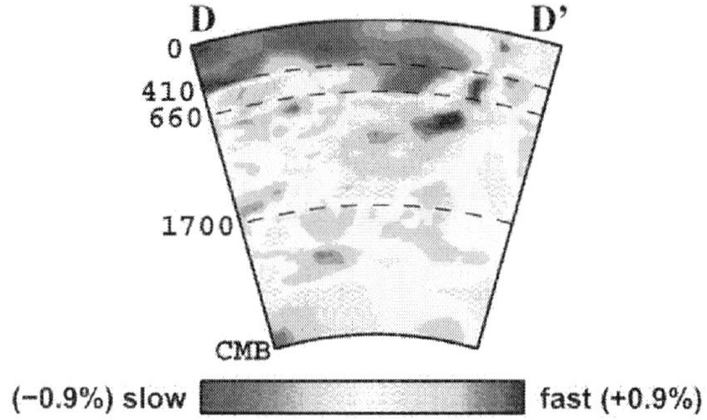

Fig. 1. Depth cross section DD' in front area of subduction zone area in Java-Sumatra, as revealed by P travel-time tomography using ISC data (Romanowicz, 2003)

I present the analysis result of the S wave, the Love and Rayleigh surface waves, to evaluate the nature of mantle below South East Asia and South China Sea, by analyzing the seismogram from the earthquakes that occurred on the surface of Java-Sumatra subduction zone that were recorded in the observatory station TATO, Taiwan. The objective of this research is to understand the S wave velocity structure beneath SE Asia using seismogram fitting method on the S and surface waves, in time domain and three components simultaneously.

The synthetic seismogram is calculated using GEMINI Program (Dalkolmo, 1993; Friederich and Dalkolmo, 1995), which is equivalent with Normal Mode method, but the difference is in the independent variable used. GEMINI uses complex frequency, instead of real frequency as in the Normal Mode.

2. Data

Table 1 presents the epicenter location of the analyzed earthquakes. The earthquake set consists of three earthquakes in North Sumatra Coast, five earthquakes in South Sumatra and Sunda Strait, and two earthquakes in South Java.

Fig. 2 indicates wave propagation from the earthquake hypocenters in the subduction zone to the TATO station. The waves pass through the western side of Southeast Asia that lies on the

No	Earthquake Code	Latitude	Longitude
1	B031501B	8.66	94.01
2	C103194B	3.03	96.27
3	C110895A	1.44	95.59
4	C090100C	1.44	96.59
5	C011601D	-4.02	101.78
6	B020893B	-4.86	101.96
7	B082400C	-6.03	102.69
8	B020399B	-6.19	104.22
9	C011502D	-6.31	105.21
10	C060394F	-10.49	112.87
11	B061594F	-10.28	113.85

Table 1. Codes and epicenter location of the earthquakes in Sumatra-Java that are analyzed in TATO observatory station

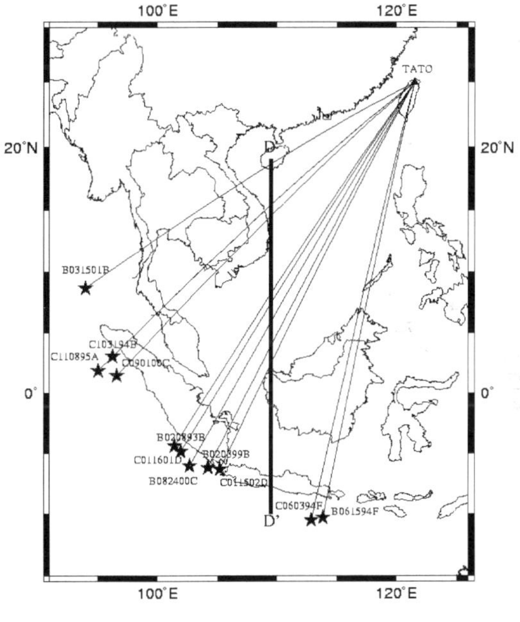

Fig. 2. Vertical projection of ray paths from epicenter to observatory station in TATO and line cross section DD′

front side of the Philippines subduction zone. The wave propagating medium between the earthquakes epicenter to TATO station traverses this non-tectonic area, which experiences a slacking strain as a result of the convergence between the India Sub-continent and the Eurasia Sub-continent plates.

3. Seismogram analysis and discussion

3.1 Seismogram analysis and fitting

Figure 3 presents the seismogram fitting of the B031501B earthquake, Nicobar archipelago that was recorded in TATO station. We can see that the PREMAN (the vertical anisotropic version of PREM (Dziewonski and Anderson, 1981)) global earth model predicts synthetic Love waveform that arrives rather earlier than the measured Love waveform, and the observation on the synthetic wave Rayleigh also shows earlier arrival time than the measured one. The corrected earth model gives a good synthetic seismogram on the Love and Rayleigh surface waves and also the SV and SH body waves. It is obtained by changing the gradient β_h in the upper mantle layers into positive and by giving a negative correction on the zero order polynomial coefficients that describe the β structure on the mantle layers up to 730 km depth.

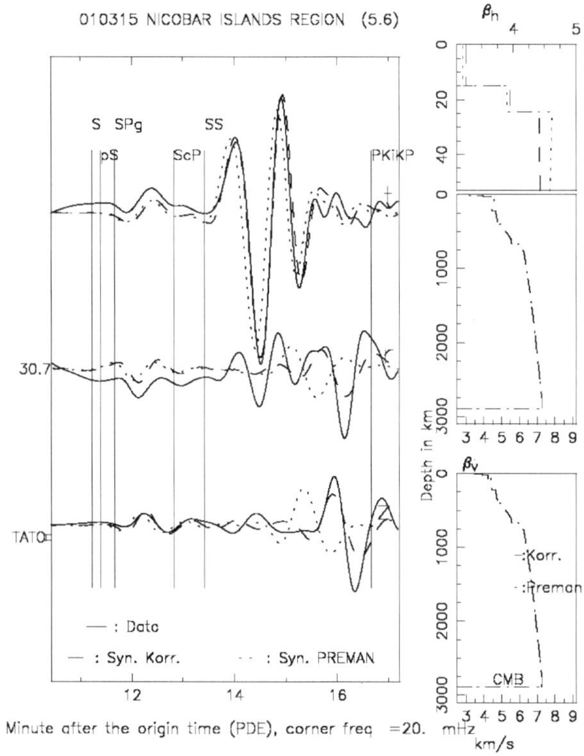

Fig. 3. The seismogram fitting of Nicobar B031501B earthquake in TATO.

Figure 4 indicates an illustration about the seismogram fitting of the C103194B earthquake, North Sumatra at TATO station. We see in the spheroidal components (r and z components),

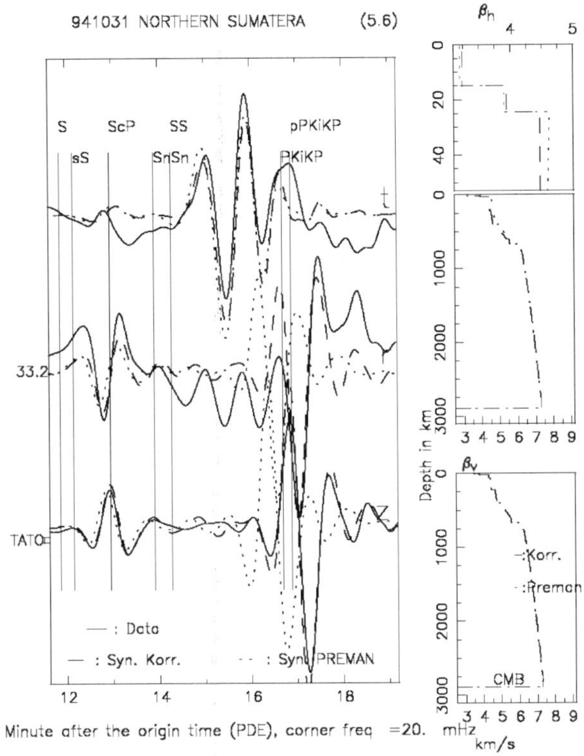

Fig. 4. The seismogram fitting of North Sumatra C103194B earthquake in TATO.

that the synthetic seismogram constructed from the PREMAN earth model arrives further earlier in the Rayleigh wave and arrives less early for S wave. It indicates that a great negative correction should be imposed on β_v in the upper mantle layers and also weaker negative correction on the mantle layers below until 730 km depth.

Figure 5 presents the seismogram analysis of the C110895A earthquake, the North Sumatra oceanic trench in TATO station. We can see that the synthetic seismogram from PREMAN on the Rayleigh arrives earlier than its measured waveform, while the Love waveform arrives a bit earlier. The synthetic seismogram of the corrected earth model agrees well with the measured data for Rayleigh wave, and also for the SV wave. In the t component, the Love waveform agrees better with the seismogram constructed from corrected model.

The seismogram analysis and fitting of the C090100C earthquake, North Sumatra trench, can be seen in Fig. 6. Since the data quality on t component is poor, the analysis is carried out just on the Rayleigh and SV wave in the r and z components. The zeroth order coefficients of β_v velocity function requires a big negative correction on the upper mantle layers, and smaller negative correction on the mantle layers below.

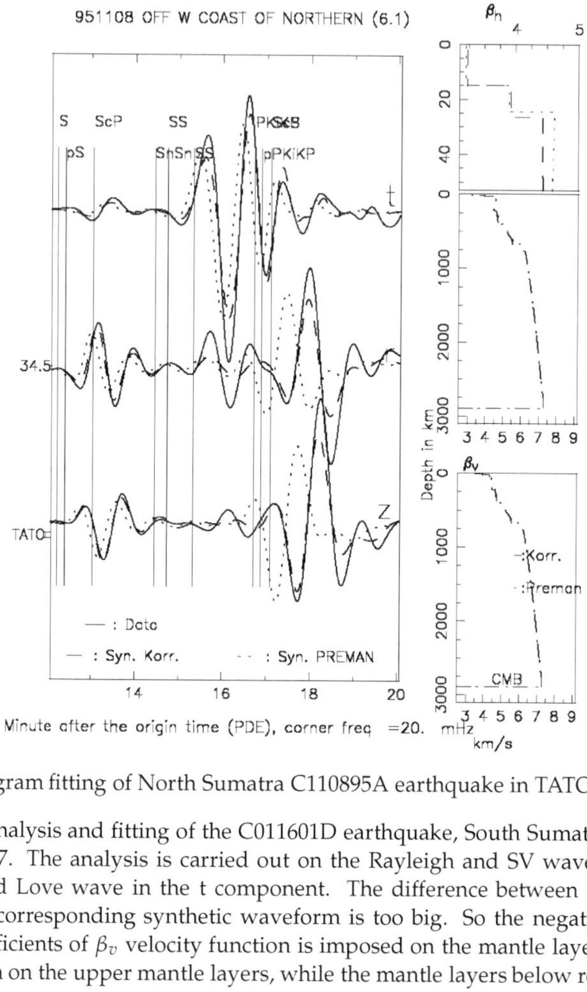

htb]

Fig. 5. The seismogram fitting of North Sumatra C110895A earthquake in TATO.

The seismogram analysis and fitting of the C011601D earthquake, South Sumatra trench, can be seen in Figure 7. The analysis is carried out on the Rayleigh and SV wave in the r and z components, and Love wave in the t component. The difference between the measured Rayleigh and the corresponding synthetic waveform is too big. So the negative correction on zero order coefficients of β_v velocity function is imposed on the mantle layers, with a big negative correction on the upper mantle layers, while the mantle layers below require a weak negative correction.

Figure 8 presents a seismogram analysis and explanation of the B020893B earthquake that occurred in the South Sumatra trench, where the seismogram was recorded in TATO station. We can see that the amplitude of the SV synthetic wave approaches the amplitude of the SV measured wave. But the amplitude of the synthetic Rayleigh wave is much higher than that of the measured Rayleigh. Therefore, a fitting is carried out in order to find a good fitting on the travel time of the maximum Rayleigh amplitude. The discrepancies in wave amplitude height were resulted from inaccuracy in the solution determination of the CMT earthquake (Dreger, 2002).

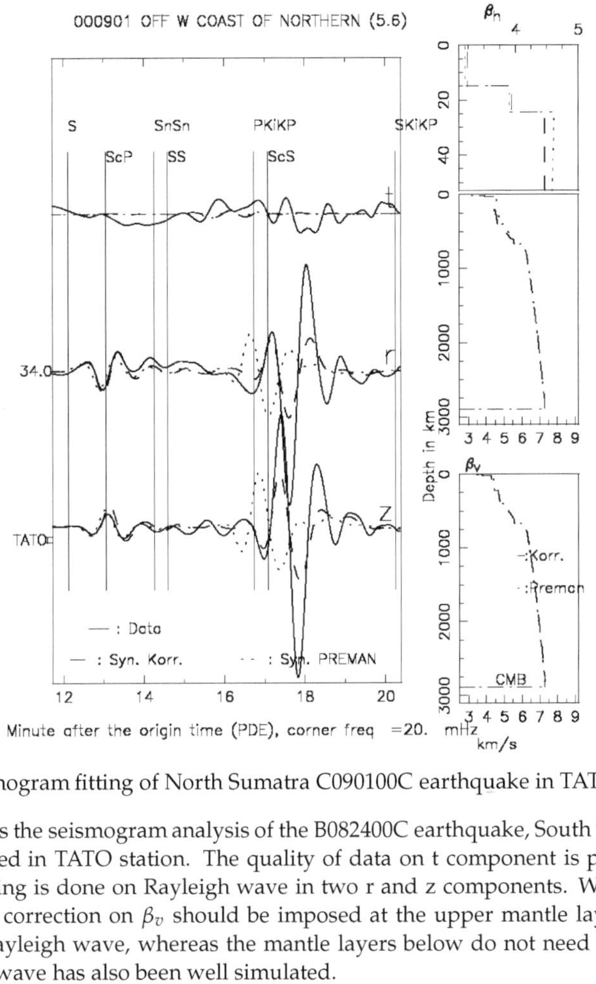

Fig. 6. The seismogram fitting of North Sumatra C090100C earthquake in TATO.

Figure 9 presents the seismogram analysis of the B082400C earthquake, South Sumatra Trench that was recorded in TATO station. The quality of data on t component is poor, so that the seismogram fitting is done on Rayleigh wave in two r and z components. We can see that a strong negative correction on β_v should be imposed at the upper mantle layers to obtain a fitting on the Rayleigh wave, whereas the mantle layers below do not need to be corrected, because the SV wave has also been well simulated.

The analysis and fitting of the B020399B seismogram that occurred in Sunda Strait which is recorded in TATO station, is illustrated in Figure 10. The synthetic seismogram that constructed from the PREMAN earth model has a different Rayleigh waveform from the measured waveform. They arrive earlier, and with the different amplitude heights distribution in the first and second maximums of Rayleigh wave; whereas the measured Rayleigh has a smaller decaying trend. The corrected earth model provides a better fitting on the pattern distribution of the maximums of the Rayleigh wave and the arrival time that approaches the measured Rayleigh wave. It is obtained by imposing a correction on β_v, and by using the positive β_h gradient on the upper mantle layers.

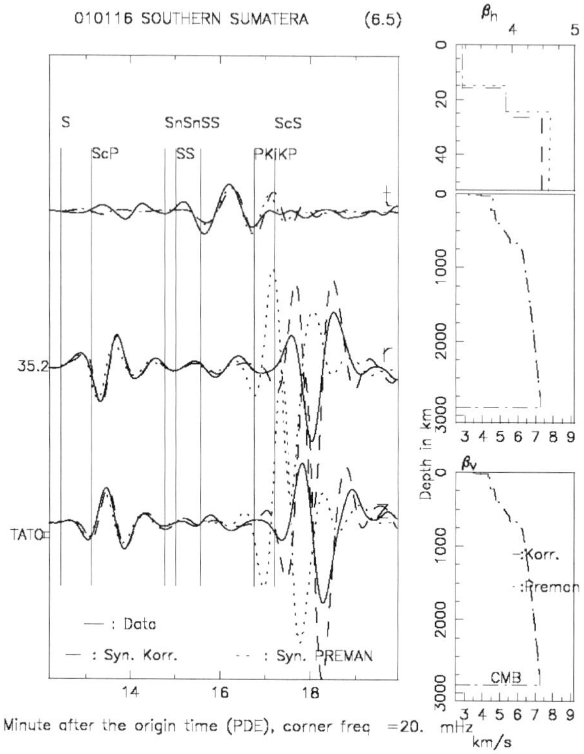

Fig. 7. The seismogram fitting of North Sumatra C011601D earthquake in TATO.

Figure 11 presents the seismogram analysis and fitting of the C011502D earthquake, Sunda strait, which is recorded in TATO station. We can see that the corrected earth model provides a synthetic seismogram that resembles more closely to the measured seismogram, than the synthetic seismogram that is constructed from the PREMAN earth model. The main negative correction is imposed on β_v on the upper mantle layers.

Figure 12 presents seismogram analysis and comparison of the C060394F earthquake, the south Java trench that occurred on June 3rd, 1994. We can see that the fitting of the synthetic seismogram calculated from the corrected earth-model approaches the measured seismogram well, both on the beginning of the Love wave (constituting the SS wave), the Rayleigh wave and the SV and SH. The big amplitude at the end part of the Love wave cannot be simulated by both synthetic waves.

Figure 13 indicates a comparison of the measured seismogram and the synthetic one from a B061594F earthquake that occurred in the South Java trench in June 15th 1994, twelve days after the earthquake that is presented in Figure 11. We can see the deviations of the synthetic seismogram constructed from the PREMAN earth model against the measured seismogram. The corrected earth model provides a synthetic seismogram that approaches

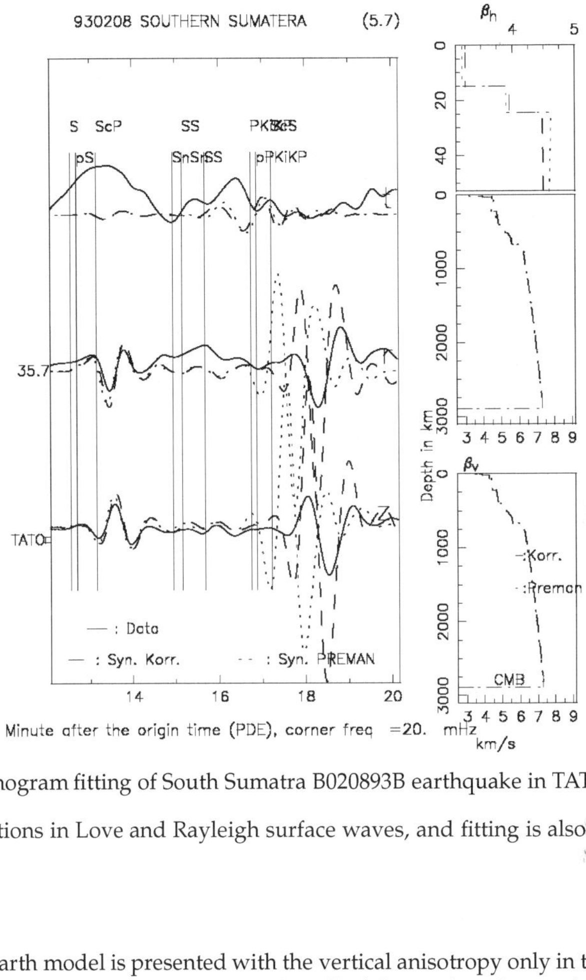

Fig. 8. The seismogram fitting of South Sumatra B020893B earthquake in TATO.

the main oscillations in Love and Rayleigh surface waves, and fitting is also achieved on the SV wave.

3.2 Discussion

The PREMAN earth model is presented with the vertical anisotropy only in the upper mantle layers. Following this research, we can see that all of the earthquakes analyzed indicate the earlier arrival time on the synthetic surface Rayleigh wave from the PREMAN model. This requires a negative correction on the velocity structure β_v, with a significant magnitude, where it indicates a greater vertical anisotropy than the one stated in the PREMAN earth model. To obtain a fitting on the S wave (SV and SH) on the seismograms of some earthquakes, we need a negative correction with a weaker magnitude on the mantle layers until 730 km depth. It indicates that the vertical nature of anisotropy also occurs in the mantle layer below the upper mantle. This anisotropy nature has not been used in the seismogram research that is based on the travel-time data. Result of this research complements the result from (Replumaz et al., 2004) and (Romanowicz, 2003) about the S wave velocity structure under South East Asia and South China Sea.

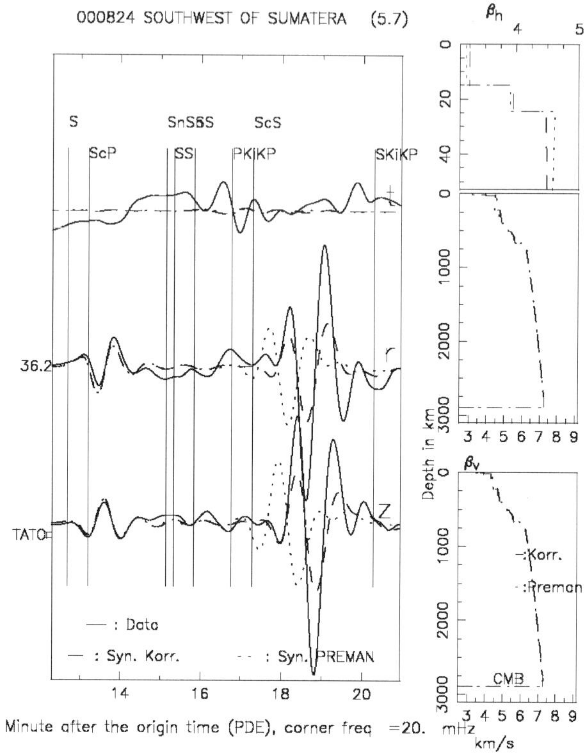

Fig. 9. The seismogram fitting of South Sumatra B082400C earthquake in TATO.

The epicenters of three earthquakes in North Sumatra coast lie close to each other, five earthquakes in Sunda Strait and South Sumatra also lie close each other, and two earthquakes in South Java Coast, indicate that the earth model beneath South-East Asia has S wave velocity anomaly which is big negative in upper mantle layers, and small negative in mantle layers till 730 km depth.

The CMT solution determination of earthquakes, that is done with CMT Routine (Dreger, 2002), relies on the isotropic earth model, and known as the 1066B earth model, in which the Green functions are calculated for various depths of the earthquake sources; and then reconstruction of the synthetic seismogram. The software in this package performs a time domain inversion of three-component seismic data for the seismic moment tensor. Source depth is found iteratively by finding the solution that yields the largest variance reduction between the observed and synthetic seismogram and the magnitude of the earthquake is obtained by Monte Carlo Method. The TDMT_invc (Dreger, 2002) uses still isotropic earth model to determine the magnitude and moment tensors of the earthquake, which after this research we found that the earth model should be described with an anisotropic earth model.

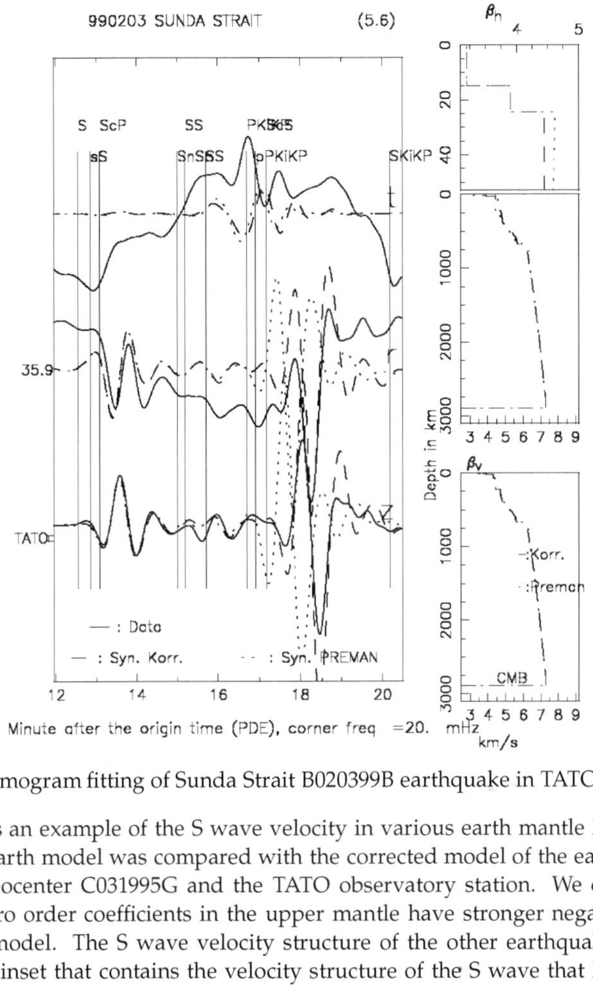

Fig. 10. The seismogram fitting of Sunda Strait B020399B earthquake in TATO.

Table 2 presents an example of the S wave velocity in various earth mantle layers, in which the PREMAN earth model was compared with the corrected model of the earth between the earthquake hypocenter C031995G and the TATO observatory station. We can see that the values of β_v zero order coefficients in the upper mantle have stronger negative terms than the PREMAN model. The S wave velocity structure of the other earthquakes can be seen by noticing the inset that contains the velocity structure of the S wave that lies on the right side of Fig.3 through Fig.13. We can see that the vertical anisotropy (η value) occurs in all of the mantle layers, not only in the upper mantle layer as stated in the PREMAN earth model. We notice further by comparing the second and third columns with the fifth and the sixth column, respectively, that the zero order coefficients of the S wave velocity generally has a negative anomaly in all layers of the mantle.

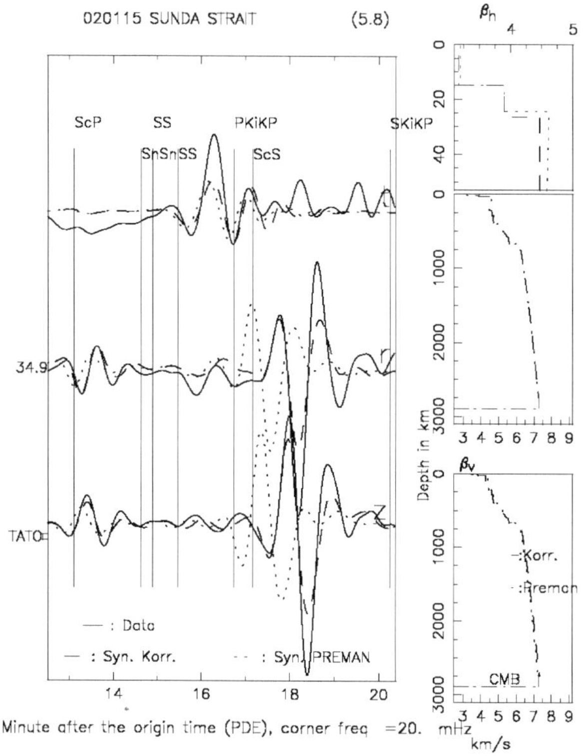

Fig. 11. The seismogram fitting of Sunda Strait C011502D earthquake in TATO.

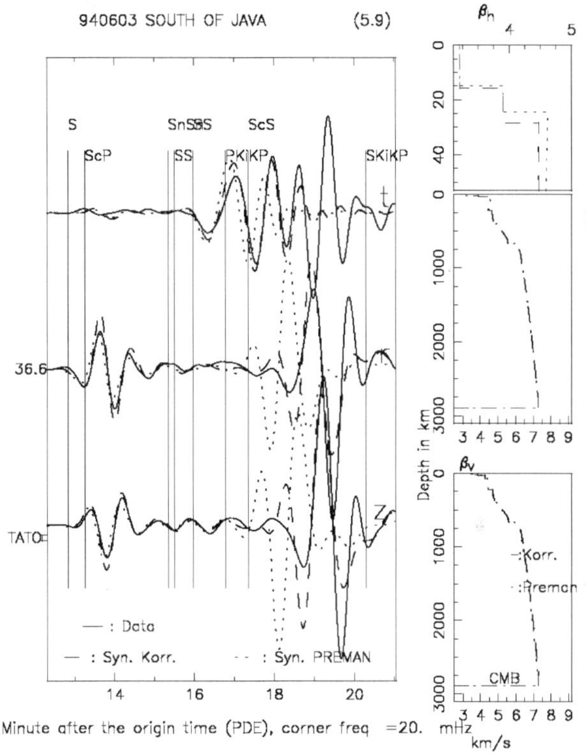

Fig. 12. The seismogram fitting of South Java trench C060394F earthquake in TATO.

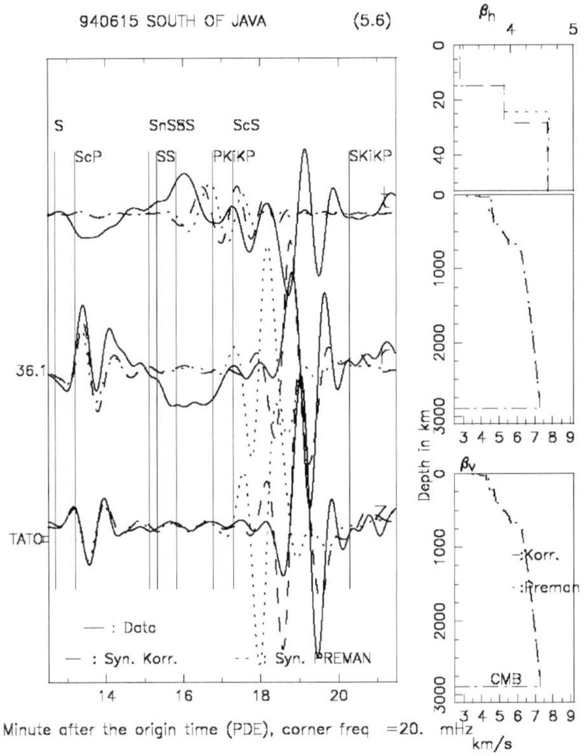

Fig. 13. The seismogram fitting of South Java trench B061594F earthquake in TATO.

	PREMAN			B020893B-TATO		
Radius km	β_v (km/s)	β_h (km/s)	η	β_v (km/s)	β_h (km/s)	η
3480.0	6.9254	6.9254	1.0	6.9454	6.9254	0.9971
3630.0	11.1671	11.1671	1.0	11.1871	11.1671	0.9982
5600.0	22.3459	22.3459	1.0	22.3559	22.3959	1.0017
5701.	9.9839	9.9839	1.0	10.0039	10.0039	1.0000
5771.0	22.3512	22.3512	1.0	22.3912	22.4412	1.0022
5971.0	8.9496	8.9496	1.0	8.9696	9.0496	1.0089
6151.0	5.8582	-1.0839	3.3687	5.7382	5.9009	1.0283
gradient	-1.4678	5.7176		-1.4678	-1.4278	
6291.0.0	5.8582	-1.0839	3.3687	5.7382	5.9009	1.0283
gradient	-1.4678	5.7176		-1.4678	-1.4278	
6346.6	3.9000	3.9000	1.0	3.8900	3.9400	1.0129
6356.0	3.2000	3.2000	1.0	3.1900	3.2400	1.0157
6371.0						

Table 2. The velocity structure of S wave between the PREMAN earth model and the corrected model of the earth for the B020893B earthquake in the TATO station; the zero-order coefficients of S wave and the velocity gradient of β_h on the upper mantle (6151 6346.6 km) are also shown

4. Conclusion

The S wave velocity of the region ahead of the subduction field and non-tectonic areas beneath South-East Asia have been investigated using seismogram analysis of the earthquakes that occurred on the subduction plane of Java – Sumatra that was recorded in TATO station, Taiwan, in the time domain and the three Cartesian components simultaneously. The data set used in this research contains all information in seismograms, unlike the travel time data set, which is just a small part of the information contained in the seismogram.

The synthetic seismogram is calculated using the GEMINI method which is equivalent to the Normal Mode. To simplify the waveform, the seismogram is low-pass-filtered, with a corner frequency of 20 mHz

The seismogram comparison indicates that the synthetic seismogram from Java-Sumatra earthquake that was recorded in TATO and calculated from the PREMAN earth model, indicates the earlier arrival-times than the measured wave on the same phase, on both Love and Rayleigh waves. A correction is imposed by changing the gradient β_h into positive, and the zero-order coefficients of β_v and β_h on the polynomial function of the S wave velocity, in every mantle layer. The negative correction is primary on the β_v. Fitting is obtained on the Rayleigh wave, a correction is also tried on the S and SS body waves. It indicates that the vertical anisotropy is stronger than the one included in the PREMAN earth model and also occurs in the mantle -layers under the upper mantle.

A big negative correction occurs in the upper layer of mantle. It suggests that the part of mantle that experiences a straining due to a tectonic process where the Indian continent impact the Asia Continent, has a negative velocity anomaly on the upper mantle layer and the layers below.

5. References

Dalkolmo, J., (1993). Synthetische Seismogramme für eine sphärisch symmetrische, nichtrotierend Erde durch direkte Berechnung der Greenschen Funktion, Diplomarbeit, Inst. für Geophys., Uni. Stuttgart.

Dreger, D.S., (2002). Time-Domain Moment Tensor INVerse Code (TDMT_INVC), The Berkeley Seismological Laboratory (BSL), report number 8511.

Dziewonski, A.M. and Anderson, D.L. (1981). Preliminary reference Earth model, Phys. of the Earth and Plan. Int., Vol. 25, 297 – 356.

Engdahl, E.R., Van Der Hilst, R.D., Buland, R.P., (1998). Global teleseismic earthquake relocation with improved travel times and procedures for depth determination, Bull. Seism. Soc. Am., Vol. 88, 722 – 743.

Friederich, W. and Dalkolmo, J., (1995). Complete synthetic seismograms for a spherically symmetric earth by a numerical computation of the green's function in the frequency domain, Geophys. J. Int., Vol. 122, 537 – 550.

Grand, S.P., Van Der Hilst, R.D., Widiyantoro, S., (1997). Global seismic tomography; a snapshot of convection in the Earth, GSA Today, Vol. 7, 1 – 7.

Kennett, B.L.N., (1991). IASPEI 1991, Seismological Tables, Research School of Earths Sciences, Australian National University.

Li, C., Van der Hilst, R.D., and Toksoz, N.M., (2006). Constraining spatial variations in P-wave velocity in the upper mantle beneath SE Asia, Phys. of the Earth and Plan. Int., Vol. 154, 180 – 195.

Romanowicz, B., (2003). 3D Structures of the Earth Lower Mantle, C.R. Geoscience, Vol. 335, 23 – 35

Replumaz, A, Kárason, H, van der Hilst, R. D., Besse, J. and Tapponnier, P., (2004). 4-D evolution of SE Asia's mantle from geological reconstructions and seismic tomography, Earth and Planetary Science Letters, Vol. 221, 103 – 115

11

Tectonic Background of the Wenchuan Earthquake

Yunsheng Wang, Runqiu Huang, Yonghong Luo,
Hongbiao Xu, Shitian Wang, Liangwen Jiang and Yusheng Li
State Key Lab. of Geohazard Prevention, Chengdu University of Technology, Chengdu China

1. Introduction

Since the occurrence of the Wenchuan Earthquake on May 12, 2008, several papers on its seismotectonics analysis have been published (Lei et al. 2009; Yue, 2010; Chen et al. 2009; Tang et al. 2009; Zhu et al. 2009). Although their opinions were not uniform, most authors believe that the source of the earthquake was the Central Longmen Mountain Range fracture with a length of about 230km. However, this hypothesis cannot explain the following phenomena: 1) the focus, of which the depth was provided by the State Seismological Bureau, cannot be projected to the central fracture but is located on the front range fracture; 2) the epicenter is not the center of the isoseismal contour; 3) the surface rupture along the front range fracture and its thrusting displacement are similar to the surface rupture and displacement along the central fracture, and 4) the area in which the aftershocks (Ms≥4.0)occurred have a shape in the form of a '√'.

The ambiguity results from the complicated tectonic environment of northwestern Sichuan. People generally pay more attention to the large boundaries of the triangular block in northwestern Sichuan (SCNWTB in Fig 1), but neglect the effect of the Minshan block (MSB) in the geostress conditions. The present authors think the "bottlle neck" geostress concentration in the Minshan block is the main controlling factor for the occurrence of the Wenchuan earthquake. The Minshan block is a sub-block of the northwest Block of Sichuan, it is bordered by Longmen Mountain Range fractures ((2) in Fig 1)in the south, the Huya fracture (5) in the east, the Maqin-Lueyang fracture in the north and the Mounigou Valley fracture (6) in the west. The block has been recognized earlier as an important tectonic element in the northwestern region of Sichuan (Tang et al. 2009; Jiang et al. 2004; Zhao et al. 1994a; Tang et al. 1991; Qian et al. 1999; Zhou et al. 2000). It not only includes the Minshan uplifted block in a narrow sense, but also the middle segment of the Longmen Mountain Range structural belt (Fig.1). Although the positions of the eastern and western block boundaries are still controversial, their existence and their recent activation are widely accepted. Analysis of seismological setting and the deformation data shows that the Minshan block has been activated by the north Mounigou Valley fracture ((6) in Fig 1) in the west, the Huya fracture (5) in the east, the back range fracture in the south and the Maqin-Lueyang fracture (1) in the north.

The whole middle segment of the Longmen Mountain Range structures has been strongly pushed along the southern boundary of the Minshan block since the Mesozoic. The Paleo-

Peng(zhou)-Guan(xian) complex has been thrusted over the lower Pleistocene, especially since the Quaternary. During this overthrusting, nappes in Longmenshan structural belt have been formed. The landform of the block shows a great contrast in elevation and relief. As a result of the uplifting of the Minshan block, strong erosion caused the deposition of a thick Quaternary accumulation in the Chengdu plain and formation of the Longquan mountain. Due to the repeated Minshan block lock-up→earthquake→geo-stress release (stick-slip) mechanism, the middle segment of the Longmen Mountain Range structural belt is more active than the southwestern and northeastern segments. The earthquake history of northwestern Sichuan reveals that the migration of seismicity in northwest Sichuan is around the Minshan block, and the seismic activities on the northern margin of the Minshan block created the conditions for the Wenchuan earthquake.

The aim of the Chapter is to present a new idea through the tectonic background analysis of Wenchuan earthquake, so that people pay more attention to the important role of the Minshan Block activities in the northwestern seismic tectonics.

2. The geological setting

2.1 The triangular block in northwestern Sichuan

The triangular block is bordered by the Longmen Mountain Range fractures in the south, the Maqin-Lueyang fracture in the north and the Xianshuihe fracture in the west (SCNWTB

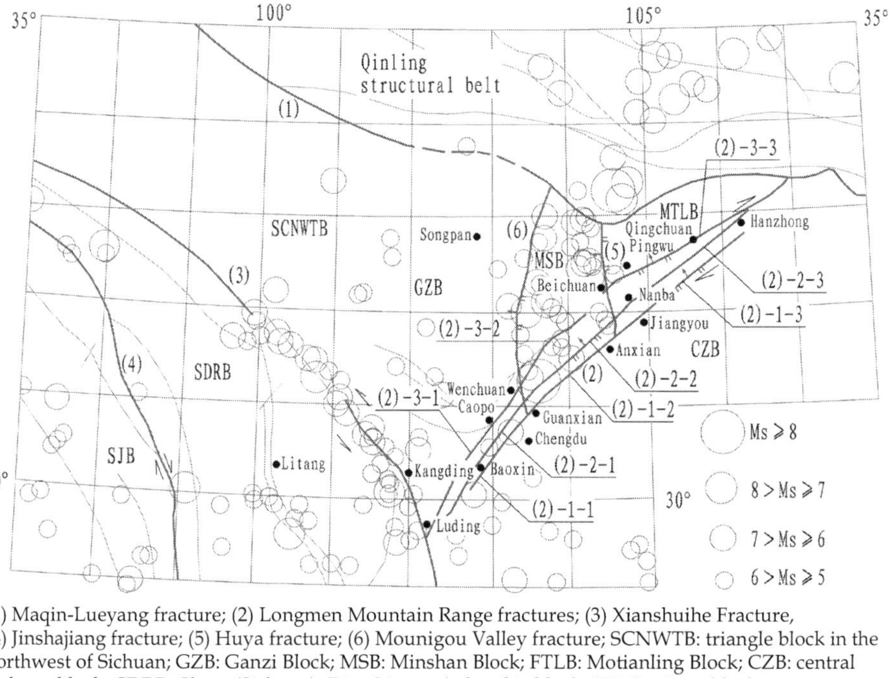

(1) Maqin-Lueyang fracture; (2) Longmen Mountain Range fractures; (3) Xianshuihe Fracture, (4) Jinshajiang fracture; (5) Huya fracture; (6) Mounigou Valley fracture; SCNWTB: triangle block in the northwest of Sichuan; GZB: Ganzi Block; MSB: Minshan Block; FTLB: Motianling Block; CZB: central Sichuan block; CDRB: Chuan(Sichuan) -Dian(Yunnan) rhombic block; SJB: Sanjiang block

Fig. 1. Tectonic framework in the northwestern Sichuan

in Fig.1). The block used to be an ocean (geosyncline) and was folded and uplifted in the late Triassic. Pre-Cambrian rocks are outcropping in Pingwu county and Maowen county, they are composed of metamorphic intermediate-basic and intermediate-acid volcanic rocks, volcanic clastic rock, intercalated with little siliceous rock. Lower Proterozoic is outcropping along the boundary of the Block. Slightly metamorphic Devonian and Carboniferous clastic rocks and carbonates are exposed in the Pingwu-Maowen region; mainly composed of carbonate, intercalated with little volcanic rock in the south part of the Block. The Permian exposed near the northern boundary of the block (Animaqin) consists of flysh, intermediate-basic volcanic rocks and carbonates. The Triassic system is widely exposed in the block, and is mainly composed of flysh. The triangular block can be divided into three sub-blocks, they are Ganzi block (GZB), Minshan block (MSB) and Motianling block (FTLB) (Fig.1). According to geophysical survey (Wang Xuben, 2000), the lithosphere of the block can be divided into three layers vertically: Upper crust, low-velocity layer in the crust, and the lower lithosphere (Fig.2).

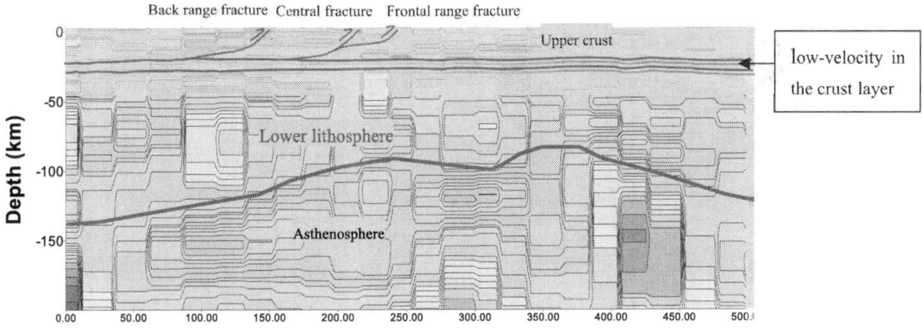

Fig. 2. Inversion result of the MT profile in northwestern Sichuan

2.2 The boundary fractures of the triangular block

The Xianshuihe Fracture is the most active fault with large magnitude (Ms6-7) earthquakes and 30-40 years recurrence period, while the Longmen Mountain Range fracture and the Maqin-Lueyang fracture earthquakes occur with super magnitude(Ms8)and low frequency(more than 2000 years).

The Xianshuihe fracture starts in Donggu, Ganzi, runs southeastward, via Luhuo, Daofu, Qianning, Kangding, Moxi, and ends in Tianwan. The strike of the fracture is NW in the north of Kangding, it is NNW in the south of Kangding, showing an arc shape. The dipping direction is SW, with medium- steep angles (45-80°). According to the satellite image and geological data, the fault is large in scale and clearly linear. Valleys such as of the Xianshuihe River are developed along this fracture, the total displacement along the fracture can reach 5000m. Since its formation in the Jurassic, it showed strong activity during the uplift of the Tibet plateau. The seismic activity along the fracture is high. Since 1725, 36 earthquakes with magnitude Ms≥5.0 have been recorded, of which 13 had a Ms≥6.0 and 6 had a Ms≥7.0. The strongest recorded earthquake with Ms7.9 of Luhuo happened on 6th of Feb., 1973, with a focal source depth of about 10-20km.

The Maqin-Lueyang fracture: starts at the Tuosuohu lake, and leads via Maqin, Maqu, Nanping, Kangxian, to Lueyang, with a length of over 2000km. The strike ranges from 290°, 270°, 70°, with a southward arc. 6 earthquakes with a magnitude larger than Ms7.0 happened along this fracture(see references: Gu Gongshu (1983); Liu Guangshun (1996); Seismic bureau of Qinghai Province (1999); Yi Guixi (2002); Xu Xiwei et al. (2005)). In Yangbuliang, Jiuzhai, Carboniferous-Permian metamorphic carbonate rocks have thrusted onto Triassic clastic rocks, forming nappes. There is basic rock and super-basic rock outcrop along the fracture in Langmusi.

The Longmen Mountain Range fracture: Longmen Mountain Range fracture belts are composed of three deep fractures. From the southeast to the northwest: the front range fracture, the central fracture and the back range fracture. The strike of the three fractures is NE-SW, with dipping direction towards the northwest. Since the Cenozoic, these fractures have experienced thrusting from the northwest to the southeast with some right-slip component. The front range fracture is composed of the Dachuan-Shuangshi fault, the Guanxian-Anxian fault, and the Jiangyou fault. The central fracture is composed of the Yanjing-Wulong fault, the Yingxiu -Beichuan fault and the Beichuan-Linyansi fault. The back range fracture is composed of the Longdong-Gengda fault, the Wenchuan-Maoxian fault and the Pingwu-Qinchuan fault(Fig.3).

2.3 The Minshan Block

The Minshan Block is an important tectonic element in the northwest region of Sichuan. It not only includes the Minshan uplift in a narrow sense, but also the middle segment of Longmen Mountain Range structural belt (Fig.2, 3).

Although the west and east boundaries are still controversial, their existence and new activities are widely accepted. Analysis of seismic geological setting and the deformation data shows the Minshan uplift to be activated by the Mounigou Valley fracture in the west, Huya fracture and Leidong fracture in the east, back range fracture in the south and Maqin-Lueyang fracture in the north. Also the whole middle segment of the Longmen Mountain Range structures are especially strong active along south boundary since the Mesozoic, especially since Quaternary, namely, the Paleo-Peng(zhou)-Guan(xian) complex has been thrusted onto lower Pleistocene, nappes have been formed. There is great contrast in elevation and drop on the landform in the block. In addition strong erosion caused a thick Quaternary accumulation along the Chengdu plain and Longquan mountain was formed as a result of the uplifting of the Minshan block.

The northern part of the Minshan block (MSB in Fig. 3) is narrow (50 km wide) and it widens (90km) towards the south. The SN length is about 200km (Fig.3); the three dimensional shape is narrow in the lower, deeper part and wider near the surface (Fig.4-9). The rigidity of the Block is stronger than that of the adjacent areas because the Vs (≥3.2) of the Minshan Block is larger than that of the adjacent areas (Fig.5).

The main part of the Minshan block is composed of series of peaks such as Gonggaling, Hongxingyan, Xuebaoding, Xueguzhai, Maoheshan and Jiudingshan, whose summits are at about 4000-5000m a.s.l. The landform is intensively dissected and deep gorges are well developed. The Songpan plateau is situated to the west of the Minshan block and is slightly dissected while a peneplain(planation surfacce) with an elevation of about 4000m is preserved.

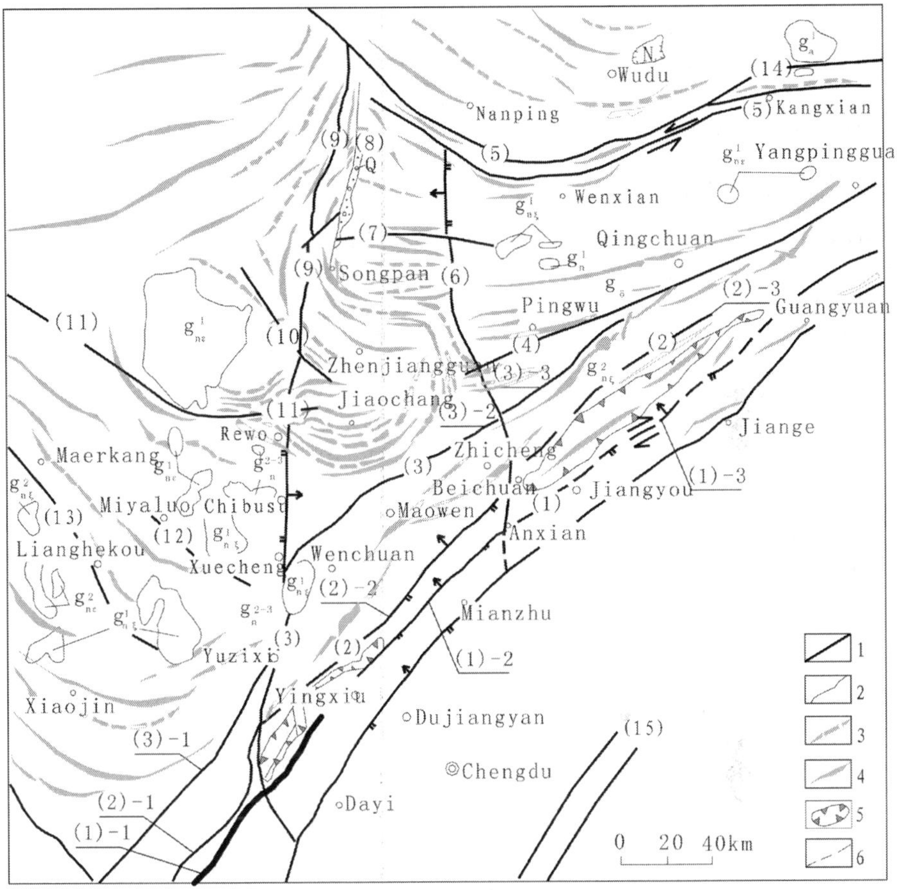

1. regional main deep fractures; 2. common fractures; 3. synclines; 4. anticlines; 5. nappes; 6. predicted fracture; (1) Guanxian-Anxian fracture; (2) Yingxiu-Beichuan fractures; (3)Maowen fracture; (4) Pingwu-Qingchuan fracture; (5) Maqin-Lueyang fracture; (6) Huya fracture; (7) Xueshan fracture; (8) Minjiang fracture; (9) Mounigou Valley fracture; (10)Songpinggou fracture; (11) Aba-Heishui-Jiaochang arc fracture, (12) Miyaluo-Lixian fracture; (13)Maerkang-Lianghekou fracture; (14)Wudu-Chengxian fracture; (15) Longquan Mountain Range fracture; (1)-1 Dachuan-Shuangshi fracture; (1)- 2 Guanxian-Anxian fracture; (1) -3 Jianyou fracture; (2)-1 Yanjing-Wulong fracture; (2)-2 Beichuan-Yingxiu fracture; (2)-3 Beichuan-Linanshi fracture; (3)-3 Gengda-Longdong fracture; (3)-2 Maowen-Wenchuan fracture; (3)-3 Pingwu-Qingchuan fracture

Fig. 3. Skeleton of the northwest block of Sichuan

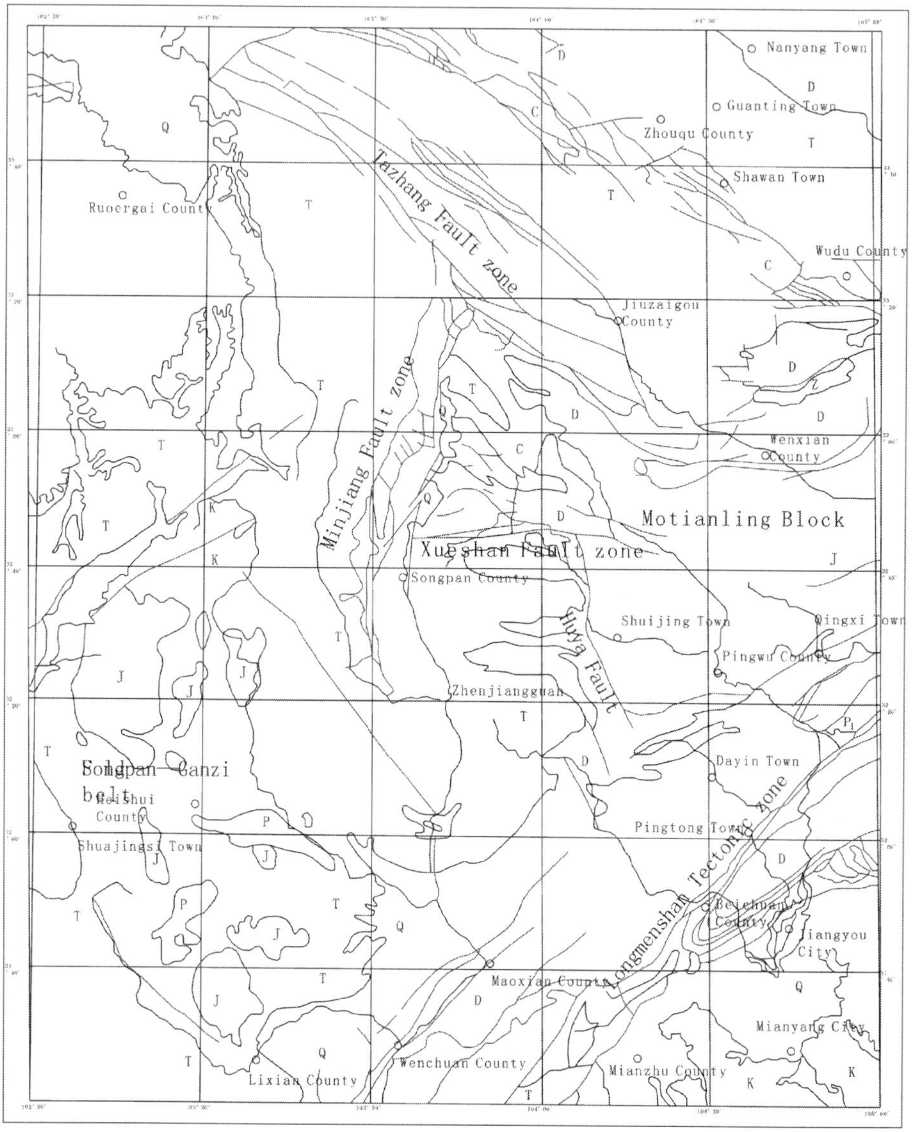

S: Silurium System; D: Devonian System; D-T: Devonian-Triassic System; C: Carboniferous System; T: Triassic System; J: Jurassic System; K: Cretaceous System; Q: Quaternary System; γ : PreCambrian;

Fig. 4. The geological map of Minshan block

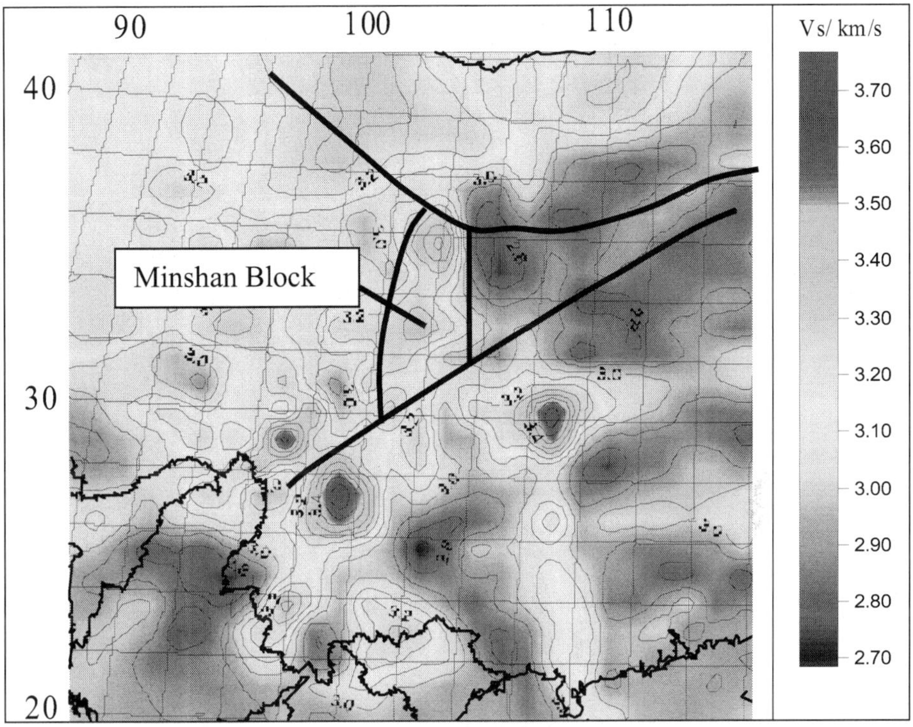

Fig. 5. Vs distribution of surface waves in the northwest of Sichuan (at depth of 8km. From Wang Xuben, personal communication) **The coarse black line is the boundary fracture**

According to the activities and the distribution of Quaternary and landform characteristics, Minshan Block can be divided into three segments: the Gonggaling to Zhenjiangguan, Zhenjiangguan to Maoxian and Maoxian to Guanxian segments.

The Gonggaling to Zhenjiangguan segment is composed of a series of peaks with an elevation of about 4500m and the mountain range is with SN strike, the highest peak being Hongxingyan (5010m) with outcropping Devonian-Triassic limestone, dolomite, metamorphic sandstone and slate. Its boundary fractures are active and a thick Quaternary accumulation is developed in the Zhangla basin.

The Zhenjiangguan to Maoxian segment is composed of peaks with an elevations of about 4500m and deeply eroded valleys. The summits follow a SN strike direction, and have Tertiary metamorphic sandstone, phylitte, slate and Devonian Weiguan Group schist, Silurian Maoxian Group phyllite, and slate and quartzite as its bedrock (Fig.4).

The Maoxian to Guanxian segment is located in the area with superimposed SN and NE structures. The strata are striking is NE direction and the general elevations is over 2000m. The highest peak is Jiudingshan (4989m), the next highest peak is Qianfoshan (3033m), and the area has 2000-4000m in elevation differences with respect to the Chengdu plain. The landform is deeply dissected and the tectonic deformation is intensive. The bedrock consists of Precambrian granite, granodiorite, Silurian schist and quartzite, Triassic sandstone and mudstone.

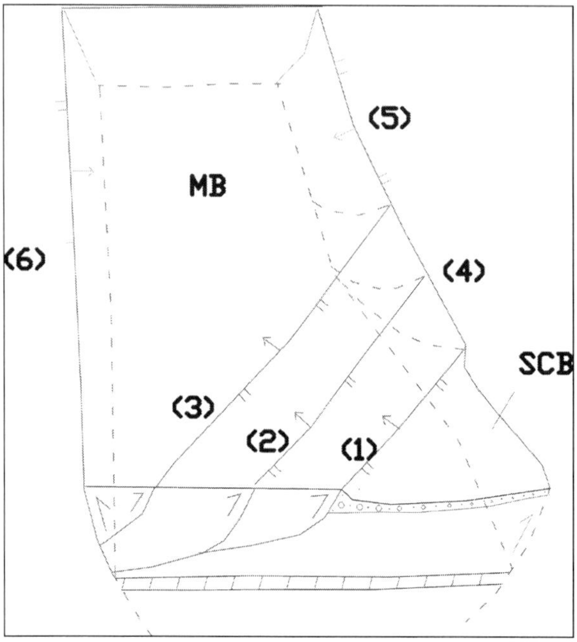

(1) front range fracture, (2) central fracture, (3) back range fracture, (4) Leidong fracture, (5) Huya fracture, (6) Mounigou fracture, MB Minshan block, SCB Sichuan basin

Fig. 6. 3D structural diagram of the Minshan block

Fig. 7. Western boundary fracture of the Minshan block, (camera facing south) The fault zone is composed of cataclastic rock and fault gouge Carbonization is obvious, fault zone width 20-30m

Fig. 8. East boundary of Minshan block, (camera facing north) The fault zone is composed of cataclastic rock and fault gouge Carbonization is obvious, fault zone width 120m.

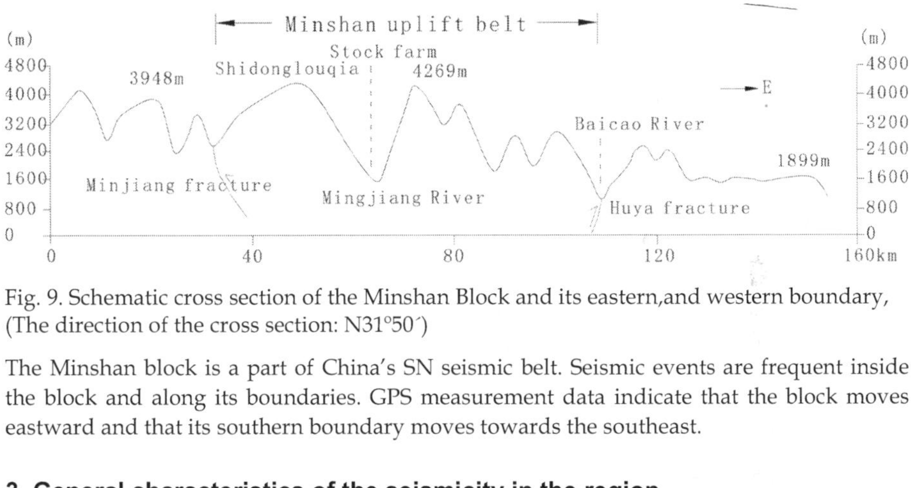

Fig. 9. Schematic cross section of the Minshan Block and its eastern,and western boundary, (The direction of the cross section: N31°50′)

The Minshan block is a part of China's SN seismic belt. Seismic events are frequent inside the block and along its boundaries. GPS measurement data indicate that the block moves eastward and that its southern boundary moves towards the southeast.

3. General characteristics of the seismicity in the region

About 100 earthquakes with magnitude over 4.7 have occurred in the region, since the Wudu, Wen County magnitude 7.0 earthquake which occurred in 186 BC. 18 Earthquakes with a magnitude over 6.0 and 7 earthquakes with a magnitude over 7.0 have occurred since 186 BC (Fig.9, Table.1). The magnitudes of Serial number 1-7 are determined according to the historical earthquake description of local chronicles and field survey of historical earthquakes. As there were no instrumental records, the value is not accurate. The distribution of seismic activity has obvious features of zoning (Fig.10). The area can be divided into three division zones and 8 subdivisions (Table.2).

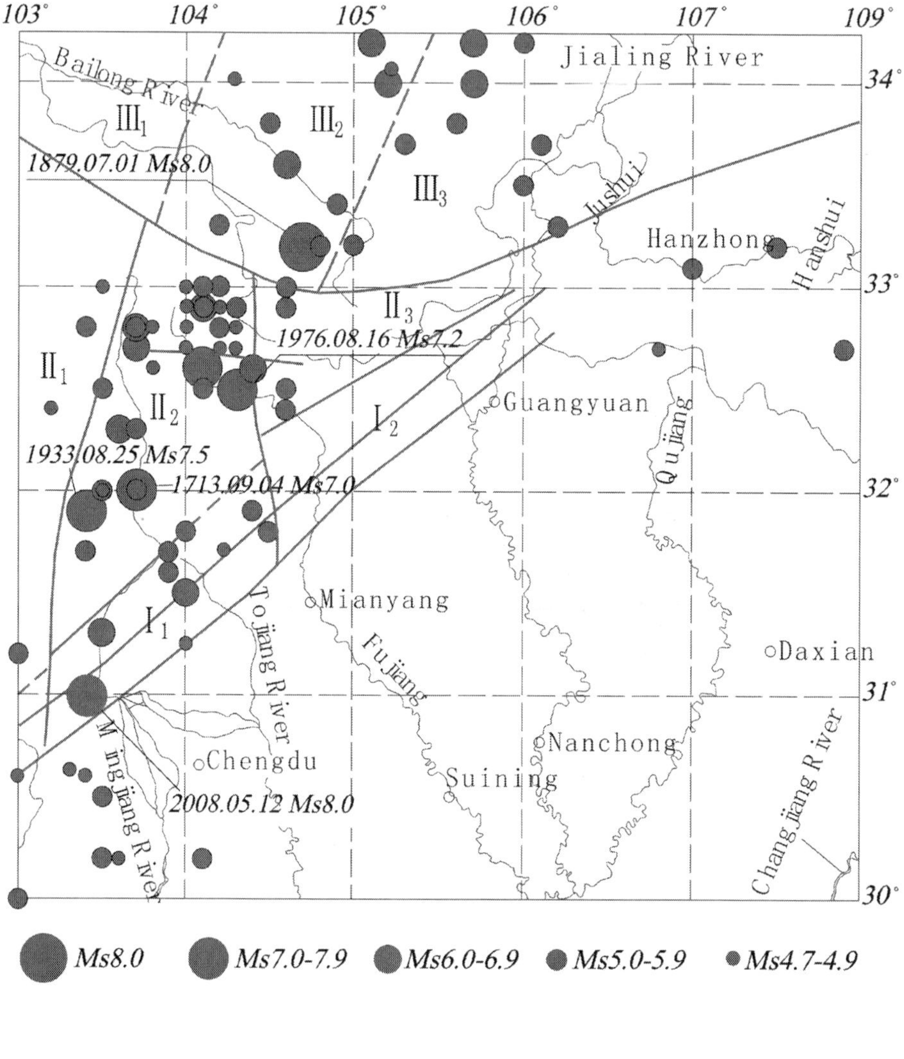

Fig. 10. Historical earthquakes in northwestern Sichuan (from Hu Xingping et al 2008 modified)

Serial number	date	Epicentral coordinate			Magnitude (Ms)	Depth (km)
		°N	°E	Reference sites		
1	186.2.22 BC	33.4°	104.8°	Wudu, Wenxian	7	_
2	1630.1.16	32.6°	104.1°	Songpan-Pingwu	6.75	_
3	1657.4.21	31.5°	103.6°	Wenchuan	6	_
4	1713.9.4	32.0°	103.7°	Diexi, Maoxian	7	_
5	1879.6.19	33.2°	104.7°	Wudu, Wenxian	6	_
6	1879.7.1	33.2°	104.7°	Wudu, Wenxian	7.5	_
7	1881.7.20	33.6°	104.7°	Zhouqu, Gansu	6.5	_
8	1933.8.25	32.0°	103.7°	Maoxian, Diexi	7.5	20 km
9	1938.3.14	32.3°	103.6°	Zhenjiangguan	6	32 km
10	1941.10.8	32.1°	103.3°	Heishui county	6	20 km
11	1958.2.8	31.8°	104.0°	Maoxian county	6.2	_
12	1960.11.9	32°47′	103°40′	Zhangla, Songpan	6.75	_
13	1970.2.4	30°36′	103°12′	Dayi county	6.25	15 km
14	1973.8.11	32°55′	103°55′	Huanglong,Songpan,	6.2	20 km
15	1976.8.16	32.7°	104.2°	Songpan- Pingwu	7.2	9 km
16	1976.8.22	32.6°	104.15°	Songpan- Pingwu	6.7	15 km
17	1976.8.23	32.5°	104.1°	Songpan-Pingwu	7.2	17 km
18	1989.9.22	31.58°	102.47°	Lianghekou, Xiaojin	6.3	14 km

Notes: the magnitude of Serial number 1-7 is according to the description of annals of local history

Table 1. Strong earthquake catalogue of northwestern Sichuan (Ms≥6.0)

Division	Subdivision
Seismic activity zone of Longmenshan Mountain structural belt(I)	The weak seismic activity sub-region in the north section(I_1) The strong seismic activity sub-region in the south and intermediate section(I_2)
Seismicity zones inside the inverted-triangle- shaped fault block in the northwest of Sichuan(II)	The weak seismic activity sub-region of western inside the fault block(II_1) The strong seismic activity sub-region of Minshan block(II_2) The weak seismic activity sub-region of eastern inside the fault block(II_3)
Seismic activity zones of western of Qinling tectonic seismic belt(III)	The weak seismic activity sub-region in the west section(III_1) The strong activity sub-region in the middle curved tectonic section(III_2) The strong seismic sub-region in the east section(III_3)

Table 2. The seismic activity zoning

3.1 Spatial distribution of seismic activity

The seismicity in this region is mainly concentrated at the western side of "tectonic bottle neck zoning " of Wudu, Wen County - Pingwu, Qingchuan, in the area of 103° ~ 105°E, which is a nearly S-N banded zone. The main seismogenic structures were the boundary faults in the north and south parts and the faults within the Minshan block. However, the earthquake frequency and maximum earthquake magnitude inside the fault block are higher than those along the Maqin-Lueyang fracture and the Longmen Mountain fault zone (Fig.10). Compared with the southern boundary zone, the seismic intensity in Maqin-Lueyang fracture is far larger (Table.3). The main causative faults inside the block triggered by a nearly E-W regional tectonic stress-strain field are:

1. the Huya fault, striking nearly N-S, it is the source of reverse fault earthquakes;
2. the Songpinggou fault, in NW-SE, it is the source of left-lateral strike-slip earthquakes ;
3. the Dongmengou fault in NEE-SWW, it is the source of right-lateral strike-slip earthquakes.

3.2 Migration pattern of the large magnitude earthquake sources

According to the historical seismic data (Table.1), the general pattern in this region is that the large magnitude seismicity migrates from the boundary faults to the triangular-shaped fault block in northwestern Sichuan in nearly N-S direction, then migrates to the boundary faults. There are some differences between the migration patterns before 1900 and after 1900.

1. The migration laws of strong seismic activities before 1900 is: earthquake in north boundary fault (186 BC, February 22, Wudu, Wen County magnitude 7.0 earthquake)→(to the south) earthquake inside the fault block (January 16, 1630, Songpan and Pingwu magnitude 6.7 earthquake)→(to the south) earthquake in the Longmen Mountain fault in the southeast boundary (April 21,1657, Wenchuan magnitude 6.5 earthquake) →(to north) earthquake inside the fault block (September 4,1713, Diexi magnitude 7.0 earthquake) →(to the north) earthquakes in the north boundary fault (June 19,1879, Wudu magnitude 6.0 earthquake, and July 1,1879, Wudu, Wen County magnitude 7.5 earthquake, and Zhouqu, Wudu magnitude 6.5 earthquake in 1881). Generally speaking, the seismic activity began in the north boundary fault, finally returned to north boundary fault again after the north-south round-trip before 1900. Seismic activities mainly occurred inside the fault block or along the north boundary fault.
2. The migration laws of strong seismic activities after 1900 is: earthquake inside the fault block(August 25, 1933, Diexi magnitude 7.5 earthquake) →(internal adjustment) earthquakes inside the fault block(two adjustable strong seismic activities: March 14, 1938, Zhenjiangguan magnitude 6.0 earthquake, and October 8,1941, Heishui magnitude 6.0 earthquake) →(to south)earthquake of Longmen Mountain in southeast fault block (February, 1958, Maoxian magnitude 6.2 earthquake)→(to north)earthquake inside fault block(November 9, 1960, Zhangla magnitude 6.7 earthquake)→(to south) earthquakes in southeast boundary fault block(February 24, 1970 Dayi magnitude 6.25 earthquake)→(to north)earthquakes inside fault block (August 11, 1973, Songpan, Huanglong magnitude 6.2 earthquake, and 1976 August 16~August 23 Songpan, Pingwu magnitude 7.2 swarm earthquake)→(to south)(internal adjustment)earthquake

inside fault block(in the southwest area of Lianghekou in Xiao Jin), and this belongs to the adjustable phase of seismicity. Obviously, strong seismicity has presented the trend of migration to internal of fault block, and southward. Seismic activities mainly occurred inside the fault block or along the southeast boundary fault.

In summary, these migration patterns suggest that the triangular-shaped Minshan fault block is transferred along the deep fault zones on both sides of the boundary faults in an irregular process. As the northern boundary fault zone is the main impedance boundary for eastward movement, every cycle of eastward fault-seismic activity started at the northern zone, and then migrated into the fault block and towards the southeastern block boundary fault.

3.3 Periodicity of large magnitude seismicity

Analysis of the historical seismic data in this region shows that their seismicity has been alternating active and quiet periods since 1920 (Fig.11).

1924-1941 was the first active period in the 20th century, which lasted about 18 years; largest earthquake during this period with magnitude 7.5 occurred in Diexi on August 25, 1933. This high activity period was followed, by the first quiet period in the region, which lasted about 10 years(1941-1952).

From 1952-1978, the second active period occurred in the region, which lasted about 26 years; its strongest earthquakes are the 2 Songpan-Pingwu magnitude 7.2 earthquakes on August 16 and 23, 1976. This period was followed by the second quiet period in the region, which lasted about 10 years (1978-1989). From 1989 to the present, we experience the third active period, which may last until 2012(±8). Wenchuan earthquake magnitude 8.0 occurred during this period, whose epicenter is south boundary of the Minshan block (Table 3).

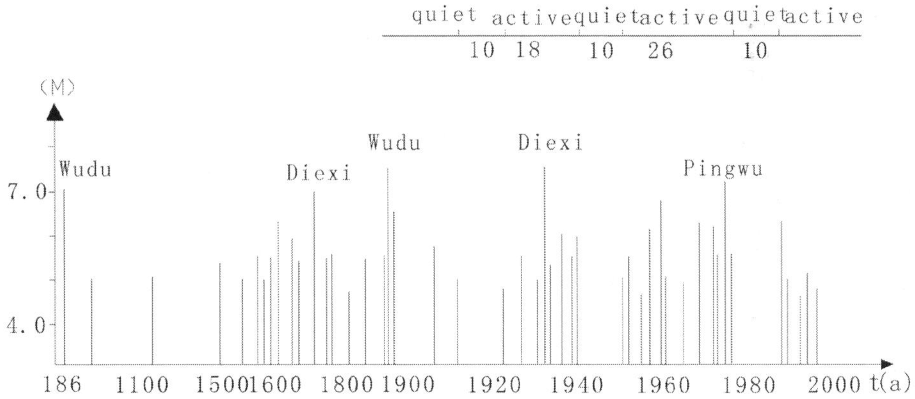

Fig. 11. M-t diagram of the strong earthquakes in northwestern Sichuan

The names of fault zones	The times of earthquakes		Occurring interval, and times
	Magnitude between 6.0 and 6.9	Magnitude over 7.0	
The north boundary fault zone	2	2	Before 1900, 4 times
The Minshan block	7	4	Before 1900, twice After 1900,9 times
The south boundary fault zone	3	0	Before 1900, once After 1900, twice

Table 3. Location and period of occurrence of strong earthquakes in northwestern Sichuan

3.4 Recurrence interval of strong earthquakes (Ms≥ 7.0)

Table.4 shows that the time interval between strong earthquakes in the region (the recurrence interval) is shortening in a slightly exponential way.

The names of earthquakes	Earthquake magnitude	The time of occurrence	The time interval between two strong earthquakes
Wudu, Wen County earthquake	7.0	February 22, 186 BC	
Diexi earthquake	7.0	September 4, 1713	1898 years
Wudu, Wen County earthquake	7.5	July 1, 1879	166 years
Diexi earthquake	7.5	August 25, 1933	54 years
Songpan, Pingwu earthquake	7.2	August 16~23, 1976	43 years

Table 4. The recurrence interval between strong earthquakes (Ms≥7.0)

From this time series, it may be assumed that the time interval until the occurrence of the next over magnitude 7.0 earthquake is about 40 years. In fact, it only took 32 years from Pingwu earthquake to Wenchuan earthquake. The accelerating tendency of strong seismicity in the region is obviously closely related to the triangular tectonic system controlling strong earthquakes.

3.5 Fractal characteristics of the seismic magnitude/frequency pattern

Aki(1981)shows that the G-R relation(logN=a-bM, Gutenberg-Richter, 1954) is equivalent to the definition of fractal distribution(Chen Chunzai, 1997). D. L. Turcotte(1989) suggested that that there is a quantitative relationship between the value b and the fractal dimension, D_f, from the point of seismic wave energy and seismic distance, that is $D_f=2b$, and then revealed the fractal characteristics of the G-R relation, i.e. the D_f stands for the frequency of the earthquake occurring. The equation of logN=a-bM show that, if the value of D_f is smaller, the region is more prone to large magnitude earthquakes.

We conducted a special fractal study of the southern and middle part of the Longmen Mountain (I), the Minshan Block (II_2), and the tectonic seismic belt of west Qinling (III). The results show that the seismicity fractal dimension (D_f) of the zone (I) is the highest: 1.2105;

that (D_f) value of the zone (II_2) is the lowest: 0.913; and that the (D_f) value of the zone (III) has an intermediate value: 0.9788. This suggest that, the magnitude level of seismicity of the Minshan block is the largest; and the south and intermediate section of Longmen Mountain has the lowest number of large magnitude earthquakes.

For the region as a whole, the fractal value (D_f) is 1.0054, close to 1, meaning that the seismic activities of all magnitudes in the region were controlled by a scale-independent earthquake mechanism. This proves that an triangular block tectonic system controls the earthquakes in the region(Table 5).

Zone \ N \ Ms	≥4.7	≥5.0	≥6.0	≥7.0	b	D_f=2b	r
Longmen Mountain fault zone(I)	20	10	3	0	0.6053	1.2105	0.9998
The uplift belt of Minshan (II)	44	30	9	4	0.4565	0.9130	0.9992
Western of Qinling tectonic seismic belt(III)	22	19	6	2	0.4889	0.9778	0.9999
$\sum$	86	59	18	6	0.5072	1.0054	0.9997

Table 5. The results of fractal analysis of seismic activity in northwestern Sichuan

4. The seismotectonics analysis of the Wenchuan earthquake

As the Indian plate pushes strongly toward the Qingzang Plateau, different blocks in southwestern China have different movement directions (Fig.12), The northwestern block of Sichuan mainly moves towards the east. In the movement process, due to the Minshan block lock-up →earthquake→geo-stress release (stick-slip) cyclic process, the middle segment of the Longmen Mountain Range structural belt is more active than the southern and northern segments.

Both the geophysical survey and the geological analysis strongly suggest that the three fractures in the southern part of the Minshan block converge into an intracrust low velocity layer and form a slip plane (Fig.2). The stress accumulation in the southern part of the Minshan block caused a high stress concentration along the Yingxiu-Beichuan and Guanxian-Anxian fractures, and they became a rupture point that probably could be fractured at anytime. According to the hypocenter parameters of the 5.12 Wenchuan Earthquake (epicenter at Niumiangou with depth of focus 14 km), the initial focus is located at the front range fracture. At the surface, the bottle-neck's lock-up of the Minshan block is small in the front and large in the tail; vertically, it is small in the lower part and big near the surface (Fig.5). Because of the restriction in the northern and southern boundaries, together with the blockage in the east, the dextral strike-slip action of the block requires thrusting movement at the southern boundary to provide enough space. Thus, the movement of the

Minshan block is mainly characterized by vertical thrusting in combination with a small strike slip. The transient variation of stress and the sudden release of tremendous amounts of strain energy forced movement in the Yingxiu-Beichuan fracture (just above the focus) and induced large scale rupture toward the northeast. Because of the obstacle (at the block's east boundary) around Beichuan County, the consequent progressive failure released large amounts of strain energy. After the thrusting movement occurred along the southern boundary of the Minshan block, the lock-up effect was lost instantly. The powerful pushing from the west was immediately transmitted to the eastern (Motianling) block, which was under lower levels of stress because of the barrier action from the Minshan block. This caused continuous aftershocks in the Motianling block and the northern segment of Longmen Mountain Range which up to then were fairly calm. Because of the eastward movement of the Minshan block, the stress level at the southern segment of the western boundary fracture changed significantly. The stress adjustment caused relative movements of both sides along the boundary, which led to continuous aftershocks along the Yuzixi - Lixian-Chibusu and the '√' shape distribution of aftershocks (Fig.12) (Zhao et al. (1994b); Chen et al. (1994a); Chen et al. (1994b); Hua et al. (2009); Hu et al. (2008)). The secondary geological hazard anomaly zone has similar shape too.

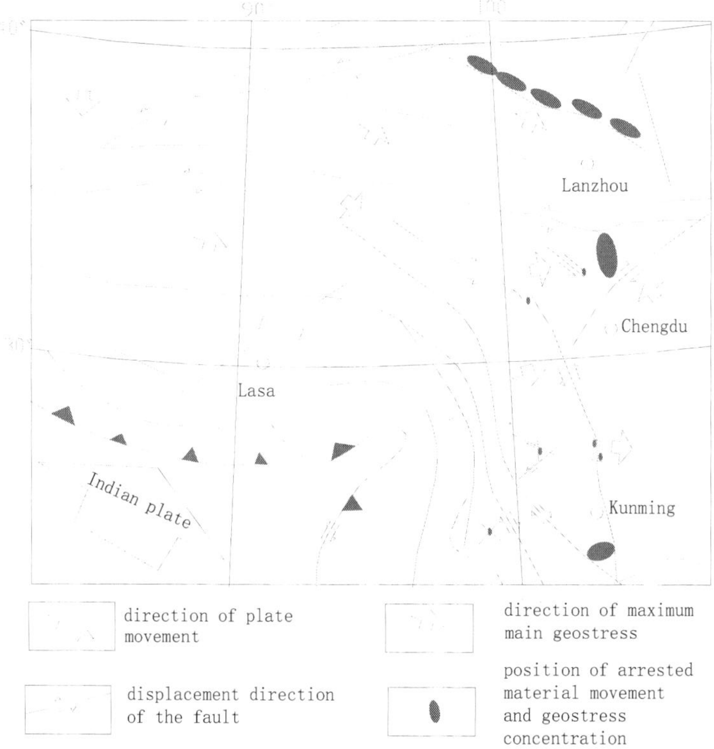

Fig. 12. Regional geostress directions in southwestern China

According to historical earthquake records, since 186 B.C the spatial distribution of earthquake in northwestern Sichuan was constrained by the Minshan Block. The earthquake record in the Minshan block demonstrates a higher magnitude and frequency than along its southern boundary, the middle-segment of the Longmen Mountain Range structural belt. Historically, there was no earthquake with a magnitude over 7.0. The occurrence of the magnitude 8.0 earthquake in Wenxian in 1879 was the consequence of stress accumulation, which led to the migration of seismicity from north to south. Earthquakes with a magnitude larger than 6.0 occurred several times along the boundaries and in the interior of the Minshan Block, and earthquakes larger than 5.0 magnitude are frequent. Except for the 5.0 magnitude earthquakes that occurred in Beichuan, Dayi and Mianzhu counties in 1999, the seismic activity along the southern boundary was historically not frequent. In other words the stress accumulation that can cause a magnitude 8.0 earthquake takes a long time to develop, more than 2000-4000 years.

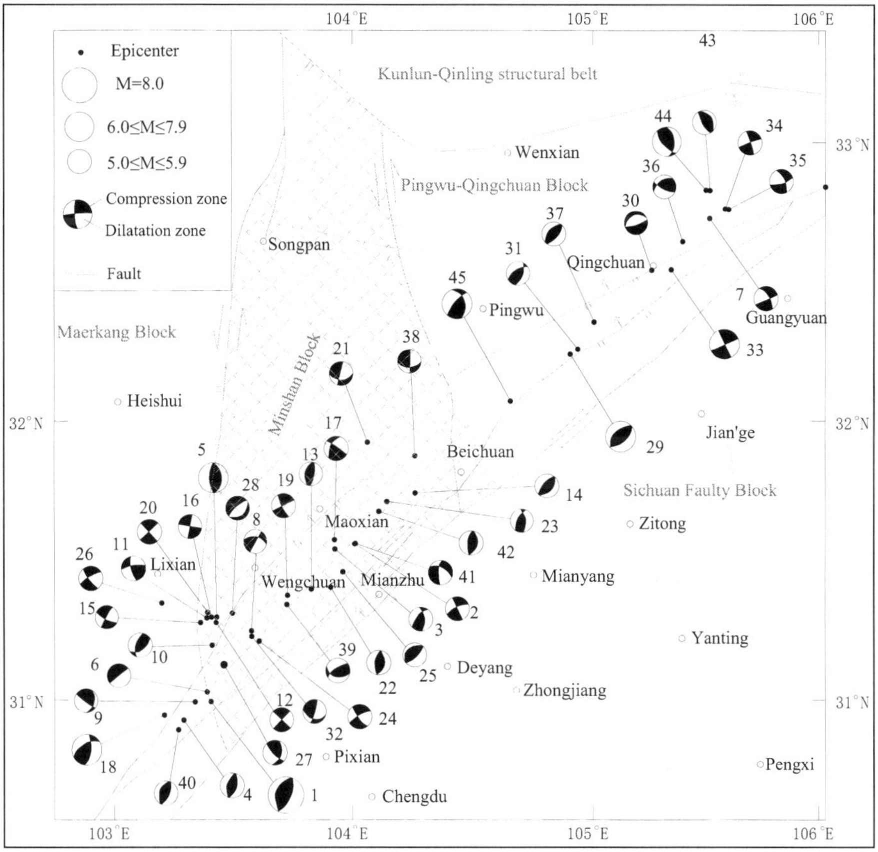

Fig. 13. The structural framework and focal mechanism solution for the Wenchuan earthquake and its aftershocks (from Hu Xingping et al 2008 modified)

5. Conclusion

On the basis of the presented analysis, we can draw the following conclusions: (1) the Minshan block is an important tectonic element with larger rigidity than that of the surrounding areas; (2) This unique tectonic framework caused a bottle-neck effect in the Minshan block. Exactly the effect which causes stress concentrations along the boundary and in the interior of the block which lead to frequent earthquakes; (3) Before the nineteenth century, the earthquakes were concentrated on the northern, western and eastern Minshan block boundaries. A magnitude 8.0 earthquake occurred in the Wenxian county in 1879 and triggered another round of large-scale reverse faulting and strike-slip movements. Against this background, the movement of the southern boundary of the Guanxian-Anxian fracture induced the 5.12 Wenchuan Earthquake; (4) The fracture just above the focus, the Yingxiu-Beichuan fracture, was forced to act strongly; (5) The release of the Minshan block lock-up lead to a high stress level in eastward direction causing continuous aftershocks in the northern segment of the Longmen Mountain Range structural belt and in the Motianling block. (6) Further aftershocks will occur in the middle and northern segment of Longmen Mountain Range structural belt and the southern segment of the Mounigou Valley fracture.

6. Acknowledgements

The research has been supported by National Natural Science Foundation of China (NSFC), (Grant No. 2008CB425801, 41072231, and by the China Geological Survey Bureau (Grant No.1212010914010). We also particularly thank Niek Rengers for comments and suggestions to earlier versions of the manuscript.

7. References

Aki, K.(1981) A probabilistic synthesis of precursory phenomena, in Earthquake Prediction: An International Review, Maurice Ewing Ser., vol.4, edited by D. W. Simpson and P. G. Richards, pp. 566-574, AGU, Washington, D.C.

Chen Chunzai, Jin Youyu(1997) Application of fractal theory in metallogenic prognosis, Mineral resources and geology, 11(4), 272-276.

Chen G, Ji F, Zhou R(2007) Primary Research of Activity Segmentation of Longmenshan Fault Zone Since Late-Quaternary. Seismology and Geology 29(3): 657-673.

Chen Z, Lin B, Bai W, et al. (2009) The mechanism of generation of May 12, 2008 Ms8.0 Wenchuan earthquake. Chinese J. Geophys 52(2):408-417.(in Chinese)

Chen S, Deng Q, Zhao XL, et al. (1994a) Evolution history, deformed mechanism and earthquake of middle segment nappes and related structures(1). Seismology and geology 16(4): 404-412.

Chen S, Deng Q, Zhao XL, et al. (1994b) Evolution history, deformed mechanism and earthquake of middle segment nappes and related structures(2) Seismology and geology 17(4): 404-412.

Cheng GW, He XB, Chen GR, et al. (2010) Change in Sediment Load of the Yangtze River after Wenchuan Earthquake. Journal of Mountain Science 7:100-104.

Cheng GW, Wang XD, He XB et al. (2008) Outburst Risk of Barrier Lakes in Sichuan, China. Journal of Mountain Science 5:189-193.

Gu Gongshu(1983) Index of China Earthquakes, Beijing: Science Publishing House, pp307.

Hu X, Yu C, Tao K, et al. (2008) Focal mechanism solutions of Wenchuan earthquake and its strong aftershocks obtained from initial P wave polarity analysis. Chinese Journal of Geophysics 51(6):1711-1718.

Hua W, Chen Z，Li Z(2009) A study of Wenchuan earthquake triggering and spatial distribution of aftershocks. Seismology 29(1):33-39.

Jiang L, Wang S, Li Y, et al. (2004) Research on the Formation Mechanism of Strong Seismic Belts and the Characteristics of Seismic Activity in the East Part of Inverted Triangle-Shaped Fault Block in the Northwest Plateau of Sichuan. Advance in Earth Sciences 19:217-222.

Kirby E, Whipple KX, Burchfiel BC, et al. (2000) Neotectonics of the Min Shan, China: Implications for mechanisms driving Quaternary deformation along the eastern margin of the Tibetan Plateau. GSA Bulletin 112(3):375-393.

Lei J, Zhao D, Su J, et al. (2009) Fine seismic structure under the Longmenshan fault zone and the mechanism in the large Wenchuan earthquake. Chinese J · Geophys 52(2):339-345.(in Chinese)

Liu Guangxun(1996) Eastern Kunlun Active Fault Zone and Its Seismic Activity, Earthquake Research In China, 12(2), 119-126.

Qian H, Zhou R, Ma SH(1999) South Segment of Minjiang Fault and Diexi Earthquake in 1993. Earthquake Research in China 15(4):333-338.

Seismic Bureau of Qinghai Province, Institute of Crustal dynamics, CEA(1999) Eastern Kunlun Active Fracture Zone[M], Beijing: Geological Publishing House

Tang H, Li D, Hu X(2009) Faulting Characteristics of Wenchuan Earthquake and Evaluation Theory of Regional Crustal Stability for Engineer. Journal of Engineering Geology 17(2):145-152.

Tang R, Wen D, Huang Z, et al.(1991) Quaternary activities of the main active fractures in Longmenshan region. Earthquake Research in China 7(3):64-71.

Tang W, Liu Y, Chen ZH, et.al. (2004) The preliminary study of the tectonic activities along the boundary faults around the Minshan Block, western Sichuan. Sedimentary Geology and Tethyan Geology 24(4):31-34.

Turcotte, D. L.(1989). Fractals in geology and geophysics. *Pure and Applied Geophysics* 131:171-96.

Yang J, Deng T, Wang Y, et al.(2007) Geomorphology characteristics of Minjiang upstream. Seismology and Geology 1(3):68-75.

Wang Xuben, Wang Yunsheng et al.(2000) Geophysical-geological comprehensive research of Songpan- Shaoyang section[R], Chengdu University of Technology.

Wang X，Yun, Zhu Y，et al.(2008) Preliminary result of magnetotelluric sounding in the Longmen thrust belt of West Sichuan, China，Journal of Chengdu University of Technology (Science & Technology Edition) 35(4): 398-403.

Yi Guixi, Wen Xueze, Xu Xiwei et al.(2002) Study on Recurrence Behaviors of Strong Earthquakes for Several Entireties of Active Fault Zones in Sichuan-Yunnan Region, Earthquake Research In China, 18(3), 267-276

Xu Xiwei(2005) Features of Active Tectonics And Recurrence Behaviors Of Strong Earthquakes In The Western Sichuan Province Aand Its Adjacent Regions, Sesimology And Geology, 27(3), 446-461.

Yue ZQ(2010) Features and mechanism of coseismic surface ruptures by Wenchuan earthquake of May 12, 2008. In Xie Furong (Eds.), Rock Stress and Earthquake, 2010 Taylor and Francis Group, London, ISBN 987-0-415-60165-8.

Zhao X, Deng Q, Chen SH (1994) Tectonic Geomorphology of the Minshan Block in Western Sichuan, Southwestern China. Seismology and geology 15(4):429-439.

Zhou R, Pu X, He Y, et al. (2000) Recent activity of Minjiang fault zone, Block of Minshan block and their relationship with seismicity of Sichuan. Seismology and Geology 22(3):285-294.

Zhao Q, Chen Z, Liu Y et al. (2005) The Crustal movement in the Qinghai-Tibet plateau and its southeast foreland from GPS surveying. Progress In Geophysics 20,(2): 524-527.

Zhu S (2009) A Study on Preparation Mechanism Ms 8.0 of Wenchuan Earthquake and Its Implication to Earthquake Prediction. Journal of Institute of Disaster-Prevention Science and Technology 11(1):10-19.

12

Non-Linear Analysis of Point Processes Seismic Sequences in Guerrero, Mexico: Characterization of Earthquakes and Fractal Properties

E. Leticia Flores-Marquez and Sharon M. Valverde-Esparza
Instituto de Geofísica, UNAM, Cd. Universitaria, México D.F.
Mexico

1. Introduction

The purpose of this work is to reveal the efficiency of some statistical non-linear methods so as to characterize a seismic zone linked to subduction in Mexico. The Pacific plate subducting into the North American plate produces an important number of earthquakes (EQs), whose magnitudes exceed $M_n = 5$. This region comprises the following States: Jalisco, to the northwest, Michoacan, Guerrero, and Oaxaca, to the southeast; it extends along roughly 1350 km (Figure 1). Therefore, the characterization of this region - in all scopes - is very important. Here, we focus on the application of non-linear methods in the Guerrero State, because it displays an important number of EQs (their magnitudes rise up to 6) and it has a different slip inclination to the rest of the subduction zone, and some authors (Singh et al., 1983; Pardo and Suarez, 1995) have considered that there are some lags of seismicity. The assumptions of the non-linear methods analyzed in this work are: that EQs are stochastic point processes; that the Fano Factor (FF) reveals the fractality of EQs; and that the NHGPPP adjusts to extreme events. The application of these methods to the Guerrero seismic sequence allows us to explain the phenomenological behaviour in the subduction zone.
Traditionally, studies to characterize earthquakes' processes focus on the tectonics mechanism, basically following deterministic approaches. Recently, some studies have investigated the time scale properties of seismic sequences with non-linear statistical approaches so as to understand the dynamics of the process. The deep comprehension of the correlation time structures governing observational time-series can provide information on the dynamical characterization of seismic processes and the underlying geodynamical mechanisms (Telesca et al., 2001).
Scale-invariant processes provide relevant statistical features for characterizing seismic sequences. Since 1944, Gutenberg and Richter have found that earthquake magnitude size follows a power-law distribution. Other scale-invariant features were determined in Kagan (1992, 1994) and Kagan and Jackson (1991). A theory to explain the presence of scale-invariance was proposed by Bak et al. (1988); they introduced the idea of self-organized criticality (SOC), beginning from a simple cellular automaton model, namely a sand pile (Turcotte, 1990; Telesca et al., 2001).

Sieh (1978) and Stuart and Mavko (1979) proposed that earthquakes are due to a stick-slip process involving the sliding of the crust of the earth along faults. When slip occurs at some location, the strain energy is released and the stress propagates in the vicinity of that position. As such, the SOC concept is well-suited for rationalizing observations of the occurrences and magnitudes of earthquakes (Bak and Tang, 1989). An important part of the relaxation mechanism of the crust of the earth is submitted to inhomogeneous increasing stresses accumulating at continental-plate borders (Sornette and Sornette, 1989). The use of scaling laws concerning earthquakes has been especially used to develop models of seismogenesis, and the efforts of the characterization of EQs in Guerrero state on the part of some authors have been devoted to the shape of waves' propagation in order to reduce the uncertainty in magnitude determination and location (Singh et al., 1983; Pardo and Suarez, 1995).

In addition, many authors using several statistical techniques for specific volcanoes have carried out some studies of volcanic eruption time-series. Most of them have been developed within the scope of statistical distributions. Some of the earliest (Wickman, 1965, 1976; Reyment, 1969; Klein, 1982) employed stochastic principles to analyze eruption patterns. Further studies included the transition probabilities of Markov chains (Carta et al., 1981; Aspinall et al., 2006; Bebbington, 2007), Bayesian analysis of volcanic activity (Ho, 1990; Solow, 2001; Newhall and Hoblitt, 2002; Ho et al., 2006; Marzocchi et al., 2008), homogeneous and non-homogeneous Poisson processes applied to volcanic series (De la Cruz-Reyna, 1991; Ho, 1991), a Weibull renewal model (Bebbington and Lai, 1996a, b), geostatistical hazard-estimation methods (Jaquet et al., 2000; Jaquet and Carniel, 2006), a mixture of Weibull distributions (Turner et al., 2008) and, finally, non-homogeneous statistics to link geological and historical eruption time-series (Mendoza-Rosas and De la Cruz-Reyna, 2008). An exhaustive list of the available literature on this subject is made in Mendoza-Rosas and De la Cruz-Reyna (2009).

Along the same research lines, several distributions have been used to model seismic activity. Among these, the Poisson distribution - which implies the independence of each event from the time elapsed since the previous event - is the most extensively used, since in many cases and for large events a simple discrete Poisson distribution provides a good fit (Boschi et al., 1995).

Like some random phenomena, such as noise and traffic in communication systems (Ryu and Meadows, 1994), biological ion-channel openings (Teich, 1989), trapping times in amorphous semiconductors (Lowen and Teich, 1993a,b), seismic events occur at random locations in time. A stochastic point process is a mathematical description which represents these events as random points on the time axis (Cox and Isham, 1980). Such a process may be called fractal if some relevant statistics display scaling, characterized by power-law behaviour - with related scaling coefficients - that indicates that the represented phenomenon as containing clusters of points over a relatively large set of time scales (Lowen and Teich, 1995). Kagan (1994) and Telesca et al. (1999, 2000a,b, 2010) maintain that an earthquake's occurrence might be characterized by clustering properties with both short and long timescales with temporal correlation among the seismic events.

In this paper, we discuss the estimating of the fractality of a point process modelling a seismic sequence, corresponding to the Guerrero coast (the most seismically active area of the southern coast of Mexico), analyzing the performance of the Fano factor. Afterwards, we look at the extreme-value theory applied to NHGPPP so as to quantitatively evaluate the probabilities of extreme EQ occurrences. This work is organized as follows: first, we present the theoretical concepts of stochastic point processes, fractal analysis by Fano factors and NHGPPP; then we present the EQ data series of the Guerrero region; and finally, we show the results of the analysis of this data when treated as stochastic point processes.

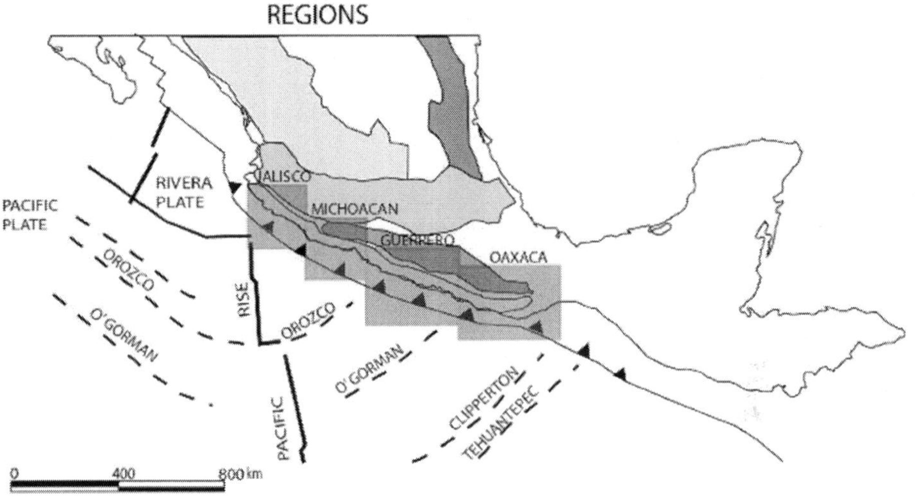

Fig. 1. Four seismicity regions dividing southern Mexico along the Mexican subduction zone, based on the seismicity and shape of the subduction (modified from Singh et al., 1983).

2. Point process

A stochastic point process was described by some authors (Telesca et al., 2001; Cox and Isham, 1980; Lowen and Teich, 1995) in terms of a mathematical description which represents the events as random points on the time axis (Cox and Isham, 1980). Such a process may be called fractal if some relevant statistics display scaling, characterized by power-law behaviour - with related scaling coefficients - that indicates that the represented phenomenon as containing clusters of points over a relatively large set of timescales (Lowen and Teich, 1995).

In this work, any earthquake sequence is assumed to be a realization of a point process, with events occurring at some random locations in time, and it is completely defined by the set of event times - or equivalently - by the set of inter-event intervals. Over a continuous time process, events can occur anywhere on the time axis. In a discrete time point process, the occurrence of events occurs at equally spaced increments. The continuous time point process is a simple Poisson process. If the point process is Poissonian, the occurrence times are uncorrelated; for this memoryless process, the inter-event interval probability density function $P(t)$ behaves as a decreasing exponential function $P(t) = \lambda e^{-\lambda t}$, for $t \geq 0$, with λ as the mean rate of the process.

If the point process is characterized by fractal behaviour, the inter-event interval probability density function $P(t)$ generally decreases as a power-law function of the inter-event time, $P(t) = kt^{-(1+\alpha)}$, with α the so-called fractal exponent (Thurner et al., 1997). The exponent α measures the strength of the clustering and represents the scaling coefficient of the decreasing power-law spectral density of the process $S(f) \propto f^{-\alpha}$ (Lowen and Teich, 1993a,b). The power spectral density furnishes information about how the power of the process is concentrated at various frequency bands (Papoulis, 1990) and it provides information about the nature of the temporal fluctuations of the process.

In recent studies, some authors (Bodri, 1993; Luongo et al., 1996) have focused their attention on the observational evidence of time-clustering properties in earthquake sequences of different seismic areas, demonstrating the existence of a range of time scales with scaling behaviour. The method that they used - the Cantor dust method (Mandelbrot, 1983) - consists of dividing the time interval *T*, over which *N* earthquake occur, into a series of *n* smaller intervals of length $t = T/n$ with n = 2, 3, 4, . . . and computing the number *R* of intervals of length *t* which contain at least one event. If the distribution of events has a fractal structure (Smalley et al., 1987) then $R \approx t^{1-D}$, where *D* is the fractal dimension, which has sub-unitary values: the clustering is higher as D approaches to 0, while a value of 1 corresponds to an uniform distribution (events equally spaced in time). But the parameter *R* does not give information about the temporal fluctuations, because it is not directly correlated to the power spectral density *S(f)* of the process itself.

3. Fractal analysis (Fano factor)

The self-organized critical systems reach the critical steady state with temporary fluctuations in their events characterized by the energy they release. To detect the presence of clustering of events in a time series, several methods can be used among which is the Fano factor calculation, which estimates the value of the fractal exponent α of the study process.

According to such authors as Telesca et al. (2004), assuming a sequence of events is the result of a point process defined by the set of occurrence times. You can use a statistical measure such as the Fano factor $FF(\tau)$ to characterize the process.

For fractal process, that displays clustering properties, *P*(*t*) generally behaves as a power-law function of the inter-event time t with exponent $(1+\alpha)$, were α is called fractal exponent, which characterizes the clustering of the process.

The representation of a point process is given by dividing the time axis into equally spaced contiguous counting windows of duration τ, and producing a sequence of counts $\{N_k(\tau)\}$, with $N_k(\tau)$ denoting the number of earthquakes in the kth window:

$$N_k(\tau) = \int_{t_{k-1}}^{t_k} \sum_{j=1}^{n} \delta\left(t - t_j\right) dt \tag{1}$$

The sequence is a discrete-random process of natural numbers.

The $FF(\tau)$ (Thurner et al., 1997) is a measure of correlation over different timescales. It is defined as the variance of the number of events in a specified counting time τ divided by the mean number of events in that counting time, that is:

$$FF(\tau) = \frac{\left\langle N_k^2(\tau) - N_k(\tau) \right\rangle^2}{\left\langle N_k(\tau) \right\rangle} \tag{2}$$

where $\langle\ \rangle$ denotes the expectation value.
The *FF* varies as a function of counting time τ. The exception is the Homogeneous Poisson Point Process (HPP). For an HPP, the variance-to-mean ratio is always unity for any counting time τ. Any deviation from unity in the value of $FF(\tau)$ therefore indicates that the point process in question is not a homogenous Poisson in nature. An excess greater than the unit reveals that a sequence is less ordered than an HPP, while values below the unit signify sequences that are more ordered.
The $FF(\tau)$ of a fractal point process with $0<\alpha<1$ varies as a function of counting time τ as:

$$FF(\tau)=1+\left(\frac{\tau}{\tau_0}\right)^{\alpha} \tag{3}$$

The monotonic power-law increase is representative of the presence of fluctuations on many timescales (Lowen and Teich, 1995); τ_0 is the fractal onset time and it marks the lower limit for significant scaling behavior in the $FF(\tau)$ (Teich et al., 1996). Therefore a straight-line fit to an estimate of $FF(\tau)$ vs. τ on a doubly logarithmic plot can also be used to estimate the fractal exponent. However, the estimated slope of the FF saturates at unity so that this measure finds its principal applicability for processes with $\alpha<1$.

4. The NHGPPP analysis

Within self-organized critical systems there are a great number of small events; however, the main changes of the system are associated with extreme events. The theory of extreme values is an area of statistics that is devoted to developing statistical models and techniques for estimating the performance of the unusual. These rare events are those which are far from the bulk of the distribution. However, there is no formal definition of extreme events in many cases, being defined as those events that exceed some threshold of magnitude, though they can also be defined as the maximum or minimum of one variable over a certain period. From a statistical standpoint, the problem of extreme value theory is a problem of extrapolation. The basic idea that leads to such extrapolation is that of finding a good parametric model for the tail of the distribution of the data generated by the process that can then be adjusted for extreme observations.
Overall, there are two approaches to the topic of Extreme Value Theory (EVT), a group of older models, known as Block top models and a new group of models known as "Peaks Over Threshold" (POT). The latter group corresponds to a pre-fixed high threshold models (Coles, 2001; Beguería, 2005). EVT focused on peak values above a value *u*, with these values being distributed as a Generalized Pareto Distribution.
The method characterizes the exceeding of a threshold based on the assumption that the occurrence of excesses on a strict threshold of a series characterized by an independent identically distributed random variable has a Poisson behaviour, and that the excesses have an exponential distribution or - more generally - a Generalized Pareto (GP) (Davison and Smith, 1990; Coles, 2001).
The distribution of excess F_u represents the probability of exceeding the threshold "*u*", in at most an amount of "*y*", which is conditioned by the information that has already exceeded the threshold (Cebrian, 1999, Lang et al. 1999; McNeal, 1999).

Definition: a distribution function with two parameters is known as the Generalized Pareto Distribution (GPD).

$$G_{k,a}(y) = \begin{cases} 1-\left(1-\dfrac{ky}{a}\right)^{\frac{1}{k}}, & k \neq 0 \\ 1-e^{\frac{y}{a}}, & k = 0 \end{cases} \tag{4}$$

Where $a > 0$ and k is arbitrary, the range of y is: $0 < y < a/k$ if $k \leq 0, 0 < y < \infty$, if $k > 0$.

The $k < 0$ case is just a re-parameterization of one or more forms of the Pareto distribution, but the extension $k \leq 0$ was proposed by Pickands (1975). The case $k = 0$ is interpreted as the limit when $k \to 0$, (i.e. the exponential distribution).

4.1 Properties for stability threshold

Property 1. If Y is a GPD $u > 0$ a threshold, then the conditional distribution for excesses over a threshold - the conditional distribution $Y - u$ given $Y > u$ - is also distributed as a Generalized Pareto Distribution.

Property 2. If N has a Poisson distribution with conditional on N, where N is the number of the exceedances of a threshold and $Y_1, Y_2, ..., Y_N$ are independent random variables identically distributed as a GP, then $\max(Y_1, Y_2, ..., Y_N)$ for each N follows a Generalized Extreme Value Distribution. Thus, the exceedances satisfy a Poisson process, with excess distributed as a GPD that implies the Classical Distribution of Extreme Values.

Both properties characterize the GPD in the sense that does not exist another family that has these properties.

The excesses of a variable with GPD also follow a GPD - by Property 1 - allowing it to obtain the value(s) of (the) threshold(s) that rise to the extreme values, which also represents a distribution whose parameter values are constant.

Davison and Smith (1990) apply this idea to the expected value of the excess over a threshold u in the case where the GPD is a linear function of the threshold (Diaz, 2003; Beguería, 2005; Lin, 2003).

If $k > -1$, $u > 0$ and $a - uk > 0$ then

$$E(x-u, x > u) = \frac{a-uk}{1+k} \tag{5}$$

On the values above the threshold at which the GPD is adequate, the mean of excess of the sample is,

$$x_u = \frac{\sum_{i,x_i>u} (x_i - u)}{N_u} \tag{6}$$

This should be approximately a linear function of u, where N_u is the number of the exceedance above a predetermined threshold (McNeil and Saladin, 1997; Martínez, 2003, Lin 2003).

Consider the graph of the mean excesses z_u (the sum of positive differences in the magnitude of the fixed threshold and the EQ magnitude that exceeds that threshold, by the number of excesses) against the threshold u:

$$z_u = \frac{\sum_{i,x_i>u} x_i}{N_u} \tag{7}$$

If the assumption that it behaves as a GPD is correct, then the plot should follow a straight line with the intercept $\frac{a}{1+k}$ and slope $\frac{k}{1+k}$. Therefore, it is enough to fit a straight line so as to obtain both parameters (a, k). This is a relatively simple method for corroborating the linear relationship between the mean excess and the threshold u (Davison and Smith, 1990; Coles, 2001; Beguería, 2005, Lin 2003).
The parameters of the GPD for each year are given in Table 2, and the average parameters are calculated and compared in terms of how well they fit with the data annually and how well they make a global settlement for all years. Once obtained, the parameters can be estimated as a GPD function that adjusts the excesses.
Finally the probability of the excesses is obtained by using a non-homogeneous Poisson Pareto process, and for this we need the rate of occurrence and the GPD as a function of the intensity of the NHGPPP, namely:

$$\theta(e) = \frac{N_u}{t}\left[1 - \frac{k(e-u)}{a}\right]^{\frac{1}{k}} \tag{8}$$

The general methodology - as described earlier - is to obtain the variable rate of occurrence of a Non-Homogeneous Poisson Process that we will use for the analysis of the EQs that occurred in Guerrero.

5. Data processing and analysis

In the Mexican Republic - including its territorial sea - five tectonic plates converge: the North American, the Pacific, the Caribbean, the Cocos plate and the plate Rivera (Nava, 1987; Kostoglodov et al., 2001).
The subduction zone includes the entire Pacific coast between Puerto Vallarta in the state of Jalisco to Tapachula in Chiapas state. This extension has produced the largest earthquakes to have occurred this century in Mexico (Kostoglodov et al., 2001). Subduction earthquakes occur mainly in the coastal state of Guerrero. This type of earthquakes is rated as the most dangerous and they deform the ocean floor and generate tsunamis. According to the above description, it is possible to locate Guerrero in a territorial space latent to earthquakes' presence, not only with it being built on ground plate convergence but also by the presence of seismic gaps.
The Guerrero gap is one of the main concerns of researchers, being a rupture zone which for more than ninety years has not recorded an earthquake. Considering the magnitude of past earthquakes, these have ranged between 7.5 and 7.9 degrees and so - according to the

elapsed time - an earthquake is expected to peak at 8.4 degrees, which would represent a greater seismic event than occurred in 1985, which is why in this study we have provided the data epicentres located off the coast of Guerrero - in the area between 15.5 -17.5 N and 98.0-102.0 W for a period of 16 years from 1990 to 2005, as provided by the National Seismological Service (NSS, Servicio Sismologico Nacional).
Figure 2 represents the 4700 epicentres located in the study area. The figure shows that the number of small and medium earthquakes increases towards the state of Oaxaca and we can distinguish a swarm near the Guerrero gap. Note also that the EQs have a greater magnitude away from the coast.

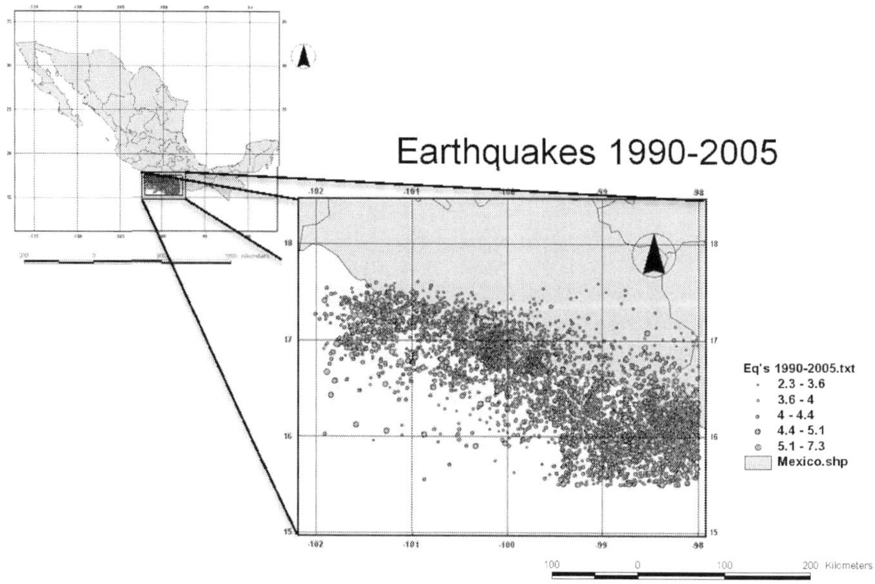

Fig. 2. Location of the 4700 epicenters in the study area (15.5-17.5 N and 98.0-102.0 W) for the 16 year period, 1990-2005, such that the size of each circle is proportional to the magnitude of the event.

5.1 Magnitudes of seismic data for years

First we made a preliminarily analysis of the data per year (Fig. 3 shows two years as an example) in order to determine the extreme events of the series.

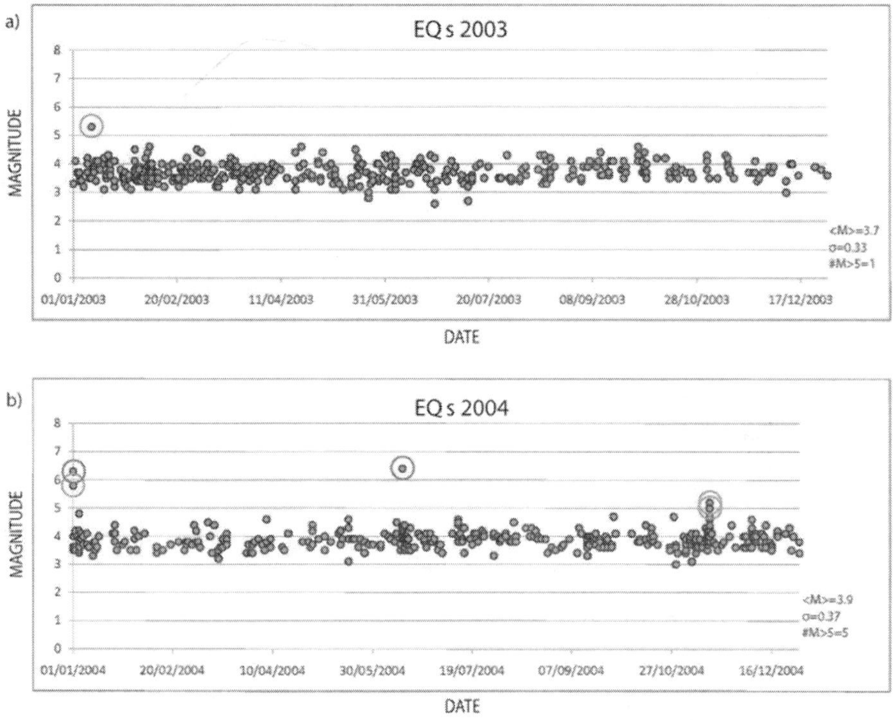

Fig. 3. An example of the EQs' series occurring per year in terms of its magnitude. The selected years are: a) 2003 b) 2004. The green open circles focus the EQs with $5.0 \leq M_n < 6.0$ and magenta circles shows EQs with $M_n \geq 6.0$.

The analysis made of the data finds that the overall average magnitude earthquakes can be considered as 3.9 ± 0.4, and the number of EQs with a magnitude greater than the threshold 5 is between 0-7 per year. However, the years 1997 and 2002 have 9 and 14 events respectively, which doubles the number of events of $M_n > 5$ occurring in the area. The influence of the type of instrumentation used to record the earthquakes is obvious because we observe that events which have a magnitude of less than two are not registered. Equally, as to the implementation of broadband seismographs in 1992, the number of records of EQ magnitudes of less than 3 increased significantly. We also observed that for the analyzed period there are no EQs with $M_n < 7.5$, so we decided to take thresholds corresponding to magnitudes: 3, 4, 5, 6, and 7 for this analysis. This data can be seen concentrated in Table 1, which also shows the dependence between the existence of large-scale EQs and the total number of events.

Year	$M_n > 5$ (year)	No. Max EQ (month)
1990	5	27
1991	5	36
1992	7	24
1993	7	32
1994	3	27
1995	3	98
1996	7	54
1997	9	78
1998	6	57
1999	0	34
2000	0	30
2001	7	201
2002	14	94
2003	1	57
2004	5	58
2005	1	44

Table 1. Extreme events and maximum number of EQs occurring per month and per year

5.2 Temporal clustering analysis of EQs

The self-organized critical systems reach this condition due to temporary fluctuations in their events, where they release much of their energy. For this reason, it is necessary to obtain the Fano factor for calculating the fractal exponent (α), so that it can detect the temporal clustering of events characteristic of the type of event detected.

The exponent of Fano is an estimate of the fractal exponent α of the power law that characterizes the density spectrum of a process with scaling properties. The value of α indicates the degree of clustering in a process according to Thurner et al. (1997). Over long time scales, the curve behaves essentially as $\sim T^{\alpha}$ and the curve can be fit by a straight line of the slope α. When $\alpha \approx 0$, the point process is a Homogeneous Poisson and the occurrence times are uncorrelated. However, if $\alpha \neq 0$, the point process is α Non-Homogeneous Poisson and has scaling properties. The value of α obtained for the studied area - as is shown in figure 4 ($\alpha = 0.6653$) - indicates, as expected, the presence of scaling behavior of the occurrence of the earthquakes.

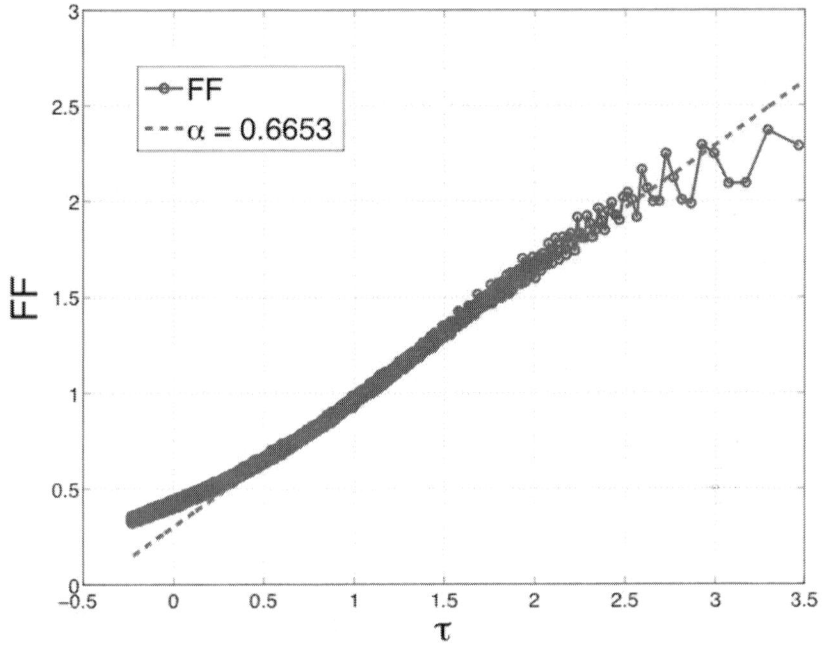

Fig. 4. *FF* curve for $\alpha = 0.6653$. The estimation of the fractal exponent α - indicative of the degree of clusterization - is carried out estimating the slope of the curve plotted in bi-logarithmic scale in the linear range (in red) of the counting times.

5.3 Setting a process of Non-Homogeneous Poisson Pareto

Since the temporal clustering analysis of the EQ series indicates that it is a Non-homogeneous Poisson process, then the methodology - discussed in section 3 - for the analysis NHGPPP is appropriated so as to apply to these data series and in order to calculate the probability of the occurrence of extreme EQs. For the property of the stability threshold it is known that the excesses can be fitted to a GPD, and to validate the method we proceed to make an analysis of the excesses.

To compute the mean excess (that is, the sum of positive differences in the magnitude of the fixed threshold and the magnitude of earthquakes that exceed the threshold, per number of excesses), and the mean exceedance (the sum of the magnitude of the earthquakes that exceed the threshold fixed by the number of the exceedance) we used equations 6 and 7, respectively.

First, you get the graph average exceedance over a threshold and check the feasibility of the linear fit of the observed data with R^2 so as to be close to the unit, in this case for the mean exceedance $R^2 = 0.9832$, which is indicative of the reliability and applicability of the proposed method. Next, we proceed to obtain the shape and scale parameters of the GPD (i.e. k and a, respectively), and for this - as shown in Figure 5 - we fit a straight line by a linear regression in the plot of the mean of the excesses against the magnitude of the fixed threshold.

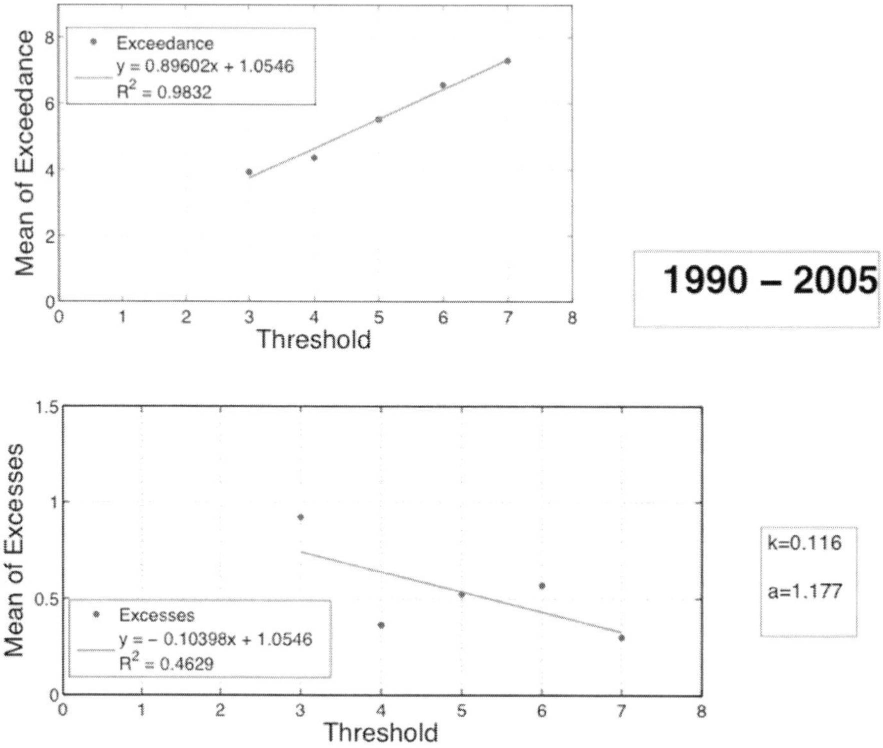

Fig. 5. Fitting of the mean exceedance for the period 1990-2005, with $R^2 = 0.9832$ for the mean of exceedance and $R^2 = 0.4629$ for the mean of excesses.

Once the linear regression is fitted, we get the shape and scale parameters. This procedure was done for earthquakes occurring by year, and the parameters of the GPD for each year of the studied period are given in Table 2; the parameters were also calculated for the whole period - repeating the procedure already explained - and obtaining k and a for the whole period. The comparison between the parameters computed annually and those for the global setting allows us to conclude that the global k and a can be used for all the data in the studied region, as it is shown in figure 6.

The obtained parameters k and a were used to fit the GPD to the excess. Figure 6 presents this comparison between the distribution obtained by k and a per year and for whole period for two selected years as an example of good fitting (year 2004) and the worst of them (year 2003).

In most of the years that were analyzed, the parameters of the GPD average reproduce the behaviour of the data; however, the years 2002, 2003 and 2005 show clear differences in the settings, as these years have atypical features of the studied area, so we will proceed to calculate the probabilities of the whole area for periods of up to 100 years, with the parameters obtained from the full term.

Year	k	a
1990	0.1765	1.5294
1991	0.2195	1.7073
1992	0.25	1.75
1993	0.0695	1.0695
1994	0.1628	1.3953
1995	0.0616	1.1677
1996	0.1381	1.409
1997	0.0891	1.042
1998	0.1635	1.3369
1999	0.1451	1.1776
2000	0.1364	1.1364
2001	0.1905	1.4286
2002	0.0516	0.8448
2003	0.087	0.837
2004	0.1236	1.236
2005	0.2318	1.6239
1990-2005	0.116	1.177

Table 2. Parameters k and a obtained of linear fitting per year and for the complete data series.

To calculate the intensity distribution of the NHGPPP and to obtain the probabilities of the EQs' occurrence, we use equation 8. Since the approach of exceedance implicitly assumes that the scale inherent to the phenomena is open, we force the magnitude scale to ends at $M_n < 9$ and we then subtract the probabilities of the EQs exceeding that magnitude from the probabilities of the lower magnitudes. As it is, the probabilities of occurrence lower than one event of $M_n < 5$ for a period of 100 years is always 1. Using the same equation (8) we compute the values of the intensity function for a threshold of $M_n > 5$ for a period of 100 years, and we obtain $P(M_n > 7) = 1$.

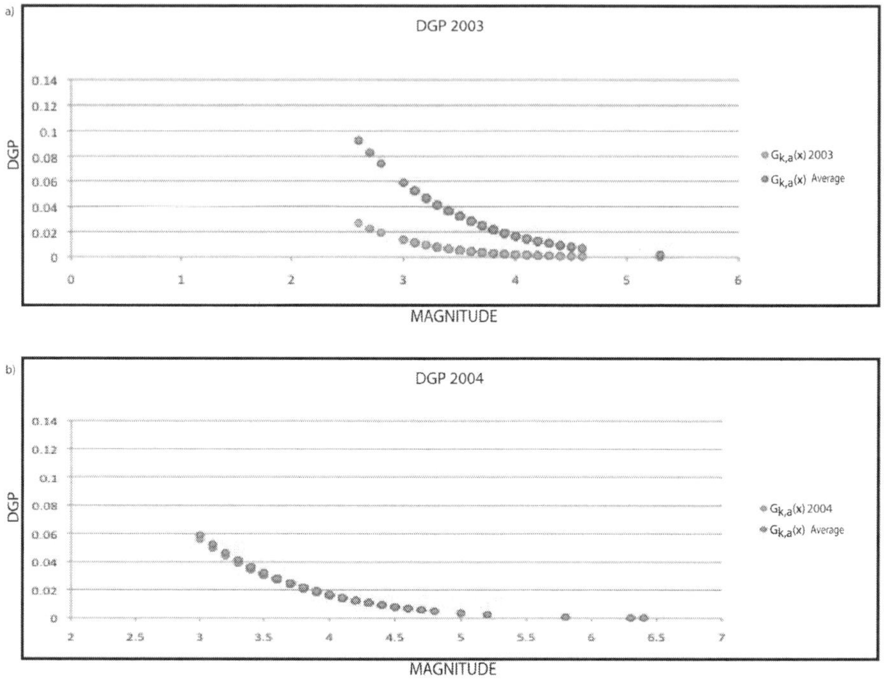

Fig. 6. Graph of the GPD fitted to the excesses using the parameters for each year compared to the excesses GPD adjusted for the period of 16 years, for the years: a) 2003 and b) 2004, providing examples of good and worst adjust.

6. Discussions

One of the goals in describing geological processes is to be able to predict their future behaviour. However, this has not been possible despite the many efforts being made in science. So, a small step is made by the characterization of such processes as earthquakes and volcanic eruptions, allowing for a step towards prediction.

As has been indicated, the data analysis was conducted over a period of 16 years - from 1990 to 2005 - looking at the earthquakes of the Guerrero state recorded by the NSS seismic web.

From the preliminary analysis of the EQ series, it was observed that the influence of the kind of instrumentation used to record EQ events is evident throughout the years, as with the implementation of broadband seismographs in 1992 where the registration of the number of EQ magnitudes of less than 3 increases considerably - and so the behaviour of the data is affected. The mean EQ magnitudes of 3.9 ± 0.4 was computed for all the data, and the number of EQ with $M_n > 5$ is between 0 - 7 per year; however, the years 1997 and 2002 have 9 and 14 events respectively, which doubles the number of events of that magnitude which occurred in the area. In addition, the number of small and medium

earthquakes increases towards the state of Oaxaca and we can distinguish a swarm near the Guerrero gap; also, we note that the EQs are of a greater magnitude away from the coast.
Following this, in order to characterize the EQ events as punctual point process series, the clusterization of this data was allowed by the Fano factor, which indicates an α value of 0.6653. This points out that the EQs follow a Non-Homogeneous Poisson Process.
Next, for the property of the stability threshold, it is known that the excesses can be fitted to a GPD and so we next proceed to an analysis of the excess, obtaining the parameters k and a to fit again to a GPD. The adjusted parameters were computed in two ways: one analysis of the EQs covered each year and the other the whole period (16 years). In most of the studied years, the adjustment of the mean of the GPD parameters reproduced the behaviour of all the data, except for the years 2002, 2003 and 2005, which show clear differences in the settings (as these years have atypical features of the area, the calculation was done with the average parameters).
The last step was to calculate the intensity distribution of the NHGPPP and to obtain the probabilities of the EQs' occurrence with the specific magnitudes in which we are interested. Because the approach of exceedance implicitly assumes an open scale of the phenomena, we assume that the magnitude scale ends at $M_n < 9$, and we compute the values of the intensity function for a threshold of $M_n > 7$ for a period of 100 years, and so we obtain $P(M_n > 7) = 1$.

7. Conclusions

Finally we can conclude that the parameters determined from the seismic data for the 16 year period allows us to calculate the fractal dimension of the data -$\alpha = 0.6653$ - it demonstrates the presence of temporal clusterization and variations in values linked to the occurrence of a high magnitude events. This fact allows us to say that the analyzed data follows a Non-Homogeneous Poisson Process and we can use a NHGPPP to characterize the distribution intensity of the data. As such, a good approximation of the probability of the occurrence of earthquakes of magnitudes greater than 7 for a period of 100 years in the area of Guerrero sees a finding of $P(M_n > 7) = 1$.

8. Acknowledgments

Authors are indebted to the National Seismological Service (SSN, Servicio Sismológico Nacional) dependent of National Autonomous University of Mexico (UNAM) for provided the EQ data used for this study, in particular to Casiano Jiménez Cruz for the corrected seismic catalogue. Sharon M. Valverde-Esparza was supported by the Consejo Nacional de Ciencia y Tecnología Scholarship.

9. References

Aspinall, W.P., Carniel, R., Jaquet, O., Woo, G., Hincks, T., 2006. Using hidden multi-state Markov models with multi-parameter volcanic data to provide empirical evidence for alert level decision-support. J. Volcanol. Geotherm. Res. 153, 112124.
Bak, P., Tang, C., Wiensenfeld, K., 1988. Self-organizing criticality. Phys. Rev. A 38, 364–374.

Bak, P., Tang, C., 1989. Earthquakes as a self-organized critical phenomenon. J. Geophys. Res. 94, 15635-15637.

Bebbington, M.S., 2007. Identifying volcanic regimes using hidden Markov models. Geophys. J. Int. 171, 921-942.

Bebbington, M.S., Lai, C.D., 1996a. Statistical analysis of New Zealand volcanic occurrence data. J. Volcanol. Geotherm. Res. 74, 101-110.

Bebbington, M.S., Lai, C.D., 1996b. On nonhomogeneous models for volcanic eruptions. Math. Geol. 28, 585-600.

Beguería, S., 2005, Uncertainties in Partial Duration Series Modelling of Extremes Related of the Choice of the Threshold Value, Journal of Hidrology, vol. 303, pp. 215-230.

Bodri, B., 1993. A fractal model for seismicity at Izu-Tokai region, central Japan. Fractals 1, 539-546.

Boschi, E., Gasperini, P., Mulargia, F., 1995. Forecasting where larger crustal earthquakes are likely to occur in Italy in the near future. Bull. Seism. Soc. Am. 85, 1475-1482.

Carta, S., Figari, R., Sartoris, G., Sassi, E., Scandone, R., 1981. A statistical model for Vesuvius and its volcanological implications. Bull. Volcanol. 44, 129-151.

Cebrián, A. C., 1999. Análisis, Modelación y Predicción de Episodios de Sequía, Tesis doctoral, Universidad de Zaragoza.

Coles, S. 2001, An Introduction to Statistical Modeling of Extreme Values, Springer-Verlang London.

Cox, D.R., and Isham, V., 1980. Point Processes. Chapman & Hall, London.

Davison, A. C., Smith, R. L., 1990. Models for Exceedances over High Thresholds, Journal of the Royal Statistical Society. Series B, vol. 52, pp. 393- 442.

De la Cruz-Reyna, S., 1991. Poisson-distributed patterns of explosive activity. Bull. Volcanol. 54, 57-67. Southern Idaho and Central Mexico. Brigham Young Universi ty Geology Studies,

Díaz Hernández, A., 2003, Teoría de Valores Extremos para Sucesiones de Variables Aleatorias Dependientes, Tesis Licenciatura (Actuario) UNAM, Facultad de Ciencias.

Ho, C.-H., 1990. Bayesian analysis of volcanic eruptions. J. Volcanol. Geotherm. Res. 43, 91-98.

Ho, C.-H., 1991. Nonhomogeneous Poisson model for volcanic eruptions. Math. Geol. 23, 167-173.

Ho, C.H., Smith, E.I., Keenan, D.L., 2006. Hazard area and probability of volcanic disruption of the proposed high-level radioactive waste repository at Yucca Moun- tain, Nevada, USA. Bull. Volcanol. 69, 117-123.

Jaquet, O., Löw, S., Martinelli, B., Dietrich, V., Gilby, D., 2000. Estimation of volcanic hazards based on Cox stochastic processes. Phys. Chem. Earth 25, 571-579.

Jaquet, O., Carniel, R., 2006. Estimation of volcanic hazard using geostatistical models. In: Mader, H.M., Coles, S.G., Connor, C.B., Connor, L.J. (Eds.), Statistics in Volcanology. IAVCEI Publications n.1. Geological Society, London, pp. 89-103.

Kagan, Y.Y., 1992. Correlations of earthquake focal mechanism. Geophys. J. Int. 110, 305-320.

Kagan, Y.Y., 1994. Observational evidence for earthquakes as a nonlinear dynamic process. Physica D 77, 160-192.

Kagan, Y.Y., Jackson, D.D., 1991. Long-term earthquake clustering. Geophys. J. Int. 104, 117-133.

Klein, F.W., 1982. Patterns of historical eruptions at Hawaiian volcanoes. J. Volcanol. Geotherm. Res. 12, 1-35.

Kostoglodov V., R. Valenzuela, A. Gorabtov, J. Mimiaga, S.I. Franco, J.A. Alvarado, R. Peláez, 2001, *Deformation in the Guerrero seismic gap, Mexico, from leveling observations*, J. Geodesy, vol. 75, pp. 19-32.

Lang, M., Ouarda, T. b. M. J., Bobée B., 1999, Towards operational guidelines for over-threshold modeling, Jounal of Hydrology vol. 225, pp. 103-117.

Lin, X. G., 2003. Statistical modeling of severe wind gust, International Congress on Modelling and Simulation, Townsville, vol. 2, pp. 620-625.

Lowen, S.B., Teich, M.C., 1993a. Estimating the dimension of a fractal point process. Proc. SPIE 2036 (Chaos in Biology and Medicine), 64–76.

Lowen, S.B., Teich, M.C., 1993b. Fractal renewal processes generate $1/f$ noise. Phys. Rev. E 47, 992–1001.

Lowen, S.B., Teich, M.C., 1995. Estimation and simulation of fractal stochastic point processes. Fractals 3, 183–210.

Luongo, G., Mazzarella, A., Palumbo, A., 1996. A fractal approach to clustering of the 1983–1984 seismicity in the Campi Flegrei Caldera, southern Italy. Fractals 4, 29–34. Biology. CRC Press, Boca Raton, FL, 1996, pp. 383–412.

Mandelbrot, B.B., 1983. The Fractal Geometry of Nature, W.H. Freeman, New York

Martínez, R. P. A., 2003, Un Modelo Estadístico para Valores extremos en Series de Datos Actuariales, Tesis de Licenciatura (Actuario) UNAM, Facultad de Ciencias.

Marzocchi, W., Sandri, L., Selva, J., 2008. BET_EF: a probabilistic tool for long- and shortterm eruption forecasting. Bull. Volcanol. 70, 623–632.

McNeil, A. J. y Saladin, T., 1997, The Peaks Over Thresholds Method for Estimating High Quantiles of Loss Distributions, Coloquio 28th international ASTIN Bulletin.

McNeil, A. J., 1999, Extreme Value Theory for Risk Managers, International Modelling and CAD II, Risk Book, pp. 99-113.

Mendoza, R. A. T., 2007, Análisis de Peligro para los Volcanes Colima, Citlaltepetl y Popocatepetl, Tesis de Maestría UNAM, Programa de Posgrado en ciencias de la Tierra, Instituto de Geofísica.

Mendoza-Rosas A. T., and De la Cruz-Reyna S., 2008. A statistical method linking geological and historical eruption time series for volcanic hazard stimations: applications to active polygenetic volcanoes. Journal of Volcanology and Geothermal Research 176 (2008) 277–290

Mendoza-Rosas A. T., and De la Cruz-Reyna S., 2009. A mixture of exponentials distribution for a simple and precise assessment of the volcanic hazard. Nat. Hazards Earth Syst. Sci., 9, 425–431, www.nat-hazards-earth-syst-sci.net/9/425/2009/

Nava, A., 1987, Terremotos, La Ciencia desde México, Fondo de Cultura Económica.

Newhall, C.G., Hoblitt, R.P., 2002. Constructing event trees for volcanic crises. Bull. Volcanol. 64, 3–20.

Papoulis, A., 1990. Probability, Random Variables, and Stochastic Processes. McGraw-Hill, New York

Pardo, M. And Suarez, G., 1995. Shape of the subducted Rivera and Cocos plates in southern Mexico: Seismic and tectonic implications. Jour. Geophys. Res., vol. 100, No. B7, pp. 12,357-12,373.

Reyment, R.A., 1969. Statistical analysis of some volcanologic data. Regarded as series of point events. PAGEOPH 74 (3), 57–77.

Ryu, B.K., Meadows, H.E., 1994. Performance analysis and traffic behavior of Xphone videoconferencing application on an Ethernet. In: W. Liu (Ed.), Proceedings of the 3rd International Conference on Composition Community Network, pp. 321-326.Phys. Earth Planet. Int. 120, 315–326.

Sieh, K.E., 1978. Prehistoric large earthquakes produced by slip on the San Andreas Fault at Pallet Creek. J. Geophys. Res. 83, 3907–3917.

Singh, S.K., Rodriguez, M., and Esteva, L., 1983. Statistics of small earthquakes and frequency of occurrence of large earthquakes along the Mexican subduction zone. Bulletin of the Seismological society of America, vol. 73, No. 6, pp. 1779-1796.

Smalley, R.F., Chatelain, J.-L., Turcotte, D.L., Prévot, R., 1987. A fractal approach to the clustering of earthquakes: applications to the seismicity of the New Hebrides. Bull. Seism. Soc. Am. 77 (4), 1368–1381.

Solow, A.R., 2001. An empirical Bayes analysis of volcanic eruptions. Math. Geol. 33 (1), 95102.

Sornette, A., Sornette, D., 1989. Self-organized criticality and earthquakes. Europhys. Lett. 9, 197–202.

Stuart, W.D., Mavko, G.M., 1979. Earthquake instability on a strike-slip fault. J. Geophys. Res. 84, 2153–2164.

Teich, M.C., 1989. Fractal character of the auditory neural spike train. IEEE Trans. Biomed. Eng. 36, 150–160.

Teich, M.C., Heneghan, C., Lowen, S.B., Turcott, R.G., 1996. Estimating the fractal exponent of point processes in bilogical systems using wavelet- and Fourier-transform methods. In: Aldroubi, A., Unser, M. (Eds.), Wavelets in Medicine and Biology. CRC Press, Boca Raton, FL, 1996, pp. 383–412.

Telesca, L., Cuomo, V., Lanfredi, M., Lapenna, V., Macchiato, M., 1999. Investigating clustering structures in time-occurrence sequences of seismic events observed in the Irpinia-Basilicata region (southern Italy). Fractals 7, 221–234.

Telesca, L., Cuomo, V., Lapenna, V., Vallianatos, F., 2000a. Self-similarity properties of seismicity in the southern Aegean area. Tectonophysics 321, 179–188.

Telesca, L., Cuomo, V., Lapenna, V., Macchiato, M., 2000b. Analysis of time-scaling behaviour in the sequence of the aftershocks of the Bovec (Slovenia) April 12, 1998 earthquake. Phys. Earth Planet. Int. 120, 315–326.

Telesca, L., Cuomo, V., Lapenna, V. and Macchiato, M., 2001, Statistical analysis of fractal properties of point processes modelling seismic sequences, Phys. Earth Planet. Int., 125, 65-83.

Telesca L, Colangelo G, Lapenna V, Macchiato M. 2004. On the scaling behavior of rain event sequence recorded in Basilicata region (Southern Italy). *Journal of Hydrology* 296: 234-240.

Telesca, L. and Chen, C.-C., 2010, Nonextensive analysis of crustal seismicity in Taiwan, Nat. Hazards Earth Syst. Sci., 10, 1293-1297.

Thurner, S., Lowen, S.B., Feurstein, M.C., Heneghan, C., Feichtinger, H.G., Teich, M.C., 1997. Analysis, synthesis, and estimation of fractal-rate stochastic point processes. Fractals 5, 565–596.

Turner, M. B., Cronin, S. J., Bebbington, M. S., and Platz, T.: Developing probabilistic eruption forecasts for dormant volcanoes: a case study from Mt Taranaki, New Zealand, Bull. Volcanol., 70, 507–515, 2008.

Turcotte, D.L., 1990. Fractal and chaos in geology and geophysics. Cambridge University Press, Cambridge, 1990, p. 221.

Wickman, F.E., 1965. Repose period patterns of volcanoes, 5. General discussion and a tentative stochastic model. Ark. Mineral. Geol. 4, 351-367.

Wickman, F.E., 1976. Markov models of repose-period patterns of volcanoes. In: Merriam, DF (Ed.), Random Processes in Geology. Spŕinger-Verlag, Berlín, pp. 135–161.

13

Rock-Fluid Interaction Along Seismogenic Faults Infered from Clay Minerals in Okitsu Mélange, the Cretaceous Shimanto Belt, SW Japan

Yoshitaka Hashimoto and Umihiko Kaji
Kochi Univeristy
Japan

1. Introduction

Rock-fluid interactions along seismogenic faults are significant issues, because they are strongly related to seismogenic mechanisms and also to modifications of the seismogenic fault itself. Various mechanisms for seismogenesis have been proposed, such as frictional melting (Sibson, 1975; Spray, 1992), thermal pressurization (Mase and Smith, 1987; Melosh, 1979; O'Hara et al., 2006; Sibson, 1977; Wibberley and Shimamoto, 2005), acoustic fluidization (Melosh, 1979; Otsuki et al., 2003), elastohydrodynamic lubrication (Brodsky and Kanamori, 2001), and silica gel lubrication (Di Toro et al., 2006). Some of these are related to frictional heating. Heating signatures from natural faults have been well-studied on the basis of the remaining grains in pseudotachylyte along faults (Ikesawa et al., 2003; Ujiie et al., 2007), the vitrinite reflectance anomaly (O'Hara et al., 2006), borehole logging (Kano et al., 2006; Mishima et al., 2006; Tanaka et al., 2006), the thermal decomposition of paramagnetic minerals (Mishima et al., 2006), and the distribution of minor elements (Ishikawa et al., 2008). In addition to the thermal effects, some of the seismogenic mechanisms are also strongly related to rock-fluid interactions. Studies focusing on rock-fluid interactions along fossil seismogenic faults have been conducted at some major fault zones. These include the Nojima fault in Japan (an intra-crustal seismogenic fault), where bulk rock chemistry analysis was used (Tanaka et al., 2007), the Chi-Chi fault in Taiwan (an active subduction plate boundary fault), which was studied on the basis of its clay characteristics (Hashimoto et al., 2008; Hashimoto et al., 2007), fossil faults such as the Mugi mélange, in the Shimanto Belt, Japan, again using bulk rock chemistry (Hashimoto et al., 2009), and an out of sequence thrust in the Shimanto Belt, Japan, using minor element distributions (Honda et al., 2011; Yamaguchi et al., 2011).

In this study, we focused on clay minerals within the fossil seismogenic fault along the subduction interface, in order to understand rock-fluid interactions at the fault. Clay minerals are commonly produced along faults, possibly by alteration of fine-grained abraded host rock materials due to rock-fluid interaction. The characteristics of clay minerals along seismogenic faults, in comparison with those of host rocks, provide clues to help understand rock-fluid interactions at the time of seismogenesis or related phenomena.

The studied fault is a fossil seismogenic fault along a subduction interface, in the Okitsu melange, the Cretaceous Shimanto Belt, SW Japan. The Shimanto Belt is the most studied on-land accretionary complex in the world, with lithology, age, thermal structure, and deformation structures available. These studies have revealed that the Shimanto Belt includes a deformation along its subduction interface from underthrusting to underplating, and that the Shimanto Belt is experienced at the seismogenic depth on the basis of the thermal model for seismogenic zones (Hyndman and Wang, 1993; Oleskevich et al., 1999). At the northernmost boundary fault of the Okitsu melange, the first pseudotachylyte within the sedimentary rocks was reported (Ikesawa et al., 2003), indicating that the fault was formed by melt lubrication along the subduction interface.

We conducted an X-ray diffraction (XRD) analysis on the host and fault rocks along the fossil seismogenic fault, and examined mineralogy, iron and magnesium substitution in chlorite, illite crystallinity, and semi-quantification of illite and chlorite to determine the clay characteristics for seismogenic fault rocks, in comparison with those of the host rocks. Finally, characteristic rock-fluid interactions in seismogenic faults, due to melt lubrication along the subduction interface, are discussed.

2. Geological setting of Okitsu melange and the northern boundary fault zone with pseudotachylytes

The Shimanto Belt is an ancient accretionary complex exposed on land from the Kanto region to the Okinawa islands, Japan, almost parallel to the Nankai Trough (Fig. 1). The Shimanto Belt is divided into two units on the basis of age, the northern Cretaceous Unit and the southern Tertiary Unit (Taira et al., 1988) (Fig. 1A). On Shikoku Island, the 4th largest island in Japan, the Shimanto Belt is bounded by the Butsuzo Tectonic Line from the Chichibu Belt, a Jurassic accretionary complex (Fig. 1B), and is further classified by its lithology as a melange unit and a coherent unit (Taira et al., 1988) (Fig. 1B). A melange unit is composed of chaotic rocks representing blocks in matrix textures. Most of the melange in the Shimanto Belt is tectonic in origin (e.g., Kimura and Mukai, 1991; Onishi and Kimura, 1995; Hashimoto and Kimura, 1999). A coherent unit is composed mainly of an alternation of sandstone and mudstone with weaker deformations (Taira et al.1988).

The study area is in the Okitsu mélange, Shikoku Island, SW Japan. The Okitsu melange is located at the southern end of the Cretaceous Unit (Fig. 1). The main lithology of the Okitsu melange is sandstone and black shale showing tectonic melange textures, minor basalts, cherts, red shale, and tuff. Foliations are well developed within the shale matrices. The mélange foliations strike ENE–WSW and dip steeply to the north. The radiolarian age of the Okitsu melange is Cenomanian to Turonian from cherts, and Santonian to Campanian from black shale (Taira et al., 1988). The paleo-maximum temperature of the mélange is about 270 (± 30)°C, based on vitrinite reflectance (Sakaguchi, 1999).

The northern boundary fault bounds the Okitsu mélange from Nonokawa formation, a coherent unit in the north (Fig. 1B). The fault zone is about 5–10 m wide (Fig. 2), and mainly develops within the Okitsu melange. Basalt blocks, and black shale matrices mixed with tuffaceous shale in some parts, are included within the fault zone, which strikes ENE–WSW, almost parallel to the melange foliations (Fig. 2).

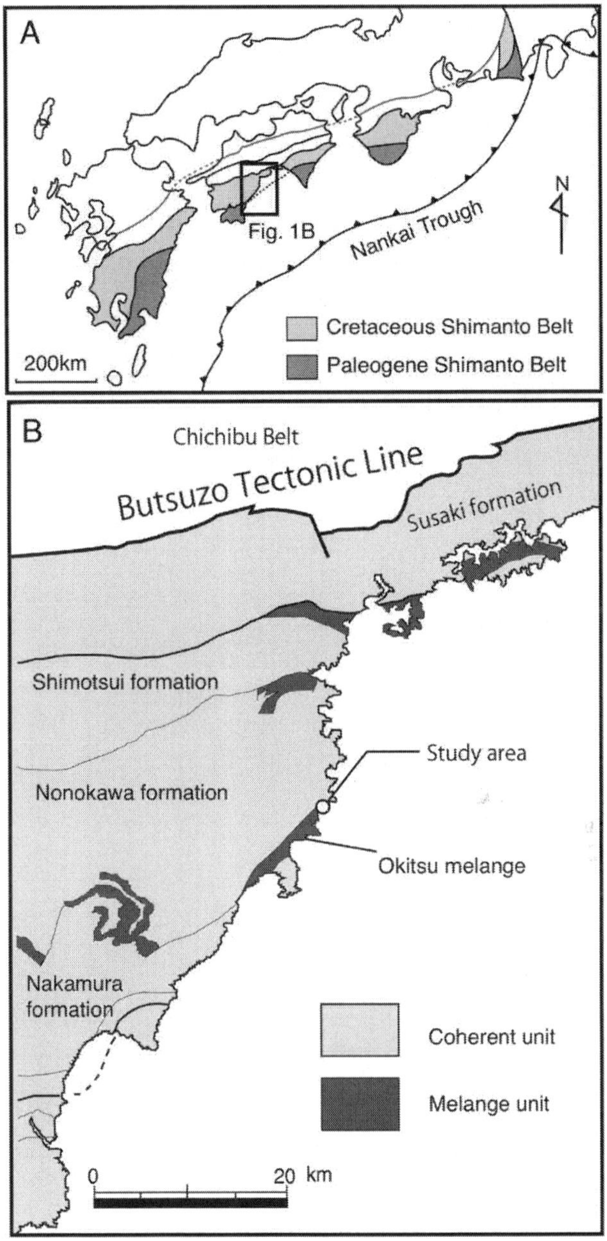

Fig. 1. A) Distribution of the Shimanto Belt along SW Japan. B) Distribution of mélange and coherent units in SW Shikoku Island. Study area is also shown.

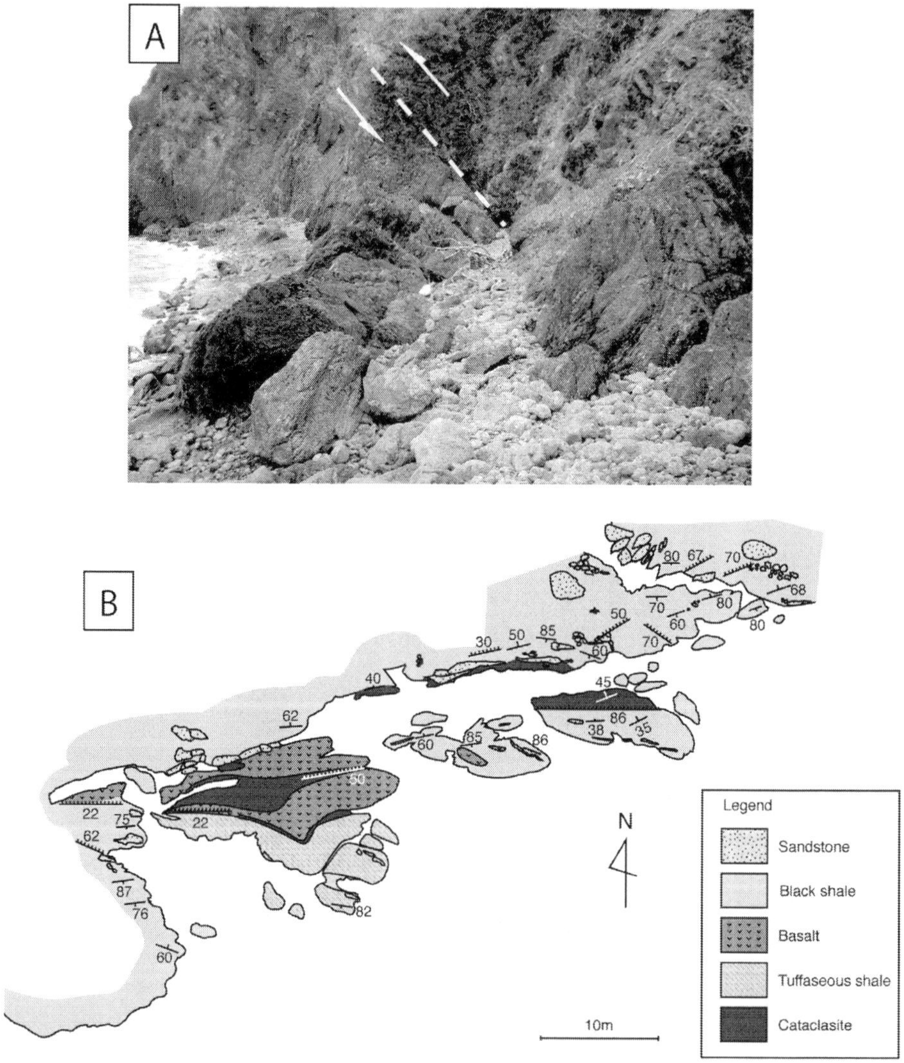

Fig. 2. A) A photo of the study area from ENE to WSW. B) Route map of the northern boundary fault zone between Okitsu melange and Nonokawa formation.

Cataclasites of 2–3 m in thickness are well observed along the fault zone (Fig. 2). The cataclasites include pseudotachylyte, as reported by Ikewasa et al. (2003). They estimated a temperature rise for the fault of at least 450°C, based on the mineral assemblage of the pseudotachylyte.

We focused on two kinds of fault rocks. One is the tectonic melange, as the host rock, and the other is the cataclasites containing pseudotachylyte, as fossil seismogenic fault rocks. The latter type are developped within the host rocks of the tectonic melanges, and thus a comparison of clay minerals in the host rocks and the cataclasites provides the characteristic modification of clay minerals along seismogenic faults.

3. Occurrence of tectonic mélange and Cataclasites with pseudotachylyte

The occurrence of tectonic melanges and cataclasites with pseudotachylyte in outcrop scale is represented in Fig. 3.

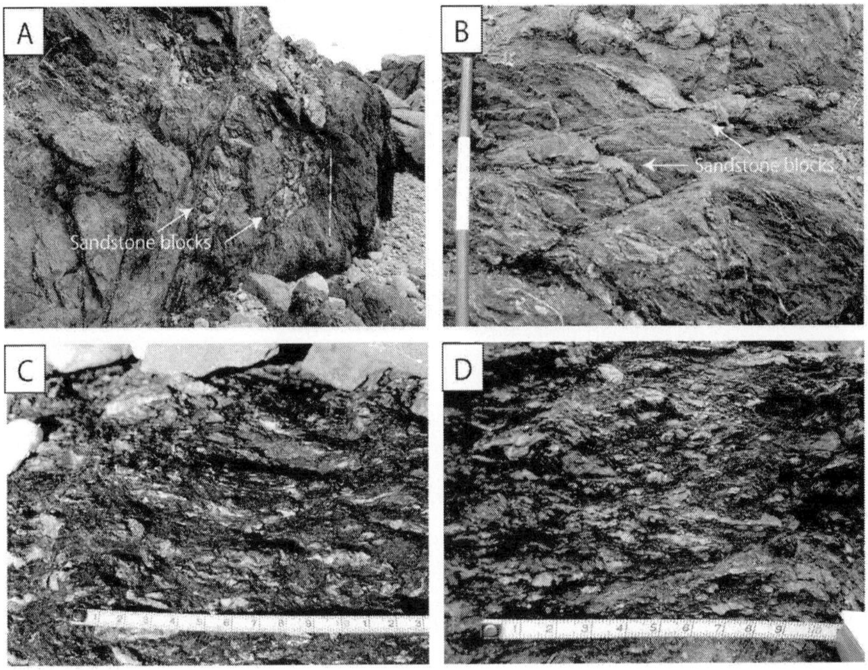

Fig. 3. Occurrences of tectonic melanges (host rocks) and cataclasites with pseudotachylyte in outcrop scale. A) A photo of tectonic melanges. Color bar indicates 1 m length. B) A close-up photo of tectonic melanges. White interval represents 10 cm. C) A photo of cataclasites showing quartz grains surrounded by black matrices. D) A photo of cataclasite. Very thin faults are developed within the cataclasites.

The melanges show blocks in matrix textures, as commonly reported from other tectonic melange zones. The blocks are asymmetrically shaped, indicating that the shear deformation is strongly related to the texture formation (Figs. 3A and B). The melange blocks of sandstone range from a few cm to about 2-3 m in diameter in outcrop scale (Figs. 3A and B).

Foliations are well developed in the shale matrices, representing composite planar fabrics, with the interval between foliations on the scale of mm. Micro-faults, with a thickness of less than 1 cm and a displacement of less than 1 m, can also be observed cutting into the melange fabrics in outcrop scale (Figs. 3A and B). Most of the micro-faults are accompanied by quartz and calcite veins. Some mineral veins in the study area are ankerite (Fe–Mg carbonate). The relationship between the micro-faults and cataclasites containing pseudotachylyte is unknown.

The cataclasites with pseudotachylyte are composed of relatively small grains (less than a few cm diameter) of quartz and calcite surrounded by black material (Figs. 3C and D). These blocks also have an asymmetric shape, and the long axis of the blocks is aligned in the same direction as the melange fabrics (Figs. 3C and D). In some parts, very thin (less than 1 mm), continuous faults are observed within the cataclasites, although these thin faults are obscure (Fig. 3D).

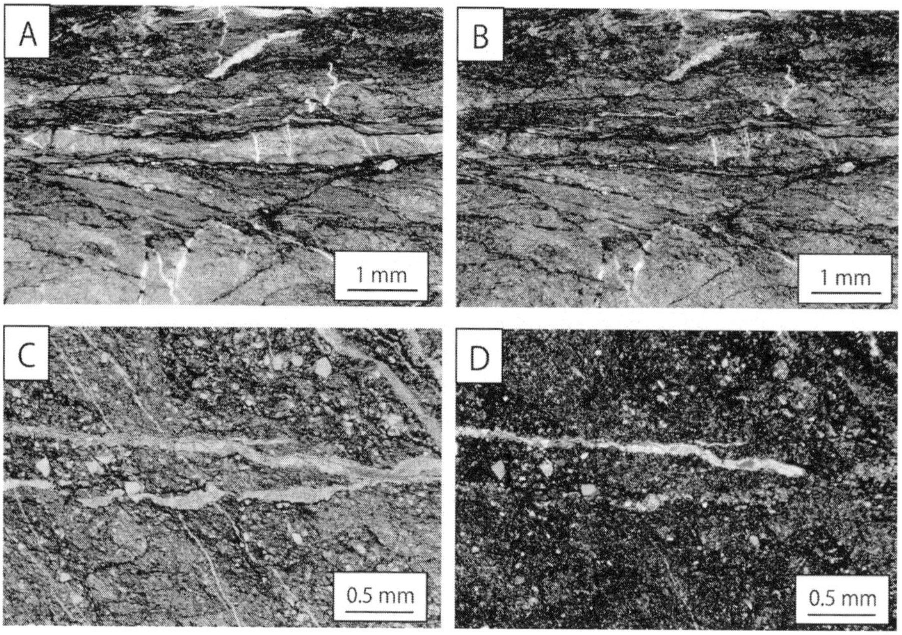

Fig. 4. Micro-textures of tectonic melanges. A) Black seams (Pressure solution cleavages) are well developed in shale matrices. B) A photo of Fig. 4A under cross poralized light. C) Very weak anastomosed pressure solution cleavage in sandy shale matrices. D) A photo of Fig. 4C under cross polarized light.

At the microscopic scale, the occurrence of tectonic melanges and cataclasites with pseudotachylyte is also distinctive.

Microscopic occurrence of tectonic melanges is characterized by a weak pressure solution cleavage within shale matrices (Fig. 4). The pressure solution cleavages develop along melange foliations, also representing composite planar fabrics. In coarser grained areas, pressure solution cleavages are weakly observed, showing anastomosed networks of pressure solution cleavages (Figs. 4C and D).

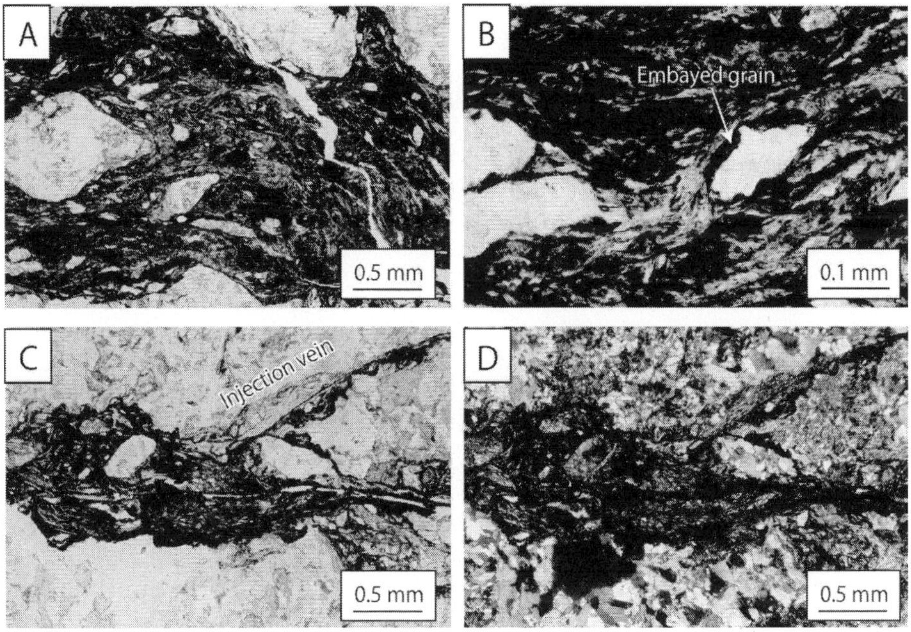

Fig. 5. Micro-texture of cataclasites with pseudotachylyte. A) Quartz grains are surrounded by shale matrices. Thin faults and micro-folding are identified within the shale matrices. B) Embayed grains surrounded by shale matrices. C) Injection vein from main fault surface (Horizontal). The boundary with host rocks is embayed. D) A photo of Fig. 5C under cross-polarized light.

In cataclasites with pseudotachylyte, highly fractured grains surrounded by shale matrices are observed (Figs. 5A and B). The grains are composed mainly of quartz aggregates, with grain size ranging from tens of μm to a few mm. The shale matrices represent highly deformed fabrics with lighter and darker brownish materials (Figs. 5A and B). Very thin and sharp faults are recognized within the shale matrices of the cataclasites (Fig. 5A and B). The shape of grains is embayed (Fig. 5B), and injection veins from the main shear surface can also be seen (Figs. 5C and D). Along the main shear surface, highly fractured cataclasites are observed. The boundary between host rocks and cataclasites shows embayed texture.

Injection veins from the main shear surface also have an embayed boundary with the host rocks (Figs. 5C and D).

4. Method

We analyzed the clay and other minerals using an X-ray diffractometer (MultiFlex, RIGAKU) for randomly oriented and oriented samples. Randomly oriented samples were analyzed for bulk rock samples. The oriented samples were prepared using <1.4 μm grains as clay size fraction. Oriented samples were further analyzed using an ethylenglicoled treatment. The XRD analysis was conducted under the following conditions: 45 kV, 40 mA of Cu kα radiation, step size of 0.01°, and a 2θ range of 2-35°.

From the XRD charts obtained, we examined bulk and clay mineralogy, iron and magnesium substitution in chlorite, and illite crystallinity, and performed a semi-quantification of illite and chlorite in the samples. The peak intensities were obtained using MacDiff 4.2.5. Twenty samples each of tectonic melanges and cataclasites were analyzed.

Illite crystallinity is expressed by a width of the illite 001 peak at half of the peak height above the background for an oriented, < 2 μm fraction of sample (Kubler, 1969). The width is controlled by X-ray-scattering-domain size and percentage of expandable layers (Srodon and Eberl, 1984; Eberl and Velde, 1989). A smaller scattering domain and/or more expandable layers would lead to a wider peak.

The chlorite in the study is Fe–Mg chlorite from both the host melanges and cataclasites with pseudotachylyte, based on the bulk rock analysis. Chlorite is composed of silicate and hydroxide layers, both layers having three sites for positive ions. The substitution of iron and magnesium in the chlorite layers was estimated from the XRD charts, following the method of Moore and Reynolds (1989). I(003)/I(005) gives the symmetry of the Fe distribution (the D value in Moore and Reynolds (1989)) and [I(002) + I(004)]/I(003)′ gives the total number of Fe atoms in six octahedral sites (the Y value in Moore and Reynolds (1989)). I(003)′ is calculated from the following equation (Brown and Brindley, 1980):

$$I(003)' = \frac{I(003)(114)^2}{(114 - 12.1D)^2} \tag{1}$$

Reference frame in the configuration described above for the number of iron and magnesium from 0 to 3 at intervals of 0.5 in the silicate and hydroxide layers (the total number of patterns is 49), respectively, were calculated by NEWMOD (Reynolds Jr, 1985). The results of the NEWMOD calculations in the I(003)/I(005) vs. [I(002) + I(004)]/I(003)′ space are shown in Fig. 7 as dotted lines.

For the semi-quantitative analysis of illite and chlorite, we used the Mineral Intensity Factor (MIF) method (Moore and Reynolds, 1989). To obtain the MIF value, we computed the mineral reference intensities for illite and chlorite also using the NEWMOD (Moore and Reynolds, 1989). As the MIF value for chlorite depends on its composition, we used the result from the examination of iron-magnesium substitution in chlorite described above. We used an illite composition of 0.1 Fe and 0.75 K as a reference mineral. Values of $\mu^* = 14$ and sigma* = 25 were used in the NEWMOD calculations, as suggested by Moore and Reynolds (1989).

sample	Chlorite	illite	quartz	calcite	anorthite	ankerite
ok060501-2	X	X	X		X	
ok060501-3	X	X	X	X	X	
ok060902-1	X	X	X	X	X	
ok060902-4	X	X	X		X	
ok060902-5	X	X	X		X	
ok060902-6	X	X	X	X	X	
ok060902-7	X	X	X	X	X	
ok060902-14	X	X	X	X	X	
ok061004-18	X	X	X		X	
ok061004-19	X	X	X	X	X	
ok060501-7	X	X	X	X	X	
ok060616-7	X	X	X	X	X	
ok060620-1	X	X	X	X	X	
ok060902-8	X	X	X	X	X	
ok060919-6	X	X	X		X	
ok060919-7	X	X	X		X	
ok060919-8	X	X	X		X	
ok060919-9	X	X	X		X	
ok060919-10	X	X	X		X	
ok060919-11	X	X	X		X	

Table 1. Minerals in host melanges

sample	Chlorite	illite	quartz	calcite	anorthite	ankerite
ok060428-2	X	X	X		X	
ok060620-4	X	X	X		X	
ok060620-13	X	X	X		X	
ok060902-13	X	X	X		X	
ok061015-7	X	X	X		X	
ok061015-8	X	X	X		X	
ok061015-9	X	X	X		X	
ok061015-10	X	X	X	X	X	
ok061015-11	X	X	X	X	X	
ok061015-12	X	X	X	X	X	
ok060501-8	X	X	X	X	X	
ok060616-4	X	X	X	X	X	
ok060902-9	X	X	X	X	X	
ok060902-12	X	X	X	X	X	
ok060919-12	X	X	X	X	X	
ok061004-23	X	X	X	X	X	
ok061004-24	X	X	X	X	X	
ok061004-25	X	X	X	X	X	X
ok061004-26	X	X	X		X	
ok061004-27	X	X	X	X	X	X

Table 2. Minerals in cataclasites with pseudotachylyte

5. Results

In this section, we describe the results of our analysis for clay and other mineralogy, iron and magnesium substitution in chlorite, illite crystallinity, and the semi-quantification of illite and chlorite.

5.1 Clay and other mineralogy

All samples, from both the host melanges and the cataclasites, included quartz, anorthite, illite, and chlorite (Tables 1 and 2), and some of the samples also contained calcite. Ankerite (Fe–Mg carbonates) were found in a number of samples from cataclasites with pseudotachylyte (Tables 1 and 2). On the basis of the bulk powder analysis for XRD, the chlorite are Fe–Mg chlorite. The results from samples analyzed by ethylenglicoled treatment suggest that smectite is not present in any sample (Fig. 6).

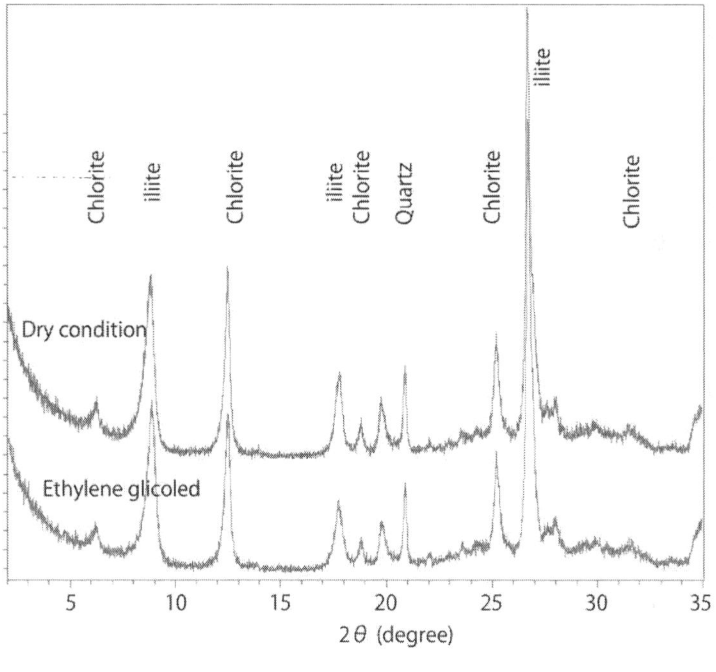

Fig. 6. Examples of XRD charts for oriented samples of cataclasites with pseudotachylyte.

5.2 Iron and magnesium substitution in chlorite

Iron and magnesium substitution in chlorite is shown in Fig. 7. The x-axis indicates I(003)/I(005), related to the D value, and the y-axis indicates [I(002) + I(004)]/I(003)', related to the Y value. In this parameter space, the amount of iron in the hydroxide and silicate layers is shown by dotted lines. For host tectonic melanges, the iron content in both hydroxide and silicate layers is relatively higher than that in the cataclasite samples. The iron content in the host tectonic melanges is distributed over a wide area (Fig. 7). In contrast, these plots show that the iron in cataclasites with pseudotachylyte is concentrated in a smaller area around smaller iron content of both the hydroxide and silicate layers.

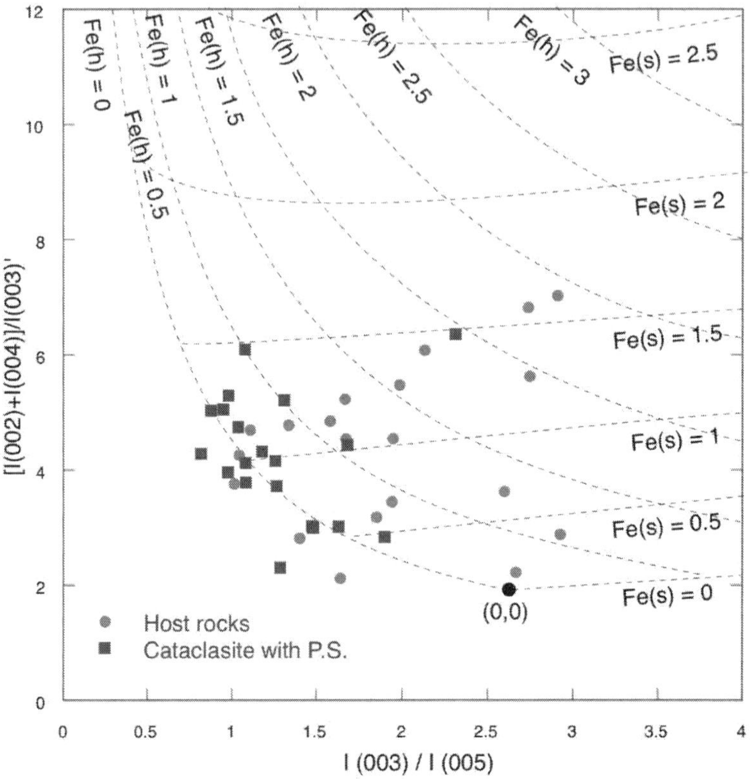

Fig. 7. A diagram of the Fe–Mg substitution of chlorite.

5.3 Illite crystallinity

The illite crystallinity of the tectonic mélange and cataclasites samples is shown in Fig. 8. The host melange samples have an illite crystallinity ranging from about 0.3 to 0.46, with an average value of about 0.4. Illite crystallinity in the cataclasites with pseudotachylyte varies from 0.41 to 0.58, averaging around 0.46. The larger values of crystallinity suggest that the illite in cataclasites is less crystallized than in the host melanges.

5.4 Semi-quantification between illite and chlorite

The semi-quantification of illite and chlorite was conducted using the MIF method, as described above. The results are shown in Fig. 8. The illite ratio to chlorite for host rocks ranges from about 20 wt% to about 70 wt%, with an average value of 46 wt%. The same ratio in cataclasites with pseudotachylyte varies from 40 wt% to 90 wt%, with an average of 65 wt% (Fig. 8). The illite to chlorite ratio is larger in cataclasites than the host melanges, suggesting that the amount of illite increases in cataclasites if the amount of chlorite is constant, or the amount of chlorite decreases relative to the amount of illite.

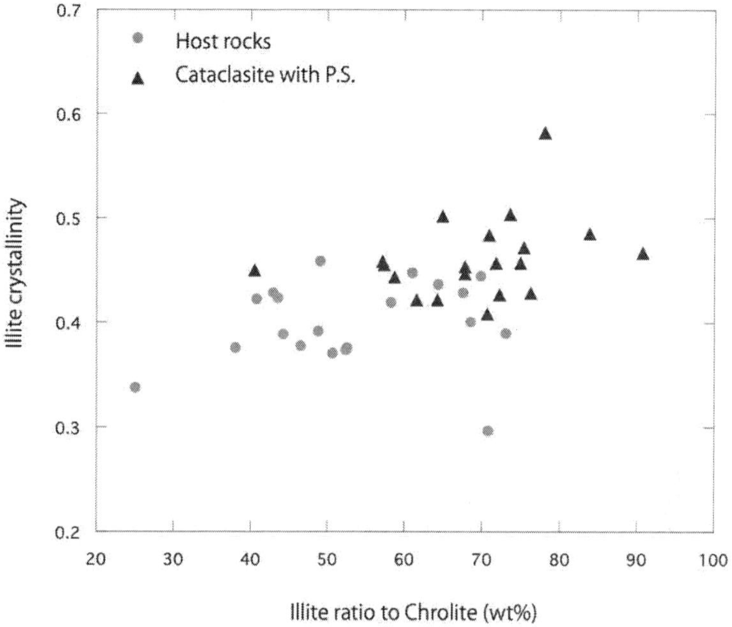

Fig. 8. Illite ratio to chlorite (wt%) vs. illite crystallinity for host rocks and cataclasites with pseudotachylyte

6. Discussions

6.1 Iron consumption in chlorite in cataclasites with pseudotachylyte

In chlorite from cataclasites with pseudotachylyte, the iron content is decreased in both the hydroxide and silicate layers, compared with host melanges, as described above (Fig. 7). The same trend has been reported in the Taiwan Chelung-pu fault (Hashimoto et al., 2007; Hashimoto et al., 2008); however, the initial state of the host rocks in Taiwan is different from those studied here. Although host rock samples from Chelung-pu fault displayed greater iron content than the Okitsu samples, fault rocks from both areas showed a decreased level of iron in their chlorite.

The decrease of iron in chlorite from fault rocks can be controlled by the temperature of the source fluid (Ohta and Yajima, 1988), the pH of the fluid (Malmstrom et al., 1996; Ross, 1969), or lithology (host rocks), as discussed in Hashimoto et al. (2008). Because the same trend of a decrease of iron levels in fault rocks was observed from different host rocks, lithology control is less significant. Ohta and Yajima (1988) reported that higher temperature is related to higher iron content in chlorite in hydrothermal environments, which is the opposite trend to that observed in our analysis, as fault rocks are expected to experience higher temperatures due to frictional heating. Therefore, temperature rise cannot explain the decrease in iron from chlorite in fault rocks. The pH of the source fluid is thus the most likely cause of the decreased iron content in chlorite from fault rocks. The change in pH of the fluid can be caused by radical reactions; that is, the reactions between water and newly created surfaces of minerals due to the breakage of mineral grains. Laboratory experiments on radical reactions showed that the pH of a fluid could be decreased or increased by the breakage of quartz, feldspar, and micas (Kameda et al., 2003; Saruwatari et al., 2004). The change in pH of fluid depends on a complex series of interactions with the broken minerals, and is thus difficult to understand quantitatively outside the laboratory.

The decrease of iron content in ultracataclasites with pseudotachylyte has also been observed in isocon diagrams from a bulk chemical analysis at another fault zone (the Mugi melange, Cretaceous Shimanto Belt, Shikoku, SW Japan; see Hashimoto et al., 2009). This suggests that the characteristic pH of the source fluid can be related to iron consumption from host rocks. The consumed iron from host rocks is observed within minerals precipitated from related fluid. Ankerite veins (Fe–Mg carbonates) are concentrated in seismogenic faults, such as the Okitsu fault described in this study.

6.2 Higher illite crystallinity and illite content in cataclasites with pseudotachylyte

The illite crystallinity of cataclasites with pseudotachylyte had a higher value than that of host melanges. Illite crystallinity is commonly used as an index of paleo-maximum Temperature (e.g., Guithrie et al., 1986; Awan and Kimura, 1996), with higher illite crystallinity indicating a lower paleo-maximum temperature. The cataclasites with pseudotachylyte are expected to have been subjected to a higher paleo-maximum temperature than the host rocks, due to frictional heating. Ikesawa et al. (2003) estimated the minimum temperature from frictional heating to be about 450°C on the basis of the composition of pseudotachylytes. Therefore, the higher illite crystallinity in cataclasites cannot be interpreted by the paleo-maximum temperature. Theoretically, illite crystallinity is controlled by X-ray-scattering–domain size and percentage of expandable layers (Srodon and Eberl, 1984; Eberl and Velde, 1989). As the illite in this study did not include expandable layers within it, the effect of the percentage of expandable layers on the crystallinity is negligible. A wider peak (higher illite crystallinity) indicates a smaller scattering domain. Therefore, the higher value of illite crystallinity for cataclasites with pseudotachylyte indicates that the particle size of illite is smaller than that of host melanges. Possible mechanisms for making smaller illite particles include comminution during cataclastic deformation, and a smaller size of authigenic illite formation related to pseudotachylyte formation.

The semi-quantitative analysis to examine the ratio of illite and chlorite indicates that the proportion of illite increases in cataclasites with pseudotachylyte, compared with that in host mélanges. This result suggests either an increase of illite or a decrease of chlorite in cataclasites. The chemical analysis and mineralogical observations of pseudotachylyte in sedimentary rocks indicates that the melt mainly originates from clay minerals, and quartz and feldspar grains that are resistant to melting (Ujiie et al., 2007). This resistance is due to differences in melting temperatures between clay and other minerals in sedimentary rocks. As it might be difficult to melt chlorite selectively, the increase in the illite ratio might be related to the authigenic illite through pseudotachylyte formation.

Smectite concentrations have been reported from the seismogenic fault zone in the Chelung-pu fault, Taiwan (Kuo et al., 2009). It was found that the smectite in the fault zone does not include an illite-smectite mixed layer, and this was interpreted as meaning that the smectite can be formed by alteration of glass (pseudotachylyte) (Kuo et al., 2009). The Chi-Chi earthquake occurred in 1999, and the Taiwan Chelung-pu fault Drilling Project (TCDP) was conducted in 2004. Within that 5-year interval, the alteration of glass to form smectite could have been progressed. Such alteration to form smectite from glass matrices in pseudotachylyte is also expected in the Okitsu examples. The formation of smectite might also be a significant rock-fluid interaction along a seismogenic fault.

The authigenic smectite is transformed to illite due to diagenetic processes. Smectite-illite transition proceeds with temperature, and the illitization is almost complete at 150°C (Moore and Vrolijk, 1992). Therefore, the smectite from pseudotachylyte glass could be transformed to illite. The illitization from authigenic smectite can also be related to a smaller size of illite, as identified by the larger value of illite crystallinity in cataclasites with pseudotachylyte. The linear relationship between the illite to chlorite ratio and illite crystallinity can be seen in Fig. 8, indicating that a larger illite ratio is associated with a smaller illite grain size. The relationship might also be explained by the illitization of authigenic smectite.

During illitization, the interlayer of smectite is dewatered. The water should migrate to the fault zone after a few years of seismogenesis.

Frictional behaviors of smectites and illites have been reported from laboratory experiments in a sliding velocity range from 0.1 to 200 μm (Saffer and Marone, 2003). While smectite indicates velocity weakening at low normal stress, illite represents only velocity strengthening behavior in the wide range of experiments. The velocity strengthening behavior in illite is not supported by the hypothesis that the smectite-illite transition is related to seismogenesis at the seismic front, as suggested by Hyndman et al (1999). Rather, illitization makes faults aseismic (Saffer and Marone, 2003). The illite concentration in cataclasites with pseudotachylyte in this study suggests that the rock-fluid interactions along a seismogenic fault, such as smectite formation from the glass of pseudotachylyte, and illitization from the authigenic smectite, are processes that modify the fault to an aseismic fault.

7. Conclusion

The iron content in chlorite in cataclasites with pseudotachylyte is smaller than that of tectonic melanges (host rocks), as observed in other seismogenic fault rocks. The decrease of iron in chlorite suggests that a specific pH fluid reacted with fault rocks due to radical reactions.

Illite crystallinity in cataclasites is higher than in host rocks. In addition, the relative amount of illite is increased in cataclasites compared with host rocks. These results imply that the smectite altered from glass in pseudotachylyte transformed into illite along a seismogenic fault.

8. Acknowledgment

We thank Prof. Higashi, M. for the help in XRD analysis. We also appreciate Dr. Kameda, J. for his constructive discussions.

9. References

Awan, M. A. & Kimura, K. (1996). *Thermal structure and uplift of the Cretaceous Shimanto Belt, Kii Peninsula, Southwest Japan: An illite crystallinity and Illiteq bo, lattice spacing study*, Island Arc, v. 5, no. 1, p. 69-88.

Brodsky, E. E. & Kanamori, H. (2001). *Elastohydrodynamic lubrication of faults*, Journal of Geophysics Research, v. 106, no. 16, p. 357-16.

Brown, G. & Brindley, G. W. (1980). X-ray diffraction procedures for clay mineral identification, Crystal structures of clay minerals and their X-ray identification, v. 5, p. 305-359.

Di Toro, G., Hirose, T., Nielsen, S., Pennacchioni, G. & Shimamoto, T. (2006). *Natural and experimental evidence of melt lubrication of faults during earthquakes*, Science, v. 311, no. 5761, p. 647-649.

Eberl, D. D. & Velde, B. (1989). *Beyond the Kubler index*, Clay Minerals, v. 24, no. 4, p. 571-577.

Guthrie, J. M., Houseknecht, D. W. & Johns, W. D. (1986). *Relationships among vitrinite reflectance, illite crystallinity, and organic geochemistry in Carboniferous strata, Ouachita Mountains, Oklahoma and Arkansas*, AAPG Bulletin, v. 70, no. 1, doi: 10.1306/9488561A-1704-11D7-8645000102C1865D.

Hashimoto, Y. & Kimura, G. (1999). *Underplating process from melange formation to duplexing: Example from the Cretaceous Shimanto Belt, Kii Peninsula, southwest Japan*, Tectonics, v. 18, no. 1, p. 92-107.

Hashimoto, Y., Ujiie, K., Sakaguchi, A. & Tanaka, H. (2007). *Characteristics and implication of clay minerals in the northern and southern parts of the Chelung-pu fault, Taiwan, Tectonophysics*, v. 443, no. 3-4, p. 233-242.

Hashimoto, Y., Tadai, O., Tanimizu, M., Tanikawa, W., Hirono, T., Lin, W., Mishima, T., Sakaguchi, M., Soh, W. & Song, S. R. (2008). *Characteristics of chlorites in seismogenic fault zones: the Taiwan Chelungpu Fault Drilling Project (TCDP) core sample*, eEarth, v. 3, no. 1, p. 1-6.

Hashimoto, Y., Nikaizo, A. & Kimura, G. (2009). *A geochemical estimation of fluid flux and permeability for a fault zone in Mugi melange, the Cretaceous Shimanto Belt, SW Japan*, Journal of Structural Geology, v. 31, no. 2, p. 208-214.

Honda, G., Ishikawa, T., Hirono, T. & Mukoyoshi, H. (2011). *Geochemical signals for determining the slip-weakening mechanism of an ancient megasplay fault in the Shimanto accretionary complex*, Geophysical research letters, v. 38, no. 6, doi:10.1029/2011GL046722.

Hyndman, R. D. & Wang, K. (1993). *Thermal constraints on the zone of major thrust earthquake failure: The Cascadia subduction zone*, Journal of Geophysical Research, v. 98, no. B2, p. 2039-2060.

Ikesawa, E., Sakaguchi, A. & Kimura, G. (2003). *Pseudotachylyte from an ancient accretionary complex: Evidence for melt generation during seismic slip along a master d collement?*, Geology, v. 31, no. 7, p. 637-640.

Ishikawa, T., Tanimizu, M., Nagaishi, K., Matsuoka, J., Tadai, O., Sakaguchi, M., Hirono, T., Mishima, T., Tanikawa, W. & Lin, W. (2008). *Coseismic fluid-rock interactions at high temperatures in the Chelungpu fault*, Nature Geoscience, v. 1, no. 10, p. 679-683.

Kameda, J., Saruwatari, K. & Tanaka, H. (2003). *H2 generation in wet grinding of granite and single-crystal powders and implications for H2 concentration on active faults*, Geophysical research letters, v. 30, no. 20, doi:10.1029/2003GL018252.

Kano, Y., Mori, J., Fujio, R., Ito, H., Yanagidani, T., Nakao, S. & Ma, K. F. (2006). *Heat signature on the Chelungpu fault associated with the 1999 Chi-Chi, Taiwan earthquake*, Geophys. Res. Lett, v. 33, L14306, doi:10.1029/2006GL026733.

Kimura, G. & Mukai, A. (1991). *Underplated units in an accretionary complex: Melange of the Shimanto Belt of eastern Shikoku, southwest Japan*, Tectonics, v. 10, no. 1, p. 31-50.

Kubler, B. (1969). *Crystallinity of illite, Detection of metamorphism in some frontal parts of the Alps:* Fortschr. der Mineralogie, v. 47, p. 39-40.

Kuo, L. W., Song, S. R., Yeh, E. C. & Chen, H. F. (2009). *Clay mineral anomalies in the fault zone of the Chelungpu Fault, Taiwan, and their implications*, Geophysical research letters, v. 36, no. 18, doi:10.1029/2009GL039269.

Malmstrom, M., Banwart, S., Lewenhagen, J., Duro, L. & Bruno, J. (1996). *The dissolution of biotite and chlorite at 25°C in the near-neutral pH region*, Journal of contaminant hydrology, v. 21, no. 1-4, p. 201-213.

Mase, C. W. & Smith, L. (1987). *Effects of frictional heating on the thermal, hydrologic, and mechanical response of a fault*, Journal of Geophysical Research, v. 92, no. B7, p. 6249-6272.

Melosh, H. J. (1979). *Acoustic fluidization-A new geologic process*, Journal of Geophysical Research, v. 84, p. 7513-7520.

Mishima, T., Hirono, T., Soh, W. & Song, S. R. (2006). *Thermal history estimation of the Taiwan Chelungpu fault using rock-magnetic methods*, Geophysical research letters, v. 33, no. 23, doi:10.1029/2006GL028088.

Moore, D. M. & Reynolds, R. C. (1989). X-ray Diffraction and the Identification and Analysis of Clay Minerals. Oxford University Press, Wellington, pp. 348.

Moore, J. C. & Vrolijk, P. (1992). *Fluids in accretionary prisms*, Reviews of Geophysics, v. 30, no. 2, p. 113-135.

O'Hara, K., Mizoguchi, K., Shimamoto, T. & Hower, J. C. (2006). *Experimental frictional heating of coal gouge at seismic slip rates: Evidence for devolatilization and thermal pressurization of gouge fluids*, Tectonophysics, v. 424, no. 1-2, p. 109-118.

Ohta, E. & Yajima, J. (1988). *Magnesium to Iron Ratio of Chlorite as Indicator of Type of Hydrothermal Ore Deposits*, Mining Geology, Special Issue, v. 12, p. 17–22.

Oleskevich, D. A., Hyndman, R. D. & Wang, K. (1999). *The updip and downdip limits to great subduction earthquakes: Thermal and structural models of Cascadia, south Alaska, SW Japan, and Chile*, Journal of Geophysical Research, v. 104, no. B7, p. 14965-14991.

Onishi, C. T. & Kimura, G. (1995). *Melange fabric and relative convergence in subduction zone*, Tectonics, v. 14, p. 1273-1289.

Otsuki, K., Monzawa, N. & Nagase, T. (2003). *Fluidization and melting of fault gouge during seismic slip: Identification in the Nojima fault zone and implications for focal earthquake mechanisms*, J. Geophys. Res, v. 108, doi:10.1029/2001JB001711.

Reynolds Jr, R. C. (1985). NEWMOD, a computer program for the calculation of one-dimensional diffraction patterns of mixed-layered clays, R. C. Reynolds, Jr., 8 Brook, Dr., Hanover, New Hampshire, 03755.

Ross, G. J. (1969). *Acid dissolution of chlorites: release of magnesium, iron and aluminum and mode of acid attack*, Clays and Clay Minerals, v. 17, p. 347-354.

Saffer, D. M. & Marone, C. (2003). *Comparison of smectite-and illite-rich gouge frictional properties: application to the updip limit of the seismogenic zone along subduction megathrusts*, Earth and Planetary Science Letters, v. 215, no. 1-2, p. 219-235.

Saruwatari, K., Kameda, J. & Tanaka, H. (2004). *Generation of hydrogen ions and hydrogen gas in quartz-water crushing experiments: an example of chemical processes in active faults*, Physics and chemistry of minerals, v. 31, no. 3, p. 176-182.

Sibson, R. H. (1975). *Generation of pseudotachylyte by ancient seismic faulting*, Geophysical Journal of the Royal Astronomical Society, v. 43, no. 3, p. 775-794.

Sibson, R. H. (1977). *Fault rocks and fault mechanisms*, Journal of the Geological Society, v. 133, no. 3, p. 191-213.

Spray, J. G. (1992). *A physical basis for the frictional melting of some rock-forming minerals*, Tectonophysics, v. 204, no. 3-4, p. 205-221.

Srodon, J. & Eberl, D. D. (1984). *Illite*, Reviews in Mineralogy and Geochemistry, v. 13, no. 1, p. 495-544.

Taira, A., Katto, J., Tashiro, M., Okamura, M. & Kodama, K. (1988). *The Shimanto belt in Shikoku, Japan evolution of Cretaceous to Miocene accretionary prism*, Modern Geology, v. 12, no. 5, p. 5-46.

Tanaka, H., Chen, W. M., Wang, C. Y., Ma, K. F., Urata, N., Mori, J. & Ando, M. (2006). *Frictional heat from faulting of the 1999 Chi-Chi, Taiwan earthquake*, Geophysical research letters, v. 33, no. 16, L16316, doi:10.1029/2006GL026673.

Tanaka, H., Omura, K., Matsuda, T., Ikeda, R., Kobayashi, K., Murakami, M. & Shimada, K. (2007). *Architectural evolution of the Nojima fault and identification of the activated slip layer by Kobe earthquake*, Journal of Geophysical Research, v. 112, no. B7, B07304, doi:10.1029/2005JB003977.

Ujiie, K., Yamaguchi, H., Sakaguchi, A. & Toh, S. (2007). *Pseudotachylytes in an ancient accretionary complex and implications for melt lubrication during subduction zone earthquakes*, Journal of Structural Geology, v. 29, no. 4, p. 599-613.

Wibberley, C. A. J. & Shimamoto, T. (2005). *Earthquake slip weakening and asperities explained by thermal pressurization*, Nature, v. 7051, p. 689-692.

Yamaguchi, A., Cox, S. F., Kimura, G. & Okamoto, S. (2011). *Dynamic changes in fluid redox state associated with episodic fault rupture along a megasplay fault in a subduction zone*, Earth and Planetary Science Letters, v. 302, no. 3-4, p. 369-377.

14

Some Complexity Studies of Electroseismic Signals from Mexican Subduction Zone

L. Guzmán-Vargas[1], R. Hernández-Pérez[1], F. Angulo-Brown[1] and A. Ramírez-Rojas[2]
[1]*Instituto Politécnico Nacional*
[2]*Universidad Autónoma Metropolitana - Azcapotzalco*
México

1. Introduction

The analysis of complex signals associated to geoelectric activity is important not only for earthquake prognosis but also for understanding non linear processes related to earthquake preparation. Previous studies have reported alterations, such as the emergence of correlated dynamics in geoelectric potentials prior to an important earthquake (EQ). One important feature of geoelectric signals is the absence of regularity patterns with fluctuations apparently influenced by noise. In past decades, earthquake prediction methods have attracted the attention of researchers from different areas of science. The search for effective seismic precursors has not been successful. However, despite some pessimism, in many seismically actives zones around the world there exist research programs for the study of possible precursory phenomena of earthquakes(Cicerone et al., 2009; Hayakawa, 1999; Hayakawa & Molchanov, 2002; Lomnitz, 1990; Telesca & M., 2005; Uyeda et al., 2000; Varotsos et al., 2003a;b;c; 2004; 2005). In particular, one of the techniques used in the search of earthquake precursors since more than three decades ago consists in monitoring the so-called electric self-potential field. The main motivation to explore this kind of signals is that it is expected that before the occurrence of an earthquake (Varotsos, 2005), the stress (pressure) gradually varies in the focal area, which affects various physical properties, for example the static dielectric constant (Varotsos, 1980; 1978). In addition, this stress variation may change the relaxation time for the orientation of the electric dipoles formed due to lattice defects (Lazaridou et al., 1985). It may happen that, when the stress (pressure) reaches a critical value (Varotsos & Alexopoulos, 1984b), these electric dipoles exhibit a cooperative orientation (collective organization), thus leading to emission of transient electric signals termed Seismic Electric Signals, SES (Uyeda et al., 2000; Varotsos & Alexopoulos, 1984a). This generation mechanism of signal emission is named pressure stimulated polarization currents (PSPC) (Varotsos & Alexopoulos, 1986). It is expected that precursory electric signals associated with large earthquakes should exhibit anomalous changes and, in some cases, fractal complex organization (Varotsos, 2005; Varotsos et al., 2003c; 2004). Additionally, several other physical mechanisms have been proposed as possible causes of electromagnetic (EM) precursory signals before EQ's, such as electrokinetic effects (EK) (Haartsen & Pride, 1997; Ishido & Mizutani, 1981; Mizutani et al., 1976), piezoelectric effects (PE) (Gershenzon et al.,

1993) and electromagnetic induction effects (Gershenzon et al., 1993; Honkura et al., 2000; Iyemori et al., 1996; Matsushima et al., 2002). A very recent review about PSPC, EK, PE and other possible generation mechanisms of signal emissions can be seen in Uyeda et al. (Uyeda et al., 2008). We have measured the ground electrical potential (the self-potential) in several sites along the Mexican coast, near the Middle American trench, which is the border between the Cocos and the American tectonic plates. In some previous articles we have reported more detailed descriptions of that region and some studies of possible precursory electric phenomena associated to several earthquakes of magnitude larger than six (Flores-Márquez et al., 2007; Muñoz Diosdado et al., 2004; Ramírez-Rojas et al., 2007). Recent studies focused on fractal and non linear properties of physical and biological times series have revealed that this organization is strongly related to a complex interaction of multiple components and mechanism across multiple scales. In particular, published studies about the complexity of ground electric self-potential behavior have pointed out that changes inthe fractal organization have been observed in a period prior to an important earthquake. However, a clear evidence with statistical support about the mechanisms involved in these changes, has not been presented, although some important suggestions and discussions have been proposed to address this problem (Gotoh et al., 2003; 2004; Ida et al., 2005; 2006; Smirnova et al., 2004; Telesca & Lapenna, 2006; Varotsos et al., 2008). One important feature of geoelectric signals is the absence of regularity patterns. These fluctuations are embedded into noise activity produced by the combined contribution of many high-dimensional processes, which due to the central-limit theorem, are Gaussian-distributed. The direct application to this kind of signals of nonlinear methods such as power spectrum, detrended fluctuation analysis (DFA) and fractal dimension method reveals that different correlation levels are present in the vicinity of a main shock. Very often the double log plot of scaling exponents obtained from the aforementioned methods present a crossover behavior between different scales. On the other hand, a long term relaxation-EQ-preparation-main shock-relaxation process has been reported before some large EQ's (Varotsos, 2005). From this point of view, one could expect that a relaxed surface layer of earth's crust corresponds to white noise in geoelectric signals and the EQ-preparation process corresponds to a background white noise mixed with a kind of correlated geoelectric signals expressed through a crossover behavior. However, this idea must be taken as a speculative hypothesis which requires a more profound attention. Thus, in this context is very important to incorporate a variety of methods to statistically distinguish and evaluate these complex dynamics.

In this chapter, we report some complexity studies of geoelectric signals during a two year period from Jun 1st. 1994 to May 31st 1996 in two sites (Acapulco and Coyuca stations) located in southern Mexico. In particular, our study is related to an $Ms = 7.4$ earthquake occurred on September 14, 1995 with epicentral distance of 110 km from Acapulco and 200 km from Coyuca, respectively. Previous studies have reported changes in the correlation dynamics observed prior to this earthquake (Guzmán-Vargas et al., 2008; Ramírez-Rojas et al., 2008; 2007; 2004; Telesca et al., 2009). However, the possible existence of seismic precursors associated with this event has not been deeply explored. Here, we perform a systematic study of DFA exponents and sample entropy to evaluate the level of irregularity and correlations of geoelectric time series. We observe important changes in the entropy a few months before the occurrence of the earthquake mentioned above. On the other hand, we use a procedure to statistically estimate two DFA-scaling exponents and the crossover scale which are representative of changes in the underlying dynamics prior to the main shock. Moreover, we perform a pattern synchrony analysis based on the computation of the cross-sample entropy between the geoelectric signals from two channels, which represents a modern approach to the study of geoelectric signals. The chapter is organized as follows. In Sec. 2, a brief description of the entropy and the detrended fluctuation analysis methods are presented. We also describe the geolectric time series. In Sec. 3, we present the results and discussions. Finally, some concludings remarks are presented.

2. Methods and data

2.1 Entropy methods

The entropy of a single discrete random variable X is a measure of its uncertainty. In the case of a stochastic process, the mean rate of creation of information is measured by the Kolmogorov-Sinai (KS) entropy (Eckmann & Ruelle, 1985). However, the KS entropy is not applicable to finite length real world series because only entropies of finite order can be computed numerically and KS is underestimated as the order becomes large. An alternative procedure to estimate the entropy of a signal was given by Grassberger et al. (Grassberger & Procaccia, 1983). They proposed the K_2 entropy to characterize chaotic systems which is a lower bound of the KS entropy. Later, based on K_2 definition, Pincus introduced the *Approximate Entropy* ($ApEn$) to quantify the regularity in time-series (Pincus, 1991; 1995). Briefly, $ApEn$ is constructed as follows: given a time series $X_i = x_1, ..., x_N$ of length N. First, m-length vectors are considered: $u_m(i) = x_i, x_{i+1}, ..., x_{i+m-1}$. Let $n_{im}(r)$ represent the number of vectors $u_m(j)$ within r of $u_m(i)$. $C^i_m(r) = n_{im}(r)/(N-m+1)$ is the probability that any vector $u_m(j)$ is within r of $u_m(i)$. Next, the average of C^i_m is constructed as $\Phi^m(r) = 1/(N-m+1)\sum_{i=1}^{N-m+1} \ln C^m_i(r)$. Finally, $ApEn$ is defined as $ApEn(m,r) = lim_{N\to\infty}\left[\Phi^m(r) - \Phi^{m+1}(r)\right]$: which, for finite N, it is estimated by the statistics $ApEn(m,r,N) = \Phi^m(r) - \Phi^{m+1}(r)$. In words, the statistics $ApEn(m,r,N)$ is approximately equal to the negative average natural logarithm of the conditional probability that two sequences that are similar for m points remain similar at the next point, within a tolerance r (Richman & Moorman, 2000). It is obtained that a low value of $ApEn$ reflects a high degree of regularity. Even though the implementation and interpretation of $ApEn$ is useful to distinguish correlated stochastic processes and composite deterministic/stochastic models (Pincus, 1995), it has been found there is a bias in $ApEn$ because the algorithm counts each sequence as matching itself (Richman & Moorman, 2000). The presence of this bias causes $ApEn$ to lack two important expected properties: (a) $ApEn$ is heavily dependent on the time-series length and is uniformly lower than expected for short series and, (b) it lacks relative consistency in the sense that if the value of $ApEn$ for a time-series is higher than that of another, it does not remain so if the test conditions change (Pincus, 1995). Therefore, the development of an alternative method was desirable to overcome the limitations of $ApEn$. Based on K_2 and $ApEn$ methods, Richman and Moorman (Richman & Moorman, 2000) introduced the so-called Sample Entropy (S_E), to reduce the bias in $ApEn$. One of the advantages of S_E is that does not count self-matches and is not based on a template-wise approach (Richman & Moorman, 2000). $S_E(m,r,N)$ is precisely defined as

$$S_E(m,r,N) = -\ln\frac{U^{m+1}}{U^m}, \tag{1}$$

that is, the negative natural logarithm of the conditional probability (U) that two sequences similar for m points remain similar at the next point, within tolerance r, without counting the self-matches. S_E results to be more robust than $ApEn$ statistics when applied to short time series from different stochastic processes over a wide range of operating conditions. For instance, a lower value of S_E indicates a more regular behavior of a time-series whereas high values are assigned to more irregular, less predictable, time series (Costa et al., 2005). It applies to realworld time series and, therefore, has been widely used in physiology and medicine (Costa et al., 2005).

2.2 Cross sample entropy

Entropy can also be calculated between two signals, and this mutual entropy characterizes the probability of finding similar patterns within the signals. Therefore, the cross-entropy technique was introduced to measure the degree of asynchrony or dissimilarity of two time series (Pincus, 1995; Pincus & Singer, 1996).

When calculating the cross-entropies, the patterns that are compared are taken in pairs from the two different time series $\{u(i)\}$ and $\{v(i)\}$, $i = 1, \ldots, N$. The vectors are constructed as follows:

$$\mathbf{x}_m(i) = [u(i), u(i+1), u(i+2), \ldots, u(i+m-1)],$$
$$\mathbf{y}_m(i) = [v(i), v(i+1), v(i+2), \ldots, v(i+m-1)],$$

with the vector distance defined as

$$d[\mathbf{x}_m(i), \mathbf{y}_m(j)] = max\{|u(i+k) - v(j+k)| : 0 \leq k \leq m-1\}.$$

With this definition of distance, the S_E algorithm can be applied to compare sequences from the *template* series to those of the *target* series to obtain the Cross Sample Entropy (C_E). It is usual that the two time series are first normalized by subtracting the mean value from each data series and then dividing it by the standard deviation. This normalization is valid since the main interest is to compare patterns.

It is quite possible that no vectors in the target series can be found to be within the distance r to the template vector and then the value of C_E is not defined. One important property of C_E is that its value is independent of which signal is taken as a template. In particular, the Cross Sample Entropy is used to define the *pattern synchrony* between two signals, where synchrony refers to pattern similarity, not synchrony in time, wherein patterns in one series appear (within a certain tolerance) in the other series. Moreover, C_E assigns a positive number to the similarity (synchronicity) of patterns in the two series, with larger values corresponding to greater common features in the pattern architecture and smaller values corresponding to large differences in the pattern architecture of the signals (Veldhuis et al., 1999). When no matches are found, a fixed negative value is assigned to C_E to allow a better displaying of the results.

The conceptual difference between pattern synchrony, as measured by the C_E, and correlations, as measured by the cross-correlation function, can be expressed as follows: let us suppose that we have two time series $\{x(k)\}$ and $\{y(k)\}$. The C_E deals with patterns: a sequence of data points of a certain length m is taken from the template time-series $\{x(k)\}$

and this pattern is searched for in the target time-series $\{y(k)\}$ within a tolerance r. However, the C_E does not collect the time-stamp of the matching sequence in the time series $\{y(k)\}$, but counts the number of sequence matches of lengths m and $m+1$. On the other hand, the objective of the cross-correlation function is to find the time lag τ for which the whole time series $\{x(k)\}$ resembles $\{y(k)\}$, but the time series are not decomposed in sequences of points. Therefore, the C_E analysis is complementary to the cross-correlation and spectral analysis since it operates on different features of the signals (see the *Appendix* of Ref. Pincus & Singer (1996)).

2.3 Multiscale entropy analysis

Recently, Costa et al. (Costa et al., 2002) introduced the multiscale entropy analysis (MSE) to evaluate the relative complexity of normalized time series across multiple scales. This procedure was proposed to give an explanation to the fact that, in the context of biological signals, single-scale entropy methods (S_E and $ApEn$) assign higher values to random sequences from certain pathological conditions whereas an intermediate value is assigned to signals from healthy systems (Costa et al., 2002). It has been argued that these results may lead to erroneous conclusions about the level of complexity displayed by these systems (Costa et al., 2005). The MSE methodology shows that long-range correlated noises as the output of healthy systems are more complex than uncorrelated signals from some pathological conditions. Briefly, the the MSE method consists of: given a time series $X_i = x_1, ..., x_N$, a coarse-grained procedure is applied (Costa et al., 2005). A scale factor τ is introduced to perform a moving average given by $y_j = 1/\tau \sum_{i=(j-1)\tau+1}^{j\tau} x_i$, with $1 \leq j \leq N/\tau$. Note that the length of the coarse-grained time series is given by N/τ, that is, for scale one the original time series is obtained. To complete the MSE procedure the S_E algorithm is applied to the coarse-grained time series for each scale. Finally, the entropy value is plotted against the scale factor. Typically, under MSE analysis, the entropy values for a random noise monotonically decreases whereas for long-range correlated noise ($1/f$-noise) the entropy remains constant for several scales, indicating that $1/f$-noise is structurally more complex than uncorrelated signals (Costa et al., 2005).

2.4 DFA method

The power spectrum is the typical method to detect correlations in a time series. For example, consider a stationary stochastic process with autocorrelation function which follows a power law

$$C(s) \sim s^{-\gamma}, \tag{2}$$

where s is the lag and γ is the correlation exponent, $0 < \gamma < 1$. The presence of long-term correlations is related to the fact that the mean correlation time diverges for infinite time series. According to the Wiener-Khintchine theorem, the power spectrum is the Fourier transform of the autocorrelation function $C(s)$ and, for the case described in Eq. 2, we have the scaling relation,

$$S(f) \sim f^{-\beta}, \tag{3}$$

where β is called the spectral exponent and is related to the correlation exponent by $\gamma = 1 - \beta$. When the power spectrum method is used to estimate the presence of correlations in real nonstationary time series, as in the case of heartbeat interval signals, it may lead to unreliable results. In past decades, alternative methods have been proposed to the assessment of correlations for stationary and nonstationary time series. A method which is very appropriated to the assessment of correlations in stationary and nonstationary time series is the detrended fluctuation analysis (DFA). This method was introduced to quantify long-range correlations in the heartbeat interval time series and DNA sequences (Peng et al., 1995a;b). The DFA is briefly described as follows: First, we integrate the original time series to get, $y(k) = \sum_{i=1}^{k} [x(i) - x_{ave}]$, the resulting series is divided into boxes of size n. For each box, a straight line is fitted to the points, $y_n(k)$. Next, the line points are subtracted from the integrated series, $y(k)$, in each box. The root mean square fluctuation of the integrated and detrended series is calculated by means of

$$F(n) = \sqrt{\frac{1}{N} \sum_{k=1}^{N} [y(k) - y_n(k)]^2}, \tag{4}$$

this process is taken over several scales (box sizes) to obtain a power law behavior $F(n) \sim n^{\alpha}$, with α an exponent, which reflects self-similar and correlation properties of the signal. The scaling exponent α is related to the spectral exponent β by means of $\alpha = (\beta + 1)/2$ (Peng et al., 1995a). It is known that $\alpha = 0.5$ is associated to white noise (non correlated signal), $\alpha = 1$ corresponds to $1/f$ noise and $\alpha = 1.5$ represents a Brownian motion. This exponent is also related to the autocorrelation function exponent by $\alpha = 1 - \gamma/2$ where the autocorrelation function is $C(\tau) \propto \tau^{-\gamma}$ with $0 < \gamma < 1$ (Makse et al., 1996).

2.5 Data

The time series considered in this study were collected during a two year period, from June 1994 to May 1996, in two electroseismical stations located at Acapulco (16.85 N, 99.9 W) and Coyuca (18.35 N, 100.7 W), both located in the South Pacific coast in Mexico (Ramírez-Rojas et al., 2004). The electrical signals consist of the electric self-potential fluctuations V between two electrodes buried 2 m into the ground and separated by a distance of 50 m. Each pair of electrodes was oriented in one direction: North-South and East-West, as it is indicated by VAN methodology (Varotsos & Alexopoulos, 1984a;b). Two time series were simultaneously recorded at each electroseismic station (N-S and E-W channels). Due to technical adjustments, two different sampling rates were used in different time intervals along the mentioned period, $t = 4$ s in Coyuca and $t = 2$ s in Acapulco (Yépez et al., 1995). In Figure 1 representative time series of potential differences for one year period (Jan. 1st. to Dec. 31st. 1995) in Acapulco station are presented. During the period of study, two EQs with $M > 6$ occurred with epicenters within 250 km of the two monitoring stations. The first EQ occurred on September 14, 1995 with $M = 7.4$ and epicenter with coordinates (16.31 N, 98.88 W), with focal depth of 22 Km; the hypocenter was at $d = 112$ km from Acapulco and $d = 146.6$ km from Coyuca. The second EQ occurred on February 24, 1996 with $M = 7.0$ and epicenter with coordinates (15.8 N, 98.25 W), with focal depth of 3 Km; and hypocenter at $d = 220.02$ km from Acapulco and $d = 250.01$ km from Coyuca. As can be seen from Figure 2, the two earthquakes had epicenters located closer to the Acapulco station. The analyzed noisy time series were not preprocessed and non significant nonstationary features affecting the correlation properties

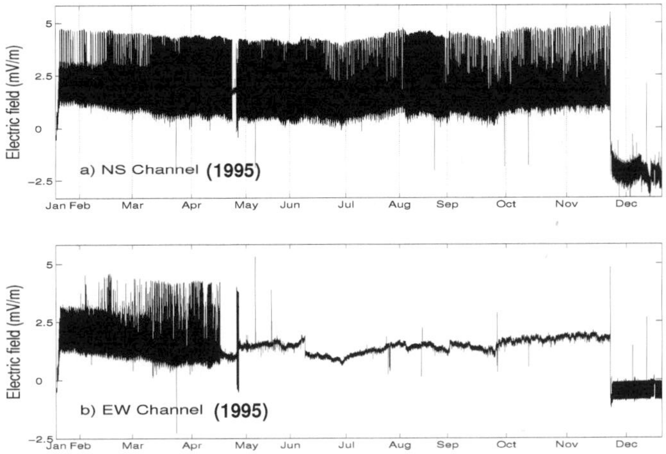

Fig. 1. Representative geoelectric time series from Acapulco station for one year period (Jan. 1st to Dec. 31st., 1995). **(a)** N-S channel and **(b)** E-W channel.

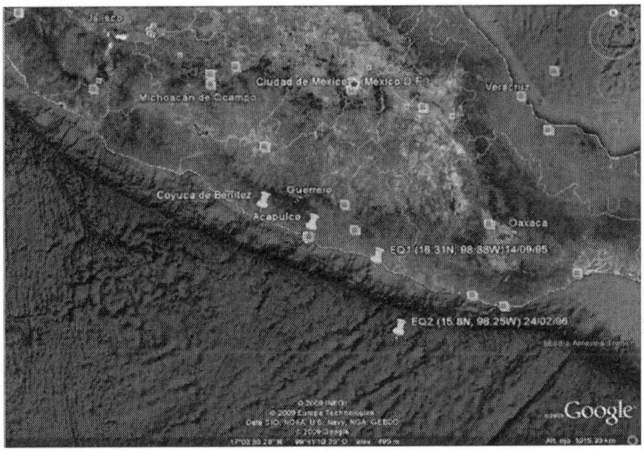

Fig. 2. Location of the monitoring stations and the epicenters of the earthquakes occurred during the studied time period.

of a signal mentioned by Chen et al. (Chen et al., 2002), were present in a remarkable way in our data. When comparing these two signals, different kind of fluctuations can be identified. An important question here is to evaluate the level of irregularity across multiple scales and its relation with the presence of long range correlations. We evaluate the changes in the variability by means of S_E, which estimates the amount of new information arriving at any time, the cross sample entropy and the presence of correlations by using the DFA method.

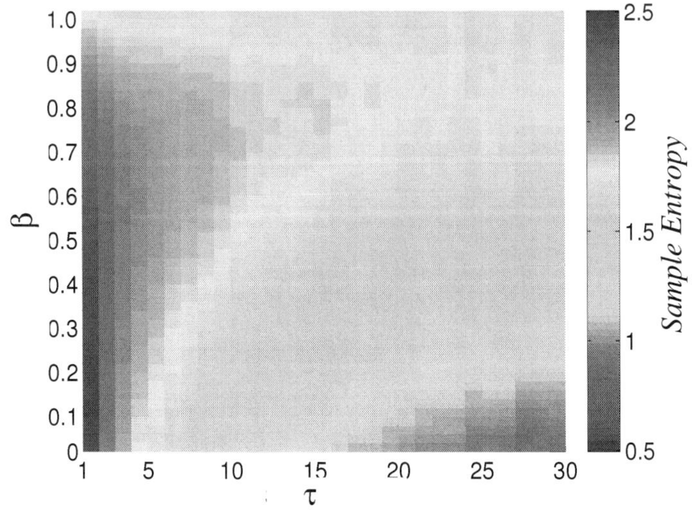

Fig. 3. Plot of MSE analysis for $1/f^{\beta}$-noises with $0 \leq \beta \leq 1$, that is, for noises with power-law correlations. We used the Fourier filtering method to generate time series of 32000 points. In this plot, each point represents the average of 10 independent realizations. The value of S_E is given according to the color panel. Note that as the spectral exponent increases the entropy value remains high even for large time scales .

3. Results and discussion

3.1 MSE results

First, in order to get a better estimation of entropy values for Gaussian noises with power law correlations, we performed simulations of noises with power spectrum of the form $1/f^{\beta}$ with $0 \leq \beta \leq 1$. We generated time series with 32000 points by means of the Fourier filtering method (Makse et al., 1996). We applied the MSE analysis to the generated data for several values of β in the interval $0 \leq \beta \leq 1$ and a range of time scales. In Fig. 3, the results for entropy are presented according to the color panel. Notice that for $\beta = 0$ and $\beta = 1$, the main results described in (Costa et al., 2002) are recovered. We observe that as the spectral exponent β increases , that is, as long-range correlations are present, S_E decreases moderately but at the same time remains constant for several time scales. This behavior indicates that, in the context of simulated signals, the amount of new information arriving at any time is "regulated" by the presence of correlations.

In order to apply the MSE procedure to the geoelectric time series we considered non overlapped time windows of 5,400 data points each, corresponding approximately to 3 hours of records Guzmán-Vargas et al. (2009). First, the data points of the original signal are divided by its standard deviation and S_E is calculated for each time scale according to the MSE method. We repeated the MSE procedure for the corresponding shuffled version of each window. In all the cases presented here, we used the following values for parameters r and

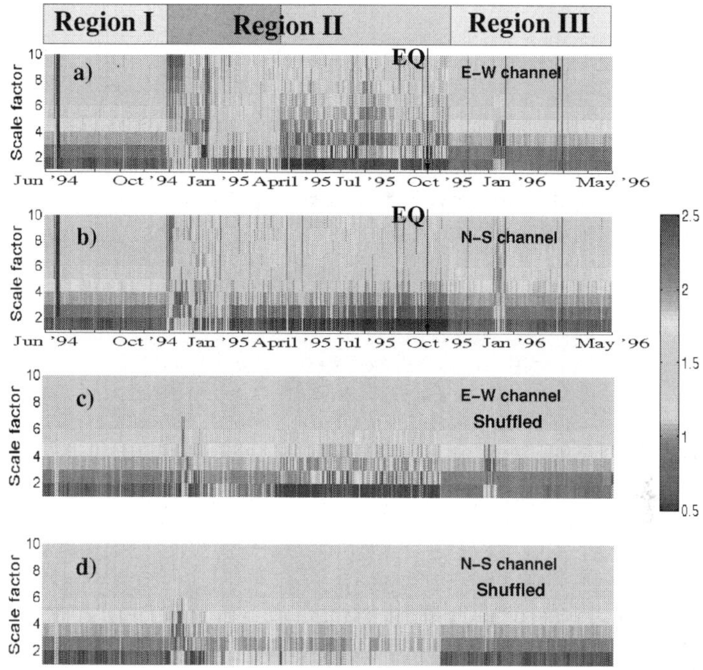

Fig. 4. MSE analysis of geoelectrical time series from Acapulco station. **(a)** MSE results for E-W channel, three main regions can be identified according to the changes of S_E for different scale factors. Note that Region II is mostly characterized by a high entropy value even for large time scales. **(b)** Entropy results for N-S channel. In this case, Region I and III also display white noise profile whereas Region II shows high regularity for short scales. **(c)**, **(d)** As in **(a)** and **(b)** but for randomized data. Note that in these shuffled cases the data display mostly white noise profile (Guzmán-Vargas et al., 2009).

m: $r = 0.15$ and $m = 2$. In Fig. 4, the results of S_E for both channels of Acapulco station are presented. The color panel represents the values of S_E in the interval 0.5 to 2.5.

For the period from June 1994 to October 1994 we define Region I and we observe that during this period and in both channels, S_E shows a high value for scale 1 and rapidly decreases as the scale factor increases as it occurs with white noise dynamics (Figs. 4(a) and 4(b)). We also identify region III from November 1995 to May 1996 where entropy values show a similar profile as in region I, that is, mostly white noise dynamics. For the period from November 1994 to October 1995, we define region II which is characterized by a complex behavior. For E-W channel, we observe that for a short interval at the beginning of this period, S_E shows a low value for scale one and a small increment for large time scales is observed, followed by a new short period with complex behavior. After this transient behavior, the entropy is small for short scales, that is, a high regularity in the original data is observed . Interestingly, for the period comprising April 1995 to October 1995, the entropy profile reveals that S_E remains

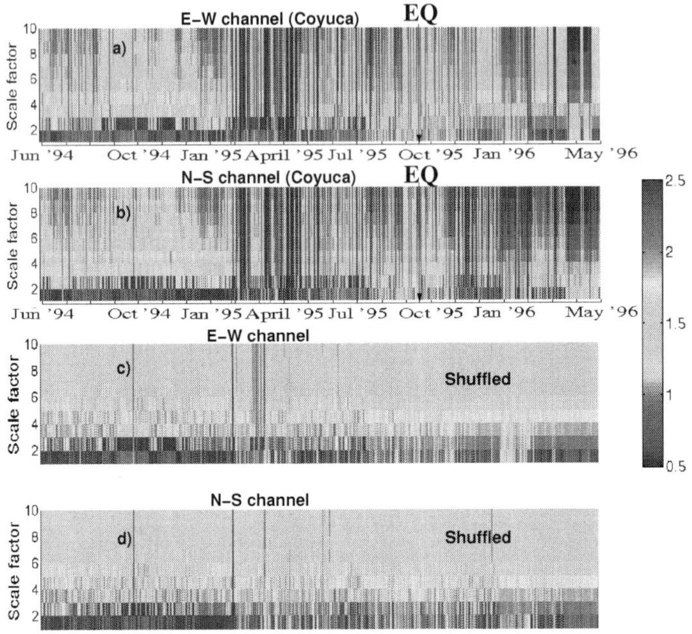

Fig. 5. MSE analysis of geoelectric time series from Coyuca station. **(a)** MSE results for E-W channel, we observe that entropy value is high for short scales and rapidly decreases such that for scales larger than $\tau = 6$ it shows high regularity, except for short periods with a low entropy at short scales. **(b)** Entropy results for N-S channel. **(c)**, **(d)** As in **(a)** and **(b)** but for randomized data. Note that in these cases the data display mostly white noise profile (Guzmán-Vargas et al., 2009)

high even for scale $\tau = 5$, indicating a presence of complex dynamics probably related to the presence of long range correlations (Fig. 4(a)).

For N-S channel (Fig. 4(b)), the entropy is small for scale one and shows a small increment as the scale factor increases, that is, more regularity in the fluctuations is present in the original time series. This behavior is observed for almost the whole period in region II, except for a short interval at the beginning where a transient very similar to the one identified in E-W channel is observed.

In Figs. 4(c) and 4(d) results for the corresponding shuffled versions are presented. For E-W data, we observe that for almost the whole two year period a pattern similar to white noise is present, except for a high value, corresponding to scale one, which is identified in the period of complex dynamics. For N-S channel, the entropy shows a profile similar to white noise. For Coyuca Station and for both channels, we observe that S_E-values are high for short scales indicating a high variability in the signals (see Figs. 5(a) and 5(b)). Another important features observed in both channels are the presence of multiple short periods with a low entropy value across multiple scales and that entropy values rapidly decrease as the scale factor increases, indicating more regularity for large scales. When these results are compared to their corresponding surrogate sequences, the entropy profile is similar to white noise and the short periods with low entropy values are changed to uncorrelated dynamics (see Figs. 5(c) and 5(d)).

3.2 DFA results

To obtain further insights in the evaluation of the complex dynamics observed in some periods of the records and its relation with the presence of correlations, we apply the DFA method (Peng et al., 1995b). The DFA is applied to segments of the same length as in the case of entropy calculations (Guzmán-Vargas et al., 2009) . Representative cases of $F(n)$ vs. n for some periods during 1995 (from Region II defined in Fig. 4(a)) are shown in Fig. 6. As we can see in these plots, two different scaling exponents can be defined to describe correlations. To get a better estimation of α-values and the crossover point, we consider the following procedure: given the fluctuation values $F(n)$, a sliding pointer is considered to perform linear regression fits to the values on the left and to the elements on the right. At each position of the pointer, we calculate the errors in the fits (e_l and e_r) and we monitored the total error defined by $e_t = e_l + e_r$. We define two stable exponents when e_t reaches its minimum value and the position of the crossover point is within the interval $6 \leq n \leq 500$. The results of DFA exponents for two regimes (separated by the crossover point $n_\times$) from Acapulco and Coyuca are presented in Figs. 7 and 8.

For both channels in Acapulco station, as it occurred in MSE analysis, we identify three different regions which are characterized by different correlation dynamics. For region I, we see that α_1 and α_2 are quite similar each other with values around 0.5 which indicates a white noise behavior (see Fig. 7(a),(b)). For region III, defined from November 1995 to May 1996, we observe that the signals also display mostly white noise dynamics. Interestingly, for region II, that is, for a period comprising November 1994 to October 1995, the dynamics can be described by two values distinctly different, both of them higher than 0.5 and close to 1, indicating long-term correlations. A more detailed observation of the scaling exponents within this region in E-W channel (Fig. 7(a)) reveals that, from November 1994 to March 1995, α_1 is close to the Brownian motion value ($\alpha_{BM} = 1.5$) whereas α_2 oscillates and stabilizes around the white noise value. In the immediate period from April until October 1995, both scaling exponents are close to 1, indicating the presence of power-law correlations. For N-S channel, a more remarkable crossover behavior is identified for the whole region II. In this case, for short scales $\alpha_1 \approx 1.5$ and for large scales $\alpha_2 \approx 0.5$.

Also, for this period and both channels, we find that these two scaling exponents are splitted by the average crossover point $\overline{n}_\times \approx 14$, which corresponds to 28 seconds, that is, approximately a half minute in time scale. We also performed the same crossover analysis to the data from Coyuca station (see Fig. 8).

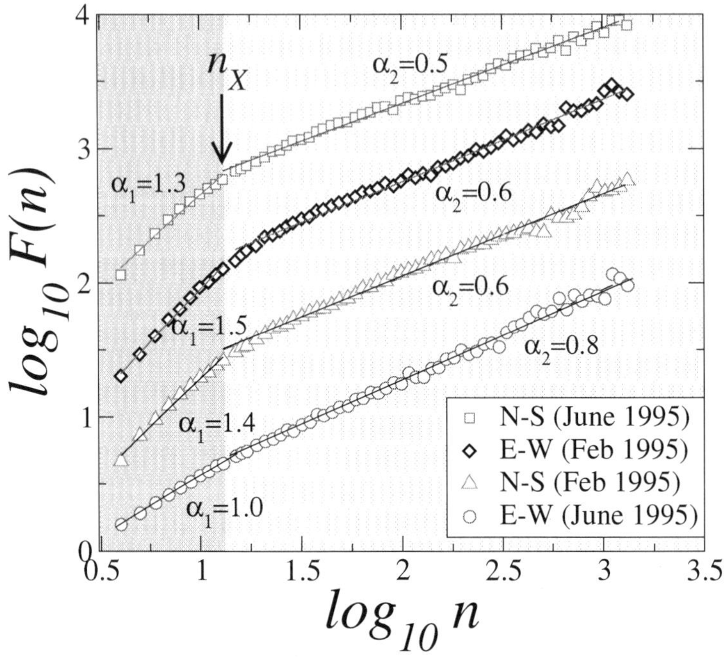

Fig. 6. Representative plots of $\log F(n)$ vs. $\log n$ for segments from Acapulco station during 1995. We observe that two scaling regimes can be defined to describe correlations. We calculated α_1 and α_2 according to the procedure described in the text. We find that there is an approximate typical characteristic time at which the crossover is present in these four cases. Notice that the data from June 1995 (E-W channel, open circles) show a weak croosover with both scaling exponents close to 1, indicating long-term correlations whereas data from N-S channel (open squares) lead to a clear crossover with a value close to a random walk ($\alpha_1 \approx 1.3$) for short scales and uncorrelated fluctuations ($\alpha_2 \approx 0.5$) over large scales (Guzmán-Vargas et al., 2009).

3.3 Cross sample entropy results

3.3.1 Simulated signals

Figure 9 shows the C_E profile for the simulated signals with power spectral density of the form $f^{-\beta}$, with $0 \leq \beta \leq 1$. For each value of the spectral exponent, ten independent realizations were performed and averaged to obtain the displayed results. As can be seen, C_E stays well-defined for longer sequences when longer-range correlations become present in the signal (increasing β). Specifically, we observe that for values of β close to the white noise fluctuations ($\beta = 0$), the pattern synchrony shows a high value and persists for a sequence length of around 8 samples whereas for values of β close to one, the C_E is sligthly lower than for the uncorrelated case but it persists for a larger sequence length such that for $\beta = 1$ it is around 12 samples.

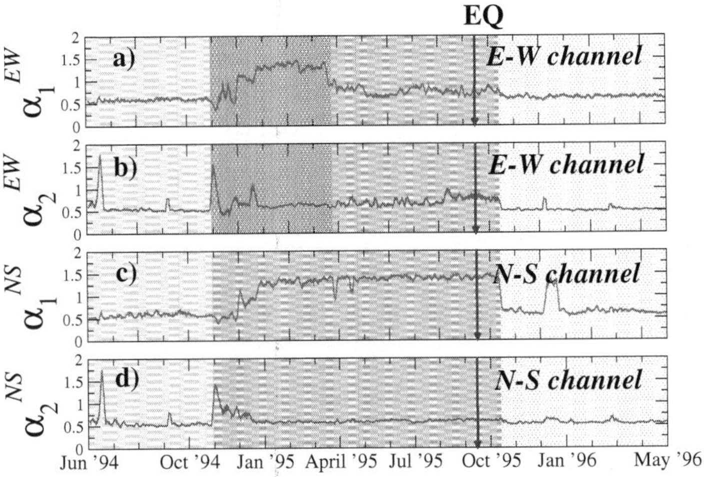

Fig. 7. Time evolution of averaged DFA-exponents for Acapulco station. The results of E-W channel are presented in Figs. 7(a) and 7(b). We also identify three main regions as in entropy results (Fig. 4(a)). For regions I and III defined in Fig. 4(a), we observe that both scaling regimes are quite similar each other with a value close to white noise behavior. In contrast, Region II display significant alterations in both scaling exponents. At the beginning of this region both exponents show an increment such that α_1 is close to the Brownian motion value, after this period both exponents are close to one, indicating the presence of long-term correlations. For N-S channel, Regions I and III also show values close to white noise level in both short and large scales, except because the presence of a few peaks. Remarkably, Region II reveals a clear crossover with $\alpha_1^{NS} \approx 1.5$ and $\alpha_2^{NS} \approx 0.5$ (Guzmán-Vargas et al., 2009).

In addition, Figure 9 shows the results of the C_E between signals with different spectral exponents β_1 and β_2, for different sequence lengths. As can be seen, the pattern synchrony between signals with correlations of longer range (for $\beta \to 1$) persists for longer sequences.

3.3.2 Acapulco data

The results of the C_E calculation for the original and shuffled data from the Acapulco monitoring station are shown in Figure 10.

We observe regularity in the C_E profile for region I. Moreover, the C_E profile for the original data is not significantly different to the one obtained for the shuffled data, except that the C_E reaches systematically higher values for the shuffled data. For the original data there is a period of time within June 1994 and towards the end of the region I for which the pattern synchrony remains for long sequences.

Moreover, in order to assess the effect of the data shuffling on the C_E calculation, we obtain the distribution of the maximum sequence length for which the C_E is well-defined (for which there is pattern synchrony), in each calculation window. In other words, for each calculation window we obtain the value of the longest data-points sequence (pattern) for which there is pattern synchrony, such that for longer sequences the value of C_E is not well-defined. The

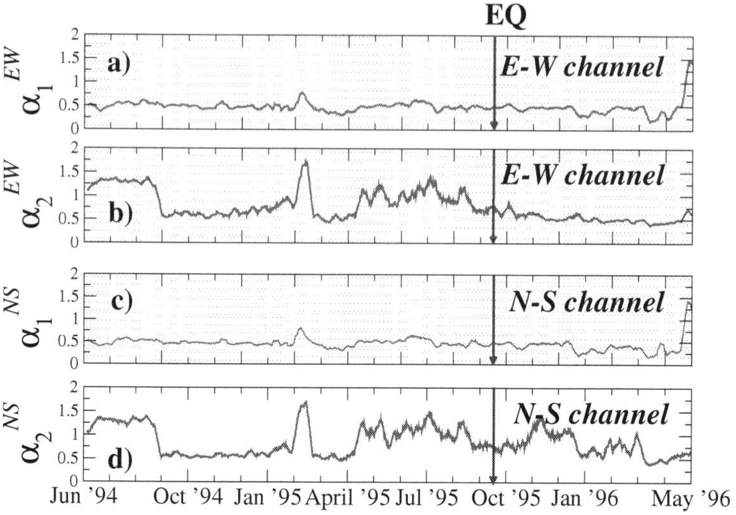

Fig. 8. Time evolution of DFA-exponents for Coyuca station. For both channels and for short scales (Figs. 8(a) and 8(c)), we observe that the exponents are close to the white noise value: $\alpha_1^{EW} = 0.52 \pm 0.26$ and $\alpha_1^{NS} = 0.49 \pm 0.20$. In contrast, for large scales the averaged exponents in both channels are bigger than 0.5 (Figs. 8(b) and 8(d)) (Guzmán-Vargas et al., 2009).

majority of the calculation windows shows presence of pattern synchrony for sequences up to 7 data-points, although for the original data we observe that there are calculation windows for which the pattern synchrony is present for longer sequences. This suggests that the pattern synchrony between the channels in this region resembles the one exhibited by white noise-like signals. This result is connected to previous works Guzmán-Vargas et al. (2008; 2009), on which we have found that for region I the signals in each channel exhibit a variability and correlations profile similar to the one for white noise.

On the other hand, we observe more variability in the C_E profile for region II. In particular, notice the significant variation of the C_E that occurred between January and April 1995. Also, notice that there is certain variability of the C_E profile towards the end the region. From our previous studies on correlations and variability for the signals in separate channels Guzmán-Vargas et al. (2008; 2009), the geoelectrical signals for region II exhibit long-range correlations behavior; and the present results suggest that not only the channel signals individually exhibit long-range correlations, but also there is pattern synchrony between channels that persists longer than for the other regions.

3.3.3 Coyuca data

The C_E results for the geoelectrical signals collected by the Coyuca monitoring station are shown in Figure 11. We observe regularity in the C_E profile for region I. Notice that the original data from Coyuca in region I exhibits pattern synchrony for longer sequences than for the

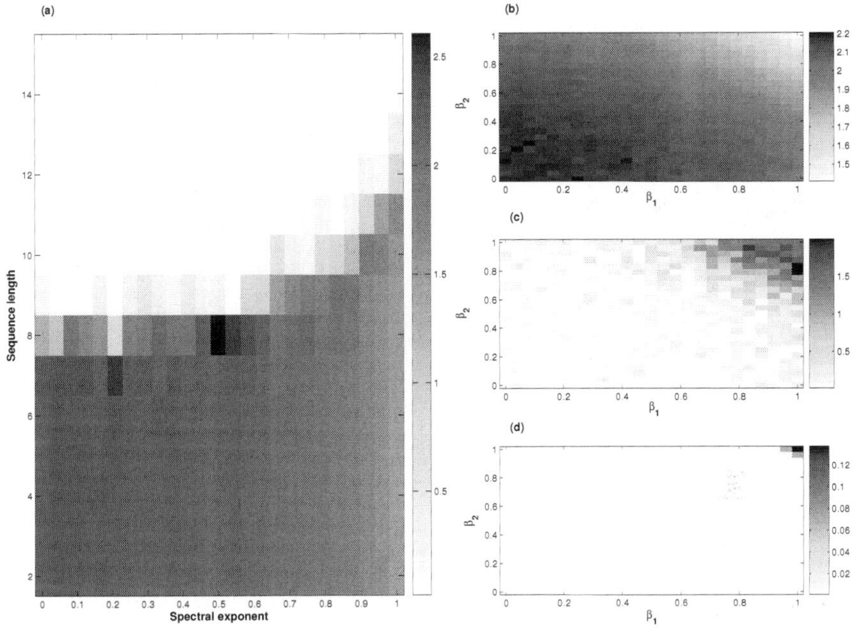

Fig. 9. Cross Sample Entropy analysis for synthetic $1/f^{\beta}$-signals, with $\beta \in [0, 1]$. (a) Shows the C_E between two signals with the same β; while the figures at the right show the results of C_E between signals with different spectral exponents β_1 and β_2, for the sequence lengths of (b) 5, (c) 10, and (d) 15 samples (Hernández-Pérez et al., 2010).

Acapulco station, where the C_E profile for the original data resembles the one obtained for the shuffled data.

On the other hand, for region II we notice the significant variation of the C_E that occurs mainly between April and June 1995. Moreover, the variability of C_E continues for the remaining part of the region. Comparing to the results for Acapulco station (see Figure 10), it can be seen that this signature occurred later for the Coyuca station, which was farther away from the EQ epicenter than Acapulco.

Finally, for region III we see that the C_E profile at the beginning of the region shows pattern synchrony for long sequences, with some gaps towards the middle and the end of the region, on which the C_E is defined for shorter sequences. Again, we observe that the C_E profile for the shuffled data still shows pattern synchrony on a non-negligible number of cases for relatively long sequences, with a more even distribution. Comparing to the results from Acapulco, for the Coyuca station we observe that the effect of the shuffling reduced less the persistence of the pattern synchrony than for the Acapulco station. Again, this difference could be due to the different local properties of the crust.

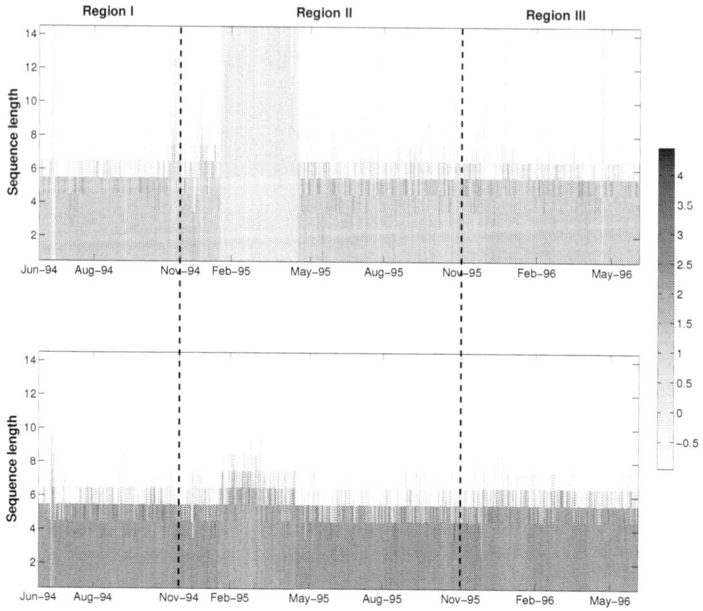

Fig. 10. Cross Sample Entropy analysis for geoelectrical time series from the Acapulco station: original (top) and shuffled data (bottom) (Hernández-Pérez et al., 2010).

3.4 Discussion

The MSE, C_E and DFA analyses suggest the existence of a relaxation−EQ−preparation−main shock−relaxation process along the June 1994 – May 1996 period. This process is approximately expressed for the sequence of white noise and correlated fluctuations, in the range of short and large scales. According with our findings, both scales showed important alterations along the period of observation. Remarkably, we observed correlated dynamics a few months before the main shock, especially in Acapulco station which is the nearest station to the epicenter (notice that the epicentral distance of Acapulco and Coyuca stations were 110 km and 200 km, respectively). These alterations were observed by means of MSE and DFA analyses; both methods consistently reveal that the changes in the geoelectrical potential observed prior to the main shock can be characterized by a complex and correlated behavior. In fact, MSE analysis incorporates a qualitatively visual manner to detect correlated fluctuations and it can be used as a complementary tool to characterize a complex behavior in noisy geolectric time series. From this point of view, is a very important task to identify the transition from white noise to correlated fluctuations, that is, the time at which a correlated signal is added to the white noise signal leading to the apparition of complex fluctuations and crossovers in the correlation scaling exponents. A more detailed observation of this transition

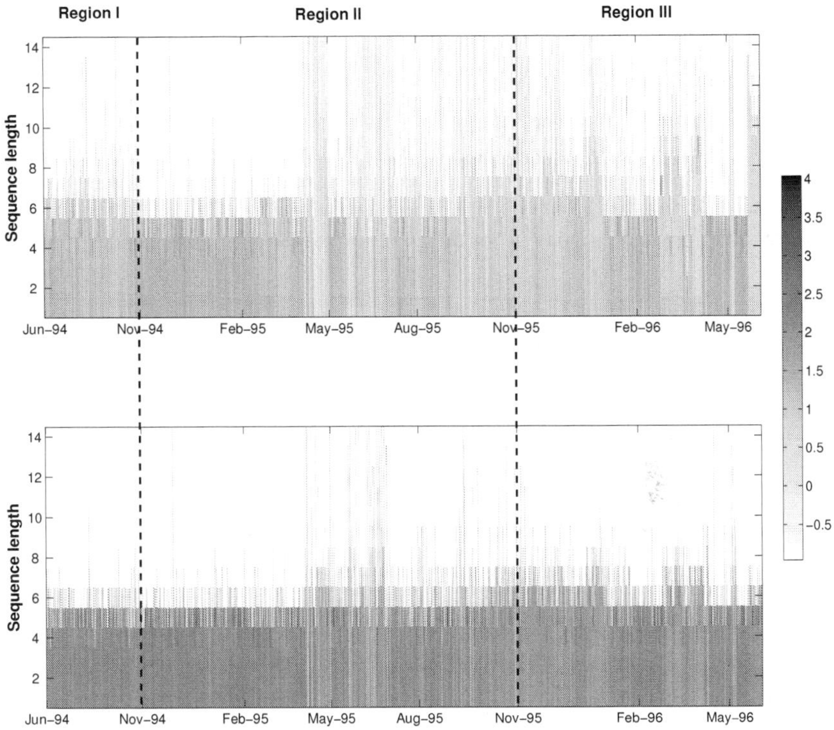

Fig. 11. Cross Sample Entropy analysis for geoelectrical time series from the Coyuca station: original (top) and shuffled data (bottom) (Hernández-Pérez et al., 2010).

located at the beginning of region II (Figs. 7a,b) reveals that, for short scales, α_1 decreases below the white noise level ($\alpha_{WN} = 0.5$) and immediately increases displaying fluctuations in the range of correlated behavior with values close to 1.5. In contrast, for large scales, the transition occurs in the opposite direction, that is, the white noise level is altered to a value close to the Brownian motion and, after a transient period, it stabilizes around a value slightly bigger than 0.5. We have identified this transient period for E-W channel from November 1994 to March 1995 (see Fig. 4a). We remark that this transition can be understood as a sequence of erratic fluctuations ranging from anticorrelated to correlated dynamics. The fact that seemingly the precursory behavior of geolectric signal are more clear in Acapulco station agrees with the empirical threshold proposed by Hayakawa et al. (Hayakawa et al., 2007) for ultra-low-frequency (ULF) geomagnetic signals given by $0.02R \leq M - 4.5$, where R is the distance between the station and the epicenter and M is the EQ magnitude. We also remark that this kind of long duration anomalies has been reported for geolectrical signals changes (lasting 56 days) for an $M = 6.6$ EQ in Japan (Uyeda et al., 2000).

4. Conclusions

We have explored geoelectrical signals from two sites in southern Mexico, to evaluate the changes in variability and correlations by using MSE, C_E and DFA methods. We have found different entropy values and correlation levels for these signals. In particular, the Acapulco station displays three different patterns of complex dynamics along the two year period which are clearly identified in E-W channel. This behavior can be interpreted as the geolectric expression of a relaxation-EQ preparation-mainshock-relaxation long-term process. The results for Coyuca station reveal that, for short scales, the entropy values and DFA exponents are close to the white noise behavior whereas, for large scales, these quantities reflect regularity resembling a random walk. The results of both stations are qualitatively compatible with previous reports based on spectral analysis (Ramírez-Rojas et al., 2004). The conceptual difference between the Sample Entropy and, as measured by the S_E, and correlations, as measured by the autocorrelation function, can be expressed as follows for a time series $\{x(k)\}$. The S_E deals with patterns: a sequence of data points of a certain length m is taken from $\{x(k)\}$ and this pattern is searched for in whole time-series looking for matches within a tolerance r. However, the S_E does not collect the time-stamp of the matching sequence in the time series, but counts the number of sequence matches of lengths m and $m + 1$. On the other hand, the objective of the autocorrelation function is to investigate the degree of dependence of future values of the time-series on present ones along the whole time series, but the time-series is not decomposed in sequences of points. Therefore, the S_E analysis is complementary to the autocorrelation and spectral analysis since it operates on different features of the signals (see the *Appendix* of Ref. Pincus & Singer (1996)). Based on this, our results on the computation of the entropy and the correlation features (DFA) are complementary since they reveal different properties of the geoelectric signals in periods with different features as captured by each monitoring station.

In summary, MSE, C_E and DFA-correlation analyses reveal important information about the complex behavior of these fluctuations and the consistent use of both methods are important complementary tools in the search of possible geoelectric precursory phenomena of earthquakes

5. Acknowledgments

This work was partially supported by CONACYT (project No. 49128-F-26020), COFAA-IPN, SIP-IPN and EDI-IPN.

6. References

Chen, Z., Ivanov, P. C., Hu, K. & Stanley, H. E. (2002). Effect of nonstationarities on detrended fluctuation analysis, *Phys. Rev. E* 65(4): 041107.

Cicerone, R. D., Ebel, J. E. & Britton, J. (2009). A systematic compilation of earthquake precursors, *Tectonophysics* 476(3-4): 371 – 396.

Costa, M., Goldberger, A. L. & Peng, C.-K. (2005). Multiscale entropy analysis of biological signals, *Phys. Rev. E* 71. 021906.

Costa, M., Goldberger, A. L. & Peng, C.-P. (2002). Multiscale entropy analysis of physiologic time series, *Phys. Rev. Lett* 89. 068102.

Eckmann, J.-P. & Ruelle, D. (1985). Ergodic theory of chaos and strange atractors, *Rev. Mod. Phys.* 57: 617.

Flores-Márquez, E. L., Márquez-Cruz, J., Ramírez-Rojas, A., Gálvez-Coyt, G. & Angulo-Brown, F. (2007). A statistical analysis of electric self-potential time series associated to two 1993 earthquakes in mexico, *Natural Hazards and Earth System Sciences* 7: 549–556.

Gershenzon, N. I., Gokhberg, M. B. & Yunga, S. L. (1993). On the electromagnetic field of an earthquake focus, *Phys. Earth Planet. Inter.* 77. 13-19.

Gotoh, K. H. Hayakawa, M., Smirnova, N. (2003). Fractal analysis of the gromagnetic data obtained at izu peninsula, japan in relation to the nearby earthquake swarm of june-august 2000, *Natural Hazards and Earth System Sciences* 3: 229–236.

Gotoh, K., Hayakawa, M., Smirnova, N. A. & K., H. (2004). Fractal analysis of seismogenic ulf emissions, *Phys. Chem. Earth* 29: 419–424.

Grassberger, P. & Procaccia, I. (1983). Estimating the kolmogorov entropy from a chaotic signal, *Phys. Rev. A* 28. 2591.

Guzmán-Vargas, L., Ramírez-Rojas, A. & Angulo-Brown, F. (2008). Multiscale entropy analysis of electroseismic time series, *Natural Hazards and Earth System Sciences* 8(4): 855–860.

Guzmán-Vargas, L., Ramírez-Rojas, A., Hernández-Pérez, R. & Angulo-Brown, F. (2009). Correlations and variability in electrical signals related to earthquake activity, *Physica A: Statistical Mechanics and its Applications* 388(19): 4218 – 4228.

Haartsen, M. W. & Pride, S. R. (1997). Electroseismic waves from point sources in layered media, *J. Geophys. Res.* 102. 24, 745-769.

Hayakawa, M. (1999). *Atmospheric and Ionospheric Phenomena Associated with Earthquakes*, Terra Sc. Publ., Tokyo.

Hayakawa, M., Hattory, K. & Ohta, K. (2007). Monitoring of ulf (ultra-low-frequency) geomagnetic variations associated with earthquakes, *Sensors* 7. 1108-1122.

Hayakawa, M. & Molchanov, O. A. (2002). *Seismo Electromagnetics: Lithosphere-Atmosphere-Ionosphere Coupling*, Terra Sc. Publ., Tokyo.

Hernández-Pérez, R., Guzmán-Vargas, L., Ramírez-Rojas, A. & Angulo-Brown, F. (2010). Pattern synchrony in electrical signals related to earthquake activity, *Physica A* 389(6): 1239 – 1252.

Honkura, Y., Isikara, A. M., Oshiman, N., Ito, A., Ucer, B., Baris, S., Tuncer, M. K., Matsushima, M., Pektas, R., Celik, C., Tank, S. B., Takahashi, F., Nakanishi, M., Yoshimura, R., Ikeda, Y. & Komut, T. (2000). Preliminary results of multidisciplinary observations before, during and after the kocaeli (izmit) earthquake in the western part of the north anatolian fault zone, *Earth Planets Space* 52. 293-298.

Ida, Y., Hayakawa, M., Adalev, A. & Gotoh, K. (2005). Multifractal analysis for the ulf geomagnetic data during the 1993 guam earthquake, *Nonlinear Processes Geophys.* 12: 157–162.

Ida, Y., Hayakawa, M. (2006). Fractal analysis for the ulf data during the 1993 guam earthquake to study prefracture criticality, *Nonlinear Processes Geophys.* 13: 409–412.

Ishido, T. & Mizutani, H. (1981). Experimental and theoretical basis of electrokinetic phenomena in rock-water systems and its application to geophysics, *J. Geophys. Res.* 86. 1763-1775.

Iyemori, T., Kamei, T., Tanaka, Y., Takeda, M., Hashimoto, T., Araki, T., Okamoto, T., Watanabe, K., Sumitomo, N. & Oshiman, N. (1996). Co-seismic geomagnetic variations observed at the 1995 hyogoken-nanbu earthquake, *J. Geomag. Geoelectr.* 48. 1059-1070.

Lazaridou, M., Varotsos, C., Alexopoulos, K. & Varotsos, P. (1985). Point defectparameters of lif, *J. Physics C: Solid State* 18: 3891–3895.

Lomnitz, C. (1990). *Fundamentals of Earthquake Prediction*, John Wiley and Sons, New York, NY.

Makse, H., Havlin, S., Schwartz, M. & Stanley, H. E. (1996). Method for generating long-range correlations for large systems, *Phys. Rev. E* 53: 5445–5449.

Matsushima, M., Honkura, Y., Oshiman, N., Baris, S., Tuncer, M. K., Tank, S. B., Celik, C., Takahashi, F., Nakanishi, M., Yoshimura, R., Pektas, R., Komut, T., Tolak, E., Ito, A., Iio, Y. & Isikara, A. M. (2002). Seimo-elecromagnetic effect associated with the izmit earthquake and its aftershocks, *Bull. Seismol. Soc. Am.* 92. 350-360.

Mizutani, H., Ishido, T., Yokokura, T. & Ohnishi, S. (1976). Electrokinetic phenomena associated with earthquakes, *Geophys. Res. Lett.* 3. 365-368.

Muñoz Diosdado, A., Pavía-Miller, C. G., Angulo-Brown, F. & Ramírez-Rojas, A. (2004). Spectral and multifractal study of electroseismic time series associated to the mw=6.5 earthquake of 24 october 1993 in mexico, *Natural Hazards and Earth System Sciences* 4. 703-709.

Peng, C. K., Havlin, S., Stanley, H. E. & Goldberger, A. L. (1995a). Long-range anti-correlations and non- gaussian behavior of the heartbeat, *Phys. Rev. Lett.* 70: 1343–1346.

Peng, C. K., Havlin, S., Stanley, H. E. & Goldberger, A. L. (1995b). Quantification of scaling exponents and crossover phenomena in nonstationary heartbeat time series, *Chaos* 5: 82–87.

Pincus, S. M. (1991). Approximate entropy as a measure of system complexity, *Proc. Natl. Acad. Sci.* 88. 2297.

Pincus, S. M. (1995). Approximate entropy (apen) as a complexity measure, *Chaos* 5: 110.

Pincus, S. & Singer, B. H. (1996). Randomness and degrees of irregularity, *Proceedings of the National Academy of Sciences* 93(5): 2083–2088.

Ramírez-Rojas, A., Flores-Márquez, E. L., Guzmán-Vargas, L., Gálvez-Coyt, G., Telesca, L. & Angulo-Brown, F. (2008). Statistical features of seismoelectric signals prior to m7.4 guerrero-oaxaca earthquake (mexico), *Natural Hazards and Earth System Sciences* 8(5): 1001–1007.

Ramírez-Rojas, A., Flores-Márquez, E. L., Guzmán-Vargas, L., Márquez-Cruz, J., Pavía-Miller, C. G. & Angulo-Brown, F. (2007). A comparison of ground geoelectric activity between three regions of different level of seismicity, *Natural Hazards and Earth System Sciences* 7: 591–598.

Ramírez-Rojas, A., Pavía-Miller, C. G. & Angulo-Brown, F. (2004). Statistical behavior of the spectral exponent and the correlation time of electric self-potential time series associated to the ms=7.4 september 14, 1995 earthquake in mexico., *Phys. Chem. Earth.* 29: 4–9.

Richman, J. S. & Moorman, J. R. (2000). Physiological time-series analysis using approximate entropy and sample entropy, *Am. J. Physiol. Heart Circ. Physiol.* 278: H2049.

Smirnova, N., Hayakawa, M. & K., G. (2004). Precursory behavior of fractal characteristics of the ulf electromagnetic fields in seismic active zones before strong earthquakes, *Phys. Chem. Earth.* 29: 445–451.

Telesca, L. & Lapenna, V. (2006). Measuring multifractality in seismic sequences, *Tectonophysics* 423: 115–123.

Telesca, L. Lapenna, V. & M., M. (2005). Multifractal fluctuation in earthquake related geoelectrical signals, *New J. Phys.* 7: 214.

Telesca, L., Lovallo, M., Alejandro, R.-R. & Angulo-Brown, F. (2009). Scaling instability in self-potential earthquake-related signals, *Physica A: Statistical Mechanics and its Applications* 388(7): 1181 – 1186.

Uyeda, S., Nagao, T. & Kamogawa, M. (2008). Short-term earthquake prediction: Current status of seismo-electromagnetics, *Tectonophysics* . doi:10.1016/j.tecto.2008.07.019.

Uyeda, S., Nagao, T., Orihara, Y., Yamaguchi, T. & Takahashi, I. (2000). Geolectric potential changes: Possible precursors to earthquakes in japan, *Proc. Natl. Acad. Sci. U.S.A.* 97: 4561–4566.

Varotsos, P. (1980). Determination of the dielectric constant of alkali halide mixed crystals, *Physica Status Solidi B* 100: 133–138.

Varotsos, P. A. (1978). An estimate of the pressure dependence of the dielectric constant inalkali halides, *Physica Status Solidi B* 90: 339–343.

Varotsos, P. A. (2005). *The Physics of Seismic Electric Signals*, TerraPub, Tokyo.

Varotsos, P. A., Sarlis, N. V. & Skordas, E. S. (2003a). Attempt to distinguish electric signals of a dichotomous nature, *Phys. Rev. E* 68(3): 031106.

Varotsos, P. A., Sarlis, N. V. & Skordas, E. S. (2003b). Electric fields that "arrive" before the time derivative of the magnetic field prior to major earthquakes, *Phys. Rev. Lett.* 91(14): 148501.

Varotsos, P. A., Sarlis, N. V. & Skordas, E. S. (2003c). Long-range correlations in the electric signals that precede rupture: Further investigations, *Phys. Rev. E* 67(2): 021109.

Varotsos, P. A., Sarlis, N. V., Skordas, E. S. & Lazaridou, M. S. (2004). Entropy in the natural time domain, *Phys. Rev. E* 70(1): 011106.

Varotsos, P. A., Sarlis, N. V., Skordas, E. S. & Lazaridou, M. S. (2008). Fluctuations, under time reversal, of the natural time and the entropy distinguish similar looking electric signals of different dynamics, *J. Appl. Phys.* 103. 014906.

Varotsos, P. A., Sarlis, N. V., Tananaka, H. K. & Skordas, E. S. (2005). Some properties of the entropy in the natural time, *Phys. Rev. E* 71(1): 032102.

Varotsos, P. & Alexopoulos, K. (1984a). Physical properties of the variations of the electric field of the earth preceding earthquakes i, *Tectonophysics* 110. 73-98.

Varotsos, P. & Alexopoulos, K. (1984b). Physical properties of the variations of the electric field of the earth preceding earthquakes ii. determination of epicenter and magnitude, *Tectonophysics* 110: 99âĂŞ125.

Varotsos, P. & Alexopoulos, K. (1986). *Stimulated current emission in the earth and related geophysical aspects. In: Amelinckx, S., Gevers, R., Nihoul, J. (Eds.), Thermodynamics on Point Defects and their Relation with Bulk Propierties*, North Holland, Amsterdam.

Veldhuis, J. D., Iranmanesh, A., Mulligan, T. & Pincus, S. M. (1999). Disruption of the young-adult synchrony between luteinizing hormone release and oscillations in follicle-stimulating hormone, prolactin, and nocturnal penile tumescence (npt) in healthy older men, *Journal of Clinical Endocrinology and Metabolism* 84(10): 3498–3505.

Yépez, E., Angulo-Brown, F., Peralta, J. A., Pavía-Miller, C. G. & González-Santos, G. (1995). Electric fields patterns as seismic precursors, *Geophys. Res. Lett.* 22: 3087–3090.

15

Hydrological and Geochemical Changes Related to Earthquakes - Examples: Three Great Earthquakes of the XX Century in the Southern Apennines (Italy)

Raimondo Pece[1] and Giuseppe Tranfaglia[2]
[1]*Dip. di Scienze della Terra, Università di Napoli Federico II, Naples*
[2]*Italian National Institute for Environmental Protection and Research, Rome*
Italy

1. Introduction

Moderate and strong (M≥5) earthquakes that take place in active tectonic crustal structures as South Apennines Chain, generate temporary or permanent environmental changes (geophysical, geochemical, geomorphic, hydrogeological and structural coseismic features). The repeated occurrence of these features leaves a geological signature in the recent stratigraphy and topography of an area, unequivocally related to the intensity of the local seismicity.

Moreover, many anomalous behaviours of aquifers have been noticed before, during and after a seismic event: sudden increase/decrease of spring flows, changes of mountain streams (streamflow level and/or flow rate) and water table level in wells, and also increase of the emanation of deep gases.

The study of the geochemical and hydrodynamic characteristics of aquifers and of their changes is considered a valid contribution to the knowledge of the natural processes connected to earthquakes, mostly because the changes in the water-rock interaction are caused by the seismic stresses in the area where the tectonic deformation leads to the seismic event.

Hydrological anomalies concomitant and/or preceding seismic events has been widely known for at least 2000 years. Only in the last three decades such variations have been related to the characteristics of seismic sources. Also in Italy possible correlation between earthquakes and hydrological and geochemical anomalies have been recognised. Many other documents of the eighteenth and nineteenth centuries report information regarding variations of water levels and spring flows in occurrence with earthquakes.

Various interpretative models indicate how the fluctuations of water level in wells and the variations in the flow of springs and/or mountain rivers can be used as earthquake precursors and, moreover, to characterise the mechanisms of the seismic source, because they are significantly influenced by the deformation field associated to the earthquake. In particular, King and Muir-Wood (1993) proposed a model of the deformation associated to

dislocation phenomena in complex systems of faults with different mechanisms and orientations. According to this model the coseismic dislocation during strong earthquakes produces a notable deformation of the superficial crust (surface layers) which influence directly the surface aquifers. Therefore, the hydrological regime and the spatial variation of anomalies are correlated to the spatial variation of the volumetric deformation produced by the phenomena of the coseismic dislocation in a wide area.
Here we resume the hydrological changes associated with some of the big earthquakes that occurred in the XX century in the Southern Apennines (Italy): 1930, 1980 and 1984. For the 23 July 1930 (M_S=6.7) in Eastern Irpinia, 23 November 1980 (M_S=6.9) in Irpinia-Lucania and 7 May 1984 (M_S=5.8) in Southern Abruzzo earthquakes (figures 1a,b and 2a,b,c). We collected an abundance of well founded hydrological information, that cannot be ascribed to environmental or anthropical causes. We analysed also hydrometric and pluviometric data monitored from the hydrological network of Italian Hydrographic Survey (IHS), looking for significant changes in wells, springs and mountain streams. Hydrological data relative to wells, rivers and springs indicate that many changes can be correlated with the earthquakes. For these earthquakes, the pre- and co-seismic stresses and the tectonic deformations have been studied in order to find a possible model of interaction between stress state and hydrological anomalies.
In this chapter a description of Southern Apennines, the three earthquakes and the types of hydrological effects for each earthquake are reported.

2. The southern Apennines

The Apennines are a Neogene and Quaternary thrust and fold belt located in the hanging-wall of the west-plunging Adria plate (Cinque et al., 1991; Doglioni et al., 1996). The Southern Apennines are a complex curved structure elongated from the Abruzzi-Molise to the Calabria-Basilicata border.
Several kilometers of vertical displacement occurred on the Tyrrhenian margin mainly along southwest-dipping normal and oblique slip faults. The extensional movements, due to their progressive shift toward the eastern sectors of the still uplifting Apennines, created deep tectonic basins elongated north-west.

Year	Month	Day	Hour-Min	Epicentral Area	I_0	M
1688	6	5	15.3	Sannio	XI	6.7
1694	9	8	11.4	Irpinia-Basilicata	X-XI	6.9
1702	3	14	5	Sannio-Irpinia	IX-X	6.3
1732	11	29	7.4	Irpinia	X-XI	6.6
1805	7	26	21	Molise	X	6.6
1930	7	23	0.08	Irpinia	X	(*) 6.7
1962	8	21	18.19	Irpinia	IX	(*) 6.2
1980	11	23	18.34	Irpinia	X	(*) 6.9
1984	5	7	17.49	Southern Abruzzo	VIII	(*) 5.8

Table 1. Major earthquakes in studied area. I_0 epicentral intensity; M macroseismic magnitude according to CPTI04 (2004), (*) surface wave magnitude.

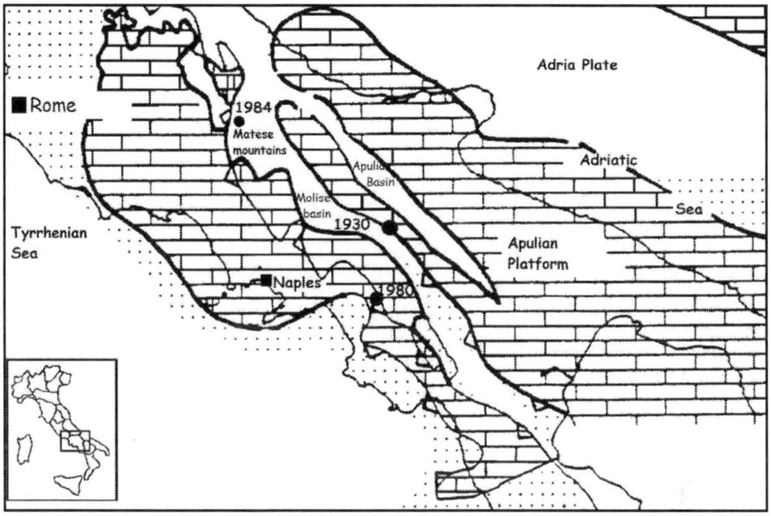

Fig. 1. (a) Sketch map of the Southern Apennines.

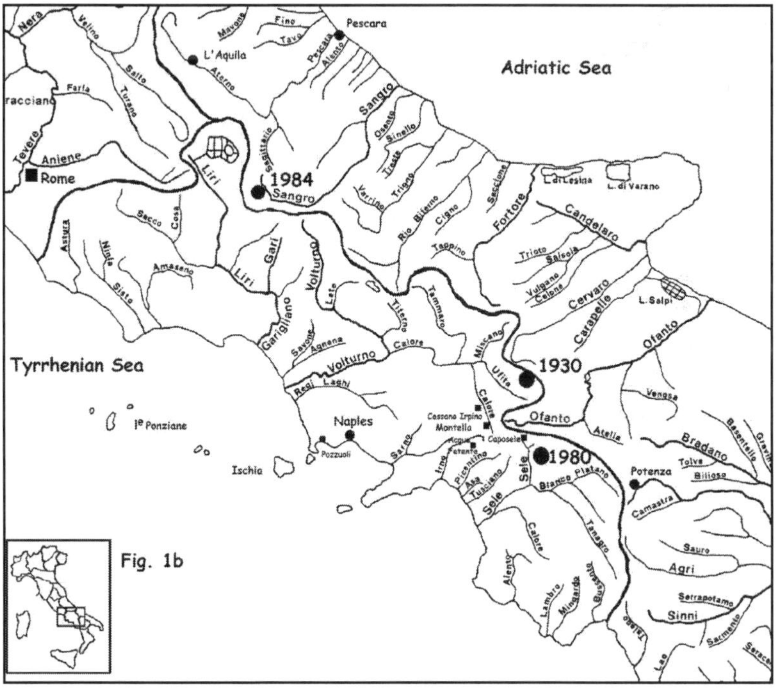

Fig. 1. (b) Hydrographic sketch map of the Southern Apennines with the epicenters of the three earthquakes

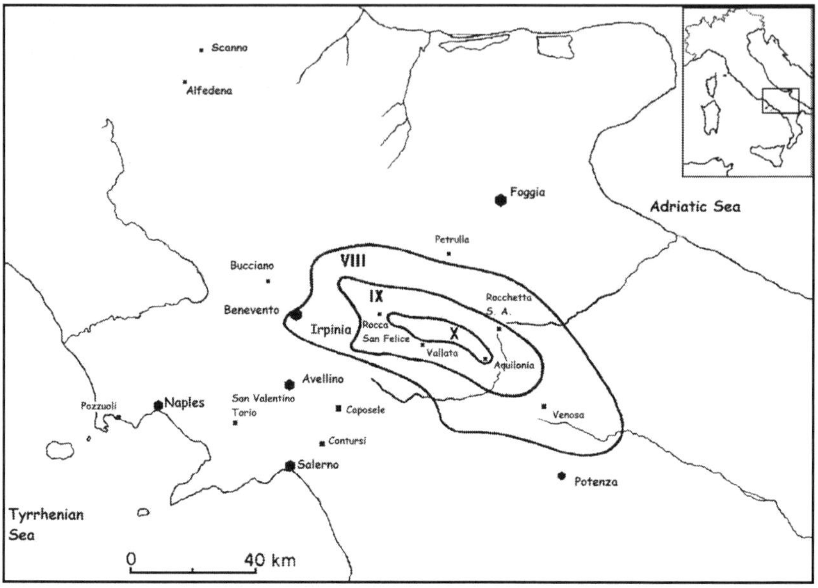

Fig. 2. (a) Isoseismal map for the 1930 earthquake

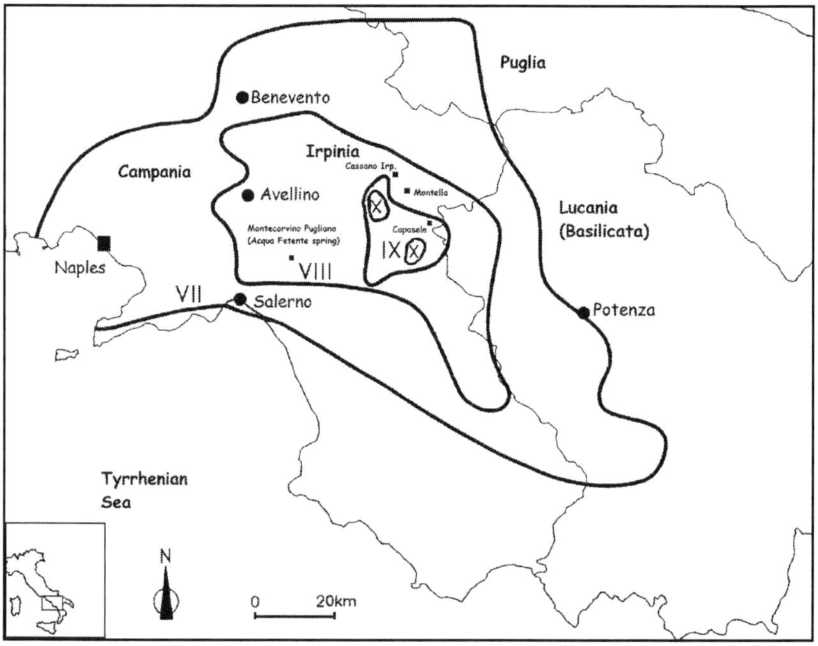

Fig. 2. (b) Isoseismal map for the 1980 earthquake

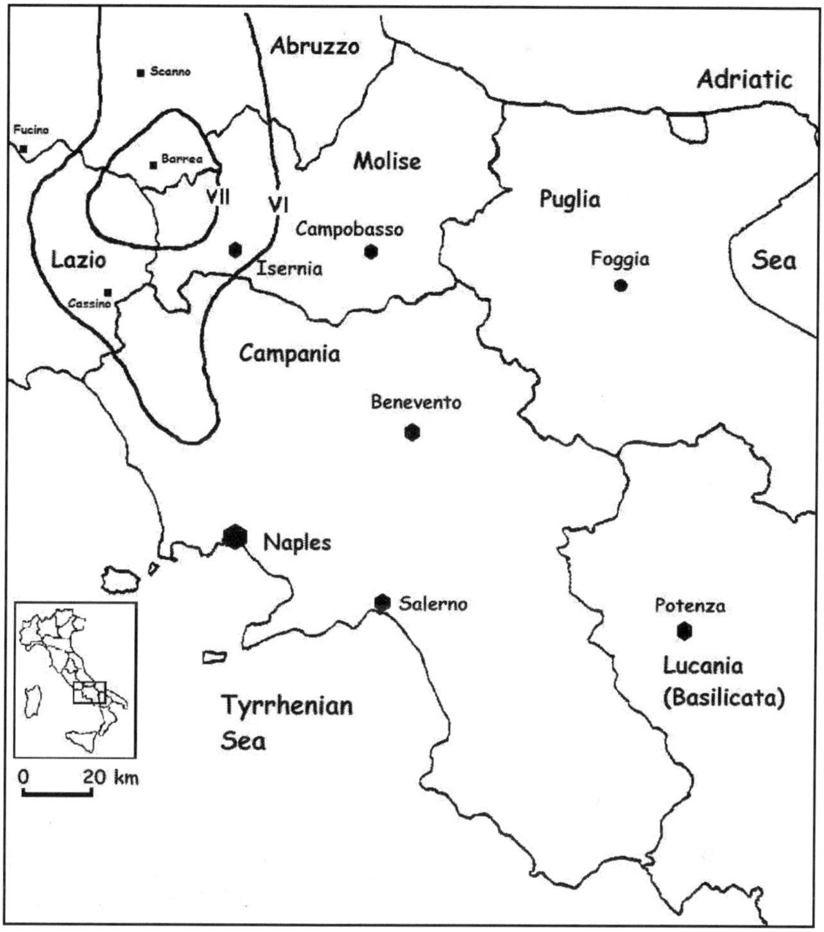

Fig. 2. (c) Isoseismal map for the 1984 earthquake

Studies of active tectonics and paleoseismicity confirm that extensional tectonics is still active in the southern Apennines, with slip-rates of several tens of millimetres per year, mostly for active faults from late Holocene until now (Westaway, 1992, 1993; Pantosti et al., 1993). The present-day tectonic setting of the mountain belt is governed by a system of Quaternary faults responsible for frequent moderate to strong crustal earthquakes, with typical hypocentral depths of 7-20 km. Table 1 lists the major earthquakes in the last centuries in studied area.

3. The earthquake of July 1930

The 23 July 1930 earthquake happened at 00:08 GMT with the greatest intensity of X MCS (Mercalli-Cancani-Sieberg scale) and a magnitude M_S=6.7, followed by many aftershocks, some of which with intensity of VII MCS. The epicentre of the main shock (figure 2a) was located at 41°05'N and 15°37'E in Irpinia (Freeman, 1930). The earthquake affected a very wide area, 36,000 km^2, comprising the regions of Campania, Puglia and Basilicata. The main shock was particularly destructive, resulting in 1425 fatalities, about 10,000 injured and more than 100,000 homeless people, 22 villages destroyed and about 40,000 dwellings damaged (Spadea et al., 1985). The epicentral zone of I_{MCS}=X is elliptical and extends over an area of 180 km^2 with the major axis of 34 km parallel to the Apennine trend (WNW-ESE). The area of the greatest effects, primary (surface faulting) and secondary effects (slope movements, ground cracks, hydrological anomalies) ($I_{MCS} \geq$ VIII) is also elliptical, extending over about 6000 km^2.

Many foreshocks and aftershocks accompanied the main event. At least two foreshocks preceded it at 23:30 on 22 July and at 00:30. The aftershocks with destructive effects occurred until 1931, also with intensity I_{MCS}>VI (Spadea et al., 1985). Surface wave magnitudes (M_S) in the range 6.2-6.7 have been estimated for the 23 July 1930 earthquake. Whereas Westaway (1992) determined a seismic moment of Mo=3.2x10^{25} dyne-cm, Jimenez et al. (1989), on the basis of seismograms recorded at Jena (Germany), calculated Mo=2x10^{25} dyne-cm. The fault plane orientation was WNW-ESE (Apennine chain trend), the fault length was 32.6 km and the depth 15 km, estimated on the basis of the equivalent ray of the major isoseismal lines (Gasperini et al., 1999).

Hydrological changes were observed in the whole macroseismic field, mostly in the far field, near the main carbonate aquifers in a widespread karstic environment (Esposito et al., 2009). They include flow increases both in springs and wells, turbid water and drying up of springs, appearance of new springs and variations in the chemical parameters of waters. From the data collected by the IHS (Annales 1925-1940) many anomalies were analysed by evaluating the shape and timing of hydrological changes in 151 sites both in the Tyrrhenian and in Adriatic watersheds (48 wells, 88 stream gauge stations, 15 springs). Spring flow at Madonna del Carmine increased from 10 l/min to 40 l/min after the earthquake, and at Monte della Guardia increased from 5 l/min to 16 l/min at the end of August (Esposito et al., 2009).

At Solfatara, a volcanic crater in Pozzuoli (near Neaples), at a distance of about 100 km from the epicentre, variations in endogenous activity were observed for about 20 days after the earthquake. Majo (1931) reports a temporary decrease in fumarolic gases, diffuse H_2S emanation from the soil, strong gas bubbling in a mud pool, and a notable temperature increase in monitored points. The possible influence of seismicity on gas release from depth can be demonstrated on the occasion of a small earthquake felt locally on 12 August 1930 near Pozzuoli. Table 2 reports the temperatures in °C measured in various sites inside the Solfatara crater (a) before 23 July, (b) during the period 28 July - 8 August and (c) during the period 13 - 26 August. In a nearby site, at Stufe di Nerone, a considerable increase in CO_2 and temperatures was also observed.

Site	Town	a	b	c
Fangaia (mud pool)	Pozzuoli	99.5	104.5	99.4
Bocca Grande (main Fumarole)	Pozzuoli	162.5	163.8	162.0
Pietra Spaccata	Pozzuoli	98.0	101.5	98.2
Stufe di Nerone	Bacoli	92.0	98.0	93.0

Table 2. Temperatures in °C measured in various sites inside the Solfatara crater.

Another site with fumaroles and mud pools, Ansanto Valley, situated about 20 km from the epicenter, presented an increase in gas emission and mud boiling, together with light flashes (Alfano, 1931). This is a very sensitive site since such phenomena also occurred in this site for other earthquakes in the southern Apennines (Italiano et al., 2000).

Near Venosa (the Vulture volcanic complex, not far from the epicentral zone) an increase in soil temperature was measured; Oddone (1932) imputed it to chemical reaction produced by a water table uplift in layers with Fe and S.

Effects on fumarole activity at distances of about 100 km from the epicenter seems hard to explain, but correlations between seismicity and volcanic phenomena are well known (Wakita et al., 1985; Hill et al., 2002; Husen et al., 2004).

Hill et al. (2002) pointed out that earthquakes and volcanoes are linked through plate tectonics and large earthquakes are capable of triggering eruptions within a matter of minutes or days at nearby volcanoes. In USA, a series of earthquakes as large as M=6.3 on 25-28 May 1980, caused turbidity and temporary increases in the discharge of hot springs in the Long Valley caldera of east-central California. These earthquakes had other obvious effects on the hydrothermal system, including emptying and refilling of boiling pools and temporary increases in fumarolic activity (Sorey and Clark, 1981). In central Japan anomalies in gas compositions were observed at fumaroles (at an epicentral distance of 9 km) and three mineral springs (at epicentral distances of 50, 71 and 95 km) about 1-3 months prior to an inland earthquake of M=6.8 on 14 September 1984 (Sugisaki and Sugiura, 1986).

Husen et al. (2004) report changes in geyser eruption behavior in Yellowstone National Park at very large distances (more than 3000 km from the epicenter) for Denali fault earthquake (Alaska), M=7.9. They interpreted these changes as being induced by dynamic stresses associated with the arrival of large-amplitude surface waves. They reported also an increase of seismic activity in Yellowstone Park and suggest that this seismicity were triggered by the redistribution of hydrothermal fluids and locally increased pore pressure.

It is plausible that such effects would occur in Southern Italy which is affected by young active tectonics with frequent strong earthquakes and many volcanically active areas (Pece et al., 1999).

Many anomalous behaviours of aquifers have been noted before, during and after a seismic event: sudden increases/decreases in spring flows, changes in piezometric levels in water wells, and increases in the emanation of deep gases (Gordon, 1970; Sorey and Clark, 1981; Whitehead et al., 1984; Wakita et al., 1985; Igarashi et al., 1992; Briggs, 1994; Curry et al., 1994; Rojstaczer and Wolf, 1992, 1994; Quilty and Roeloffs, 1997; Schuster and Murphy, 1996; Italiano et al., 2000; Thorson, 2001; Montgomery and Manga, 2003; Husen et al., 2004).

Characterizing the behaviour of aquifers and detecting anomalies in the late 1930s may be easier than in subsequent years since water resources were less exploited at that time. They are: pre- and co-seismic decreases in stream flows and water levels in wells; post-seismic increases in most of the discharges; only in some cases are they pre-seismic.

Here we illustrate the features of 7 types of the hydrological changes that we consider anomalous and connected with the 1930 earthquake. The first category of anomalous behaviour consists of decreases in stream flows before the earthquake, followed by increases after the seismic event. The four figures 3a show the data collected daily at two stream gauges (located very near great springs) on the Tyrrhenian side and at two stream gauges on the Adriatic side. In the three figures 3b the water levels measured with a 3-day frequency in one well in the Adriatic watershed and one on the Tyrrhenian side are reported, as well as the flow rate of Sanità Spring at Caposele (Sele river in figure 1b). In figure 3a.1 the anomaly consists of a sharp decrease in stream flow a few days before the seismic event, even if high rainfall preceded this decrease. The increase after the seismic event seems imputable to an anomalous discharge of the tributary springs that lasted for more than 10 days. In figure 3a.2, after the decreasing summertime trend, with a minimum reached on 24 July, there is a notable post-seismic increase from 25 July to 12 August due to

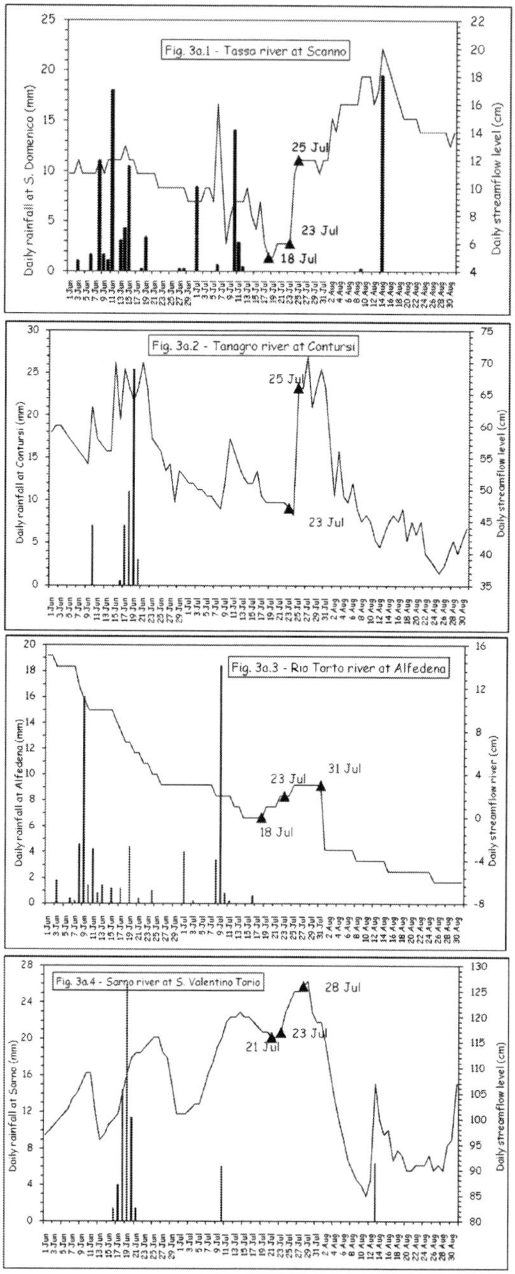

Fig. 3a.1 - 3a.2 - 3a.3 - 3a.4 (see text)

contributions from numerous large springs. In figure 3a.3 the anomaly is a temporary increase of a few cm after the earthquake.

Of great interest is the post-earthquake behaviour of 3 springs that contribute to the Sarno river (figure 3a.4). The measurements carried out at San Valentino Torio, where the total contribution of the 3 springs is measured, indicate a stream flow increase with a maximum of 127 cm on 29 July (6 days after the earthquake) followed by a decrease to a minimum of 88 cm on 12 August, a minimum level never reached before.

These types of variations have been observed for many earthquakes all over the world. In the USA, Whitehead et al. (1984) observed many significant hydrologic changes after an earthquake on 28 October 1983 in Idaho (M=7.3). Discharge measured at 10 springs and 48 stream gauging stations of the Big Lost River and surrounding watersheds increased in some instances by more than 100%. The Loma Prieta earthquake (17 October 1989) with Mw=6.9 produced hydrogeological effects reported by several authors who analysed the records of many gauging stations. Briggs (1994) analysed the hydrological effects of this earthquake in Waddell Creek watershed near Santa Cruz (California) at about 38 km from the epicenter. Numerous new springs appeared, and many inactive springs resumed flow; the springs maintained an exponential recession with minimal rain interference until they ceased flowing abruptly. As a consequence, post-seismic discharge near the mouth of Waddell Creek rose to 12.5 times the pre-earthquake discharge, followed by a gradual recession which was obscured by rain runoff beginning after about 50 days. Also Curry et al. (1994) observed very significant and unexplainable increases in the San Francisco peninsula and Santa Cruz Mountains watersheds immediately after the main shock of the Loma Prieta earthquake. For the watersheds of S. Lorenzo and Pescadero, Rojstaczer and Wolf (1992, 1994) observed that stream flows increased at most gauging stations within 15 minutes after the earthquake. Groundwater levels in the upland parts of watersheds were locally lowered by as much as 21 m within weeks to months after the earthquake.

As regard the 1930 earthquake, levels in water wells exhibited a general post-seismic increase. At Petrulla (figure 3b.1) and Bucciano (figure 3b.2) June and July rainfall did not influence the summer decreasing trend, and the increase lasted throughout August; note that at the Petrulla well the increase started 3 days before the earthquake. Figure 3b.3 shows the flow rate of Caposele spring at 22.5 km from the fault. A discharge increase of 150 liters/sec (about 3%) was measured a few hours after the seismic event, compared to the measurement on 16 July 1930, a week before the earthquake (Celentani Ungaro, 1931).

Schuster and Murphy (1996) describe an analogous hydrogeological effect for the Draney Peak earthquake, M_W=5.9 in Idaho-Wyoming (USA), on 3 February 1994: a marked increase in groundwater flow (from 4,527 to 5,695 l/min) occurred at the spring for the Auburn Fish hatchery, 5 km NE of the epicentre.

Also for the Idaho earthquake (28 october 1983, magnitude = 7.3), Whitehead et al. (1984) analyzed water levels in 69 wells: those near the epicentre generally increased rapidly after the earthquake, by as much as 3 metres. Igarashi et al. (1992), for the 2 February 1992 Tokyo Bay earthquake (M=5.9), reported possible precursor water level changes detected by the long-term groundwater observation sites. Three observation wells, about 90-110 km away from the hypocenter, showed anomalous changes: a rise and fall in water levels of 3-10 cm which began simultaneously 1-1.5 days before the earthquake. They excluded that rainfall or pumping could produce this change. The water level fall began to recover about 6 hours before the earthquake, followed by a coseismic rise of about 20 cm.

In all figures 3 the rainfall is shown. Analysis of the yearly rainfall from 1925 to 1940 shows that 1930 had slightly less than average rainfall. Moreover, the epicentral area was less rainy than

the mountainous part of the Apennines and watersheds on the Tyrrhenian side. The absence of rain on the days preceding and following the event shows that the increase in the level of the aquifer was totally due to variations in spring flow rates that flow down to the river-beds.
It is difficult to assess the anomalous variations (negative or positive). In some instances the stream flow data are sufficient to permit estimates of the total "excess" stream flow derived from a particular seismic event. Using the extensive USGS hydrological network it was estimated that the Hebgen Lake earthquake (17 August 1959; M=7.5) apparently produced about 0.3 km^3 of water, the Borah Peak earthquake (28 October 1983; M=7.3) about 0.5 km^3

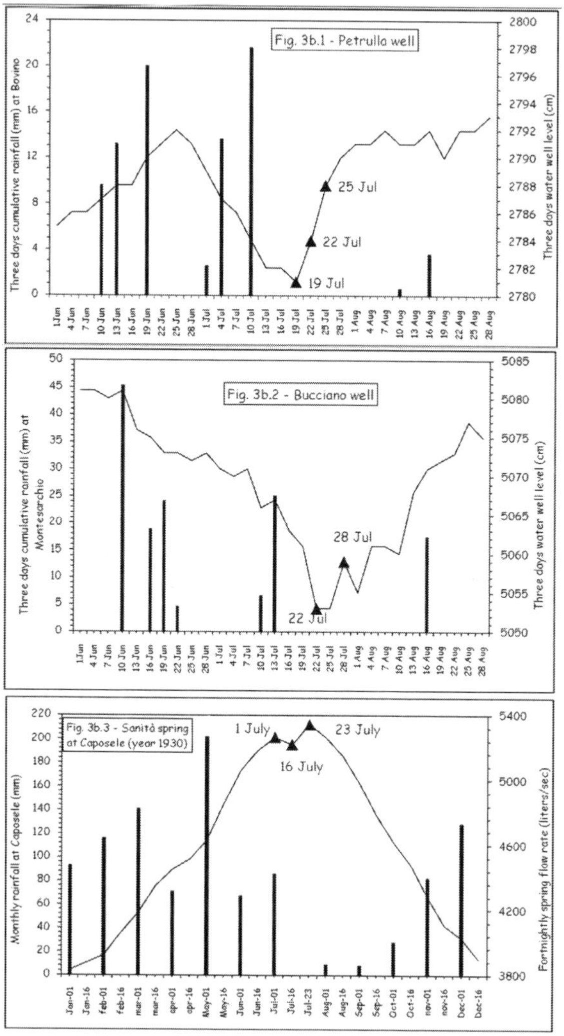

Fig. 3b.1 - 3b.2 - 3b.3 (see text)

of water, and the Loma Prieta earthquake (17 October 1989; Mw = 6.9) only about 0.01 km^3 of water (King and Muir-Wood, 1993; Rojstaczer and Wolf, 1994).
We performed an evaluation of the stage-discharge rating curves for 11 streams for which sufficient data were available (table 3). By assuming that the daily values collected in 1930 were constant in the 24 hour time frame, we calculated the average discharge in the entire anomalous period (Qav) and, obviously, the total discharge (Qtot) in this period. This permits a rough quantification of excess discharge (about 0.035 km^3 for these 11 streams) which does not appear to be correlated with the distance from the epicentre.
Gordon (1970), following the Meckering earthquake (western Australia) of 14 October 1968 (mainshock M_l=6.9), reported an increase (of about 11 cm) in water level in three boreholes 110 km west of the epicentre, which started 90 minutes prior to earthquake motion and

Site	Anomalous period (days)	Qav (m^3/d)	Q tot (m^3)	Epicentral distance (km)
Aterno river at Molina	12	162174	1946084	179.0
Tasso river at Scanno	28	62610	1565255	154.0
Sagittario river at Capo Canale	6	317486	1269946	167.5
Pescara river at Maraone	9	1498499	13486494	200.0
Lavino river at Scafa	23	74608	1715990	175.0
Zittola river at Montenero	25	9480	236995	129.0
Rio Torto river at Alfedena	15	50574	758614	107.5
Trigno river at Trivento	13	134317	1746116	107.0
Biferno river at Guardialfiera	18	536461	9656293	96.0
Fortore river at Stretta di Occhito	11	202098	2223075	70.0
Ofanto river at Rocchetta S. Antonio	8	104047	832375	14.0

Table 3. Excess discharge of the 23 July 1930 earthquake. Qav and Qtot are, respectively, the average and the total discharge in the entire anomalous period.

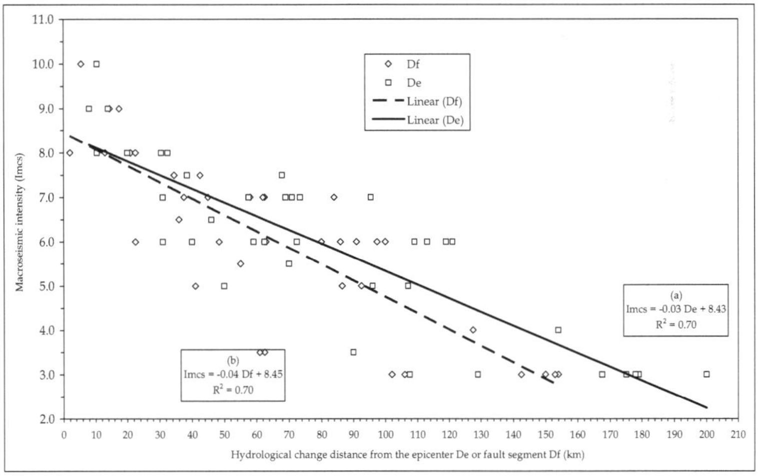

Fig. 4. Distance to epicentre (a) and to fault segment (b) versus intensity (I_{MCS}). A clear negative linear regression is visible.

lasted about six hours. For the 18 November 1755 Cape Anne historic earthquake in New England (USA) with an epicentral intensity MM=VIII, Thorson (2001) reported hydrological responses up to 275 km from the epicentre, consisting in coseismic, abrupt, long-term changes in the flow rate and chemistry of water wells from five towns in Connecticut.
The anomalies were evaluated to determine whether there were patterns of hydrologic change related to epicentral or fault distance. Figure 4 shows that: (a) most of the phenomena lie between 30-120 km from the epicentre, whereas the maximum distance was 200 km; (b) most hydrological changes occurred within 30-110 km from the fault rupture segment. The maximum distance of such variations from the fault rupture was 155 km. Note that few hydrological anomalies occurred near the fault or near the epicentre (<30 km).

4. The earthquake of November 1980

The 23 November 1980 earthquake happened inside a seismically active band extending from southern Abruzzo to Lucania (Basilicata). This Irpinia-Lucania seismogenetic zone is characterised by a complex seismotectonic structure, buried under the thick alloctonous terrains that constitute the Apennine Chain.
On the basis of geological, structural and geophysical evidences, the seismogenetic band has been subdivided in more seismogenetic zones characterised by a frequent seismicity, both historical as the earthquakes in 1561, 1694, 1851, 1857, 1930 with I≥IX MCS and recent, main shock-aftershock type, as the earthquakes in 1980, 1990, 1991 with I≥VIII and 5≤M≤6.9.
About 800 villages were damaged in Campania and Lucania. The Imax was X MSK. This great intensity and the nature of the geological structures caused very numerous effects. Surface fractures were observed in all the area from the epicentre till to the VIII isoseismal line. About 200 landslide phenomena were observed in an area more than 20.000 Km^2 wide around the epicentre.
The study of the hydrological effects has been carried out on measurements of water levels, spring flows measured each day, streamflow levels measured each day, continuous registration of water level in rivers (Esposito et al., 2001)
We considered also many other similar, but less continuous, measurements carried out during 1975-1985 (periodicity varied from weekly to monthly) on springs and rivers which exhibited a clear anomaly. Most of them are in the high Sele river and on Matese mountains (figures 5a-f).
Before the earthquake the trends of hydrometric levels were normal or decreasing, but show a general increase starting from 24 November, lasting for about 24-48 hours. In some cases the positive anomaly lasted till to 27 November.
A detailed study of the spring flow at Caposele and Cassano Irpino, located near the epicentre, have been carried out, by considering the monthly averages in 10 years (Pece et al., 1999). The year 1980 appears to be hydrologically anomalous.
In Fig. 5a the anomaly consists of the great increase in the flow rate of a thermal spring (Acqua Fetente) between 17 and 30 November (earthquake was on 23 November) and this increase lasted many months. All this behaviour is anomalous and the rainfalls could not produce such a sudden and great increase. It is sufficient to observe what is the contribution of each rainfall event on hydrograph of the flow rate. We suppose that the pre- and co-seismic stresses modified the groundwater circulation.
In Fig. 5b there is another type of anomaly. Measurements are carried out on the mountain stream by a river gauge and the continuous record shows that a decrease is followed by an

increase recorded at 12:00 MLT each day. It is not possible that this increase is due to some rainfall (no rainfalls are present in 23 and 24 November). The response of water level to rainfalls in this site is almost contemporaneous, because the catchment area of rainfall is small (see 27 November). So the increase between 23 and 24 November appears to be imputable only to the seismic event.

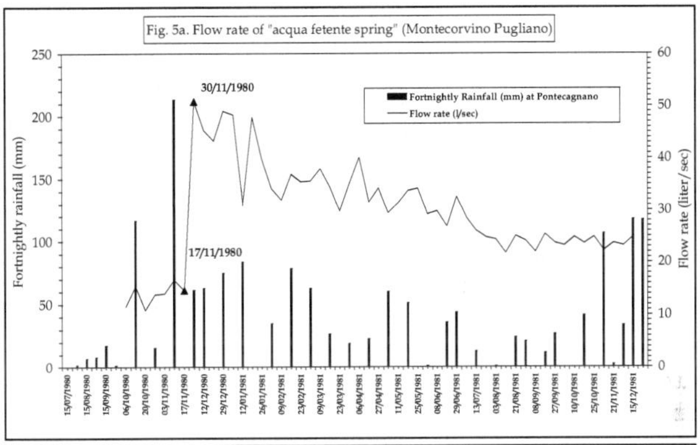

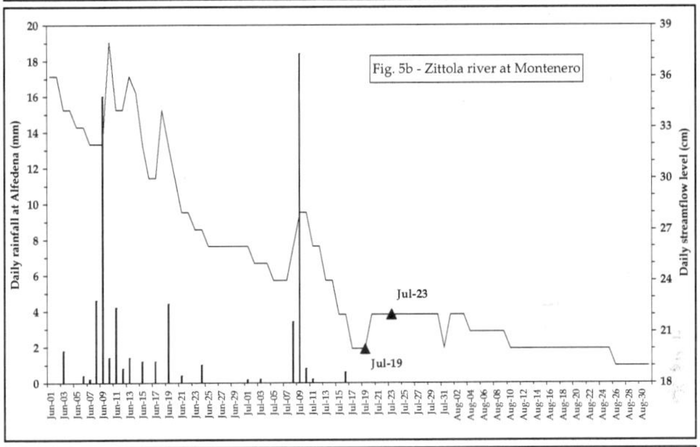

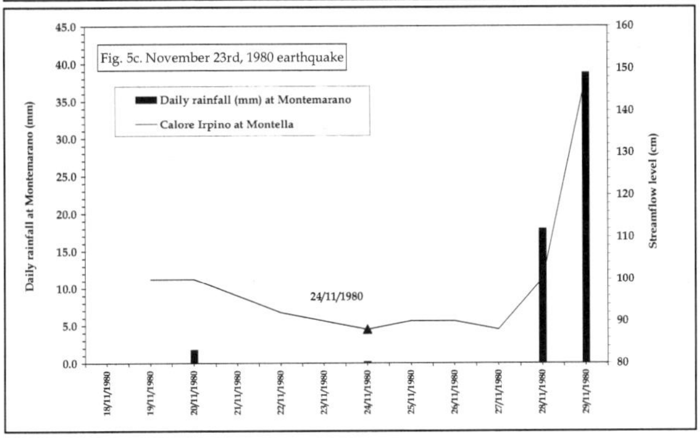

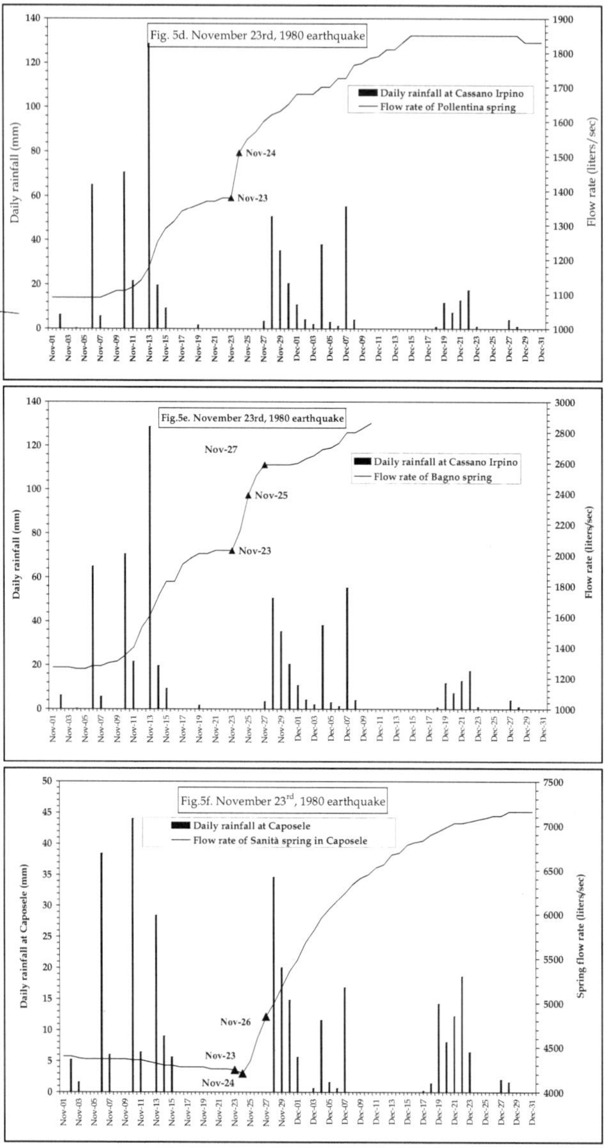

Fig. 5. Acqua Fetente spring (a); Zittola river at Montenero (b); Streamflow level of Calore Irpino at Montella (c); Cassano Irpino springs: Pollentina (d) and Bagno (e); Sanità spring at Caposele (f).

In Fig. 5c the anomaly seems to be the opposite of that shown in Fig. 5b. In fact, there is a decrease from 23 to 24 November: the increase took place on 25-26 November. Also in this case, no rainfalls occurred in those days. The decrease at this site can be explained by taking into account that near Caposele Spring the same behaviour is exhibited.

Figures 5d and 5e show the hydrological anomalies consisting of the strong increase in flow rate of Pollentina and Bagno Springs, very near each other at Cassano Irpino. It seems that this increase started about ten days before the seismic event. A further sharp increase took place soon after the earthquake.

We retain that these increases cannot be imputable to the rainfalls that preceded the seismic event. In fact, Fig. 6 shows the trends of monthly cumulated flow rates of these two springs compared to the rainfalls in the decade 1975–1985. This figure also reports the flow rate of the Caposele Spring. The lag among rainfalls and increases in the flow rates is consistent: some months, with Caposele preceding Cassano Irpino.

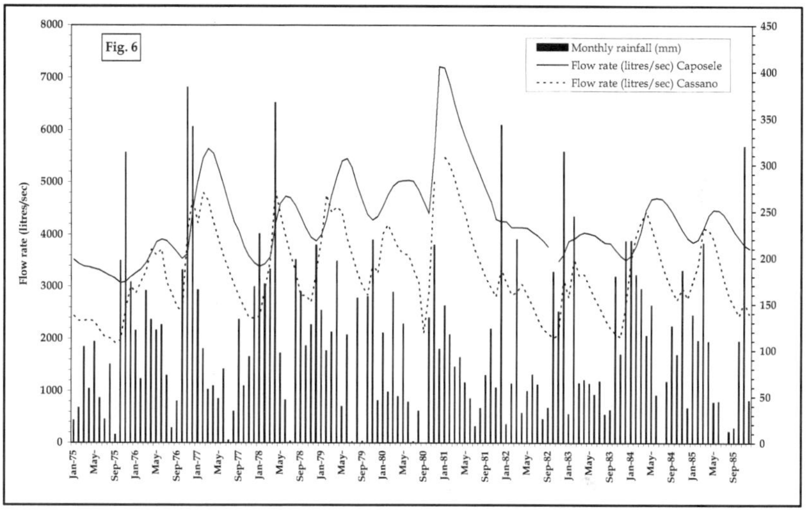

Fig. 6. (see text)

5. The earthquakes of May 1984

The seismic sequence which started in May 1984 consists of two main shocks, in 7 May (M_s=5.8 seismic moment is M_0 =0.59x10^{18} Nm and depth 10 km) and in 11 May (M_s=5.2; M_0 =0.21x10^{18} Nm, depth 10 km). The main shock fault plane was part of the Sangro fault, one of three major SW-dipping normal faults that take up much of the extension of this part of central Italy in Lazio and Abruzzo (figure 1a) (Westaway et al., 1989).

In order to evaluate the hydrological effects produced by the seismic sequence, we analyzed the data registered in 15 hydrometrical stations and the variations of levels of two lakes (Barrea and Scanno), all localised in the epicentral area and monitored by the Italian Hydrographic Survey.

Hydrological variations have been registered in 12 hydrometric sections. Among them, 3 hydrometric stations WSW in the epicentral area, and another station located west of the continuation toward north of the Val di Sangro fault did not show notable hydrological

anomalies in the days before the event of 7 May, while during all this month flow rates were comparable with the average values calculated for the period 1979-89, even if the rainfalls from January to May 1984 have been lightly higher than the average.
Nevertheless, gas outpouring and water muddying have been noticed in the evening of 5 May, in the springs at Posta Fibreno, localized about 2 km before the hydrometric sections.
Also in the hydrometric section of Sangro river at Ateleta, the hydrological anomaly was pointed out only in the value of the average streamflow level in May 1984. Consequently, for this earthquakes we have a total of 14 sites where hydrological effects have been noticed, taking into account that in 3 sites (the Rapido and Gari rivers at Cassino, the Gari river at S. Angelo in Theodice) the positive hydrological anomalies preceded the earthquake of 7 May.
Figures 7a and 7b show the anomalous behaviour in two studied sites. In Fig. 7a we show an hydrological anomaly consisting of a strong increase in the streamflow rate of the Sangro River at Barrea, not imputable to the rainfalls. Figure 7b shows the daily streamflow level of Fucino channel, "a detail" of Fig. 7a. The two sites are very near each other. It shows how during April 1984 the sharp increase of 7 May cannot be imputed to the very small rainfalls on 6 May.

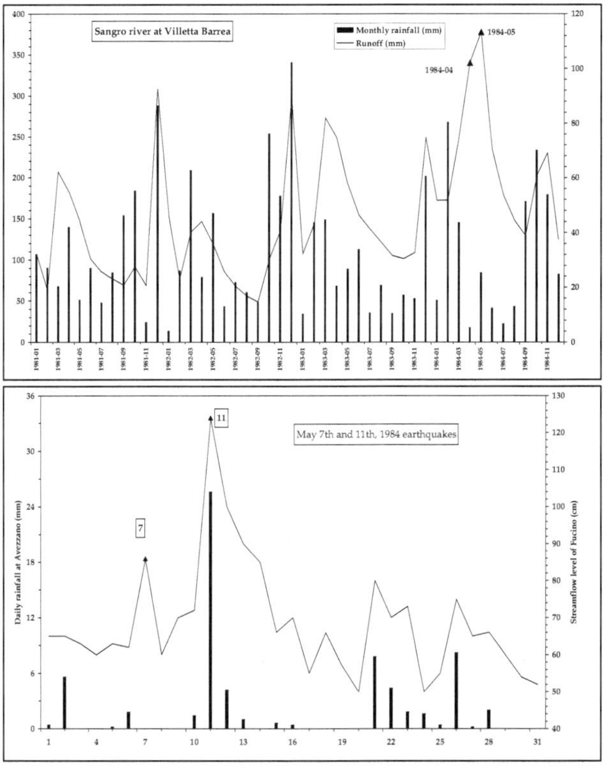

Fig. 7. Increase in the streamflow rate of the Sangro River at Villetta Barrea (a) and in daily streamflow level of Fucino channel (b).

6. Discussions and conclusions

In the Southern Apennines we found that some earthquakes produced clear forerunner signals in various areas where geochemical and hydrological parameters were controlled.
These results seem to indicate that the hydrological phenomena are associated to the changes of the stress field, during and after an earthquake.
We analyzed the relations of primary tectonic effects with the local geomorphic and structural setting.
We have also applied for some category of secondary effects a statistical test to infer the presence of trends and a regression analysis based on least-squares method was performed.
In particular, a simple bivariate scatter plot of two variables have been computed, specifically: macroseismic intensities versus epicentral or fault distances of hydrological anomalies. Figure 4 shows for the 1930 earthquake the trend macroseismic intensity-distance from the epicentre. Very similar trends have been found for the 1980 and 1984 earthquakes. Hydrological phenomena occurred throughout the macroseismic field, and were the most numerous among the induced effects. They include flow increase both in spring and well, turbid water and drying up of springs, and even creation of new springs. Some variations in chemical parameters of the waters were observed at different locations, both inside and outside the epicentral areas. The data relative to the hydrological variations' distribution versus the distance from the epicentre show that the high concentration of phenomena lie between 25-80 km, about 20% are inside the epicentral area (0-25 km); few phenomena occurred at greater distances.
A dramatic increase in the springs' flow implies the deformation of major tectonic blocks, which influences deep aquifers. Examples are shown in figures 5d,e (Cassano Irpino springs) and figure 5f (Sanità spring near Caposele) during the 23 November 1980 earthquake.
A remarkable aspect of these anomalies is that they were observed at distances of more than 200 kilometers from the earthquake epicentral area. This suggest that the impact of the Irpinia earthquake on the hydrogeologic structure of the Southern Apennines was more important, in terms of both total number of recorded anomalies and for their epicentral and fault distance.
Today there are no valid earthquake precursors, but many effects are invoked as good forerunners: geophysical changes (v_P/v_s, telluric currents, electromagnetic effects), geochemical changes (chemical composition, pH, water temperature, gases like Rn, CO_2) and hydrologic changes (piezometric levels, spring and stream flow).
Probably the simultaneous observation of all of these effects can constitute a sure forecasting. Many efforts and money are necessary for this purpose.
Taking into account that not always earthquakes are preceded by all of the above mentioned precursory phenomena, and that today's technology can provide probes for many geochemical and geophysical parameters at affordable cost, we maintain that a regional monitoring network can be installed in the Southern Apennines, in order to continuously control as many parameters as possible.
Hydrologic changes depend on both the structure of the aquifer and the strain that an earthquake induces on the area of the fault rupture.
Dobrovolsky et al. (1979) give a theoretical relation regarding earthquake magnitude, distance from the epicenter and volumetric strain. The "strain radius" R_s of a circle centered on the epicenter, in which precursor deformations and other physical phenomena occur, is given by:

$$R_s = 10^{0.43M}, \quad \text{that is } R_s \approx e^M$$

This exponential curve divides the areas where strain is lower than 10^{-8} and is greater than 10^{-8}. For strain = 10^{-8} water level changes are only 1 cm. The data of some earthquakes in Irpinia (table 1 in Onorati and Tranfaglia, 1994; tables 2, 4 and 6 in Porfido et al., 2007) are plotted in figure 8 and the strain radius is indicated.

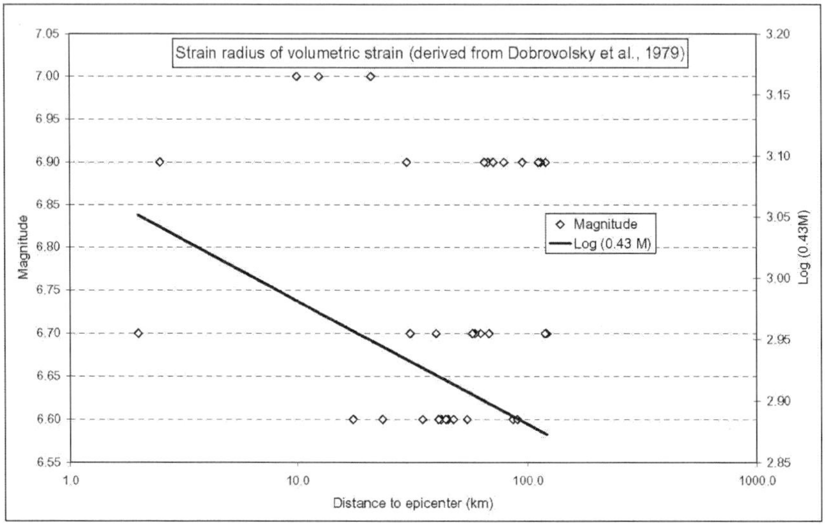

Fig. 8. (see text)

Montgomery and Manga (2003) suggest that the stream flow changes are attributable to liquefaction of valley bottom deposits. Papadopoulos and Lefkopoulos (1993) give an empirical maximum distance to the epicenter at which liquefaction can occur as a function of earthquake magnitude:

$$M = -0.44 + 3 * 10^{-8} D_e + 0.98 \log D_e ,$$

where D_e is the distance to the epicenter in cm. Figure 9 shows distance to epicentre for hydrological changes in rivers versus magnitude of Irpinia earthquakes in 1930, 1980 and 1984 (Porfido et al., 2007; Onorati and Tranfaglia, 1994). The liquefaction curve determined by the above relation is reported in figure 9. Because many hydrological changes are at distances greater than the liquefaction curve determined for valley bottom deposits, they can be caused by preseismic fracturing of carbonate aquifers in the Apennine Chain.

To define the normal hydrodynamic behaviour of an aquifer it is necessary to develop some stochastic models of the input-output type. Even the simplest stochastic model provides a lot of information on the aquifer's structure and on the connections between hydrologic variables. For this purpose an analysis of the correlation between rainfalls and water levels and flow rates has been carried out.

Four examples are reported in figures 10a-d, which shows the cross-correlograms obtained calculating the coefficients of cross-correlation for various lags. For the Pomigliano well (figure 10a), seasonal variations are evident and the coefficients of cross-correlation between the precipitations and the water table are statistically meaningful for delays until 30 months. This is in agreement with the fact that the Pomigliano well is on alluvial water table with superficial feeding due only to precipitations. For the Bucciano well (figure 10b), the cross-correlogram increases of significance until to the maximum value of 150 lags (that is about 250 days). The minor components can be due to surface feeding, but the lag=150 component is certainly due to deep feeding from the carbonate aquifer of Taburno mount. In figure 10c and figure 10d similar trends appear: seasonal correlations are very clear for Sanità Spring, fed only by carbonate aquifers of Picentini mounts; while for Tasso river the minor components indicate surface feeding, but the great lag=90 component is due to the deep feeding from Marsicano mount.

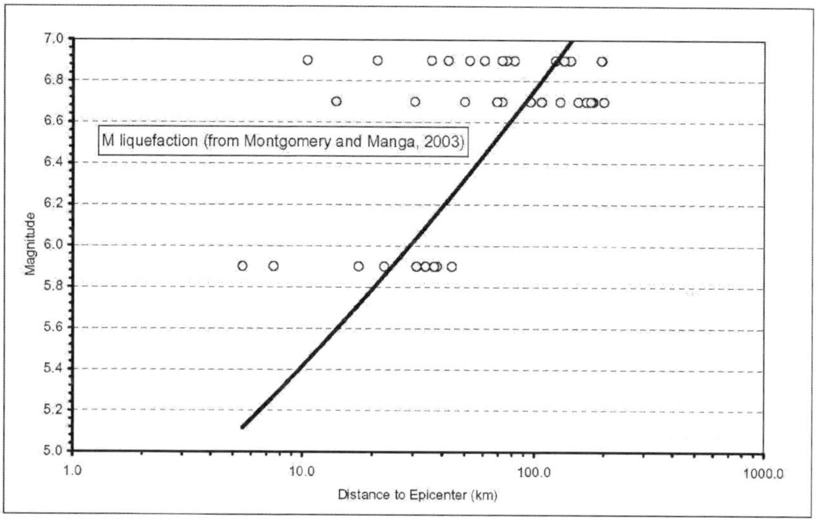

Fig. 9. (see text)

To conclude, the significant peaks can be attributed to the hydrodynamic behaviour of aquifers, and the delayed contribution from carbonate complex is evident. The delay of such peaks with respect to the start of the hydrological anomaly and/or its duration, can concur to define the space-time limits of the anomaly correlated with earthquake.

The study of the geochemical and hydrodynamic characteristics of aquifers is acknowledged to make a valid contribution to understanding the natural processes connected to earthquakes (King et al., 1981; King, 1985; Bredehoeft et al., 1987; Roeloffs et al., 1989; Kissin and Grinevsky, 1990; King et al., 1994; Quilty and Roeloffs, 1997; Ingebritsen and Sanford, 1999). Changes in the water-rock interaction are caused by the seismic stresses in the area where the tectonic deformation leads to the seismic event (Rojstaczer and Wolf, 1992, 1994; King and Muir-Wood, 1993; Quilty and Roeloffs, 1997; Roeloffs, 1998; Ingebritsen and Sanford, 1999; Manga, 2001).

Various mechanisms have been invoked to explain earthquake-related changes in water tables and in spring and stream discharges:

1. Large (as much as 20 m), near-field (probably <50 km from the epicenter) water level declines can sometimes be related to near-surface permeability enhancement due to ground motion (Rojstaczer and Wolf, 1992, 1994; Rojstaczer et al., 1995). These authors limit the validity of the relationship between seismic intensity and areas with water increases only to normal fault earthquakes.

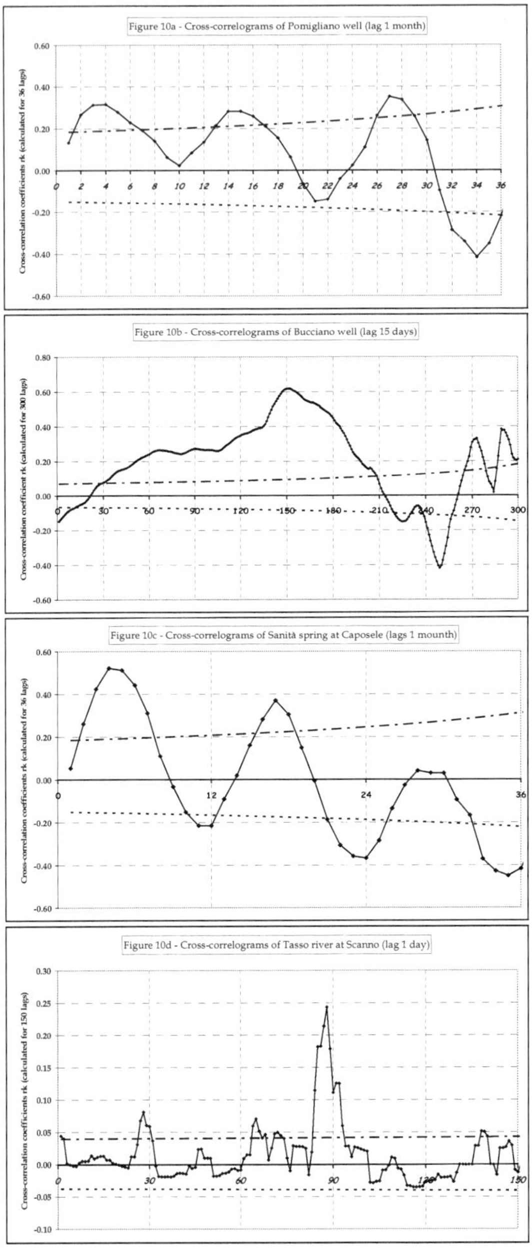

Fig. 10. Cross-correlograms of (a) Pomigliano (lag 1 month), (b) Bucciano wells (lag 15 days), Sanità spring at Caposele (lag 1 month) and Tasso river at Scanno (lag 1 day).

2. King and Muir-Wood (1993) proposed a model of deformation in complex fault systems with different mechanisms and orientations. In the model, the coseismic dislocations during strong earthquakes produce a deformation of the crust which directly influences the surface aquifers. Furthermore, in the inter-seismic periods in areas undergoing crustal extension, there is an increase in pore volume, which is then filled by percolating fluids with a consequent decrease of levels in underground waters. After a normal-faulting earthquake, the stress release produces a decrease in pore volumes and hence an increase in water levels. This hydrological behaviour can be considered a precursor anomaly. In areas undergoing inter-seismic compression, the anomalous behaviour has the opposite sign.
3. Quilty and Roeloffs (1997) analysed co-seismic changes in water level in nine wells near Parkfield, California, produced by an earthquake on 20 December 1994 (M=4.7), in order to test the hypothesis that coseismic water level changes (which for nine wells ranged from -16 to +34 cm) are proportional to coseismic volumetric strain.
4. According to Cooper et al. (1965), most of the coseismic water level oscillations observed at larger distances are resonance phenomena caused by particular fracture patterns of the formations where wells are located, that act to amplify a very small crustal strain signal.

The negative anomalies found in this work can be considered "rebound anomalies", which are the most common precursor reported by many authors and are related to increases in porosity and permeability caused by fracturing that precedes an earthquake (Roeloffs, 1988; Igarashi et al., 1992).

The total excess discharge (0.035 km^3) caused by the Irpinia 1930 earthquake (Ms=6.7) of 11 streams (table 3) is comparable with the excess discharge of about 0.01 km^3 for the Loma Prieta earthquake (Mw=6.9) (Rojstaczer and Wolf, 1992, 1994).

7. References

Alfano, G.B. (1931). Il terremoto irpino del 23 luglio 1930, *Osservatorio di Pompei 1931*, 3-57.

Bredehoeft, J.D., Riley, F.S., and Roeloffs, E.A. (1987). Earthquakes and Groundwater, *Earthquakes and Volcanoes*, 19 (4), 142-146.

Briggs, R.O. (1994). Effects of the earthquakes on surface waters in Waddell Valley, in The Loma Prieta (California) Earthquake of Oct. 17, 1989 - Hydrologic disturbances, U.S.G.S. Prof. Paper 1551-E, 21-30.

Celentani Ungaro, P. (1931). Effetti del terremoto irpino del 23 luglio 1930 - VIII sulle opere dell'Acquedotto Pugliese, *L'Ingegnere*, V, September 1931.

Cinque, A., Patacca, E., Scandone, P., and Tozzi, M. (1991). Quaternary kinematic evolution of the Southern Apennines. Relationship between surface geological features and deep lithospheric structures, *Annali di Geofisica*, 36(2), 249-260.

Cooper, H.H., Bredehoeft, J.D., Papadopulos, I.S., and Bennet, R.R. (1965). The response of well-aquifer systems to seismic waves, *J. Geophys. Res.* 70, 3915-3926.

Curry, R.R., Emery, B.A, and Kidwell, T.G. (1994). Sources and magnitudes of increased streamflow in the Santa Cruz Mountains for the 1990 water year after the earthquake. In: The Loma Prieta (California) Earthquake of October 17, 1989 - Hydrologic disturbances, U.S.G.S. Prof. Pap. 1551-E, 31-50.

Dobrovolsky I.P., Zubkov S.I., and Miachkin V.I. (1979). Estimation of the size of earthquake preparation zones, *Pageoph*, 117, 1025-1044.

Doglioni, C., Harabaglia, P., Martinelli, G., Monelli, F., and Zito, G. (1996). A geodynamic model of the Southern Apennines accretionary prism, *Terra Nova* 8, 540-547.

Esposito, E., Pece, R., Porfido, S., and Tranfaglia, G. (2001). Hydrological anomalies precursory of earthquakes in Southern Apennines (Italy), *Natural Hazards and Earth System Sciences* 1, 137-144.

Esposito, E., Pece, R., Porfido, S., and Tranfaglia, G. (2009). Ground effects and hydrological changes in the Southern Apennines (Italy) in response to the 23 July 1930 earthquake (M_s=6.7), *Natural Hazards and Earth System Sciences* 9, 539-550.

Freeman, J.R. (1930). The Italian Earthquake of July 23, 1930, *Bull. Seismol. Soc. Amer.* 20(4), 241-260.

Gasperini, P., Bernardini, F., Valensise, G., and Boschi, E. (1999). Defining seismogenic sources from historical earthquake felt report, *Bull. Seismol. Soc. Amer.* 89, 94-110.

Gordon F.R. (1970). Water level changes preceding the Meckering, Western Australia, earthquake of October 14, 1968, *Bull. Seismol. Soc. Amer.* 60(5), 1739-1740.

Hill, D.P., Pollitz, F., and Newhall, C. (2002). Earthquake-volcano interactions, *Physics Today* 11, 41-47.

Husen, S., Taylor, R., Smith, R. B., and Healser, H. (2004). Changes in geyser eruption behavior and remotely triggered seismicity in Yellowstone National Park produced by the 2002 M 7.9 Denali fault earthquake, Alaska, *Geology* 32 (6), 537-540.

Igarashi, G., Wakita, H., and Sato, T. (1992). Precursory and coseismic anomalies in well water levels observed for the February 2, 1992 Tokyo Bay earthquake, *Geoph. Res. Letters* 19(15), 1583-1586.

Ingebritsen, S.E., and Sanford, W.E. (1999). Groundwater in geologic processes, *Cambridge University Press.* Chapter 8, 215-245.

Italian Hydrographic Survey Annals (1926-1990). *Annali Idrologici, Parte I e II*, Published by Napoli, Pescara and Bari Hydrographic Departments.

Italiano, F., Martelli, M., Martinelli, G., and Nuccio, P.M. (2000). Geochemical evidence of melt intrusions along lithospheric faults of the Southern Apennines, Italy: geodynamic and seismogenic implications, *Jour. Geoph. Res.* 105(B6), 13569-13578.

King, C.Y., Evans, W.C., Presser, T., and Husk, R.H. (1981). Anomalous chemical changes in well waters and possible relation to earthquakes, *Geophys. Res. Lett* 8(5), 425-428.

King, C.Y. (1985). Earthquake hydrology and chemistry, Pure Appl. Geophys. 122, (2-4), 141-142.

King G.C.P. and Muir-Wood R. (1993). Hydrological Signatures of Earthquake Strain. *Jour. Geoph. Res.*, Vol. 98, B12, pp. 22035-22068.

King, C.Y., Basler, D., Presser, T.S., Evans, W.C., White, L.D., and Minissale, A. (1994). In search of earthquake-related hydrologic and chemical changes along Hayward fault, *Applied Geochemistry* 9, 83-91.

Kissin, I.G., and Grinevsky, A.O. (1990). Main features of hydrogeodynamic earthquake precursors, *Tectonophysics* 178, 277-286.

Majo, E. (1931). I fenomeni geofisici Flegrei susseguenti al terremoto irpino del 23 luglio 1930, *Bollettino Società Naturalisti Napoli* 43, 361-365.

Manga, M. (2001). Origin of postseismic streamflow changes inferred from baseflow recession and magnitude-distance relations, *Geophys. Res. Lett.* 28(10), 2133-2136.

Montgomery, D. R., and Manga, M. (2003). Streamflow and water well responses to earthquakes, Science 300, 2047-2049.

Oddone, E. (1932). Studio sul terremoto avvenuto il 23 luglio 1930 nell'Irpinia, in: *La meteorologia pratica, Ufficio Centrale di Meteorologia e Geofisica*, 16-26, 77-84, 116-125, 171-176.

Onorati, G., and Tranfaglia, G. (1994). L'anomalia idrologica che ha preceduto i terremoti del maggio 1984 nel Parco Nazionale d'Abruzzo, in: Terremoti in Italia - Previsione e prevenzione dei danni, *Accademia Nazionale dei Lincei,* Rome, 285-293.

Pantosti, D., Schwartz, D. P., and Valensise, G. (1993). Paleoseismology Along the 1980 Surface Rupture of the Irpinia Fault: Implications for Earthquake Recurrence in the Southern Apennines, Italy, *Jour. Geophys. Res.* 98, 6561-6577.

Papadopoulos G.A., and Lefkopoulos G. (1993). Magnitude-distance relations for liquefaction in soil from earthquakes, *Bull. Seismol. Soc. Amer.* 83(3), 925-938.

Pece, R., Tranfaglia, G., and Avino, R. (1999). Geochemical monitoring integrated in a real time hydrological network, *Il Nuovo Cimento* 22 C(3-4), 483-490.

Porfido S., Esposito E., Guerrieri L., Vittori E., Tranfaglia G., Pece R. (2007). Seismically induced ground effects of the 1805, 1930 and 1980 earthquakes in the Southern Apennines, Italy, *Ital. J. Geosci. (Boll. Soc. Geol. It.),*126(2), 333-346.

Quilty, E.G. and Roeloffs, E.A. (1997). Water-level changes in response to the 20 December 1994 earthquake near Parkfield, California, *Bull. Seismol. Soc. Amer.* 87(2), 310-317.

Roeloffs, E.A., Schulz Burford, S., Riley, F.S., and Records, A.W. (1989). Hydrologic effects on water level changes associated with episodic fault creep near Parkfield, California, *Jour. Geophys. Res.* 94(B9),12387-12402.

Roeloffs, E.A. (1998). Persistent water level changes in a well near Parkfield, California, due to local and distant earthquakes, *Jour. Geophys. Res* 103(B1), 869-889.

Rojstaczer, S.A., and Wolf, S. (1992). Permeability changes associated with large earthquakes: An example from Loma Prieta, California, 10/17/89. *Geology* 20, 211-214.

Rojstaczer, S.A., and Wolf, S. (1994). Hydrologic changes associated with the earthquake in The San Lorenzo and Pescadero drainage basins. In: The Loma Prieta (California) earthquake of October 17, 1989 - Hydrologic disturbances, U.S.G.S. Prof. Paper 1551-E, 51-64.

Rojstaczer S.A., Wolf, S., and Michel, R. (1995). Permeability enhancement in the shallow crust as a cause of earthquake-induced hydrological changes, *Nature* 373, 237-239.

Schuster, R.L., and Murphy, W. (1996). Structural damage, ground failure, and hydrologic effects of the magnitude (M_W) 5.9 Draney Peak, Idaho, earthquake of February 3, 1994, *Seismological Research Letters* 67 (3), 20-29.

Sorey, M.L., and Clark, M.D. (1981). Changes in the discharge characteristics of thermal springs and fumaroles in the Long Valley caldera, California, resulting from earthquakes on May 25-27, 1980, U.S.G.S. Open File Rep. 81-203.

Sugisaki, R., and Sugiura, T. (1986). Gas anomalies at three mineral springs and a fumarole before an Inland earthquake, Central Japan, *Jour. Geophys. Res.* 91(B12), 12996-12304.

Thorson, R.M. (2001). Remote aquifer response to the 18 November 1755 Cape Anne earthquake, *Seismological Research Letters,* 72 (3), 401-403.

Spadea, M.C., Vecchi, M., Gardellini, P., Del Mese, S. (1985). The Irpinia earthquake of July 23, 1930, in: Atlas of isoseismal maps of Italian earthquakes, *Consiglio Nazionale delle Ricerche ed.*.

Wakita, H., Nakamura, Y., and Sano, Y. (1985). Groundwater radon variations reflecting changes in regional stress fields, *Earthquake Prediction Research* 3, 545-557.

Westaway R., Gawthorpe R., Tozzi M. (1989). Seismological and field observation of the 1984 Lazio-Abruzzo earthquakes: implications for the active tectonics of Italy. *Geophysical Journal*, 98, 489-514.

Westaway, R. (1992). Seismic moment summation for historical earthquakes in Italy: Tectonic implications, *Jour. Geophys. Res.* 97(B11), 15437-15464, Correction (1993), *Jour. Geophys. Res.* 98(B3), 4539.

Westaway, R. (1993). Quaternary Uplift of Southern Italy, *Jour. Geophys. Res.* 98(B12), 21741-21772.

Whitehead, R.L., Harper, R.W., and Sisco, H.G. (1984). Hydrologic changes associated with the October 28, 1983, Idaho earthquake, *Pure Appl. Geophys.* 122, 280-293.

16

ESR Techniques for the Detection of Seismic Frictional Heat

Tatsuro Fukuchi
Yamaguchi University
Yoshida, Yamaguchi,
Japan

1. Introduction

The contribution of seismic frictional heat to the total earthquake energy budget is an important topic to elucidate earthquake rupture process. However, the frictional heat has been hardly estimated from fault rocks except melting-originated pseudotachylytes but calculated from seismological or frictional test data. The frictional heat calculated often gives a large component of the total energy budget although the San Andreas fault heat flow paradox suggests that the contribution of frictional heat may be rather small (Kanamori et al., 1998; Kanamori & Heaton, 2000; Lachenbruch & Sass, 1980, 1992). Thus, we had required any new technique to directly estimate frictional heat from fault rocks. Recent ESR (electron spin resonance) and magnetic studies of fault zones revealed that fault rocks may have been magnetized due to the thermal decomposition of iron-bearing paramagnetic or antiferromagentic minerals included in host rocks into ferrimagnetic ones by frictional heating (Fukuchi, 2003; Fukuchi et al., 2005, 2007; Ferré et al., 2005; Han et al., 2007). Since ferrimagnetic minerals commonly show huge ESR absorption due to their spontaneous magnetization, we can detect them as FMR (ferrimagnetic resonance) signals using the ESR technique. Detailed ESR analyses showed that the growth process of FMR signals during heating may fundamentally follow the zero-order reaction kinetics (Fukuchi, 2003; Fukuchi et al., 2005). Therefore, we can use FMR signals as an effective index of frictional heat. In this chapter, I will explain the basis and application of ESR techniques for the detection of seismic frictional heat using FMR signals. According to one-dimensional equations on frictional heating (McKenzie & Brune, 1972; Cardwell et al., 1978), the frictional heat strongly depends on the width of heat generation, which is equivalent to the thickness of the slip zone but not to the thickness of a pseudotachylyte vein. The thickness of the slip zone is considered to be commonly an order of millimeters or less (Kanamori and Heaton, 2000; Sibson, 2003). To actually estimate the frictional heat from a fault rock, we must sequentially detect FMR signals at a resolution of 1 mm or less. From this point of view, I develop the scanning ESR microscopic technique for sequential high-resolution measurements of FMR signals. I will introduce the case of the Nojima fault rocks in Japan.

2. Principles of ESR and FMR

A number of vacancies, interstitials and impurities exist in natural minerals. These point defects often trap unpaired electrons such as electrons ionized or holes formed in the valence energy band by natural radiation. Some transition elements such as Fe or Mn originally have unpaired electrons at the *d*-orbit. Classical physically, an electron with negative charge may be considered to be a rotating sphere, so that the rotation of an electron, that is, the electron spin generates a circular current. Since the circular current causes a magnetic field around the electron due to the electromagnetic induction, every unpaired electron has the magnetic moment. Paired electrons in materials show no magnetic moment due to the neutralizing effect of the pairing electron spins on the basis of the Pauli exclusion principle. The electron spin is responsible for all sorts of magnetism. In this section, I will explain the principles of ESR and FMR.

2.1 Electron spin resonance

Electron spin resonance is a spectroscopic technique to detect unpaired electrons in materials. Fig. 1 shows the principle of ESR. In case of paramagnetic materials, when no external magnetic field exists around, the unpaired electrons, that is, the internal electron spins are distributed at random and as a whole show no magnetic moment. On the other hand, when the external magnetic field exits, the spins are arranged in parallel or anti-parallel with the magnetic field and separate into two energy levels (E_1 and E_2). This phenomenon is called the Zeeman splitting or Zeeman effect. Consequently, the paramagnetic materials as a whole show the magnetic moment. The difference in energy level ΔE (=E_2-E_1) caused by the Zeeman splitting is expressed by $\Delta E=g\beta H$, where g is a spectroscopic splitting factor (g-value), β is the Bohr magneton and H is the magnetic field. If microwaves are added to the materials under the Zeeman splitting, parallel spins with the lower energy E_1 absorb microwave power and shift to the higher energy level E_2, and

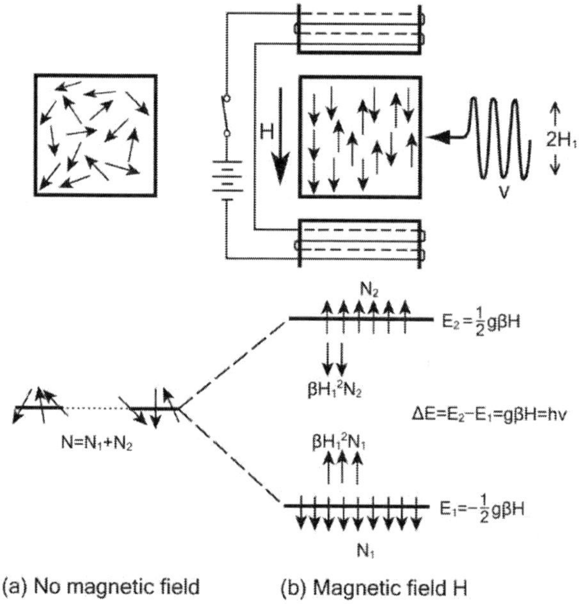

Fig. 1. Principle of electron spin resonance.

simultaneously anti-parallel spins with the higher energy E_2 emit microwaves and shift to the lower energy level E_1 when the energy $h\nu$ of microwaves are equal to the difference in energy level; $\Delta E = g\beta H = h\nu$, where h is Planck's constant and ν is the frequency of microwaves. This phenomenon is called electron spin resonance.

Since the number of the parallel spins is commonly larger than that of the anti-parallel ones, the paramagnetic materials as a whole cause the absorption of microwave power, that is, the ESR absorption. The ESR absorption is measured by sweeping the magnetic field under a fixed microwave frequency and power, and is recorded as an absorption curve using an ESR spectrometer (Fig. 2). Every paramagnetic material shows the largest absorption at the resonant magnetic field and the area of the absorption curve is proportional to the total number of the internal spins. Recent ESR spectrometers have an additional 100 kHz modulation of magnetic field to improve the S/N ratio and the rectified output is recorded as the first derivative line of the absorption curve on the recorder. We commonly call this first derivative line ESR spectrum and detect the unpaired electrons with the Zeeman energy (E_1 or E_2) as an ESR signal in the ESR spectrum (Fig. 3). Every ESR signal has an intrinsic g-value calculated from the resonant magnetic field H_r and microwave frequency ν_0 ($g=h\nu_0/\beta H_r$), so that we can identify the ESR signal by its g-value; the g-value of free electrons is 2.0023. Moreover, the linewidth and lineshape of the signal are also important physical parameters. There are two types of lineshape, the Gaussian and Lorentzian lines, in

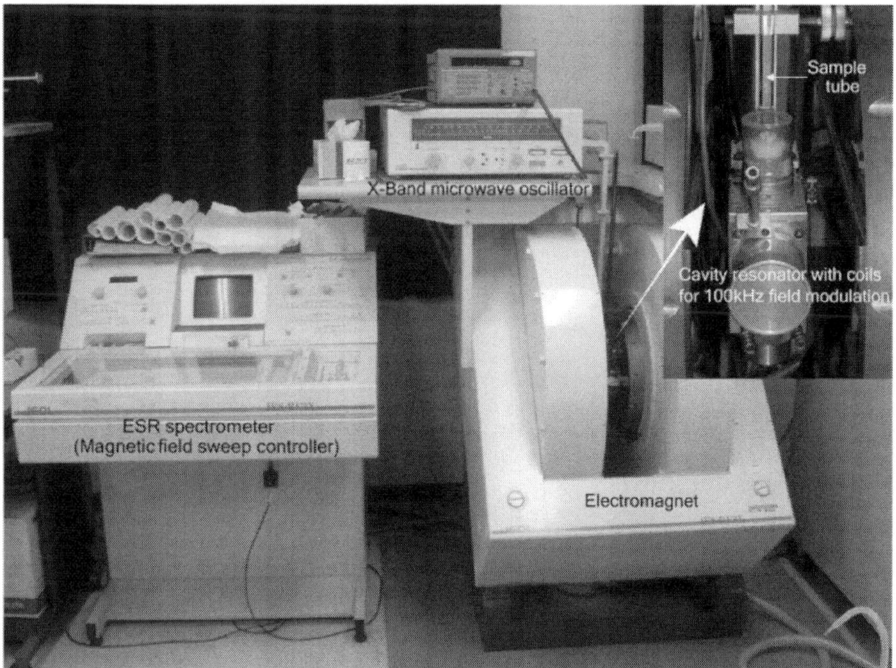

Fig. 2. X-band ESR spectrometer

the absorption curve and the first derivative line. The relationship between the peak-to-peak linewidth ΔH_{pp} and the spin-spin relaxation time T_2 are expressed by $\Delta H_{pp} = \sqrt{\pi / 2} / (\gamma T_2)$ (Gaussian) and $\Delta H_{pp} = 1 / (\sqrt{3}\gamma T_2)$ (Lorentzian), respectively, where $\gamma = g\beta/(h/2\pi)$ (Alger, 1973). On the other hand, the peak-to-peak length of the signal is proportional to the total number of the internal spins.

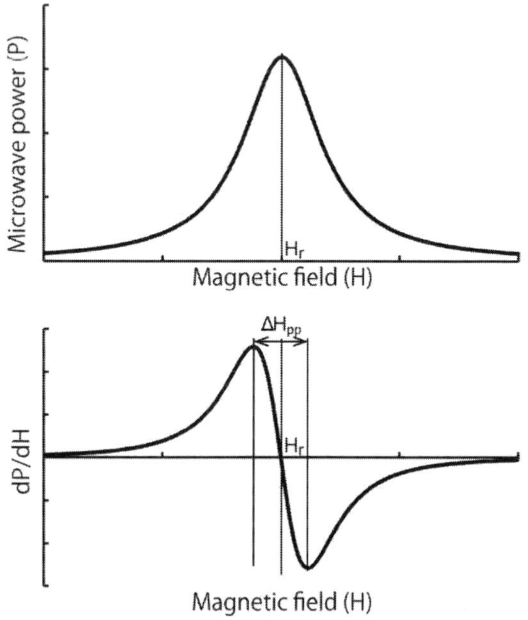

Fig. 3. ESR absorption curve and its first derivative line.

2.2 Ferrimagnetic resonance

Electron spin resonance is classified into electron paramagnetic resonance (EPR), ferromagnetic resonance, ferrimagnetic resonance, antiferromagnetic resonance (AFMR), etc. on the basis of the magnetism of the materials studied (e.g. Kittel, 2005). The principle of ferrimagnetic resonance is essentially similar to that of ferromagnetic resonance as far as the opposing magnetic moments of two sublattices in ferrimagnets precess in a magnetic field retaining their antiparallel state. In addition to this mode of precession, there is another mode that the two magnetic moments in an imperfectly antiparallel state precess on easy axes of magnetization in ferrimagnets (Kittel, 2005; Smit & Wijn, 1965). Thus, I use the term "FMR" as the abbreviation for ferrimagnetic resonance in this chapter.

In general, ferrimagnets have two sublattices consisting of different magnetic ions such as Fe^{3+} and $Fe^{3+}Fe^{2+}$ ions in magnetite. The magnetic moments of the two sublattices are opposed due to the negative exchange interaction and unequal, so that a spontaneous magnetization as a whole remains. When an external magnetic field exits, the magnetic field exerts a torque on the opposing magnetic moments and causes them to precess in one body. In case of ferromagnets, the magnetic moments are aligned in the same direction due to the positive exchange interaction, so that the external magnetic field causes the aligned moments to precess in one body. As mentioned above, in ferrimagnetic resonance there is another mode that the two magnetic moments in an imperfectly antiparallel state precess on the easy axes of magnetization. The precession in an imperfectly antiparallel state occurs in antiferromagnets as well, where the magnetic moments of two sublattices are opposed due to the negative exchange interaction and equal, so that the precession is responsible for antiferromagnetic resonance. However, the ESR absorption due to antiferromagnetic resonance is much weaker than that due to ferrimagnetic one. On the other hand, in paramagnetic resonance, every electron's magnetic moment (electron spin) is caused to precess by the magnetic field. The frequency of precession is called the Larmor frequency. Regardless of the types of magnetism, the ESR absorption occurs when the Larmor frequency is the same as the resonant frequency, that is, the frequency of microwaves added (Kittel, 2005). In magnetic resonance, the energy $h\nu$ of electromagnetic waves with the Larmor frequency coincides with the difference in the Zeeman energy levels (Fig. 1). Since the whole of the opposing moments in ferrimagnets has much larger energy level in the magnetic field than every electron's moment, ferrimagnets show much larger ESR absorption than paramagnets. Moreover, the resonant frequency in ferrimagnetic resonance and the linewidth of the ESR absorption curve strongly depend on the orientation of the material and the magnitude of the magnetic field due to the large demagnetizing field arising from the spontaneous magnetization. The ESR signals obtained from ferrimagnets are especially called FMR (ferrimagnetic resonance) signals.

3. ESR spectra of ferrimagnetic minerals

ESR has been geologically used for investigating the characteristics of paramagnetic lattice defect centers in natural minerals or ESR dating of rock-forming minerals such as quartz or feldspars. However, there are few studies on natural ferrimagnetic minerals using ESR because magnetometers are main equipments for investigating ferrimagnetic minerals in the field of geology or geophysics. ESR spectra give us important information on the magnetism of natural minerals as well as magnetometers. The g-value of the signal is an intrinsic physical parameter calculated from the microwave frequency and resonant magnetic field, which represent the energy level of unpaired electrons in minerals. The peak-to-peak linewidth and lineshape of the signal are also important physical parameters reflecting the spin-spin relaxation time. In this section, I will show ESR spectra obtained from main ferrimagnetic minerals and explain their characteristics.

3.1 Magnetite: Fe_3O_4

Magnetite is the most famous ferrimagnetic mineral. It is a cubic crystal of the spinel group and has the inverse spinel structure, which can be expressed by$Fe^{3+}(Fe^{3+}Fe^{2+})_2O_4$, indicating that each formula unit of magnetite has one Fe^{3+} in the A sublattice and one Fe^{3+} plus one Fe^{2+} in the B sublattice. Magnetite can be produced by the oxidation of iron at high temperatures in air or steam, by heating maghemite (γ-Fe_2O_3) or hematite (α-Fe_2O_3) in a reducing atmosphere, by heating siderite ($FeCO_3$) in steam or nitrogen at dull red heat, or by heating biotite ($K(Mg,Fe^{2+})_3(Al,Fe^{3+})Si_3O_{10}(OH)_2$) in nitrogen or vacuum (Deer et al., 1992). Fig.4 shows ESR spectra obtained from powder (1–10 μm) of various magnetite samples. The samples are respectively natural magnetite (≤10μm) distributed at Hanaidani Mine in

Shimane, Japan (Nichika Corporation, MJ09) (a), synthetic magnetite (~1μm) with high purity (WAKO Pure Chemical Industries, JW090103) (b), magnetite produced from natural biotite (≤10μm) distributed in the Uchinoura granite in southern Kyushu, Japan by heating at 1000°C for 1 h under 0.5 Pa (c), and magnetite produced from natural siderite (≤10μm) distributed in Guelmin Es-Semara Region, Morocco (Hori Mineralogy Ltd.) by heating at 500°C for 5 min. under 0.6 Pa (d). The spectrum c) is magnified to 50×. Measurement conditions are as follows and the same conditions are used for the other spectra in this chapter: microwave frequency; 9.43 GHz, microwave power; 1 mW, modulation width; 100 kHz 0.05 mT, sweep speed; 8 min/sweep, accumulation; 3 times, measurement temperature; room temperature. The g-value calculated from the spectra is 2.70–7.15 and the peak-to-peak linewidth of the FMR signals is 219–358 mT. The g-value tends to shift toward the lower magnetic field with increasing grain size due to the demagnetizing field. Siderite is a brownish trigonal mineral in the calcite group, while biotite is a black monoclinic mineral in the mica group. Since siderite and biotite are paramagnetic, both originally show paramagnetic signals, that is, EPR signals.

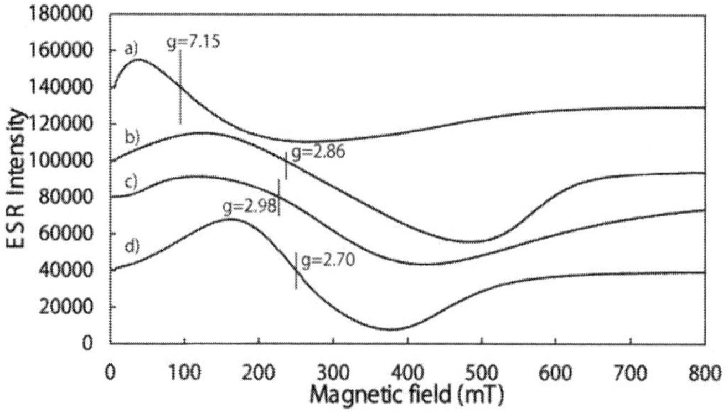

Fig. 4. ESR spectra obtained from various magnetite samples.

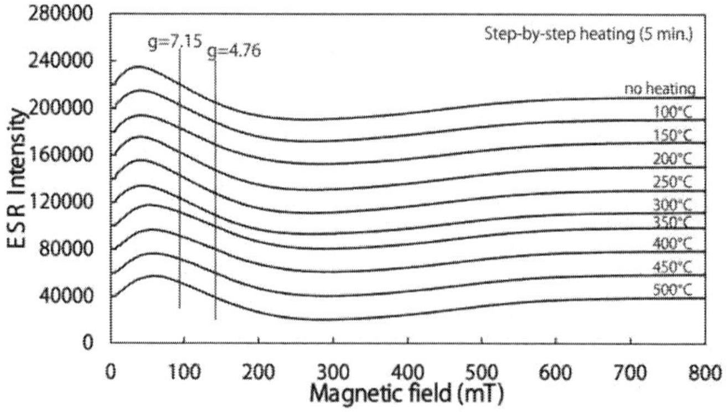

Fig. 5. Variation of the ESR spectrum of natural magnetite with step-by-step heating in air.

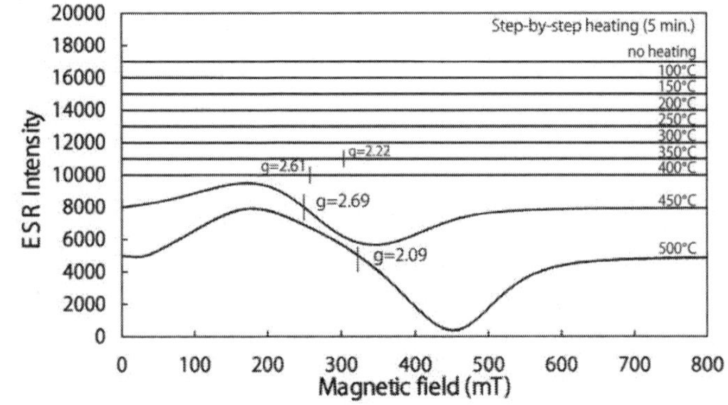

Fig. 6. Variation of the ESR spectrum obtained from natural siderite with step-by-step heating in air.

Fig. 5 shows the result from step-by-step heating experiments at 100–500°C in air. The heating duration is 5 minutes for each temperature. The natural magnetite same as the spectrum a) in Fig. 4 was used for the heating experiments. Magnetite is usually oxidized and changed into maghemite or hematite by heating in air, and the g-value of the FMR signal obtained from the natural magnetite shifts from 7.15 to 4.76 with heating. Although the signal peak becomes broader with the oxidation due to heating, its lineshape does not change so much; the peak-to-peak linewidth is in the range of 220−239 mT. This means that the oxidation may be limited to only the surface of the magnetite particles. On the other hand, Fig.6 shows ESR spectra obtained from natural siderite by step-by-step (5 min.) heating in air. The natural siderite same as the spectrum d) in Fig. 4 was used for the heating experiments. The FMR signal of magnetite produced from siderite by heating strikingly increases at 450°C, and then its lineshape is distorted at 500°C due to the production of maghemite or hematite. The FMR signal detected at 450°C in air has almost the same g-value, peak-to-peak linewidth and lineshape as that of magnetite detected at 500°C in vacuum (Fig. 4d). However, the g-value somewhat shifts from 2.69 to 2.09 with heating, while the peak-to-peak llinewidth extremely changes from 170 to 272 mT. Since the magnetite produced by thermal decomposition of siderite may consist of extremely fine crystals or amorphous particles, it is easily oxidized under an oxidizing environment.

3.2 Maghemite: γ-Fe_2O_3

Maghemite is a popular ferrimagnetic mineral and widely used as a material for magnetic recorders. It is a cubic crystal of the spinel group and has the spinel structure expressed by $Fe^{3+}(Fe^{3+}Fe^{3+}{}_{2/3},V_{1/3})O_4$ where V is a vacancy, indicating that magnetite is oxidized to maghemite by changing the valence state of two thirds of the original Fe^{2+} to Fe^{3+} while simultaneously removing one third of the original Fe^{2+} from the B sublattice. This removal occurs by diffusion producing vacancies in the spinel structure where a Fe^{2+} cation had previously resided; these vacancies account for the name cation-deficient spinel. Since ferrimagnetism of magnetite results from Fe^{2+} in the B sublattice, the removal of one third of these cations decreases saturation magnetization from 480 G (4.8×10^5 A/m) for magnetite to 420 G (4.2×10^5 A/m) for maghemite (Butler, 1992). Pure maghemite is commonly metastable and irreversibly changes its crystal structure into a hexagonal one, that is, α-Fe_2O_3 (hematite) on heating to 300–500°C. However, the transformation temperature of natural maghemite into hematite is often over 500–700°C because maghemite is stabilized with impurities inside the crystal (e.g. Fukuchi et al., 2007). As for the formation of natural maghemite, besides the oxidation of magnetite, the following three processes are pointed out on the basis of the studies of ferrimagnetic minerals in soils (Butler, 1992):

1. The formation of maghemite from iron oxides or oxyhydroxides by repeated oxidation-reduction cycles during soil formation.
2. The conversion of paramagnetic iron-bearing minerals by natural burning (above ~200 °C) in the presence of organic matter.
3. The dehydration of lepidocrocite (γ-FeOOH).

Fig. 7 shows the ESR spectrum obtained from powder (≤1μm) of synthetic maghemite and its variation with step-by-step heating (5 min.) in air. The synthetic maghemite (≤1μm) with high purity (Kojundo Chemical Lab, FE006PB) was used for the heating experiments. Maghemite shows a quite different lineshape from magnetite. The signal peak around the lower magnetic field (~100 mT) obtained from maghemite is much sharper than that from magnetite (Fig.4). The g-value calculated from the spectra is in the range of 3.26–3.34 and the peak-to-peak linewidth of the FMR signals is 346–358 mT. The FMR signal of maghemite hardly changes its lineshape during heating until 500°C and thermally shows high stability although pure maghemite transforms into hematite by heating to 300–500°C.

On the other hand, Fig. 8 shows ESR spectra obtained from natural and synthetic hematites with those from synthetic lepidocrocite and goethite (α-FeOOH) and natural limonite (FeOOH(nH_2O)). The samples are synthetic hematite (≤10 μm) with high purity (WAKO Pure Chemical Industries, JW090282) (a), natural hematite (specularite, ≤10 μm) distributed in Antananarivo, Madagascar (Hori Mineralogy Ltd.) (b), synthetic goethite (~1μm) with high purity (Kojundo Chemical Lab, FEI16PB) (c), synthetic lepidocrocite with high purity (≤10 μm, Kojundo Chemical Lab, FEI17PB) (d), and natural limonite (≤10 μm) distributed in Hwanghae-do, North Korea (Nichika Corporation, MU262) (e). Small signals like spines detected between 300 and 400 mT in the spectrum b) are Mn^{2+} markers. Hematite is a trigonal mineral of the corundum structure and shows parasitic ferromagnetism, which is a

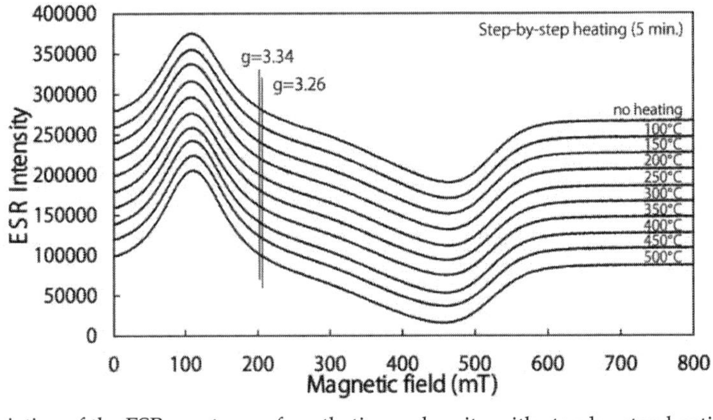

Fig. 7. Variation of the ESR spectrum of synthetic maghemite with step-by-step heating in air.

kind of antiferromagnetism. Hematite has an asymmetrical antiferromagnetic structure, so that it shows a weak spontaneous magnetization and ESR absorption (Fig. 8a). On the other hand, natural hematite (specularite) shows a quite different ESR signal from the synthetic hematite. It has the g-value of 10.8, which is much larger than that of 2.35 obtained from the synthetic hematite.

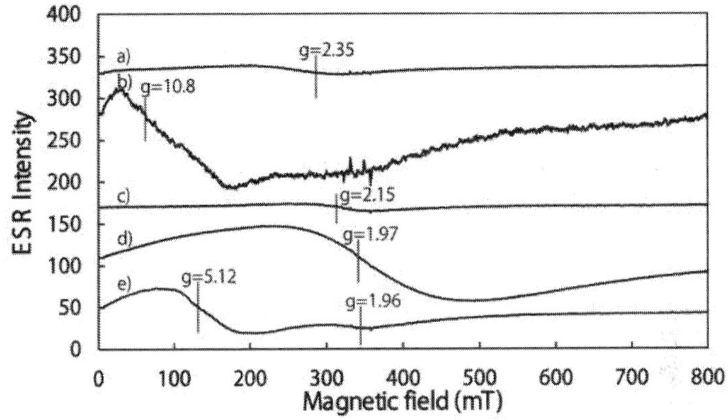

Fig. 8. ESR spectra obtained from powder of hematite, goethite, lepidocrocite and limonite.

Lepidocrocite is a monoclinic (or trigonal) paramagnetic crystal at room temperature, however shows antiferromagnetism at the Néel temperature of 77K. Therefore, lepidocrocite shows a paramagnetic signal at room temperature (Fig. 8d). A strong FMR signal of maghemite comes to be detected from heated lepidocrocite due to thermal dehydration; $2\gamma\text{-}FeOOH \rightarrow \gamma\text{-}Fe_2O_3 + H_2O$. Fig. 9 shows the variation of the ESR spectrum of lepidocrocite with step-by-step heating (5 min.) in air. The synthetic lepidocrocite same as the spectrum d) in Fig. 8 was used for the heating experiments. The FMR signal of maghemite produced from pure lepidocrocite by heating strikingly increases at 300°C. This suggests that the maghemite produced from lepidocrocite changes from a superparamagnet to a ferrimagnet between 250 and 300°C. The g-value is 2.26–2.31 and the peak-to-peak linewidth is 166–192 mT. The lineshapes of the FMR signals detected are different from those of maghemite with high crystallinity (Fig. 7). Moreover, Fig. 10 shows the result from isothermal annealing experiments at 250°C using the synthetic lepidocrocite. The lineshape and g-value of the signal obtained from the lepidocrocite characteristically change with heating time. The g-value shifts from 2.0 to 2.18 since the demagnetizing field causes the resonant magnetic field to shift toward the lower field (Fukuchi et al., 2007). The peak-to-peak linewidth also changes from 24 to 174 mT with heating time. In general, the maghemite produced from lepidocrocite has low crystallinity and shows an FMR signal with much lower g-value and peak-to-peak linewidth than maghemite with high crystallinity or one produced by the oxidation of magnetite (Figs. 5–7). Strictly speaking, such an FMR signal may be called superparamagnetic signal although it is difficult to distinguish between the superparamagnetic and ferrimagnetic signals. In this chapter, I use the term FMR signal for ESR signals derived from both superparamagnetic and ferrimagnetic maghemites. As mentioned later, the FMR signals detected from the Nojima fault rocks in Japan show almost the same characteristics as the superparamagnetic signal detected from the heated lepidocrocite.

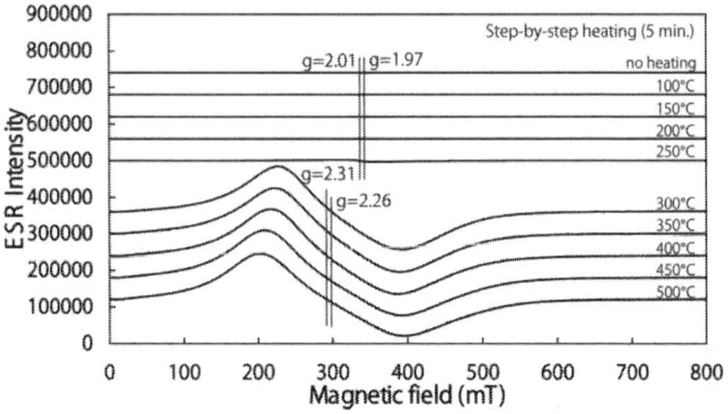

Fig. 9. Variation of the ESR spectrum of lepidocrocite with step-by-step heating in air.

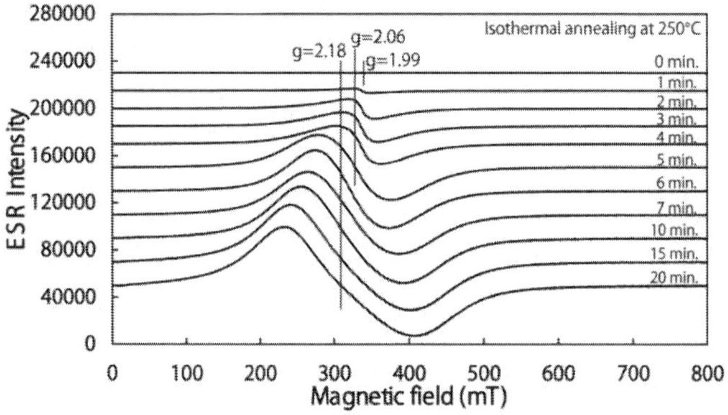

Fig. 10. Variation of the ESR spectrum of lepidocrocite with isothermal annealing at 250°C in air.

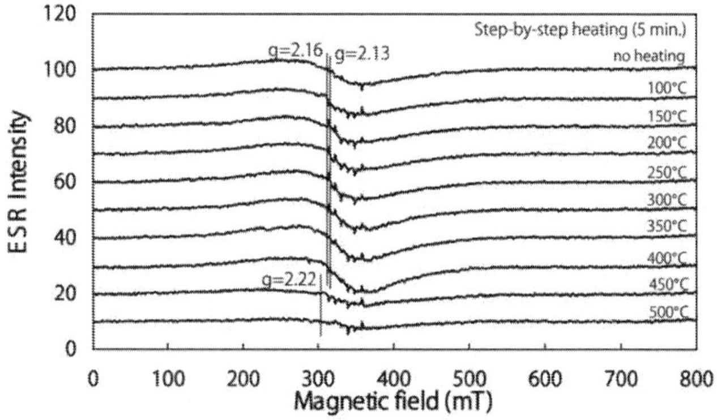

Fig. 11. Variation of the ESR spectrum of goethite with step-by-step heating in air.

Goethite is an orthorhombic mineral and a polymorph of lepidocrocite. It changes into hematite due to thermal dehydration by heating at 300–400°C; 2α-FeOOH→α-Fe_2O_3+H_2O. Fig. 11 shows the variation of the ESR spectrum of goethite with step-by-step heating (5 min.) in air using the synthetic goethite (Fig. 8c).

Goethite is an antiferromagnet however exhibits a weak ferromagnetism as well as hematite. The signal detected once increases with producing hematite, and then decays above 400°C. The g-value is in the range of 2.13–2.22 and the peak-to-peak linewidth is 92–136 mT. The characteristics of the signal are very similar to those obtained from the heated lepidocrocite although the signal intensity is very weak. On the other hand, limonite consists of cryptocrystalline goethite and lepidocrocite along with absorbed water or some hematite, so that it has a complex ESR spectrum derived from these minerals (Fig. 8e).

3.3 Pyrrhotite: $Fe_{1-x}S$ ($0 \le x < 0.125$)

Pyrrhotite has the approximate composition of FeS but always contains less iron than is indicated by this formula. Hence, it is expressed by the generic formula of $Fe_{1-x}S$, where $0 \le x < 0.125$. Pyrrhotite is a monoclinic or hexagonal crystal of the nickel arsenide (NiAs) structure and consists of several superstructures due to the presence and ordering of vacancies within the structure. In pyrrhotite, two antiparallel coupled sublattices containing iron cations exist and the number of iron cations in the opposing sublattices are unequal, so that it is responsible for the ferrimagnetism of pyrrhotite. Troillite (FeS) is antifferomagnetic and occurs mainly in meteories and lunar rocks. Pyrrhotite can be produced by the direct combination of iron and sulphur and by heating pyrite (FeS_2) in an atmosphere of H_2S at 550°C. In the Fe-S system, the pyrrhotite in equilibrium with pyrite above 400°C shows increasing iron deficit with increasing temperature (Deer et al., 1992). Fig. 12 shows ESR spectra obtained from natural pyrrhotite, synthetic troillite and synthetic pyrite. The samples used are natural pyrrhotite (≤10μm) distributed at Yanahara Mine in Okayama, Japan (Nichika Corporation, MU123) (a), synthetic troillite (≤10 μm) with high purity (Kojundo Chemical Lab, FEI06PB) (b), and synthetic pyrite (≤10 μm) with high purity (Kojundo Chemical Lab, FEI07PB) (c). The natural pyrrhotite shows very weak and broad ESR absorption, while the troillite has a similar ESR spectrum to natural hematite, that is,

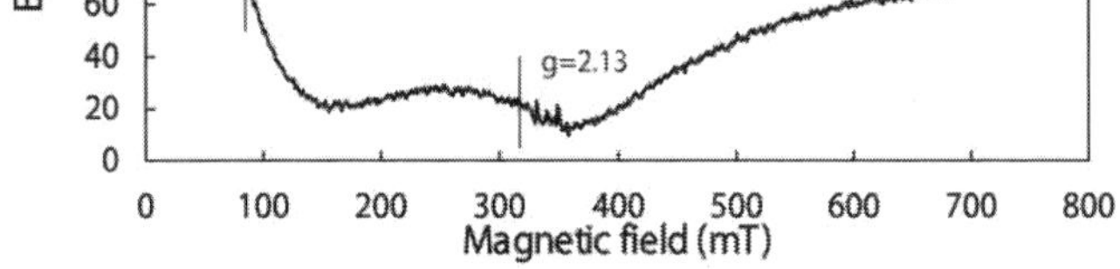

Fig. 12. ESR spectra obtained from powder of natural pyrrhotite, and synthetic troillite and pyrite.

specularite (Fig. 8b). On the other hand, pyrite is a cubic mineral and shows paramagnetism. The synthetic pyrite shows an intermediate spectrum of specularite and troillite.

4. ESR spectra of fault rocks

In general, ESR spectra obtained from natural minerals present various phases, because they reflect multiple geological events having taken place during the long geological time. Especially, faults repeatedly move, so that the fault rocks have been influenced by multiple frictional heating events and show complicated ESR spectra consisting of overlapping multiple signals. Therefore, we must separate individual signals from the whole spectrum on the basis of the physical parameters such as the g-value or peak-to-peak linewidth. In this section, I will introduce some examples of magnetized fault rocks and attempt to identify the FMR signals detected from the fault rocks by comparing their physical parameters with those obtained from known ferrimagnetic minerals.

4.1 Nojima fault

The Nojima fault is one of the most famous earthquake faults in Japan, which caused the 1995 Kobe earthquake (Magnitude 7.3). After the Kobe earthquake, the black fault rock with high magnetic susceptibility was found in the Nojima fault zone (Otsuki et al., 2003; Fukuchi, 2003). This fault rock consists of multiple sheets of a few mm wide veins, each of which was produced from granitic fault gouge during ancient earthquakes (Fig. 13). Despite no obvious melting textures, it was named "Nojima pseudotachylyte" as a descriptive name that comes from its glassy dark appearance and intrusive structure. Fig. 14 shows the ESR spectra obtained from the pseudotachylyte veins (PT-1 and PT-2) and baked fault gouge (GG-1) with the original fault gouge (NG-1) that is the source rock of the pseudotachylyte veins. Besides a

Fig. 13. Nojima pseudotachylyte veins in the Nojima fault zone.

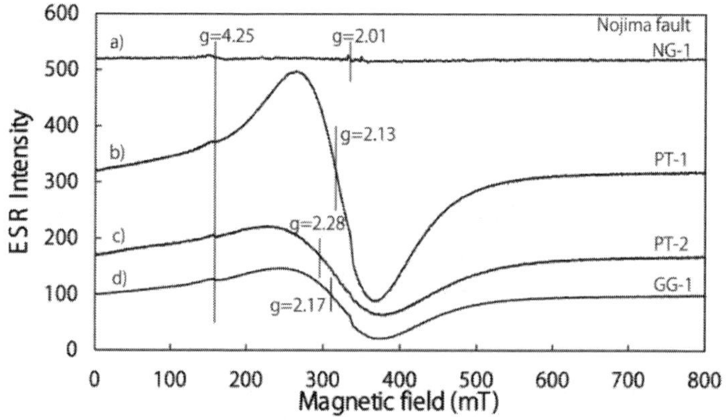

Fig. 14. ESR spectra obtained from the Nojima pseudotachylyte veins and fault gouge.

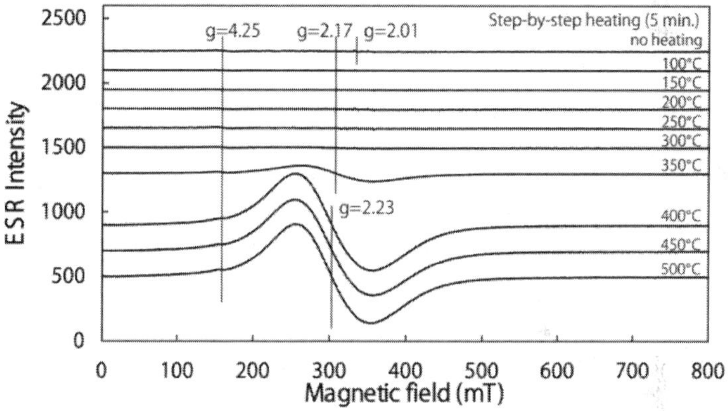

Fig. 15. Variation of the ESR spectrum obtained from the Nojima fault gouge with step-by-step heating in air.

paramagnetic Fe^{3+} ion signal at g=4.25 and other paramagnetic signals derived from quartz or clay minerals between 300–400 mT, no FMR signal is detected from the original fault gouge (NG-1). This means that there originally existed no ferrimagnetic mineral inside the fault gouge, because the FMR signals derived from magnetite and maghemite are thermally so stable, as shown in Figs. 5–7, that they cannot perfectly disappear by later frictional heating or the oxidation caused by the heating. The g-values and peak-to-peak linewidths of the signals detected from the black veins and baked gouge are 2.13–2.28 and 102–155 mT, respectively. Hence, these signals have almost the same physical parameters as the superparamagnetic signals detected from heated lepidocrocite rather than maghemite with high crystallinity or one produced by the oxidation of magnetite (Figs. 6, 7, 9 and 10).

Moreover, Fig. 15 shows the variation of the ESR spectrum obtained from the original fault gouge (NG-1) by step-by-step heating (5 min.) in air (Fig. 14a). The g-value and peak-to-peak linewidth of the signals detected by heating are 2.17–2.23 and 102–128 mT, so that they are consistent with those detected from the Nojima fault rocks and from the superparamagnetic signals of baked lepidocrocite. This strongly suggests that the magnetic source of the Nojima pseudotachylyte and baked gouge may be superparamagnetic maghemite produced by thermal dehydration of lepidocrocite, and besides that the Nojima pseudotachylyte may have been produced in an oxidizing environment.

4.2 Uchinoura shear zone

The Uchinoura shear zone is distributed in the Middle Miocene Osumi granodiorite pluton, which is located at the southern end of the Kyushu Island in Japan about 170 km away from the Nankai Trough subduction zone, and consists of a series of ENE trending faults (Fabbri et al., 2000). Along the Uchinoura shear zone, pseudotachylyte veins are exposed with cataclastic rocks such as foliated cataclasite. Fig. 16 shows a photograph of the black pseudotachylyte vein (PT) intruded into foliated cataclasite (FC). ESR spectra obtained from the Osumi granodiorite (OG), foliated calaclasite (FC) and pseudotachylyte vein (PT) are shown in Fig. 17. The Osumi granodiorite has a paramagnetic Fe^{3+} ion signal at g=4.23 and another paramagnetic signal between 300–400 mT similar to the signal of goethite and/or hematite (Fig. 8). On the other hand, the foliated cataclasite and pseudotachylyte vein have an FMR signal with the g-value of 2.57–2.88 and the peak-to-peak linewidth of 218–223 mT. These physical parameters obtained from the fault rocks are consistent with those from magnetite, and indeed we can detect similar FMR signals by heating biotite in the Osumi granodiorite over 800–1000°C in vacuum (Fig. 4c). For producing magnetite from biotite by heating in vacuum, high temperatures over 600–800°C are necessary, while biotite can be easily oxidized by heating in air and changes into hematite without producing magnetite. The ESR data indicate that the pseudotachylyte veins along the Uchinoura shear zone may have been formed by frictional heating in a reducing environment at depths.

Fig. 16. Pseudotachylyte veins distributed along the Uchinoura shear zone.

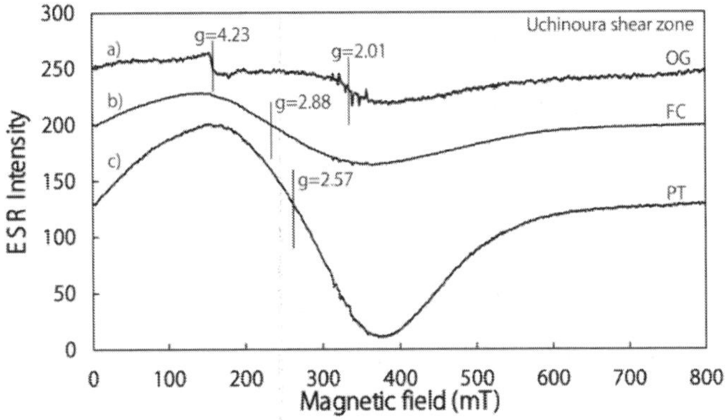

Fig. 17. ESR spectra obtained from the Osumi granite, foliated cataclasite and pseudotachylyte veins distributed along the Uchinoura shear zone.

4.3 Taiwan Chelungpu fault

The Taiwan Chelungpu fault moved in the 1999 Chi-Chi earthquake (Magnitude 7.6), which occurred in the collision zone of the Eurasian and Philippine Sea plates. After the earthquake, the Taiwan Chelungpu Fault Drilling Project (TCDP) was launched in 2002 to elucidate the rupture process caused in a subduction seismogenic zone, and continuous drill cores were collected from two main boreholes (Holes A and B) penetrating through the Chelungpu fault plane at depths (Ma et al, 2006). In the Hole B cores, there are three major fault zones at about 1136 m, 1194 m and 1243 m depths, in which a black material zone respectively exists. The black material zone is considered to have been formed by frictional heating (Hirono et al., 2006). Fig. 18 shows the black material zone in the 1194 m major fault zone. Black fault gouge exists along with the black indurated material, which may have been produced by frictional melting. Fig. 19 shows ESR spectra obtained from the black and gray gouges and black indurated material. Besides a paramagnetic Fe^{3+} ion signal (g=4.23) and an organic radical (g=2.004) (Fukuchi et al., 2007), a broad signal (g=2.27–2.44 and

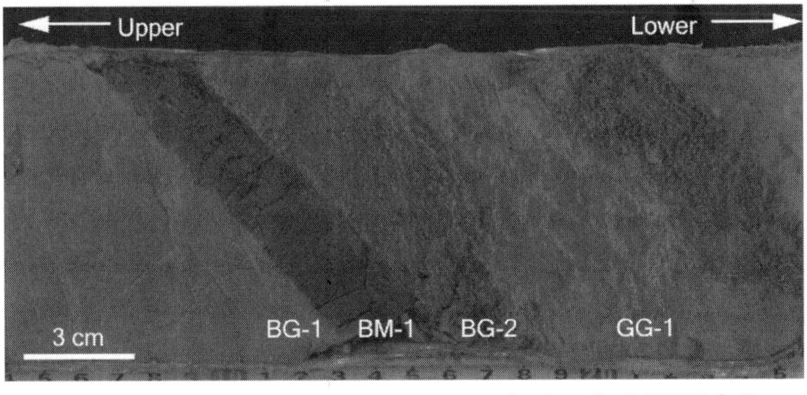

Fig. 18. Black material zone distributed at about 1194 m depth in the TCDP Hole B cores.

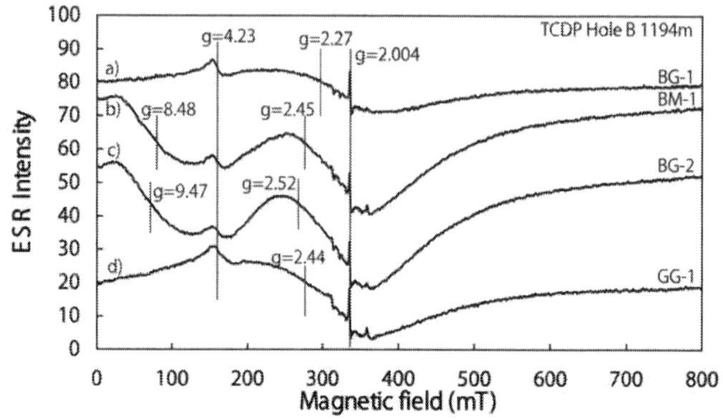

Fig. 19. ESR spectra obtained from the black material zone Taiwan Chelungpu fault zone.

ΔH_{pp}=150–173 mT) is detected from one black gouge and the gray gouge (BG1 and GG-1), while two types of large signal (g=8.48–9.47 and g=2.45–2.52) are detected from the black indurated material (BM-1) and another black gouge (BG-2). The broad signal at g=2.27–2.44 may be identified with the FMR signal of maghemite produced from lepidocrocite and/or hematite, while the lineshape of the large signal at g=8.48–9.47 is similar to that of the signal with g=10.8 obtained from specularite (Figs. 8 and 9). On the other hand, the signal at g=2.45–2.52 has the peak-to-peak linewidth of 114–123 mT, so that it may be derived from maghemite produced from lepidocrocite, goethite or hematite produced from goethite (Figs. 10 and 11). These results suggest that the black material zone may have been repeatedly subjected to frictional heating in an oxidizing environment.

5. Basic equations for frictional heat analysis

Coseismic frictional heat can be detected by measuring FMR signals in fault rocks. To calculate the frictional heat from FMR signals, we need the chemical kinetics for FMR signals besides the diffusion equations of frictional heat. As mentioned above, there are mainly two FMR signals derived from maghemite and magnetite among the FMR signals detected from natural fault rocks. Although the two FMR signals have different g-values, peak-to-peak linewidths and lineshapes, the growth processes of these signals are essentially based on the same mechanism, that is, the thermal decomposition and grain growth during heating, and are fundamentally expressed by the zero-order kinetic equation (Fukuchi, 2003). Actual fault rocks may have a mixed signal of the two FMR signals with other paramagnetic or antiferromagnetic signals, so that we must experimentally investigate the chemical kinetics on every fault rock. In this section, I will explain the basic equations necessary for calculating the frictional heat from FMR signals by inversion.

5.1 Chemical kinetics of ESR signals

There are a lot of studies on the chemical kinetics of ESR signals detected from paramagnetic minerals in connection with luminescence emitted from them. ESR signals derived from electrons or holes trapped at lattice defects in paramagnetic minerals commonly decay with time on heating and their decay processes may be expressed by the 1st, 2nd order or other kinetic model (Fukuchi, 1989, 1992; Fukuchi & Imai, 2001; Ikeya, 1993). On the other hand, the chemical kinetics of FMR signals has been studied using the FMR signal of maghemite in the Nojima fault gouge (Fukuchi, 2003; Fukuchi et al., 2005). The FMR signal of maghemite grows with the thermal decomposition of lepidocrocite and the grain growth of maghemite during heating. As shown in Fig. 10, the signal intensity increases with time on heating and its growth process follows the zero-order reaction kinetics; $dI/dt=a$, where I is the FMR signal intensity, t is time and a is the velocity constant. a is equivalent to the slope of the growth line of the FMR signal; $I=at$. When the temperature T is constant, a is expressed by the Arrhenius' equation; $a=\nu\exp[-E/RT]$, where ν is the frequency factor, E is apparent activation energy and R is the gas constant. Now we consider one-dimensional thermal conduction. When the temperature changes with distance x and time t, the velocity constant $a(x,t)$ is expressed by

$$a(x,t) = \nu \exp\left[-\frac{E}{RT(x,t)}\right] \tag{1}$$

where $T(x,t)$ is the temperature at distance x and time t. When we deal with seismic frictional heating, $T(x,t)$ means the frictional heat temperature at a distance from a fault plane (x=0) and a passing time t after earthquake rupture. If we set the FMR signal intensity as $I(x,t)$, $\partial I(x,t)/\partial t=a(x,t)$. Then, $I(x,t)$ is expressed by

$$I(x,t) = \int_0^t a(x,t)dt = \int_0^t \nu \exp\left[-\frac{E}{RT(x,t)}\right]dt \tag{2}$$

ν and E are experimentally determined from the Arrhenius plot of velocity constants measured at various temperatures (Fukuchi, 2003). In actual frictional heat analysis, we must collect the fault rock sample with a finite thickness. Therefore, when the sampling thickness is w_0, the mean FMR signal intensity F (w_0, x, t) between the distances of x and $x+w_0$ from the fault plane is expressed as follows (Fukuchi et al., 2005):

$$F(w_0,x,t) = \frac{1}{w_0}\int_x^{x+w_0} I(x,t)dx = \frac{1}{w_0}\int_x^{x+w_0}\int_0^t a(x,t)dtdx = \frac{1}{w_0}\int_x^{x+w_0}\int_0^t \nu \exp\left[-\frac{E}{RT(x,t)}\right]dtdx \tag{3}$$

5.2 One-dimensional diffusion models of frictional heat

As for the temperature during seismic frictional heating, some one-dimensional diffusion models of frictional heat have been proposed (e.g. McKenzie & Brune, 1972; Cardwell et al., 1978). According to McKenzie & Brune (1972), the equation for the diffusion of frictional heat is expressed by

$$\rho C_p \frac{\partial T}{\partial t} = k^* \frac{\partial^2 T}{\partial x^2} + Q(x,t) \tag{4}$$

where ρ is the density, C_p is the specific heat, k^* is the thermal conductivity, and Q is the frictional heat generation per unit volume and time. When we assume that $T=T_0$ and Q=0 for t<0, the solution of eq.4 can be written down as follows:

$$T(x,t) = T_0 + \frac{1}{2\rho C_p \sqrt{\pi K}} \int_0^t \int_{-\infty}^{\infty} \exp\left[-\frac{(x-x_0)^2}{4K(t-t_0)}\right] \frac{Q(x_0,t_0)}{\sqrt{t-t_0}} dx_0 dt_0 \tag{5}$$

where K is the thermal diffusivity ($K=k^*/\rho C_p$). When we further assume that the frictional heat generation is restricted to a fault plane, Q is expressed by

$$\left.\begin{aligned} Q(x_0,t_0) &= 0 & (t_0 < 0, t_0 > t_1) \\ &= \delta(x_0)\frac{\sigma_f D}{t_1} & (0 \le t_0 \le t_1) \end{aligned}\right\} \tag{6}$$

where $\delta(x_0)$ is the Dirac delta function, σ_f is the frictional shear stress, D is the displacement, and t_1 is the slip duration (McKenzie & Brune, 1972). Then, eq.5 is expressed as follows:

$$\begin{aligned} T(x,t) &= T_0 + \frac{\sigma_f D}{\rho C_p t_1}\left\{\sqrt{\frac{t}{\pi K}} exp\left[-\frac{x^2}{4Kt}\right] - \frac{x}{2K} erfc\left[-\frac{x}{2\sqrt{Kt}}\right]\right\}, & 0 \le t \le t_1 \\ &= T_0 + \frac{\sigma_f D}{\rho C_p t_1}\left[\left\{\sqrt{\frac{t}{\pi K}} exp\left[-\frac{x^2}{4Kt}\right] - \sqrt{\frac{t-t_1}{\pi K}} exp\left[-\frac{x^2}{4K(t-t_1)}\right]\right\}\right. \\ &\quad \left. - \frac{x}{2K}\left\{erfc\left[-\frac{x}{2\sqrt{Kt}}\right] - erfc\left[-\frac{x}{2\sqrt{K(t-t_1)}}\right]\right\}\right], & t > t_1 \end{aligned} \tag{7}$$

On the other hand, when we assume that the frictional heat is generated with a finite thickness, Q is expressed by

$$\left.\begin{aligned} Q(x_0,t_0) &= 0 & (t_0 < 0, t_0 > t_1) \\ &= \frac{\sigma_f D}{w t_1}\left[H\left(x_0 + \frac{w}{2}\right) - H\left(x_0 - \frac{w}{2}\right)\right] & (0 \le t_0 \le t_1) \end{aligned}\right\} \tag{8}$$

where w is the width of heat generation and H is the Heavyside step function (Cardwell et al., 1978). Then, eq.5 is expressed as follows:

$$\begin{aligned} T(x,t) &= T_0 + \frac{\sigma_f D}{2\rho C_p w t_1}\int_0^t \left\{erf\left[\frac{x+(w/2)}{\sqrt{4K(t-t_0)}}\right] - erf\left[\frac{x-(w/2)}{\sqrt{4K(t-t_0)}}\right]\right\} dt_0, & 0 \le t \le t_1 \\ &= T_0 + \frac{\sigma_f D}{2\rho C_p w t_1}\int_0^{t_1} \left\{erf\left[\frac{x+(w/2)}{\sqrt{4K(t-t_0)}}\right] - erf\left[\frac{x-(w/2)}{\sqrt{4K(t-t_0)}}\right]\right\} dt_0, & t > t_1 \end{aligned} \tag{9}$$

Since the frictional heat temperature calculated from eq.9 begins to diverge to infinity when w<0.5 mm (Fukuchi et al., 2005), then we should use eq.7 in place of eq.9.
The maximum temperature rise ΔT at $t=t_1$ and $x=0$ obtained from eq.7 is expressed as follows (McKenzie & Brune, 1972):

$$\Delta T = T_m - T_0 = \frac{\sigma_f D}{\rho C_p \sqrt{\pi K t_1}} \tag{10}$$

where T_m is the maximum temperature. Eq.10 is valid when $w \approx 0$. On the other hand, when $w \geq 0.5$ mm, the maximum temperature rise ΔT at $t=t_1$ and $x=0$ obtained from eq.9 is expressed by

$$\Delta T = T_m - T_0 = \frac{\sigma_f D}{\rho C_p w} \tag{11}$$

5.3 Frictional heat energy

The product of the frictional shear stress and displacement, that is, $\sigma_f D$ in eqs.6−11 is equivalent to the frictional heat energy per unit area φ_H. In case of large earthquakes, frictional heat is most probably generated with a finite thickness of slip zone; $w \gg 0$. Then, the relationship between φ_H and w for a fault can be expressed by Eq.11 and φ_H is proportional to w (Fig.20). When we regard T_m as a melting point, φ_H means the frictional melting energy per unit area φ_M. Therefore eq.11 can give constraints on the relationship between φ_M and w. Once frictional melting occurs in a shear zone, the temperature of melt is maintained at a melting point until the materials in the shear zone are completely molten because the subsequent frictional heat should be consumed as latent heat. In addition, once frictional melting occurs, the friction coefficient immediately drops towards zero (Di Toro et al., 2004). Thus, T_m should not be beyond the melting point and besides φ_H should not be beyond φ_M. On the other hand, φ_H increases with increasing the depth of the fault because σ_f is proportional to the normal stress σ_n when the coefficient of friction is constant over the fault plane. This means that φ_H per unit depth, that is, the frictional heat energy per unit volume ψ_H is more meaningful than φ_H. When ψ_H is constant over the fault plane, the mean value $\bar{\varphi}_H$ of φ_H for a fault is expressed using the focal depth z_0 as follows:

$$\bar{\varphi}_H = \frac{1}{z_0}\int_0^{z_0} \psi_H z dz = \frac{\psi_H z_0}{2} \tag{12}$$

Fukuchi et al. (2005) estimated the frictional heat energy for the Nojima fault using the FMR signal detected from the fault gouge in the Nojima fault 500m drill cores. The φ_H value of the

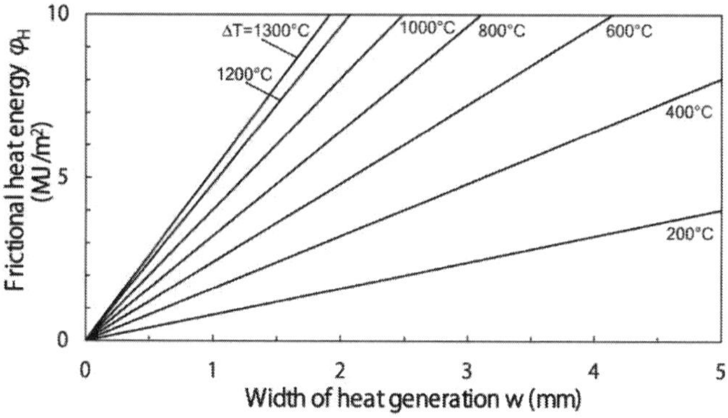

Fig. 20. Relationship between frictional heat energy and the width of heat generation.

Nojima fault at about 390 m in depth was calculated at 20.61 MJ/m^2 with the w value of 14 mm by inversion using eqs.3 and 7 or 9. However, the Nojima fault has moved frequently during the Quaternary period, so that the value obtained means the total frictional heat energy since the formation of the fault gouge. According to the geological analysis of the 500 drill cores, the total uplift along the fault plane at about 390m in depth was estimated at about 230 m (Murata et al. 2001). Therefore, the φ_H value per unit faulting may be calculated at about 0.18 MJ/m^2 when the displacement is 2.0 m compatible with that in the 1995 Kobe earthquake. Then, the ψ_H value may be estimated at about 460 J/m^3 and the $\overline{\varphi}_H$ value be calculated at 3.68 MJ/m^2 from eq.12 when z_0 is set as 16 km.

6. Scanning ESR microscopy

When we estimate the frictional heat by inversion using eqs.3 and 7 or 9, we need the sequential data of FMR signal along the fault plane. In addition, we must determine the width of heat generation, that is, the thickness of the slip zone on which the frictional heat strongly depends as shown by eq.9. The thickness of the slip zone is considered to be commonly an order of millimeters or less (Sibson, 2003). Therefore, we need the sequential data of FMR signal along the fault plane at a high-resolution of ≤1mm. However it is difficult to measure them using an ordinary ESR spectrometer for grain or powder samples (Fig.2). In this section, I will explain the scanning ESR microscopic technique for sequential high-resolution measurements of FMR signals.

In magnetic resonance, there are two physical quantities for spatially scanning, the external magnetic field and microwaves, however it is technically easier to locally measure ESR signals in the immediate vicinity of the surface of a sample by scanning localized microwave magnetic field leaking out of an aperture of the microwave cavity in a fixed external magnetic field. The scanning of localized microwaves can be carried out by shifting the sample using a mechanical X-Y stage with stepping motors controlled by a computer (Ikeya, 1991). Fig. 21 shows the TE_{111} mode cavity with a pinhole of 2.6 mm φ in diameter (Yamanaka et al., 1992). Ordinary ESR cavities have coils for 100 kHz field modulation inside, however in case of the scanning ESR microscope an external coil for 100 kHz field modulation is set above the pinhole cavity. The sample chip whose surface has been polished using 1 μm-diamond paste is put on the pinhole and the sample arm with the sample is shifted using the mechanical X-Y stage. Since the ESR spectrometer gives a first derivative line, we obtain the ESR absorption curve by integrating the first derivative curve with the magnetic field (Fig. 3). The area of the ESR absorption curve is theoretically proportional to the concentration of unpaired electrons in the sample and magnetic susceptibility. Thus, I set the value obtained by integrating the ESR absorption curve once more as the ESR absorption intensity. The FMR signal intensity is defined as the ESR absorption intensity obtained by integrating the ESR spectrum twice within the range of magnetic field where the FMR signal is detected.

The ESR absorption intensity $I_r(x,t)$ detected by the ESR cavity with a pinhole of radius r at distance x and time t is obtained by integrating the whole absorption intensity within the hemispheric domain V with the volume of $2\pi r^3/3$; the center of the hemisphere is located at distance x.

$$I_r(x,t) = \iiint_V I(x,t)dV = \iiint_V \int_0^t \nu \exp\left[-\frac{E}{RT(x,t)}\right] dt dV \tag{13}$$

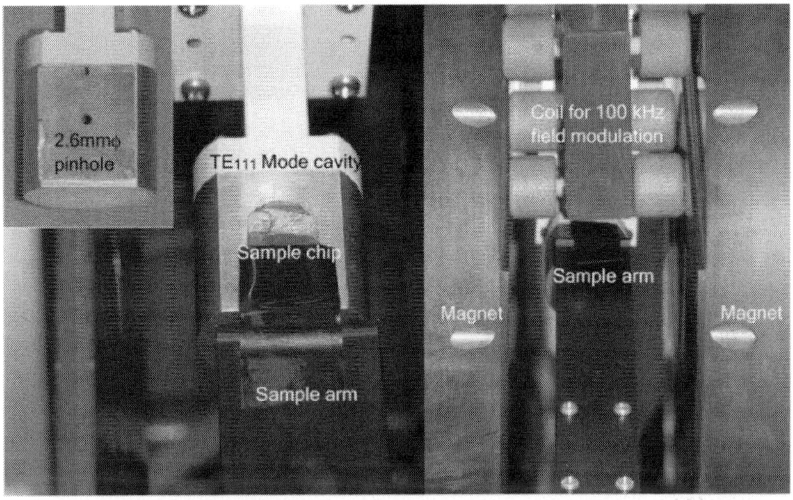

Fig. 21. TE_{111} mode cavity with a pinhole of 2.6 mm φ.

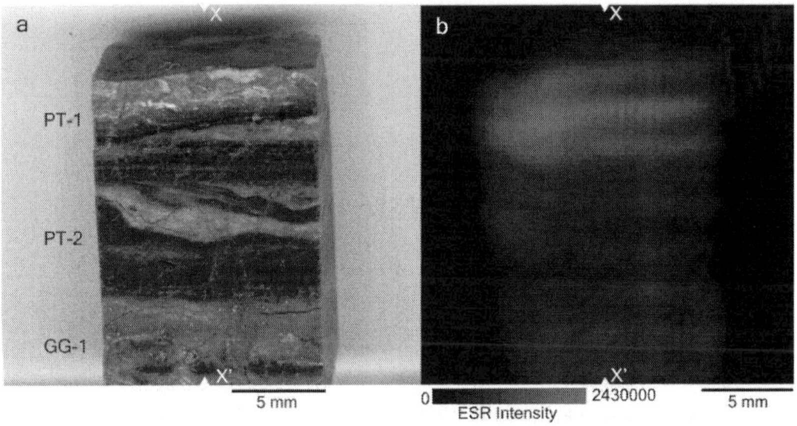

Fig. 22. A 2-D ESR map obtained from the Nojima pseudotachylyte.

Fig. 22 shows a 2-Dimensional ESR map obtained from the sample chip of the Nojima pseudotachylyte. The highest intensity is obtained from the pseudotachylyte vein (PT-1) (Fig. 14). On the other hand, Fig. 23 shows a 1-Dimensional profile obtained from the sample chip along the measuring line X-X'. Measurement conditions are as follows: microwave frequency; 9.388 GHz, microwave power; 100 mW, modulation width; 100 kHz 0.32 mT, scanning speed; 10 s/sweep (2-D) or 2.0 min./sweep (1-D), scan step; 0.25 mm, accumulation; 1 time (2-D) or 3 times (1-D), measurement temperature; room temperature. As shown in Fig. 23, the FMR signals of maghemite are sequentially detected at a resolution of 0.25 mm from the Nojima pseudotachylyte. On the other hand, the detection sensitivity of the scanning ESR microscope is much lower than the ordinary ESR spectrometer. Since the resolution of the scanning ESR microscope depends on the detection sensitivity, at this stage the limit of resolution is about 0.25 mm.

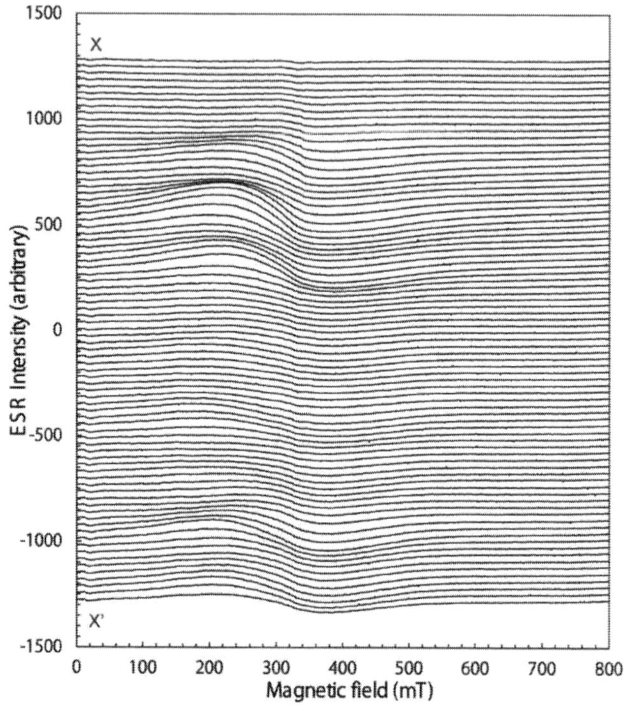

Fig. 23. A 1-D profile obtained from the Nojima pseudotachylyte.

7. Conclusion

I described ESR techniques for detecting seismic frictional heat. We can detect the frictional heat using some FMR signals derived from ferrimagnetic minerals such as maghemite or magnetite produced by heating. To actually estimate the frictional heat from fault rocks, we must carry out high-resolution measurements of FMR signals necessary for determining the width of heat generation. I believe that the scanning ESR microscopic technique makes it possible. Since faults commonly move repeatedly, the frictional heat is also generated repeatedly. Therefore, we need to separate multiplex frictional heating events to accurately estimate the frictional heat energy. Moreover, we need to determine the formation depth of fault rocks used for ESR analyses because the frictional heat commonly increases with increasing the depth of the fault. The frictional heat energy per unit volume is more meaningful than that per unit area. For these purposes, it is important to select fault rock samples with a lot of information revealed by many previous studies, for example the Nojima fault rocks.

8. Acknowledgments

I would like to thank Dr. Wonn Soh and Prof. Sheng-Rong Song for their permission to publish the ESR data on the TCDP Hole B cores in this chapter. This work was funded by Grant-in-Aid for Scientific Research (B) of the Ministry of Education, Science, Sports and Culture, Japan (No.20340139).

9. References

Alger, R. S. (1973). *Electron Spin Resonance: Techniques and Applications,* John Wiley & Sons, Inc., New York.

Butler, R. F. (1992). *Paleomagnetism,* Blackwell Scientific Publications, Boston.

Cardwell, R. K., Chinn, D. S., Moore, G. F., & Turcotte, D. L. (1978). Frictional heating on a fault zone with finite thivkness. *Geophys. J. R. Astron. Soc.,* Vol.52, pp.525–530.

Deer, W. A., Howie, R. A., & Zussman, J. (1992). *An Introduction to the Rock-Forming Minerals* (2nd Edition), Pearson Education Limited, England.

Di Toro, G., Goldsby, D. L., & Tullis, T. E. (2004). Friction falls towards zero in quartz rock as slip velocity approaches seismic rates. *Nature,* Vol.427, pp.436–439.

Fabbri, O., Lin, A., & Tokushige, H. (2000). Coeval formation of cataclasite and pseudotachylyte in a Miocene forearc granodiorite, southern Kyushu, Japan. *J. Struct. Geol.,* Vol.22, pp.1015–1025.

Ferré, E. C., Zechmeister, M. S. , Geissman, J. W., MathanaSekaran, N., & Kocak, K. (2005). The origin of high magnetic remanence in fault pseuodtahcylites: Theoretical considerations and implication for coseismic electrical currents. *Tectonophys.,* Vol.402, pp.125–139.

Fukuchi, T. (1989) Theoretical study on frictional heat by faulting using electron spin resonance. *Appl. Radiat. Isot.,* Vol.40, pp.1181–1193.

Fukuchi, T. (1992). ESR studies for absolute dating of fault movements. *J. Geol. Soc. London,* Vol.149, pp.265–272.

Fukuchi, T., & Imai, N. (2001). ESR and ICP analyses of the DPRI 500m drilling core samples penetrating through the Nojima fault, Japan. *Island Arc,* Vol.10, pp.465–478.

Fukuchi, T. (2003). Strong ferrimagnetic resonance signal and magnetic susceptibility of the Nojima pseudotachylyte in Japan and their implication for coseismic electromagnetic changes. *J. Geophys. Res.,* Vol.108, No.B6, 2312, doi:10.1029/2002JB002007.

Fukuchi, T., Mizoguchi, K., & Shimamoto, T. (2005). Ferrimagnetic resonance signal produced by frictional heating: A new indicator of paleoseismicity. *J. Geophys. Res.,* Vol.110, B12404, doi:10.1029/2004JB003485.

Fukuchi, T., Yurugi, J., &Imai, N. (2007). ESR detection of seismic frictional heating events in the Nojima fault drill core samples, Japan. *Tectonophys.,* Vol.443, pp.127–138.

Han, R., Shimamoto, T., Ando, J., & Ree, H.-H. (2007). Seismic slip record in carbonate-bearing fault zones: An insight from high-velocity friction experiments on siderite gouge. *Geology,* Vol.35, pp.1131–1135.

Hirono, T., Ikehara, M., Otsuki, K., Mishima, T., Sakaguchi, M., Soh, W., Omori, M., Lin, W., Yeh, E.-C., Tanikawa, W., Wang, C.-Y. (2006). Evidence of frictional melting from disk-shaped black material, discovered within the Taiwan Chelungpu fault system. *Geophys. Res. Lett.,* Vol.33, L19311, doi:10.1029/2006GL027329.

Ikeya, M. (1991). Electron spin resonance (ESR) microscopy in materials science. *Annu. Rev. Mater. Sci.*, Vol.21, pp.45–63.

Ikeya, M. (1993). *New Applications of Electron Spin Resonance, Dating, Dosimetry and Microscopy*, World Scientific, Singapore.

Kanamori, H., Anderson, D. L., & Heaton, T. H. (1998). Frictional melting during the rupture of the 1994 Bolivian earthquake. *Science*, Vol.279, pp.839–842.

Kanamori, H., & Heaton, T. H. (2000). Microscopic and macroscopic physics of earthquakes, in *GeoComplexity and the Physics of Earthquakes, Geophys. Monogr. Ser.*,vol.120, edited by J. B. Rundle, D. L. Turcotte, and W. Klein, pp. 147–163, AGU, Washington, D. C.

Kittel, C. (2005). *Introduction to Solid State Physics* (8th Edition), John Wiley & Sons, Inc., New York.

Lachenbruch, A. H., & Sass, J. H. (1980). Heat flow and energetic of the San Andreas fault zones. *J. Geophys. Res.* , Vol.85, pp.6249−6272.

Lachenbruch, A. H., & Sass, J. H. (1992). Heat flow from Cajon Pass, fault strength and tectonic implications. *J. Geophys. Res.* , Vol.97, pp.4995−5015.

Ma, K.-F., Tanaka, H., Song, S.-R., Wang, C.-Y., Hung, J.-H., Tsai, Y.-B., Mori, J., Song, Y.-F., Yeh, E.-C., Soh, W., Sone, H., Kuo, L.-W., Wu, H.-Y. (2006). Slip zone and energetic of a large earthquake from the Taiwan Chelungpu-fault Drilling Project. *Nature*, Vol.444, pp.473−476, doi:10.1038/nature05253.

McKenzie, D., & Brune, J. N. (1972). Melting on fault planes during large earthquakes. *Geophys. J. R. Astron. Soc.*, Vol.29, pp.65−78.

Murata, A., Takemura, K., Miyata, T., & Lin, A. (2001). Quaternary vertical offset and average slip rate of the Nojima fault on Awaji Island, Japan. *Island Arc*, Vol.10, Issue 3/4, pp.360−367.

Otsuki, K., Monzawa, N. & Nagase, T. (2003). Fluidization and melting of fault gouge during seismic slip: Identification in the Nojima fault zone and implications for focal earthquake mechanisms. *J. Geophys. Res.*, Vol.108, No.B6, 2192, doi:10.1029/2001JB001711.

Sibson, R. H. (2003). Thickness of the seismic slip zone. *Bull. Seismol. Soc. Am.*, Vol.93, pp.1169–1178.

Smit, J., & Wijn, H. P. J. (1965). *Ferrites, Physical Properties of Ferrimagnetic Oxides in Relation to Their Technical Applications* (International Edition), Tokyo Electrical Engineering College Press, Tokyo.

Yamanaka, C., Ikeya, M. & Hara, H. (1992). ESR cavities for In Vivo dosimetry of tooth enamel. *Appl. Radiat. Isot.*, Vol. 44,pp. 77–80.

17

Change of Pore Fluid Pressure Versus Frictional Coefficient During Fault Slip

Yuta Mitsui
Hokkaido University
Japan

1. Introduction

Frictional resistance on a slip interface controls slip behavior as a boundary condition within an elastic space. According to the Amonton-Coulomb principle, frictional resistance is proportional to normal stress.

In particular, in fluid-filtrated conditions, pore fluid pressure around the slip interface reduces frictional resistance as follows. Let us consider a condition that pore fluid fills around slip surfaces, confined by macroscopic normal stress σ. The conceptual model is illustrated in Figure 1. In macroscopic contact area A, solid parts support partial normal stress σ_s only by real contact area A_r and pore fluid with pressure p supports the residual normal stress. Then we obtain the equation

$$\sigma A = p(A - A_r) + \sigma_s A_r$$
$$\Rightarrow \frac{\sigma_s A_r}{A} = \sigma - p\left(1 - \frac{A_r}{A}\right) \quad (1)$$

Since just the solid part can support shear stress, the macroscopic area can resist shear only for $\sigma - p(1 - A_r/A)$. When $A_r/A \ll 1$ (this is a common feature in many materials (Dieterich & Kilgore (1994)), $\sigma - p(1 - A_r/A)$ results in $\sigma - p$. Thus usually the effective normal stress $\bar{\sigma}$ for fluid-saturated frictional surfaces is defined as $\sigma - p$. This idea was

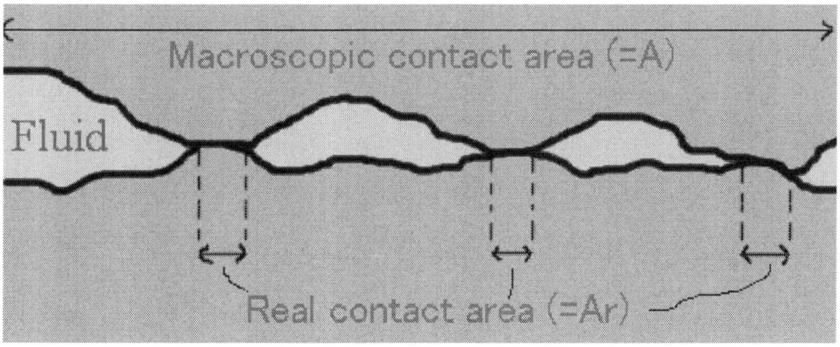

Fig. 1. Schematic illustration around a slip surface filled with pore fluid. Macroscopic and apparent contact area is A, but real contact area by solid parts is only A_r.

introduced in Soil mechanics (Terzaghi (1943)) and accepted for rocks by certain experiments (Brace & Martin (1968); Handin et al. (1963)).

Therefore, if a slip interface is fluid-infiltrated, evolution of pore fluid pressure has the same importance as frictional coefficient for governing the slip behavior. Hitherto, no one could state that pore pressure is always constant on a fluid-infiltrated slip interface. As examples of physical processes to change pore pressure, thermal pressurization due to shear heating (e.g., Lachenbruch (1980); Sibson (1973); Ujiie et al. (2010)) and pore-related pressurization due to porosity change (e.g., Brace & Martin (1968); Marone et al. (1990); Rudnicki (1986)) have been well studied.

Regarding these effects, Suzuki & Yamashita (2007) analytically derived a non-dimensional controlling parameter for slip behavior, assuming constant frictional coefficient, no diffusion of pore pressure and heat, and simple rate-dependent pore dilatation. Noda & Shimamoto (2005) presented a characteristic distance of fault slip-weakening behavior is controlled by a width of deformation zone and fluid diffusion, and Rice (2006) derived some analytical expressions for the slip-weakening behavior, owing to thermal pressurization, assuming constant slip rate and frictional coefficient. They focus on the effects of the pore pressure change on frictional resistance. On the other hand, they ignore the evolution of frictional coefficient. While the effects of the pore pressure change exceed that of frictional coefficient, their results can be regarded as constitutive laws of fault friction. But while not, they can not. It depends on circumstances.

In fact, some numerical studies including both the change of pore pressure and frictional coefficient have shown that the fluid pressurization can notably affect dynamic rupture propagation (e.g., Andrews (2002); Bizzarri & Cocco (2006)), quasi-static nucleation (e.g., Segall & Rice (2006); Shibazaki (2005)) and whole earthquake cycle (e.g., Mitsui & Hirahara (2009a;b)). These numerical studies clarified the fundamental effects of the pore pressure evolution, i.e., thermal pressurization enlarges seismic slip its recurrence intervals, and pore-related pressurization restrains seismic slip and its occurrences.

However, applying their results to actual faults is not easy, since they depend on many constitutive parameters. In order to provide a clue for obtaining a clear view, here, we will focus on an analytic representation for comparing pore pressure change with frictional coefficient change. We will obtain the condition in which the effects of the pore pressure change can exceed. Then, we will substitute the typical values of rock materials and several types of the evolution law of frictional coefficient, to get several easy relations.

2. Derivation of analytic representation

First of all, we clarify the model setup in this study. We assume a fault embedded in a poroelastic body. The frictional resistance of the fault obeys the Amonton-Coulomb principle and the Terzaghi law of effective normal stress. Frictional resistance τ_f is equal to frictional coefficient μ multiplied by effective normal stress $\bar{\sigma}$, which is macroscopic normal stress σ minus pore fluid pressure p. One more fundamental assumption is that μ does not depend on fluid pressure p and normal stress σ.

Differentiating the relation $\tau_f = \mu(\sigma - p)$ with time t, we obtain

$$\frac{d\tau_f}{dt} = -\mu\frac{dp}{dt} + \bar{\sigma}\frac{d\mu}{dt} \tag{2}$$

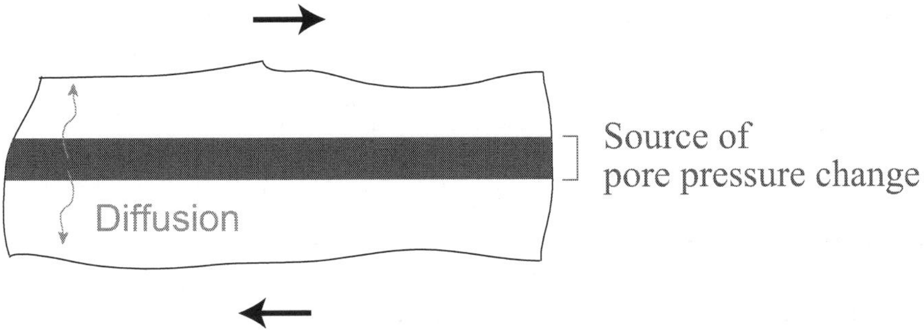

Fig. 2. Schematic illustration for the fault model with pore pressure change in this study.

The right-hand first term means the temporal alteration of friction by pore pressure change and the right-hand second term represents that by frictional coefficient. In order to compare the degree of both terms, we consider inequality with respect to the condition for exceeding of the pore pressure change. From Equation (2), the condition is written as

$$\left|\frac{d\mu}{dt}\right| \ll \frac{\mu}{\bar{\sigma}}\left|\frac{dp}{dt}\right| \tag{3}$$

Inequality (3) is a rather general representation, however, as a practical matter, we need more useful relations. In order to expand the term of the pore pressure change, we introduce a simplified poroelastic model following Segall & Rice (2006).

We assume that the pore fluid pressure first alters within the fault due to physicochemical processes, and diffuses outside via conduction processes in the poroelastic body, as illustrated in Figure 2. Based on the equation of mass conservation and the Darcy law, changing rate of a fluid mass m per unit volume in a certain bulk is given by:

$$\frac{dm}{dt} = \frac{\rho_f \kappa}{\nu}\nabla^2 p \tag{4}$$

where ρ_f is the fluid density, κ is the permeability of the poroelastic bulk and ν is the fluid viscosity.

Since a fluid mass per unit volume m is equivalent to $\rho_f \phi$, where ϕ is the porosity, we write down the following equation:

$$\frac{dm}{dt} = \rho_f \frac{d\phi}{dt} + \phi \frac{d\rho_f}{dt} \tag{5}$$

Moreover, we divide the temporal change of the porosity ϕ into elastic change and plastic change. The elastic change is given by

$$\frac{d\phi}{dt} = \phi\beta\frac{dp}{dt} + \phi\alpha\frac{dT}{dt} \tag{6}$$

introducing the pressure compressibility of the solid $\beta = (\partial\phi/\partial p)/\phi$ and the thermal expansivity of the solid $\alpha = (\partial\phi/\partial T)/\phi$, where T is the bulk temperature. Thus the form of Equation (6) is changed into

$$\frac{dm}{dt} = \rho_f \left.\frac{d\phi}{dt}\right|_{pl} + \rho_f\phi\beta\frac{dp}{dt} + \rho_f\phi\alpha\frac{dT}{dt} + \phi\frac{d\rho_f}{dt} \tag{7}$$

where the suffix $|_{pl}$ means "plastic". Likewise we divide the temporal change of the fluid density ρ_f into that by fluid pressure and temperature. The representation is as follows:

$$\frac{d\rho_f}{dt} = \rho_f \beta_f \frac{dp}{dt} - \rho_f \alpha_f \frac{dT}{dt} \quad (8)$$

where we introduce the pressure compressibility of the fluid $\beta_f = (\partial\rho_f/\partial p)/\rho_f$ and the thermal expansivity of the fluid $\alpha_f = -(\partial\rho_f/\partial T)/\rho_f$. Using the constitutive relations, Equation (7) is rewritten as

$$\frac{dm}{dt} = \rho_f \left[\left.\frac{d\phi}{dt}\right|_{pl} + \phi[\beta + \beta_f]\frac{dp}{dt} + \phi[\alpha - \alpha_f]\frac{dT}{dt} \right] \quad (9)$$

Finally, from Equations (4) and (9), we obtain the following equation for representing fluid pressurization:

$$\frac{dp}{dt} = \Lambda\frac{dT}{dt} - \frac{1}{S_t}\left.\frac{d\phi}{dt}\right|_{pl} + \varpi\nabla^2 p \quad (10)$$

where Λ is $[\alpha_f - \alpha]/[\beta + \beta_f]$, the storage capacity S_t is $\phi[\beta + \beta_f]$ and the pressure diffusivity ϖ is $\kappa/[\nu\phi(\beta + \beta_f)]$.

In addition, when dehydraction reactions occur during fault slip (Brantut et al. (2010); Hirono et al. (2008); Hirose & Bystricky (2007)), we must consider an additional term "$+c_{de}$" in the left-hand term of Equation (4), where c_{de} is the dehydration rate of fluid per unit volume of a bulk. It leads to the same additional term in the right-hand term of Equation (10):

$$\frac{dp}{dt} = \Lambda\frac{dT}{dt} - \frac{1}{S_t}\left.\frac{d\phi}{dt}\right|_{pl} + \frac{c_{de}}{S_t} + \varpi\nabla^2 p \quad (11)$$

The right-hand first term of Equation (11) corresponds to the pore pressure change due to shear heating under undrained conditions, and the second term corresponds to that owing to plastic porosity change. The third term represents the fluid pressure diffusion. Figure 3

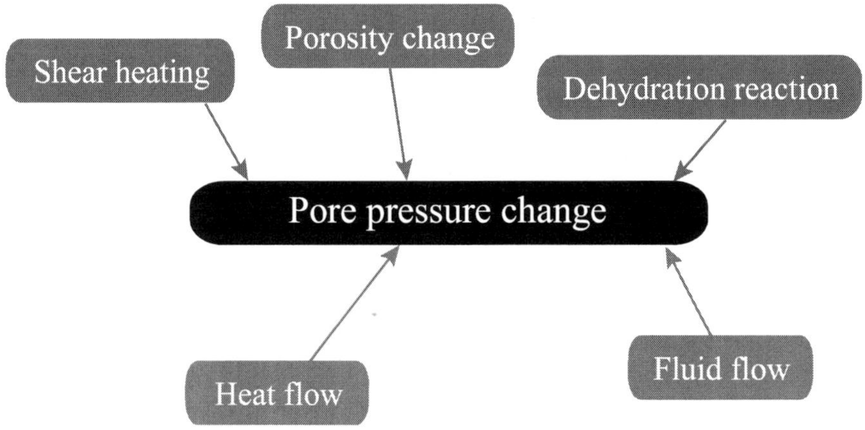

Fig. 3. A schematic presenting what factors cause the pore pressure change.

shows an illustration for the alteration processes of pore pressure. Using this equation, we can rewrite Inequality (3) as

$$\left|\frac{d\mu}{dt}\right| \ll \frac{\mu}{\bar{\sigma}}\left|\Lambda\frac{dT}{dt} - \frac{1}{S_t}\frac{d\phi}{dt}\bigg|_{pl} + \frac{c_{de}}{S_t} + \varpi\nabla^2 p\right| \tag{12}$$

2.1 Constitutive equation for thermal pressurization

The right-hand first term of Equation (11) includes both effects of shear heating and heat diffusion. Based on the energy conservation law and the Fourier law, change of the temperature T is given by

$$\rho c\frac{dT}{dt} = \tau_f \Upsilon + \lambda\nabla^2 T \tag{13}$$

where ρ, c and λ is respectively represents the density, the specific heat capacity and the thermal conductivity of the bulk composite, Υ is the shear strain rate.

How should we represent the shear strain rate Υ? This issue is in itself of consequence. One simple assumption is that Υ is roughly given by v/w, where v is the dislocation rate from a macroscopic viewpoint and w is the slip zone width (Cardwell et al. (1978); Fialko (2004)). The shear strain is assumed to be homogeneous in the slip zone. The other assumption is the Gaussian strain distribution (Andrews (2002)). Although actual processes of strain localization is much more complicated (e.g., Mandl et al. (1977); Marone et al. (2009)), here we assume $\Upsilon = v/w$ and w is constant as illustrated in Figure 4, for simplicity. p and T are the representative values at the center of the slip zone.

The temperature change can be rewritten as:

$$\frac{dT}{dt} = \frac{\tau_f v}{\rho c w} + \chi\nabla^2 T \tag{14}$$

where $\chi = \lambda/(\rho c)$ is the temperature diffusivity.

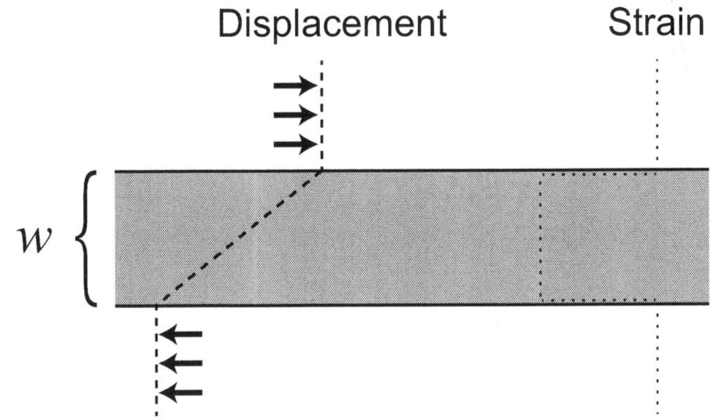

Fig. 4. A schematic for the simple assumption of the homogeneous shear strain within the slip zone.

Substituting Equation (14) into Inequality (12), we obtain an improved representation of the condition for exceeding of the pore pressure change:

$$\left|\frac{d\mu}{dt}\right| \ll \frac{\mu}{\bar{\sigma}}\left|\frac{A\tau_f v}{w} - \frac{1}{S_t}\frac{d\phi}{dt}\right|_{pl} + \frac{c_{de}}{S_t} + \varpi\nabla^2 p + \Lambda\chi\nabla^2 T\right| \tag{15}$$

where A equals $\Lambda/(\rho c)$. The new parameter A is a non-dimensional material parameter.

2.2 Constitutive equation for pore-related pressurization

The right-hand second term of Equation (10) means the porosity effect on the pore pressure change, so-called pore dilatation and compaction. Unlike thermal pressurization, the physical model of such the pore-related pressurization accompanying with shear has not been established well.

For example, Marone et al. (1990); Zhang & Tullis (1998) revealed by experiments that pore compaction (permeability decrease) evolve with shear from the initial experimental conditions. It is identical with observations of fault cores in actual faults, particularly in cases of high-porosity rocks such as sandstone (e.g., Aydin (1978); Balsamo & Storti (2010)).

However, pore compaction mechanism may not work effectively in lower-porosity rocks. For instance, Collettini et al. (2009) experimentally showed dilatational behavior with shear.

Such an effect of initial porosity is also presented experimentally by Tanikawa et al. (2010): Porosity (permeability) of initially higher-porosity rocks decreases by shear localization and that of lower-porosity rocks increases by cracking. Furthermore, Goren et al. (2010) performed granular simulations to present that initial dense packing of grains leads to initial pore dilatation, and loose packing does initial pore compaction.

Putting aside the above situation, we need certain simplified models. Since underground rocks in a seismogenic depth may have sufficiently low porosity, first we should consider the dilatation effect with shear. One simple assumption is dependence of the temporal rate of porosity increase on slip rate, as was assumed in several studies (Rudnicki & Chen (1988); Suzuki & Yamashita (2007)). In contrast, some experimental studies proposed that the temporal rate of porosity increase does not depend only on slip rate but also slip amount (Beeler et al. (1996); Marone et al. (1990)). From a practical standpoint, we adopt the former, more simple one.

In addition, beside the porosity changes with shear, chemical reaction such as pressure solution and precipitation, may occur within the fault zone. It should be time-dependent processes, indepedent of shear. Although many models have been suggested about it (e.g., Renard et al. (1999); Revil et al. (2006)), we try implementing the simplest one proposed by Gratier et al. (2003).

2.3 Typical values of material parameters

In the above relations, many material parameters appear. To evaluate them under a typical condition of underground seismogenic regions is essential for applications to fault dynamics.

First, referring to Clark (1966), we obtain the following values: the solid compressibility $\beta \sim 10^{-11}$ [Pa^{-1}], the thermal conductivity of the bulk (almost equal to the solid phase) $\lambda \simeq 2.0 \times$

10^0 [$Jm^{-1}s^{-1}K^{-1}$], the fluid compressibility $\beta_f \sim 10^{-9}$ [Pa^{-1}], the thermal expansivity of the fluid $\alpha_f \sim 10^{-3}$ [K^{-1}], and the thermal expansivity of the solid $\alpha \sim 10^{-5}$ [K^{-1}].

The fluid parameters are calculated by assuming a pressure condition of 100 [MPa] and a temperature condition of 473 [K]. There is not much difference if the fluid composition is assumed to be pure H_2O or CO_2. The notable difference of the magnitude of β and β_f means that we can look on the term $(\beta + \beta_f)$ as β_f. Likewise we are able to regard the term $(\alpha_f - \alpha)$ as α_f. Moreover, the multiplication of the bulk density ρ and the specific heat capacity c is approximately given by 3.0×10^6 [Pa K^{-1}] (Vosteen & Schellschmidt (2003)). The fluid viscosity ν is also approximated by 10^{-4} [Pa s].

The remaining parameters of the permeability κ and the slip zone width w are difficult to characterize, although they are the controlling parameters of thermal pressurization. Unfortunately, many observations have revealed that they vary in several orders according to environments (e.g., Sibson (2003)). The porosity ϕ is also an ambiguous parameter. Therefore we regard the parameters as variables in this study.

The above typical values can be substituted into Inequality (15), via $A = (\alpha_f - \alpha)/[(\beta + \beta_f)\rho c]$, $S_t = \phi(\beta + \beta_f)$, $\varpi = \kappa/[\nu\phi(\beta + \beta_f)]$ and $\Lambda\chi = A\lambda = [(\alpha_f - \alpha)\lambda]/[(\beta + \beta_f)\rho c]$. In particular, since A is a non-dimensional parameter, we can directly substitute the above typical value:

$$\left|\frac{d\mu}{dt}\right| \ll \frac{\mu}{\bar{\sigma}}\left|\frac{\tau_f v}{3w} - \frac{1}{S_t}\frac{d\phi}{dt}\right|_{pl} + \frac{c_{de}}{S_t} + \varpi\nabla^2 p + \Lambda\chi\nabla^2 T\right| \tag{16}$$

where $S_t \simeq 10^{-9}\phi$ [Pa^{-1}], $\varpi \simeq 10^{13}\kappa/\phi$ [$m^2\ s^{-1}$], $\chi \simeq (2/3) \times 10^{-6}$ [$m^2\ s^{-1}$] and $\Lambda\chi \simeq 2/3$ [$W\ m^{-1}\ K^{-1}$].

3. Application to several cases

3.1 Model of undrained and adiabatic condition without porosity change and dehydration reaction

The comparison between thermal pressurization and frictional coefficient change in cases of undrained and adiabatic condition (no fluid flow and heat flow), no porosity changes, and no dehydration reactions is an easiest exercise. This situation would be applied to actual faults during short-time slip (details are described in section 4.1). Thus, in this section, we neglect the right-hand second, third, fourth and fifth terms of Inequality (16). Inequality (16) can be simplified as:

$$\left|\frac{d\mu}{dt}\right| \ll \frac{\mu^2 v}{3w} \tag{17}$$

Note that the absolute values of the normal stress and the pore pressure vanished by the assumption.

Not only the pore pressure but the frictional coefficient μ alters during fault slip. There have been proposed so many processes and constitutive laws for the frictional coefficient (Bizzarri (2009) and references there in). By contrast, as a practical matter, simple velocity-dependent friction or slip-dependent friction are well used for boundary conditions in elastodynamic problems (e.g., Fukuyama & Madariaga (1998)).

3.1.1 Rate-strengthening friction vs thermal pressurization

When μ depends only on the slip velocity v, Inequality (17) is changed into:

$$\left|\frac{d\mu}{dv}\right| \ll \frac{\mu^2 v}{3w|dv/dt|} \tag{18}$$

Inequality (20) means that the thermal fluid pressurization must exceed in cases of constant slip velocity $dv/dt = 0$. If we perform frictional experiments for constant slip velocity using materials with pure velocity-dependent friction and confined pore fluid, friction evolution must be controlled by the thermal fluid pressurization.

In fact, several numerical and experimental researches of granular rheology have revealed that macroscopic friction of dry granular flow increases with flow rate (e.g., Hatano (2007); Jop et al. (2006)), which might be adopted as frictional characteristics of fault gouges with shear heating. In addition, several rock experiments have clarified that the frictional coefficient of rocks tend to have rate-strengthening characters with the sub-seismic slip rate in the range of 1 [¯m s^{-1}] - 1 [cm s^{-1}] (e.g., Tsutsumi & Shimamoto (1997); Weeks (1993)). Thus the "competition" between the rate-strengthening friction and the thermal pressurization of pore fluid is an important issue for fault dynamics.

In addition, for example, if the frictional coefficient has a logarithmic rate-strengthening character as many rock experiments revealed, Inequality (18) is modified as

$$\left|\frac{d\mu}{d\ln(v)}\right| \ll \frac{\mu^2 v^2}{3w|dv/dt|} \tag{19}$$

When we adopt the typical values $d\mu/d\ln(v) \simeq 0.01$ and $\mu \simeq 0.6$, we obtain

$$\left|\frac{dv}{dt}\right| \ll \frac{10v^2}{w} \tag{20}$$

The parameters in Inequality (18) are reduced to only three parameters.

3.1.2 Slip-dependent friction vs thermal pressurization

If μ depends only on the slip amount u, Inequality (17) is rewritten as:

$$\left|\frac{d\mu}{du}\right| \ll \frac{\mu^2}{3w} \tag{21}$$

Only three parameters remain in this case.

In general, earthquake breakdown processes are apparently interpreted as an initial slip-strengthening and the following slip-weakening behavior (e.g., Cocco & Tinti (2008); Ohnaka & Yamashita (1989)), both of which include many kinds of microscopic physical processes. The thermal pressurization of pore fluid might be one of a dominating process of phenomenological slip-weakening (Abercrombie & Rice (2005); Wibberley & Shimamoto (2005)). Whether apparent slip-weakening behavior in actual earthquakes is due to thermal pressurization or not, is not so easy to be judged from observations.

In order to provide a hint for this issue, we use Inequality (21). Assuming the typical values of $\mu \simeq 0.6$, we obtain

$$\left|\frac{d\mu}{du}\right| \ll \frac{0.1}{w} \tag{22}$$

Just two parameters remain: the slip-weakening rate of the frictional coefficient $d\mu/du$ and the width of the slip zone w.

It means that fault weakening by the thermal pressurization must exceed without a certain intense slip-strengthening of the frictional coefficient or intense slip-weakening owing to other mechanisms such as wearing (Matsu'ura et al. (1992)), flash heating of asperity contacts (Rice (2006)), and thermally-activated chemical reactions of rock minerals (Di Toro et al. (2004); Han et al. (2007)), dependent on the slip zone thickness w.

3.2 Model of undrained and adiabatic condition

As was discussed in Section 2.2, we assume a velocity-dependent pore dilatancy and time-dependent pore compaction.

With regard to the former, the temporal evolution of the porosity is given by

$$\left.\frac{d\phi}{dt}\right|_{pl} = Zv \tag{23}$$

where Z is a characteristic value for dilatation.

In respect to the latter, Gratier et al. (2003) provides a simple equation

$$\left.\frac{d\phi}{dt}\right|_{pl} = -\phi/X \tag{24}$$

where X is a characteristic value for compaction.

Using both models for porosity changes, Inequality (16) is changed into:

$$\left|\frac{d\mu}{dt}\right| \ll \left|\frac{\mu^2 v}{3w} - \frac{\mu Z v}{\bar{\sigma} S_t} + \frac{\mu\phi}{\bar{\sigma} X S_t} + \frac{c_{de}}{S_t}\right| \tag{25}$$

Naturally, the right-hand second term (the dilatation term) and the right-hand third term (the compaction term) in Inequality (25) have sufficient potentials to alter the conditions described by the original Inequality (17). The dehydration term also has the potential. We can further introduce inequalities under limited assumptions like (18)-(22).

For example, Inequality (20) in case of the rate-strengthening friction is modified as:

$$\left|\frac{dv}{dt}\right| \ll \left|\frac{10v^2}{w} - \frac{60Zv^2}{\bar{\sigma} S_t} + \frac{60v\phi}{\bar{\sigma} S_t X} + \frac{c_{de}}{S_t}\right| \tag{26}$$

Moreover, Inequality (22) in case of the slip-dependent friction is modified as:

$$\left|\frac{d\mu}{du}\right| \ll \left|\frac{0.1}{w} - \frac{0.6Z}{\bar{\sigma} S_t} + \frac{0.6\phi}{\bar{\sigma} S_t X v} + \frac{c_{de}}{S_t}\right| \tag{27}$$

They have no simple parameter dependencies as Inequality (20)-(22).

4. Discussion

4.1 Requirement for neglecting fluid and heat diffusion

In Section 3.1-3.2, we ignore fluid and heat flow from slip zone. It is only valid in a sufficiently short time period.

Let us define the length of the time period in which we can ignore fluid and heat diffusion. With these terms, Inequality (17) is modified as:

$$\left|\frac{d\mu}{dt}\right| \ll \left|\frac{\mu^2 v}{3w} + \frac{\mu}{\bar{\sigma}}(\varpi\nabla^2 p + \Lambda\chi\nabla^2 T)\right| \tag{28}$$

We can neglect the right-hand second term (fluid diffusion) of Equation (28) when

$$\Delta_t \ll \frac{w_{fld}^2}{4\varpi} \tag{29}$$

and the right-hand third term (heat diffusion) when

$$\Delta_t \ll \frac{w_{heat}^2}{4\chi} \tag{30}$$

where Δ_t is a time scale for consideration, w_{fld} and w_{heat} are certain characteristic lengthes for each process. Furthermore, since w_{fld} and w_{heat} can not fall below w, the conditions can be changed into

$$\Delta_t \ll \frac{w^2}{4\varpi} \tag{31}$$

$$\Delta_t \ll \frac{w^2}{4\chi} \tag{32}$$

We can use Inequality (20), (22), (26) and (27), when both (31) and (32) are true.

Usually, under actual conditions, ϖ is several order larger than χ. For instance, even assuming quite low permeability $\kappa = 10^{-21}$ [m^2] and $\phi = 0.01$ as a typical value around slip zone, the fluid pressure diffusivity is $\varpi \simeq 10^{-6}$ [m^2 s^{-1}], larger than the temperature diffusivity $\chi \simeq (2/3) \times 10^{-6}$ [m^2 s^{-1}] (see the last part of Section 2.3). It implies that Inequality (31) is a practical requirement for neglecting fluid and heat diffusion.

Besides, how large w in actual faults is another problem. Sibson (2003) reported coseismic shearing is localized in a region of less than 0.1 [m] by geological observations. A recent geochemical study on the Chelungpu fault revealed thermally-pressurized fluids might exist in the sheared bands with thickness of 0.02-0.15 [m] (Ishikawa et al. (2008)). Those studies may constrain the upper limit of w value $\sim$ 0.1 [m]. In contrast, we have no idea to constrain the lower limit of w. Too small w breaks Inequality (31) in a practical time scale Δ_t.

For example, with one particular scale $w = 0.002$ [m] and $\varpi = 10^{-6}$ [m^2 s^{-1}], Inequality (31) is turned into $\Delta_t \ll 1$ [s]. In this case, we can use (20), (22), (26) and (27) to consider whole slip processes in small earthquakes.

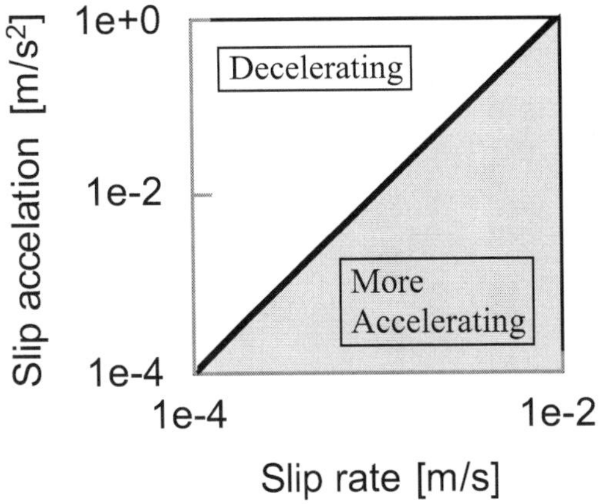

Fig. 5. The solid line displays a threshold level of the absolute value of slip acceleration as a function of sub-seismic slip rate, with $w = 1$ [cm]. When slip acceleration is above the threshold, slip will be decelerated owing to rate-strengthening friction. When it is below, slip will be accelerated due to the thermal pressurization of pore fluid.

4.2 Implications for actual earthquakes

4.2.1 Slip acceleration and deceleration with sub-seismic slip rate

As was described in section 3.1.1, frictional coefficient of rocks or fault gouges with sub-seismic slip rate may have rate-strengthening characters. Hence the "competition" between the rate-strengthening friction and thermal pressurization is a characteristic phenomenon during earthquakes especially in a slip acceleration period just before the slip rate reaches its maximum value.

Within the sufficiently short time under Inequalities (31) and (32), Inequality (20) would be a useful reference to understand fault behavior in this regime. As an example, Figure 5 presents a threshold level of slip acceleration whether slip is more accelerated by thermal pressurization or decelerated by rate-strengthening friction, when w is 1 [cm].

The Inequality and figure provide us qualitative implications about slip acceleration and deceleration during the sub-seismic slip regime. First, slip acceleration is originally loaded by external forces. Once slip rate reaches around the sub-seismic rate, the acceleration obeys the "competition", namely evolves more or less along the threshold level of the "competition" described by Inequality (20). It means that slip acceleration process might have a broadly fixed pattern within the sub-seismic slip regime. Also, with respect to slip deceleration following high-speed seismic slip, similar consideration might be able to adopted.

4.2.2 Breakdown process with seismic slip rate

After the "competition" between the rate-strengthening friction and the thermal pressurization of pore fluid, slip rate may reach a seismic slip regime. Usually this regime accompanies the phenomenological slip-weakening of frictional coefficient. We can refer to

Inequality (22) to compare such the slip-weakening of frictional coefficient with the thermal pressurization. Figure 6 shows the meaning of Inequality (22).

With respect to the slip-weakening rate, for instance, experiments by Mizoguchi et al. (2009) using fault gouge obtained from Nojima fault, southwest Japan, show $d\mu/du \sim 10^{-2}$ [m^{-1}]. Other experiments by Di Toro et al. (2004) using Arkansas novaculite, show $d\mu/du \sim 10^{-1}$ [m^{-1}]. Those results fall on the shadow zone in Figure 6, which means Inequality (22) is true, even with the largest w in geological observations: $w \sim 0.1$ [m].

It indicates that the thermal pressurization would necessarily dominate during the apparent slip-weakening behavior within the seismic slip regime, without the porosity evolution and the fluid/heat flow.

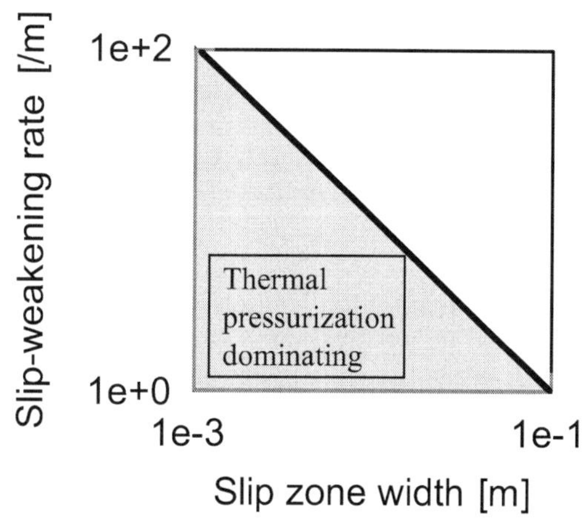

Fig. 6. The solid line represents a threshold level of slip-weakening rate of frictional coefficient $d\mu/du$ as a function of the slip zone width w. Within the shadow zone, the thermal pressurization dominates the slip-stress evolution.

5. Conclusion

We derived inequalities for comparing the change of pore pressure with that of frictional coefficient during fault slip.

The condition in which the effects of the pore pressure change on friction can exceed is represented as Ineqation (12) or (15). Substituting the typical values of rock materials for the inequality, we obtain Inequality (16). Some easy relations are further obtained by the assumptions of no porosity evolution, no dehydration reaction, and a short-time period as Inequality (31) and (32): "rate-strengthening friction vs thermal pressurization" (Inequality (20)) or "slip-dependent friction vs thermal pressurization" (Inequality (22)).

From the easy relations, we obtain a qualitative implication for slip acceleration and deceleration with sub-seismic slip rate. Slip acceleration, originally due to external forces, might have a fixed way owing to the competition between the rate-strengthening friction and the thermal pressurization. Slip deceleration does the same. In addition, the thermal pressurization would necessarily dominate the apparent slip-weakening behavior in the seismic slip regime under the undrained and adiabatic condition with no porosity evolution and no dehydration reaction.

6. References

Abercrombie, R. E. & Rice, J. R. (2005). Can observations of earthquake scaling constrain slip weakening?, *Geophys. J. Int.* 162(2): 406–424.
URL: *http://blackwell-synergy.com/doi/abs/10.1111/j.1365-246X.2005.02579.x*

Andrews, D. J. (2002). A fault constitutive relation accounting for thermal pressurization of pore fluid, *J. Geophys. Res.* 107(B12).
URL: *http://www.agu.org/pubs/crossref/2002/2002JB001942.shtml*

Aydin, A. (1978). Small faults formed as deformation bands in sandstone, *Pure Appl. Geophys.* 116: 913–930.
URL: *http://www.springerlink.com/index/10.1007/BF00876546*

Balsamo, F. & Storti, F. (2010). Grain size and permeability evolution of soft-sediment extensional sub-seismic and seismic fault zones in high-porosity sediments from the Crotone basin, southern Apennines, Italy, *Marine Petro. Geol.* 27(4): 822–837.
URL: *http://linkinghub.elsevier.com/retrieve/pii/S0264817209002141*

Beeler, N. M., Tullis, T. E., Blanpied, M. L. & Weeks, J. D. (1996). Frictional behavior of large displacement experimental faults, *J. Geophys. Res.* 101(B4): 8697–8715.
URL: *http://www.agu.org/pubs/crossref/1996/96JB00411.shtml*

Bizzarri, A. (2009). What Does Control Earthquake Ruptures and Dynamic Faulting? A Review of Different Competing Mechanisms, *Pure Appl. Geophys.* 166(5-7): 741–776.
URL: *http://www.springerlink.com/index/10.1007/s00024-009-0494-1*

Bizzarri, A. & Cocco, M. (2006). A thermal pressurization model for the spontaneous dynamic rupture propagation on a three-dimensional fault: 1. Methodological approach, *J. Geophys. Res.* 111(B5).
URL: *http://www.agu.org/pubs/crossref/2006/2005JB003862.shtml*

Brace, W. F. & Martin, R. J. (1968). A test of the law of effective stress for crystalline rocks of low porosity, *Int. J. Rock Mech. Min. Sci.* 5: 415–426.

Brantut, N., Schubnel, A., Corvisier, J. & Sarout, J. (2010). Thermochemical pressurization of faults during coseismic slip, *J. Geophys. Res.* 115(B5): 1–17.
URL: *http://www.agu.org/pubs/crossref/2010/2009JB006533.shtml*

Cardwell, R. K., Chinn, D. S., Moore, G. F. & Turcotte, D. L. (1978). Frictional heating on a fault zone with finite thickness, *Geophys. J. R. Astr. Soc.* 52: 525–530.

Clark, S. P. (ed.) (1966). *Handbook of physical constants*, The Geological Society of America, New York.

Cocco, M. & Tinti, E. (2008). Scale dependence in the dynamics of earthquake propagation: Evidence from seismological and geological observations, *Earth Planet. Sci. Lett.* 273(1-2): 123–131.
URL: *http://linkinghub.elsevier.com/retrieve/pii/S0012821X08003956*

Collettini, C., De Paola, N. & Faulkner, D. (2009). Insights on the geometry and mechanics of the UmbriaâĂŞMarche earthquakes (Central Italy) from the integration of field and laboratory data, *Tectonophysics* 476(1-2): 99–109.
URL: *http://linkinghub.elsevier.com/retrieve/pii/S004019510800396X*

Di Toro, G., Goldsby, D. L. & Tullis, T. E. (2004). Friction falls towards zero in quartz rock as slip velocity approaches seismic rates, *Nature* 427: 436–439.
URL: *http://www.ncbi.nlm.nih.gov/pubmed/14749829*

Dieterich, J. H. & Kilgore, B. D. (1994). Direct observation of frictional contacts: New insights for state-dependent properties, *Pure Appl. Geophys.* 143(1-3): 283–302.
URL: *http://www.springerlink.com/index/10.1007/BF00874332*

Fialko, Y. (2004). Temperature fields generated by the elastodynamic propagation of shear cracks in the Earth, *J. Geophys. Res.* 109(B1).
URL: *http://www.agu.org/pubs/crossref/2004/2003JB002497.shtml*

Fukuyama, E. & Madariaga, R. (1998). Rupture dynamics of a planar fault in a 3D elastic medium: Rate- and slip- weakening friction, *Bull. Seis. Soc. Am.* 88(1): 1–17.

Goren, L., Aharonov, E., Sparks, D. & Toussaint, R. (2010). Pore pressure evolution in deforming granular material: A general formulation and the infinitely stiff approximation, *J. Geophys. Res.* 115(B9).
URL: *http://www.agu.org/pubs/crossref/2010/2009JB007191.shtml*

Gratier, J.-P., Favreau, P. & Renard, F. (2003). Modeling fluid transfer along California faults when integrating pressure solution crack sealing and compaction processes, *J. Geophys. Res.* 108(B2).
URL: *http://www.agu.org/pubs/crossref/2003/2001JB000380.shtml*

Han, R., Shimamoto, T., Hirose, T., Ree, J.-H. & Ando, J.-I. (2007). Ultralow friction of carbonate faults caused by thermal decomposition, *Science* 316: 878–881.
URL: *http://www.ncbi.nlm.nih.gov/pubmed/17495168*

Handin, J., Hager, R. V., Friedman, M. & Feather, J. N. (1963). Experimental deformation of sedimentary rocks under confining pressure; pore pressure tests, *Bull. Am. Assoc. Petro. Geol.* 47(5): 717–755.

Hatano, T. (2007). Power-law friction in closely packed granular materials, *Phys. Rev. E* 75(6).
URL: *http://link.aps.org/doi/10.1103/PhysRevE.75.060301*

Hirono, T., Fujimoto, K., Yokoyama, T., Hamada, Y., Tanikawa, W., Tadai, O., Mishima, T., Tanimizu, M., Lin, W., Soh, W. & Song, S.-R. (2008). Clay mineral reactions caused by frictional heating during an earthquake: An example from the Taiwan Chelungpu fault, *Geophys. Res. Lett.* 35.
URL: *http://www.agu.org/pubs/crossref/2008/2008GL034476.shtml*

Hirose, T. & Bystricky, M. (2007). Extreme dynamic weakening of faults during dehydration by coseismic shear heating, *Geophys. Res. Lett.* 34(14).
URL: *http://www.agu.org/pubs/crossref/2007/2007GL030049.shtml*

Ishikawa, T., Tanimizu, M., Nagaishi, K., Matsuoka, J., Tadai, O., Sakaguchi, M., Hirono, T., Mishima, T., Tanikawa, W., Lin, W., Kikuta, H., Soh, W. & Song, S.-R. (2008). Coseismic fluid-rock interactions at high temperatures in the Chelungpu fault, *Nature Geoscience* 1: 679–683.
URL: *http://www.nature.com/doifinder/10.1038/ngeo308*

Jop, P., Forterre, Y. & Pouliquen, O. (2006). A constitutive law for dense granular flows., *Nature* 441: 727–730.
URL: *http://www.ncbi.nlm.nih.gov/pubmed/16760972*

Lachenbruch, A. H. (1980). Frictional Heating, Fluid Pressure, and the Resistance to Fault Motion, *J. Geophys. Res.* 85(B11): 6097–6112.
URL: *http://www.agu.org/pubs/crossref/1980/JB085iB11p06097.shtml*

Mandl, G., Jong, L. N. J. & Maltha, A. (1977). Shear zones in granular material, *Rock Mech. Roch Eng.* 9: 95–144.
URL: *http://www.springerlink.com/index/10.1007/BF01237876*

Marone, C., Cocco, M., Richardson, E. & Tinti, E. (2009). The critical slip distance for seismic and aseismic fault zones of finite width, *in* E. Fukuyama (ed.), *Fault-zone properties and earthquake rupture dynamics*, Academic press, chapter 6, pp. 135–162.

Marone, C., Raleigh, C. B. & Scholz, C. H. (1990). Frictional Behavior and Constitutive Modeling of Simulated Fault Gouge, *J. Geophys. Res.* 95(B5): 7007–7025.
URL: *http://www.agu.org/pubs/crossref/1990/JB095iB05p07007.shtml*

Matsu'ura, M., Kataoka, H. & Shibazaki, B. (1992). Slip-dependent friction law and nucleation processes in earthquake rupture, *Tectonophysics* 211(1-4): 135–148.
URL: *http://linkinghub.elsevier.com/retrieve/pii/004019519290056C*

Mitsui, Y. & Hirahara, K. (2009a). Coseismic thermal pressurization can notably prolong earthquake recurrence intervals on weak rate and state friction faults: Numerical experiments using different constitutive equations, *J. Geophys. Res.* 114(B9).
URL: *http://www.agu.org/pubs/crossref/2009/2008JB006220.shtml*

Mitsui, Y. & Hirahara, K. (2009b). Interseismic pore compaction suppresses earthquake occurrence and causes faster apparent fault loading, *Geophys. Res. Lett.* 36(20).
URL: *http://www.agu.org/pubs/crossref/2009/2009GL039932.shtml*

Mizoguchi, K., Hirose, T., Shimamoto, T. & Fukuyama, E. (2009). High-velocity frictional behavior and microstructure evolution of fault gouge obtained from Nojima fault, southwest Japan, *Tectonophysics* 471(3-4): 285–296.
URL: *http://linkinghub.elsevier.com/retrieve/pii/S0040195109001292*

Noda, H. & Shimamoto, T. (2005). Thermal Pressurization and Slip-Weakening Distance of a Fault: An Example of the Hanaore Fault, Southwest Japan, *Bull. Seis. Soc. Am.* 95(4): 1224–1233.
URL: *http://bssa.geoscienceworld.org/cgi/doi/10.1785/0120040089*

Ohnaka, M. & Yamashita, T. (1989). A Cohesive Zone Model for Dynamic Shear Faulting Based on Experimentally Inferred Constitutive Relation and Strong Motion Source Parameters, *J. Geophys. Res.* 94(B4): 4089–4104.
URL: *http://www.agu.org/pubs/crossref/1989/JB094iB04p04089.shtml*

Renard, F., Park, A., Ortoleva, P. & Gratier, J.-P. (1999). An integrated model for transitional pressure solution in sandstones, *Tectonophysics* 312(2-4): 97–115.
URL: *http://linkinghub.elsevier.com/retrieve/pii/S0040195199002024*

Revil, A., Leroy, P., Ghorbani, A., Florsch, N. & Niemeijer, A. R. (2006). Compaction of quartz sands by pressure solution using a Cole-Cole distribution of relaxation times, *J. Geophys. Res.* 111(B9).
URL: *http://www.agu.org/pubs/crossref/2006/2005JB004151.shtml*

Rice, J. R. (2006). Heating and weakening of faults during earthquake slip, *J. Geophys. Res.* 111(B5).
URL: *http://www.agu.org/pubs/crossref/2006/2005JB004006.shtml*

Rudnicki, J. W. (1986). Slip on an impermeable fault in fluid-saturated rock mass, *in* S. Das, J. Boatwright & C. H. Scholz (eds), *Earthquake source mechanics*, Vol. 6 of *AGU Geophysical Monograph 37*, pp. 81–89.

Rudnicki, J. W. & Chen, C.-H. (1988). Stabilization of rapid frictional slip on a weakening fault by dilatant hardening, *J. Geophys. Res.* 93(B5): 4745–4757.

Segall, P. & Rice, J. R. (2006). Does shear heating of pore fluid contribute to earthquake nucleation?, *J. Geophys. Res.* 111(B9).
URL: *http://www.agu.org/pubs/crossref/2006/2005JB004129.shtml*

Shibazaki, B. (2005). Nucleation process with dilatant hardening on a fluid-infiltrated strike-slip fault model using a rate- and state-dependent friction law, *J. Geophys. Res.* 110(B11).
URL: *http://www.agu.org/pubs/crossref/2005/2005JB003741.shtml*

Sibson, R. H. (1973). Interactions between Temperature and Pore-Fluid Pressure during Earthquake Faulting and a Mechanism for Partial or Total Stress Relief, *Nature* 243: 66–68.

Sibson, R. H. (2003). Thickness of the Seismic Slip Zone, *Bull. Seis. Soc. Am.* 93(3): 1169–1178.
URL: *http://bssa.geoscienceworld.org/cgi/doi/10.1785/0120020061*

Suzuki, T. & Yamashita, T. (2007). Understanding of slip-weakening and -strengthening in a single framework of modeling and its seismological implications, *Geophys. Res. Lett.* 34(13).
URL: *http://www.agu.org/pubs/crossref/2007/2007GL030260.shtml*

Tanikawa, W., Sakaguchi, M., Tadai, O. & Hirose, T. (2010). Influence of fault slip rate on shear-induced permeability, *J. Geophys. Res.* 115(B7).
URL: *http://www.agu.org/pubs/crossref/2010/2009JB007013.shtml*

Terzaghi, K. (1943). *Theoretical soil mechanics*, John Wiley & Sons.

Tsutsumi, A. & Shimamoto, T. (1997). High-velocity frictional properties of gabbro, *Geophys. Res. Lett.* 24(6): 699.
URL: *http://www.agu.org/pubs/crossref/1997/97GL00503.shtml*

Ujiie, K., Kameyama, M. & Yamaguchi, A. (2010). Geological record of thermal pressurization and earthquake instability of subduction thrusts, *Tectonophysics* 485(1-4): 260–268.
URL: *http://linkinghub.elsevier.com/retrieve/pii/S0040195110000065*

Vosteen, H. D. & Schellschmidt, R. (2003). Influence of temperature on thermal conductivity, thermal capacity and thermal diffusivity for different types of rock, *Phys. Chem. Earth* 28(9-11): 499–509.
URL: *http://linkinghub.elsevier.com/retrieve/pii/S147470650300069X*

Weeks, J. D. (1993). Constitutive Laws for High-Velocity Frictional Sliding and Their Influence on Stress Drop During Unstable Slip, *J. Geophys. Res.* 98(B10): 17637–17648.
URL: *http://www.agu.org/pubs/crossref/1993/93JB00356.shtml*

Wibberley, C. A. J. & Shimamoto, T. (2005). Earthquake slip weakening and asperities explained by thermal pressurization, *Nature* 436: 689–692.
URL: *http://www.ncbi.nlm.nih.gov/pubmed/16079843*

Zhang, S. & Tullis, T. E. (1998). The effect of fault slip on permeability and permeability anisotropy in quartz gouge, *Tectonophysics* 295(1-2): 41–52.
URL: *http://linkinghub.elsevier.com/retrieve/pii/S0040195198001140*

18

Integrated Morphometric Analysis in GIS Environment Applied to Active Tectonic Areas

Rosa Nappi and Giuliana Alessio
Istituto Nazionale di Geofisica e Vulcanologia, Sezione di Napoli, Osservatorio Vesuviano
Italy

1. Introduction

This chapter describes a methodology for constraining the structural lineaments in active tectonic areas by integrating morphological and morphometrical data derived by DEMs (Digital Elevation Models) processing, with different geophysical data, as local seismicity and ground deformation data. Furthermore, validation of the lineaments extracted from DEM is carried out by looking over geological and geomorphological maps of literature, available aerial photo and field surveys reports (Fig.1).

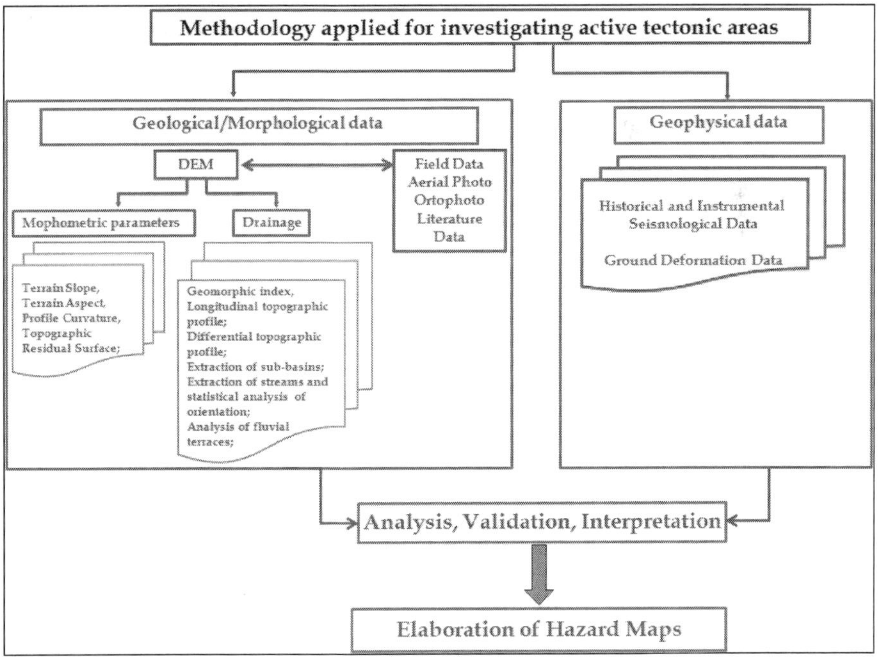

Fig. 1. Flow chart describing the methodology applied for investigating active tectonic areas.

The morphometric parameters of the terrain slope, terrain aspect, profile curvature, tangential curvature and topographic residual surface are chosen for identifying the linear continuity of the morphostructural features observed on the DEM. The criteria of lineament extraction is based on the identification of linear topographic surface features, such as valleys, ridges, breaks in slope, boundaries of elevated areas aligned in a rectilinear or slightly curvilinear shape and that distinctly differ from the patterns of adjacent features.

The geophysical data considered in this analysis include spatial distribution of local earthquakes, accurate locations of seismic sequences and relative focal mechanisms, which could confirm activity of local tectonic structures.

The areas selected for testing our methodology, located in Southern Italy, are the Agri Valley (Campania-Lucania regions), hit by the strong historical earthquake of December 26, 1857 (I_{max}=XI, M_e=7.0) and also by recent micro-seismicity; the Sannio area (Campania-Molise regions), affected in historical time by the strong earthquake of June 5, 1688 (I_{max}=XI, M_e=6.7) and in recent time by seismic sequences of moderate energy; the Campi Flegrei volcanic district (Campania region, Thyrrenian coast), characterized in the past decades by major bradyseismic crises with remarkable ground uplift and intense seismic activity, and recently by minor crises with lower deformation (Fig. 2).

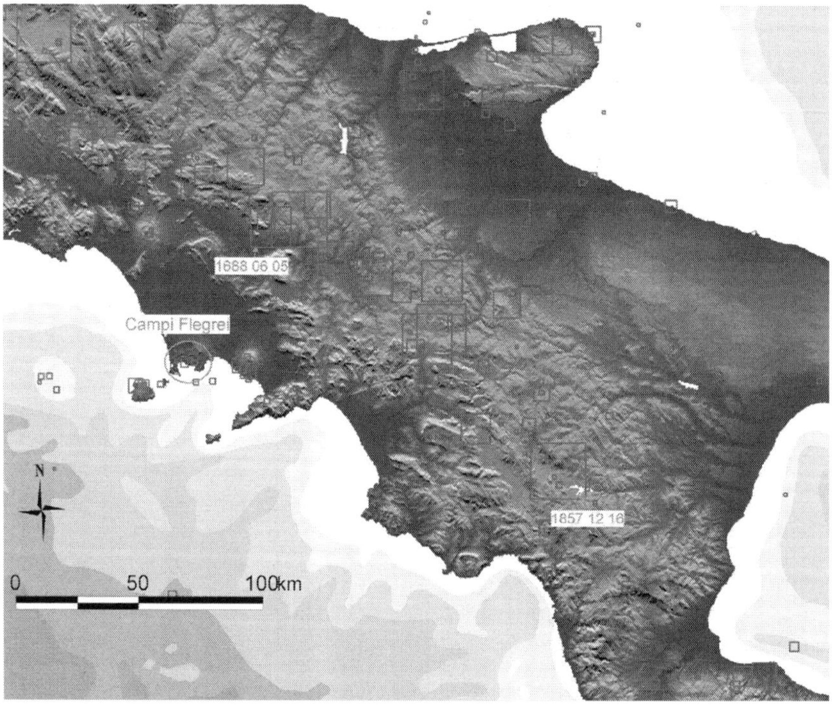

Fig. 2. Historical and recent earthquakes in the Southern Apennines of Italy (red squares) from CPTI04 catalogue; the epicenters of the 1688 Sannio earthquake (I_{max}=XI, M_e=6.7) and the 1857 Agri Valley earthquake (I_{max}=XI, M_e=7.0, CPTI, 1999) are pointed out; the Campi Flegrei volcanic area location is indicated by the orange circle.

Particulary, the Agri Valley and the Sannio area are active tectonic areas of the Southern Apennines chain and are characterized by a complex inherited tectonic setting, low-tectonic deformation rates that hide the seismogenic sources geometry, and youthfulness of the sources. Whereas the morphotectonic features of the Campi Flegrei caldera resulted from the combined action of both volcanism and regional tectonics, mainly correlated with a circular geometry of deformation (volcanism), and NW-SE, NE-SW normal faults (regional tectonics).

As regards all the above studied areas, due to the difficulty or not straightforward recognition of the induced surface deformation, besides the lack of recent surveyed structural data, the operation of the GIS system has enabled us to process and generate original informative layers, through image analysis, such as new structural lineaments.

The synthesis of our main findings has consisted in structural thematic maps of the new lineaments, which are the final results of our application, representing a contribution to understanding the potential active faults of the investigated areas, for the assessment of local geological-environmental hazard parameters.

2. Methodology

Although the interpretation of land morphology applied to tectonic deformation has been broadly studied, there are few examples of integration of digital methods for tectonic geomorphology with the classical approach (Burbank and Anderson, 2001, Keller and Pinter, 2002, Jordan et al., 2005; Nappi et. al., 2009). Quantitative analysis of the topographic features contributes to study the interaction between tectonics and surface processes, providing a basis for modelling landscape evolution. Particularly, geo-morphometrical analysis of DEMs represents a methodology for studying the morphotectonics of an area quantitatively, recognizing the main tectonic structures (e.g., Jordan et al. 2005). Geologic structures may produce linearly aligned features not typically common in nature, therefore, recognizing, measuring, and interpreting all the linear and areal surface features is of great importance in geodynamics. Intersection of bedding with the topographic surface can appear as linear features or planar features in DEMs. Linear morphological expressions of fractures include: linear valleys, linear ridgelines, and linear slope breaks.

The methodology applied consists of geo-morphometrical analysis of high resolution DEMs (20x20, 5x5 m pixel), integrated with the geological and geomorphological data derived from literature, photo-interpretation and field surveys, besides seimic data and ground deformation data analysis (Fig.1).

In particular, geomorphic analysis of topography consists of different steps: a) generation of high resolution DEMs derived from interpolation of altimetrical data in vector format; b) extraction and analysis of topographic parameters derived from DEM to identify their linear continuity associable to fault lineaments; c) statistical analysis of the azimuthal distribution of the morphostructural lineaments extracted. The morphometric parameters analyzed in our work are have been extracted following the methods of Moore et al., 1993 (*terrain aspect, terrain slope, tangential curvature, profile curvature, shaded relief, topographic residual surface*). The *terrain aspect* (At=arctan[(−dH/dy)/(dH/dx)]) measures the downhill direction of the steepest slope (i.e., dip direction) at each grid node, and identifies the linear geomorphic features (crest lines) that can be associated with footwalls of normal faults.

The *terrain slope* (S=[(dH/dx)(2)+(dH/dy)(2)](1/2)) measures the slope at any grid node on the surface and, for a particular point on the surface, is based on the direction of the steepest descent or ascent at that point, highlighting the presence of scarps and slope-breaks.

The *tangential curvature* K_T measures the topographic surface curvature in the direction perpendicular to the maximum gradient direction, or tangential to the contour lines:

$$K_T = \frac{\left(\frac{\partial^2 z}{\partial x^2}\right)\left(\frac{\partial z}{\partial y}\right)^2 - 2\left(\frac{\partial^2 z}{\partial x \partial y}\right)\left(\frac{\partial z}{\partial x}\right)\left(\frac{\partial z}{\partial y}\right) + \left(\frac{\partial^2 z}{\partial y^2}\right)\left(\frac{\partial z}{\partial x}\right)^2}{pq^{1/2}} \tag{1}$$

$$p = \left(\frac{\partial z}{\partial x}\right)^2 + \left(\frac{\partial z}{\partial y}\right)^2$$

$$q = 1 + p$$

The *profile curvature* K_p quantifies the rate of change in slope along the direction of maximum slope:

$$Kp = \frac{\left(\frac{\partial^2 z}{\partial x^2}\right)\left(\frac{\partial z}{\partial x}\right)^2 + 2\left(\frac{\partial^2 z}{\partial x \partial y}\right)\left(\frac{\partial z}{\partial x}\right)\left(\frac{\partial z}{\partial y}\right) + \left(\frac{\partial^2 z}{\partial y^2}\right)\left(\frac{\partial z}{\partial y}\right)^2}{pq^{3/2}} \tag{2}$$

$$p = \left(\frac{\partial z}{\partial x}\right)^2 + \left(\frac{\partial z}{\partial y}\right)^2$$

$$q = 1 + p$$

Negative values of both tangential curvature and profile curvature indicate that the surface is upwardly convex; positive values show that the surface is upwardly concave. A curvature value of zero indicates that the surface is flat. In our procedure, we have low-pass filtered and classified the DEM, and have identified the highest values of tangential and profile curvature for outlining basins/ridges.

The *topographic residual surface* represents a measure of relief within the landscape. This parameter indicates how suddenly channels are incised into the landforms (Hilley and Arrosmith, 2000). In areas undergoing rapid uplift, as active tectonic areas, channels will incise steeply into the surrounding soil. Therefore, high values of residual surface should correlate with rapidly uplifting areas.

The residual surface is calculated by subtracting two derived surfaces: the envelope and the subenvelope surfaces. The envelope is a surface interpolated from the stream bottoms in a landscape and the subenvelope is a surface interpolated from the ridge lines in a landscape. The final distribution of the topographic residual values depends on the points selected for the envelope and subenvelope surfaces.

The topographic residual surface for the study areas has been calculated using the algorithm of Hilley and Arrosmith, 2000, (modified). We have removed all pits from DEM using Arc/INFO "fill" command before applying the algorithm, since the reliability of the topographic residual map depends fairly strongly on data quality (Nappi et al., 2009).

The followed methodology recommends the integration of data derived from digital parameters with those extracted through classical approach. In particular, the drainage networks have been performed since they represent young features of landform and they have possibly registered the recent tectonic deformation; indeed, the drainage pattern is one of the most significant features that could be influenced by tectonic activity. Phenomena of antecedence, rivers diversions, shifts in channel pattern and longitudinal change of the channel behaviour reveal tectonic modifications. Therefore the quantitative geomorphic analysis of the drainage networks has consisted of evaluation of the following: 1) geomorphic indexes (Keller & Pinter, 2002) such as Asymmetry Factor (AF=100(Ar/At)), Transverse Topographic Symmetry Factor (T=Da/Dd), Stream Length gradient (SL=(DH/DL)L) and mountain front Sinuosity (Smf=Lmf/Ls); 2) the longitudinal topographic profile, the topographic profiles adjacent to the stream and the relative differential profile; 3) extraction of sub-basins; 4) extraction and statistical analysis of the stream orientation; 4) analysis of the river terraces.

The analysis of the geological data found in literature and field survey reports has consisted in collecting and cataloging the whole cartographic available dataset, in addition to recent publications and scientific papers, relative to the case study areas. Moreover, seismological data as ipocenters and focal mechanisms of hystorical and recent earthquakes extracted from the available seismic catalogues have been used in this study with the aim of constraining the activity of the investigated structural lineaments.

3. The case study areas

3.1 The Agri Valley basin

The Agri Valley is a NW-SE oriented Quaternary intermountain basin associated to the last stages of the Lucanian Apennines fold-and-thrust belt building, with inherited topographic setting that still represents the most striking feature (Fig.3a). It represented the epicentral area of the 1857 destructive earthquake (M_e=7.0; Imax=XI; MALLET, 1862; Gruppo di lavoro CPTI, 1999, 2004). This earthquake has been localized along the narrow topographic culmination of the chain where the historical destructive seismicity, characterized by extensional focal mechanisms, is concentrated together with the active faults of this portion of the Southern Appenines.

Different models have been suggested for the 1857 seismogenetic structure, based on: geological and structural surveys (Borraccini et al., 2002; Cello et al., 2003; Giano et al.,1997, 2000); geomorphological analysis (Di Niro et al., 1992, Di Niro and Giano, 1995; Giano, 2011); seismotectonic analysis (Benedetti et al., 1998; Burrato & Valensise, 2008; Improta et al., 2010; Maschio et al. 2005; Pantosti and Valensise, 1988; Valensise & Pantosti, 2001a), but there are still some open questions. Indeed in this region, contrasting with the events size, several tectonic and geomorphic factors concur to hide the active extensional seismogenetic faults. The inherited tectonic history of the peninsula with strong deformation readable on the topographic configuration, the low rates of present tectonic deformation, the youthfulness of the latest major change in the tectonic regime, the lithologies with strong erosional contrast (Valensise and Pantosti, 2001b), represent the reasons for the questionable recognition of the seismogenic source position. In this complex seismotectonic setting our integrated methodology turns out to be necessary for acquiring basic knowledge of new geomorphological features which could substantiate the different suggested hypotheses. Quantitative geomorphic analysis, which has been performed using a medium-resolution DEM (20x20 m), has allowed the identification of a series of NW-SE oriented features along the western side of the valley, spatially continuous and evenly distributed, even outside the

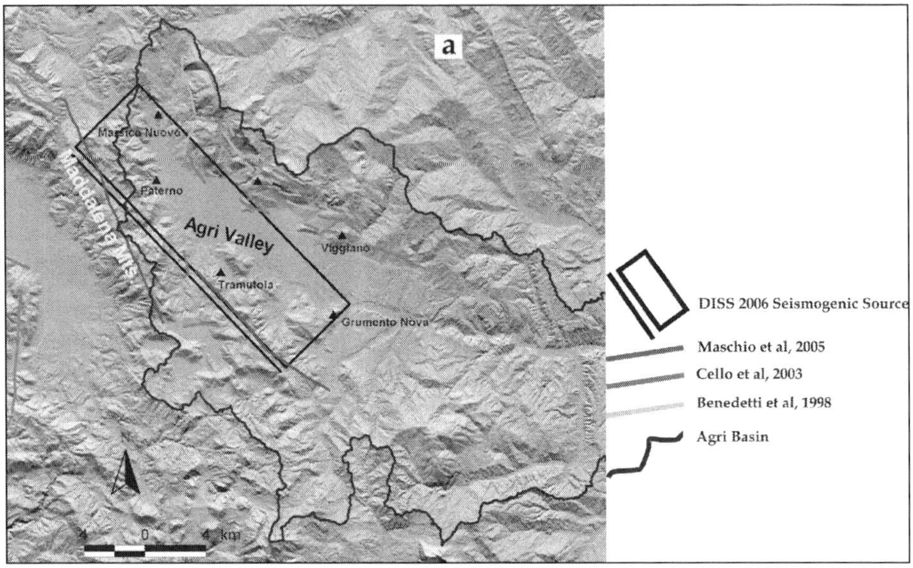
a
Marsico Nuovo
Paterno
Agri Valley
Viggiano
Tramutola
Grumento Nova
Maddalena Mts
4 0 4 km
DISS 2006 Seismogenic Source
Maschio et al, 2005
Cello et al, 2003
Benedetti et al, 1998
Agri Basin

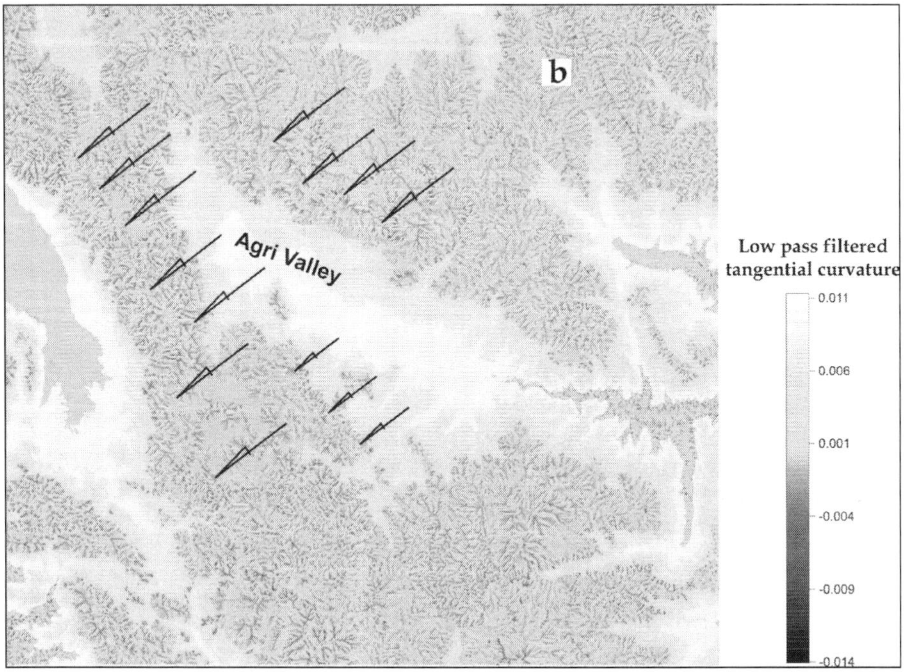
b
Agri Valley
Low pass filtered tangential curvature
0.011
0.006
0.001
-0.004
-0.009
-0.014

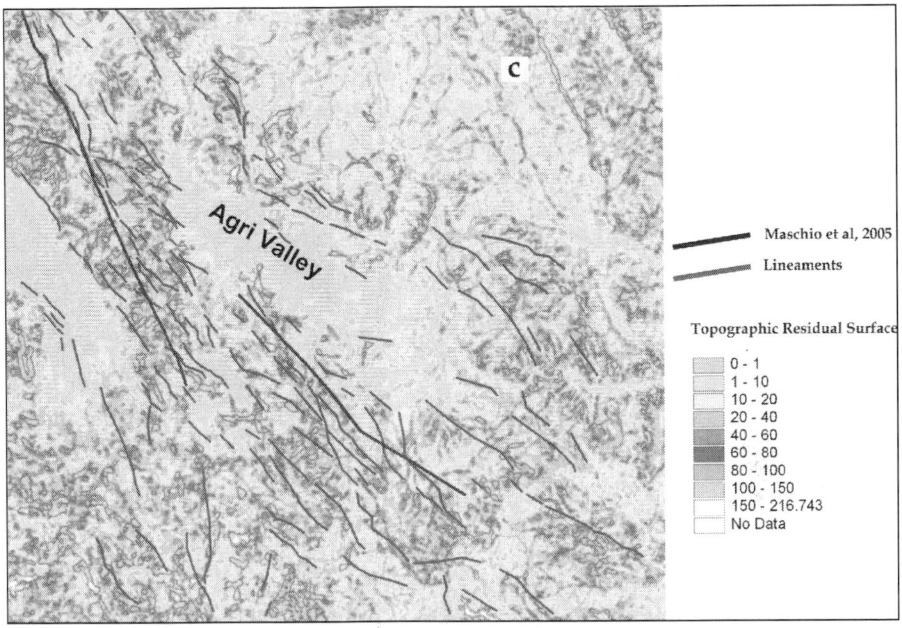

Fig. 3. a) Hypothesis from literature for the 1857 seismogenic source: black box from Valensise e Pantosti, 2001a; green line from Benedetti et al., 1998; red line from Maschio et al., 2005. Morphometric parameters extracted from DEM: b) Low pass tangential curvature with the arrows indicating the linear continuity of the parameter; c) morphostructural lineaments extracted from topographic residual surface (red lines) and the recent active lineaments derived from Maschio et al., 2005 (blue line).

basin, and coinciding with recent scarps. Whereas, on the eastern side of the valley, the lineaments coincide with the west-dipping fault system identified in the ground (Figs. 3b, 3c). It was performed also a study on the paleosurfaces of the basin (Nappi et al., 2002) wich represent relict surfaces of ancient periods of the local base-level, in geomorphic situation that was probably very different from the current conditions (Ollier, 1991). On this basis these studies point out the possible presence of zones affected by differential vertical movements and assess the geometry and kinematics of the relative structural elements. Previous studies on the paleosurfaces of the Agry Valley had been carried out (Amato e Cinque, 1999; Giano et al., 1997, 2000; Giano, 2011; Schiattarella et al., 2003) with the aim of estimating the regional tectonic uplift. On the other hand, in order to acquire information on the deformation of the topography due to active faults, further analysis and correlations based on morphological criteria have been carried out.

The spatial continuity of the paleosurfaces, the correspondence of elevation and the change of slope have been examined respect to the present topography for taking into account the probable original gradient of each surface. We have identified eight groups of remnant surfaces at different elevation of the basin, but the distribution of various orders with respect to the current topography is not the same on both sides of the basin (Fig.4).

The surfaces at highest elevations (1700-1500m a.s.l; 1400-1200m a.s.l), according to the authors (Amato e Cinque, 1999; Giano et al., 1997, 2000) are considered the most ancient and are clearly visible on the landscape. The paleosurfaces at elevation of 1700-1500 m a.s.l are only visible on the eastern side of the Agri Valley; the paleosurfaces at elevation 1400-1200 m a.s.l are well rapresented on both sides of the valley, although the distribution of several patches shows a linear continuity on the western side with lacks on the eastern side. The remnant surfaces at elevation of 1200-1100 m a.s.l are rather fragmented on both sides of the valley and are not much representative. The paleosurfaces at elevation of 1100-1000 m a.s.l. are rather fragmented on both sides of the valley and are better represented on the western side, compared to the eastern side. The remnant surfaces, of more recent age, are fragmented in small edges on the western side of the valley with evidence of faulting of paleosurfaces at elevation of 930-880 m a.s.l..Therefore, the study of paleosurfaces of the Agri Valley has evidenced that their spatial distribution is not symmetrical on both flanks of the basin. The eastern side of the valley is characterized by paleosurfaces of high elevation with uniform morphologic characteristics while the paleosurfaces of low elevation are lacking; the western side is characterized by different and fragmented high elevation paleosurfaces with constant NW-SE trend, and also by low elevation paleosurfaces. This geomorphological observation supports the landscape deformation modeled by the authors Maschio et al., (2005), Valensise & Pantosti, (2001a), DISS (2007) for the 1857 seimogenic source.

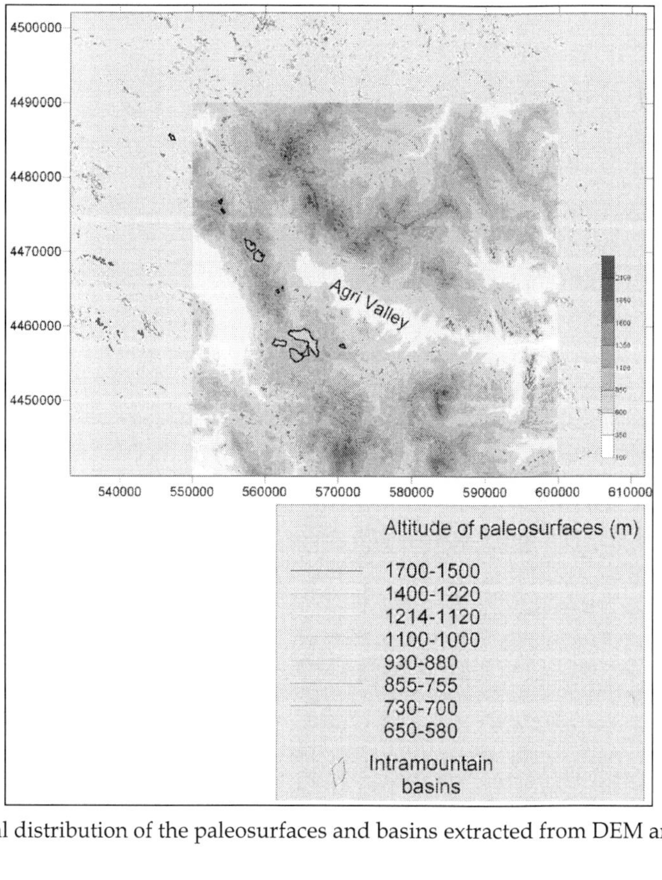

Fig. 4. Spatial distribution of the paleosurfaces and basins extracted from DEM analysis.

3.2 The Tammaro river basin

The Sannio area is located inside the Southern Apennines fold and thrust belt that has been affected by extensional deformation since middle-late Pleistocene (Cinque et al. 1993; Hippolyte et al. 1994; Westaway, 1993). The most important historical seismic event occurred in the Sannio on June 5, 1688 (M_w = 6.7, CPTI 1999); this strong earthquake destroyed many towns of the area, while in recent times low-and moderate-energy seismic sequences (M_d max = 4.1) occurred in 1990–1992 near the town of Benevento, in 1997 along the border between Campania and Molise near the Tammaro river, and in 2001 near Isernia, (Milano et al. 1999, 2005; Vilardo et al. 2003). Among the recent earthquakes, the event with maximum magnitude M_l = 4.5 occurred on March 19, 1997, and the relative motion from its focal mechanism was a dip-slip on a NW-SE trending fault (Pondrelli et al. 2006).

As regards the 1688 seismogenic fault a NW-SE trending normal fault has also been hypothesized in the DISS by Valensise & Pantosti, 2001a (Fig.5a), on the basis of the similarity among the strongest earthquakes sources occurred along the Southern Apennines chain. However, the location and geometry of the seismogenetic source of the M_w = 6.7, 1688 Sannio normal faulting earthquake is still a subject of scientific debate. This is due to several reasons: a) the possible incompleteness of the damage pattern data b) the difficult or not straightforward recognition of the induced surface deformation, c) the possible occurrence of blind or hidden faulting, and d) the low tectonic deformation rates and youthfulness of the source. The most important literature hypotheses for the 1688 source position are the following:

1. a NW-SE trending, NE-dipping, 25 km-long normal fault running along the SW margin of the Tammaro river basin (Valensise & Pantosti, 2001a; DISS Working Group, 2007);
2. a WNW-ESE trending, NE-dipping, 32 km-long normal fault located along the Calore River Valley (Di Bucci et al., 2006).

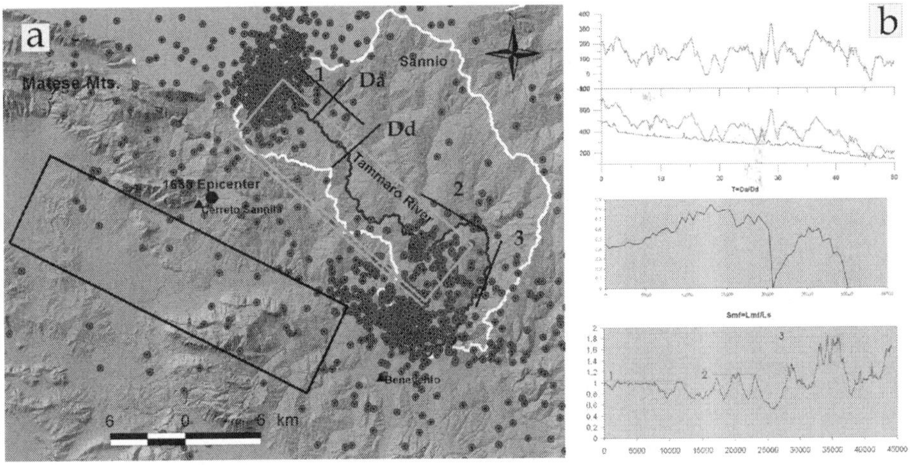

Fig. 5. a) Epicentral distribution of the instrumental seismicity from 1980 to 2006 in the Sannio area (red circles) and supposed 1688 seismogenic source positions (literature hypotheses): in orange the box of Valensise & Pantosti, 2001a, in black that of Di Bucci et al., 2006; b) from top to bottom: differential profile, elevation profile and longitudinal profile along the Tammaro river; T asimmetry factor; Smf mountain front Sinuosity evaluated along the segments 1, 2, 3 of the river, indicated on the left map.

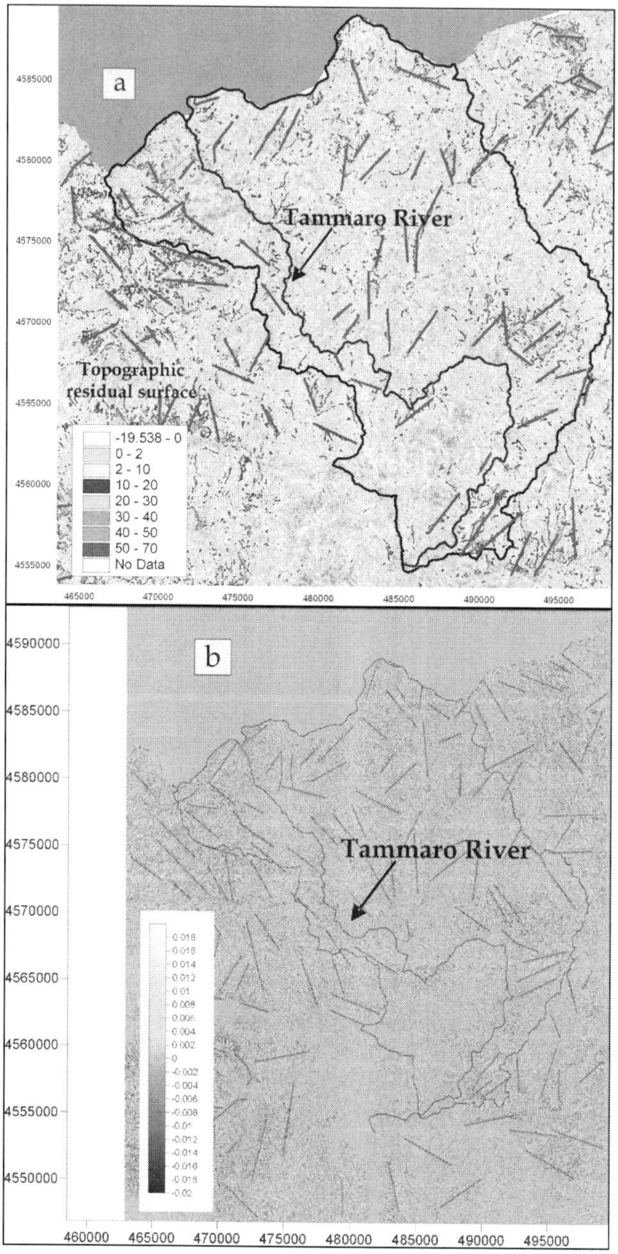

Fig. 6. The lineaments (red lines) extracted from the DEM analysis of the Tammaro basin: morphometric parameters of a) topographic residual surface; b) tangential curvature.

The methodology suggested in this chapter for investigating the seismogenetic source position has been completely applied due to the availability of high quality data. In particular it has been carried out a) detailed analysis of space-time features of the low energy seismic sequences by sub-setting each different seismic sequence in a number of events grouped on the basis of the events space clustering (Nappi et al., 2008); b) quantitative geomorphological analysis of the Tammaro river basin area for identifying the long term surface deformation possibly induced by the seismogenetic fault of the 1688 earthquake; c) integration in GIS environment of the geological and geomorphological data derived from photo-interpretation and field surveys with the morphometrical data derived by processing of a very high resolution DTM (5x5 m pixel), generated through the vectorial data of the Carta Tecnica Regionale of Campania.

The geomorphic indexes of drainage network as the mountain front Sinuosity ($1.2<Smf<2.5$) and Transverse Topographic Symmetry Factor ($T=1$ asymmetric basin) (Keller & Pinter, 2002) can be consistent with a possible recent tectonic control along the course of Tammaro river (Fig.5b), with a remarkable asymmetry of the basin towards SW. Furthermore, the longitudinal topographic profile, the topographic profile adjacent to the stream and the relative differential profile have evidenced the prevalent depositional behaviour of the early portion (20 km) of the Tammaro river (Fig. 5b).

A significant swarm of NW-SE contiguous lineaments on the SW side of the Tammaro river valley have been identified through the analysis of topographic parameters extracted by DTM (Fig 6a, b). The morphological analysis of the Ortophoto (1x1 pixel m) and the field observations have suggested a young morphological expression of such lineaments, also inferred by the existence of recent small basins along their trace, and by the identification of a structural mountain front corresponding to the above lineaments, the whole features suggesting a recent deformation activity.

3.3 The Campi Flegrei volcanic district

The Campi Flegrei volcanic district formed as a consequence of the lithospheric stretching in the central Tyrrhenian sea and Apennines belt parallel extension, since the Plio-Quaternary times (Scandone et al., 1991). The Campi Flegrei caldera is an active volcanic area located westerly of the town of Naples, characterized by high volcanic risk due to its very intense urbanization (Orsi et al., 1999).

The morphological features of this caldera resulted from the combined action of both volcanism and regional tectonics; in fact the caldera is a nested structure (Fig. 7) which originated through two major collapses related to the major eruptions of the Campanian Ignimbrite (39 ky) and the Neapolitan Yellow Tuff (15 ky) (De Vivo et al., 2001). After the Yellow Tuff eruption, volcanic activity and ground deformation have been very intense, with many different eruptive episodes. The last eruption took place in 1538 and formed the Mt. Nuovo cone (Di Vito et al., 1987). The magmatic system of the caldera is still active with intense hydrothermal activity (Chiodini et al., 2010), seismicity and ground deformation.

As regards the geodynamic history of the Campi Flegrei, the major bradyseismic crises occurred in 1969-1972 and 1982-1984, which were characterized by remarkable ground uplift and intense seismic activity; minor crises were observed in 1989, 1994, March-August 2000 and from June 2004 to October 2006, with low seismicity and moderate ground deformation (Del Gaudio et al., 2010).

Our study for the Campi Flegrei area has consisted in extraction of morphostructural lineaments based on the identification of linear topographic surface features derived from geomorphic analysis of high resolution DTM (5x5 m pixel); analysis and comparison of the lineaments spatial and statistical coherence with the structural lineaments already known from literature; correlation of the recent seismicity spatial distribution (crises of 1982-1984 and 2004-2006) with the obtained structural lineaments. Moreover, the results of the analysis have also been correlated to the local ground deformation data acquired through high precision levelling surveys over the last 20 years, and to the tiltmetric data continuously recorded locally over the last 10 years (Del Gaudio et al., 2009) with the aim of constraining the activity of the lineaments with local ground deformation data as well (Nappi et al., 2010b). In fig. 8 the morphometric parameters and the statistical analysis of the spatial distribution of the lineaments extracted are plotted. Two preferred NW-SE and NE-SW orientations are evident from the rose diagrams, in particular, from the terrain slope, terrain aspect and the topographic residual surface the lineaments with NW-SE orientation are prevalent.

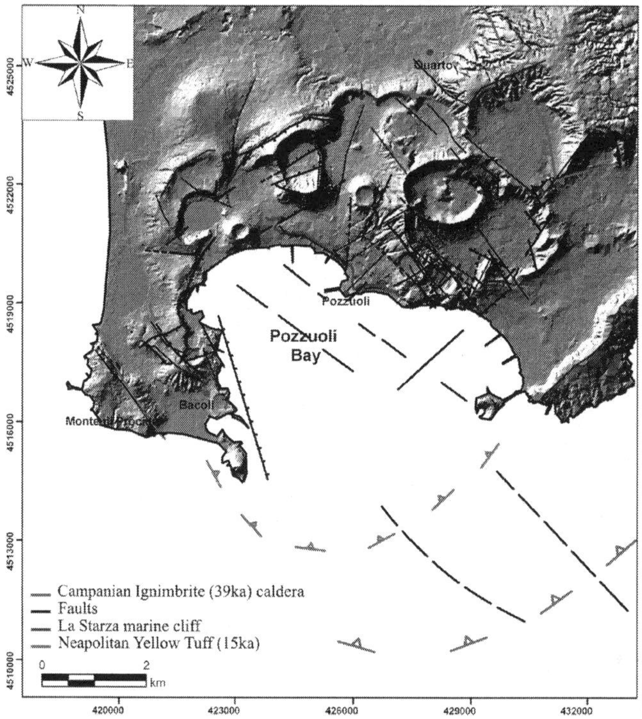

Fig. 7. Structural setting modified from Orsi et al., 1999. DTM (Digital Terrain Model) and structural data extracted from Bechtold et al., 2005.

Epicentres of the best located 1982-1984 seismic events appear clustered mostly in the central area of the maximum deformation observed (Fig. 9 left). A selection of events with $M_D \geq 2.5$ showes epicentres concentrated to the East of the Solfatara crater and hypocentres clustered above the STH and SFT (Solfatara) seismic stations with depths between 0.5 and 4 km. Epicentres of the best located 2005-2006 Volcano-Tectonic (VT) seismic events are concentrated on the eastern border of the Solfatara crater (Fig.9 right); the relative hypocenters are confined between 0.5 and 4 km and also clustered above the SFT (Solfatara) seismic station.

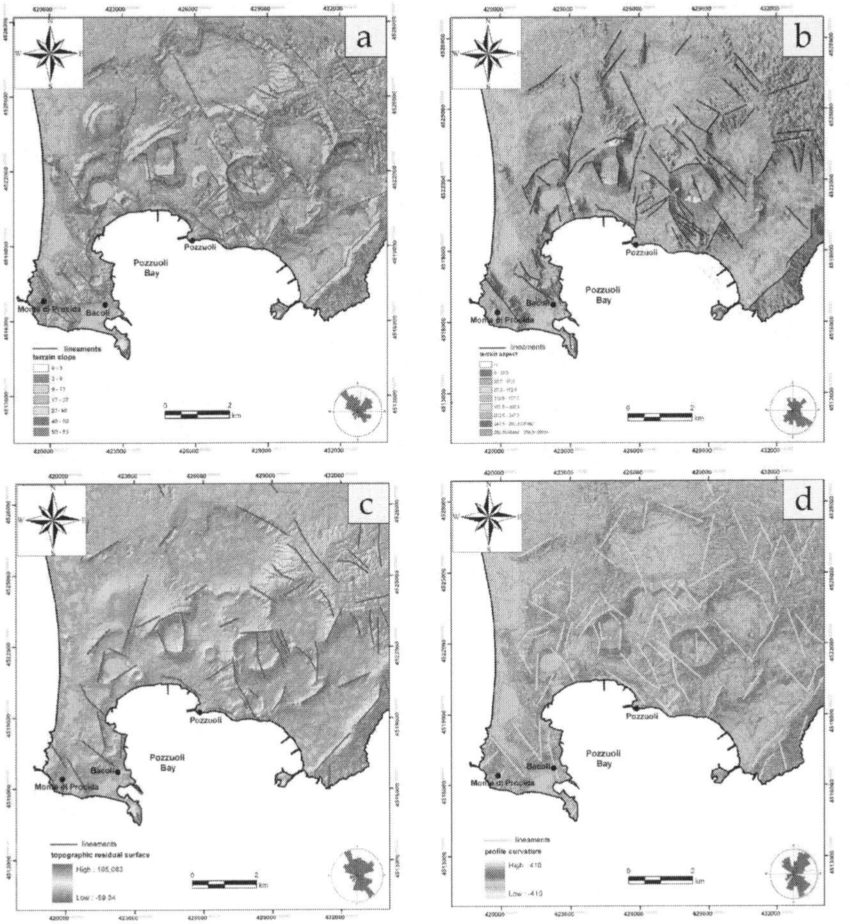

Fig. 8. The lineaments extracted from the morphometric parameters analysis; the rose diagrams represent the frequency distribution of the lineaments directions. a) the terrain slope map (lineaments in green); b) the terrain aspect map (lineaments in blue); c) the topographic residual surfaces map (lineaments in red); d) the profile curvature map (lineaments in yellow).

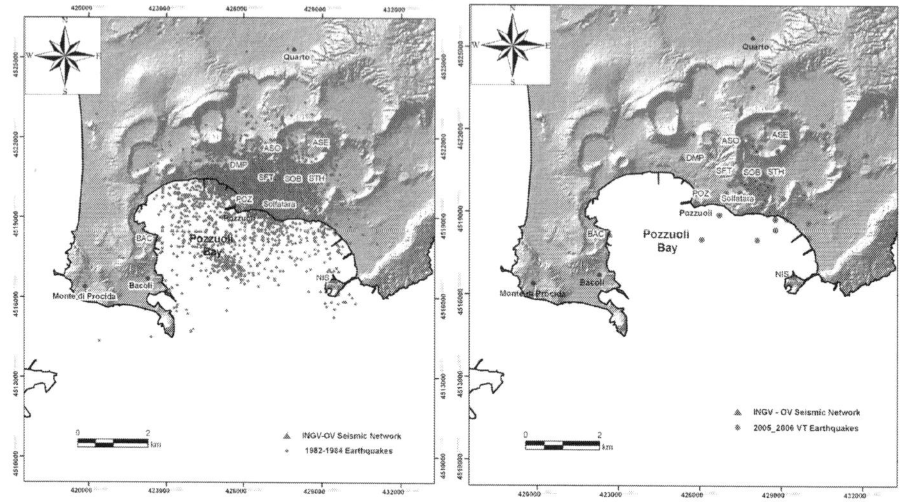

Fig. 9. The 1982-1984 seismic crises epicentres (blue circles) of the Campi Flegrei volcanic area (left map) and the epicentres of the best located 2005-2006 Volcano-Tectonic (VT) seismic events (right map).

4. Conclusions

The final step of the analysis and methodology proposed in this chapter has consisted in testing the congruence and consistency of the obtained results and of the possible interpretations, in the context of a coherent pattern of deformation taking place inside the studied areas.

The methodology applied to the Agri Valley has highlighted the agreement of the morphostructural lineaments identified on the SW side of the valley with the surface trace of faults system quoted in literature (Maschio et al., 2005). Moreover, these lineaments, jointing several intermountain sub-basins located only on the western side of the valley, are coincident with scarplets in recent deposits, in turn aligned with the main trend of the Maddalena Mts. Faults system, confirming their recent tectonic activity (Fig.3c). Throughout the study of the paleosurfaces from DEM analysis, the ancient erosional surface at 930-880 m a.s.l., on the western flank of the basin, is dislocated with throw of 25 m. The age of the paleosurfaces evaluated by Amato e Cinque, 1999; Schiattarella et al., 2003, as about 0.75 My (middle Pleistocene), suggested the recent tectonic activity of the western mountainside affected by post-middle Pleistocene deformation. The morphostructural data extracted following our methodology, relatively to the Agri Valley, accordingly support the recognized literature hypothesis of a NW-SE oriented seismogenic source, NE dipping, for the active faults system of the Agri Valley.

For the Tammaro basin area, the results obtained by applying our methodology have indicated that the strong asymmetry of the basin, not influenced by geological factors, could have been conditioned by tectonics, oriented along SW direction. The presence of NW-SE topographic lineaments extracted from DEM, on the right hand side of the valley of the Tammaro river, further constrained by morphological analyses, confirms the existence of a structural mountainside corresponding to the above lineaments. Moreover, also the presence of recent small basins along the trace of some morphostructural features extracted can be associated with recent tectonic activity.

As regards the analysis of seismological data relative to the Tammaro basin area, some recent seismic sequences of low energy, progressively activated, would concentrate particularly on some structural features exposed at surface (Fig. 10). In addition, some clusters of events (N. 5 and N. 3) concerning the same sequences are concentrated at the edge of the seismogenic source of the 1688 earthquake proposed by Valensise and Pantosti (2001a) (Fig. 5a), therefore it is possible to hypothesize that such clusters act as segment boundaries relatively to the master fault source (Nappi et al., 2008), constraining the linear dimension of the 1688 seismogenetic source.

The methodology applied to the Campi Flegrei volcanic area has allowed to identify significant structural lineaments from quantitative analysis of high resolution DEM (5x5 pixel m), extracting morphostructural features, based on their linear continuity. The statistical analysis of the spatial distribution of the lineaments extracted showes NW-SE and NE-SW preferred orientations; in particular, from the terrain slope, terrain aspect and the topographic residual surface the lineaments with NW-SE orientation are prevalent.

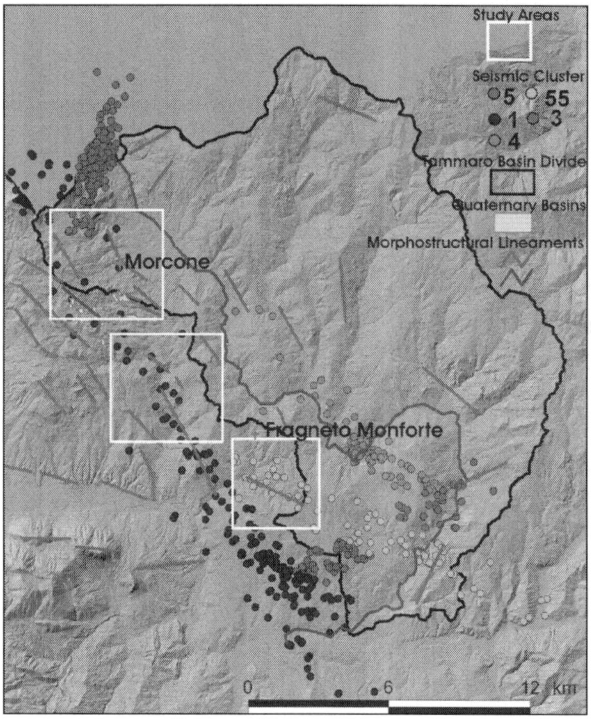

Fig. 10. Morphostructural lineaments (red and blue lines) extracted from DEM analysis; the main clusters of seismic events are mapped in group of different colors, and areas to be furtherly investigated in white boxes.

The analysis of seismological data pointed out that the epicentres of the best located 1982-1984 seismic events clustered mostly in the central area of maximum deformation but a selection of events with $M_D \geq 2.5$ have been located to the east of the Solfatara crater, with depths between 0.5 and 4 km, clustered above the STH and SFT (Solfatara) seismic stations. The S-N cross section of the 1982-84 epicentres, the S-N and W-E section of the 2005-2006 VT epicenters highlight a subvertical plane above the Solfatara crater that would suggest the activation of a seismogenic fault comparable with significant structural discontinuities NNW-SSE oriented, also identified through image analysis, located in the M. Olibano dome area (near OLB tilt station).
Moreover, the high precision levelling and tiltmetric data relative to the 2005-2006 years have clearly demonstrated asymmetrical deformation in the same area (OLB station) (Del Gaudio et al., 2009), meaning uplift of the rock volume to the SW of the fault zone and lowering to the NE fault zone side. For this reason we hypothesize a significant seismotectonic role of the fault zone located E-SE of the Solfatara crater (Fig.11), capable to accommodate the strain built up in these last unrest periods (2005-2006).

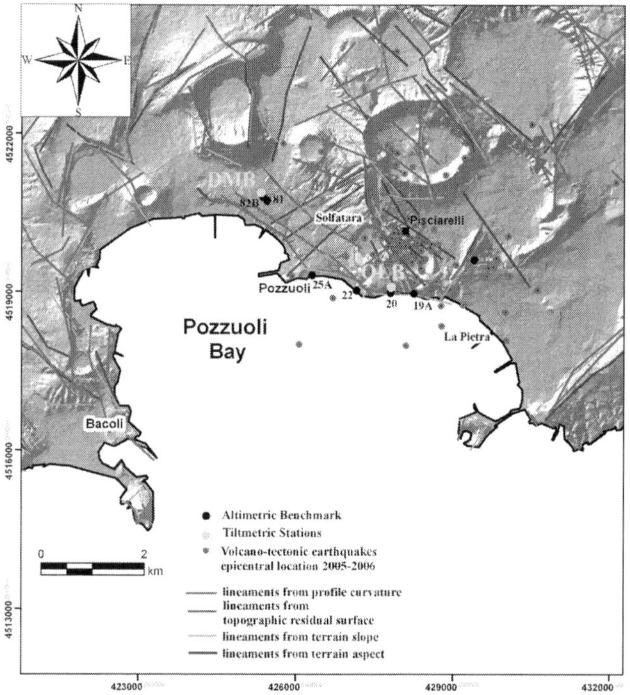

Fig. 11. The map shows different thematic layers overlapped: instrumental seismic events (VT 2005-2006) with blue circles and structural lineaments from morphometric analysis.

In conclusion, the areas selected as case study have revealed the efficiency of the methodology for investigating their complex tectonic setting, and the usefulness of generating original morphostructural data for improving scientific knowledge about active tectonic areas and associated hazards. Moreover, validation of new data with widely accepted literature data, has provided added value to the interpretations and future analysis. Accordingly, the results demonstrate the applicability of the method used so far only for active tectonic areas, even in active volcanic areas (Nappi et al., 2010a), confirming the usefulness and the need for a methodological approach of this kind.

5. References

Amato, A. & Cinque, A. (1999). *Erosional landsurface of the Campano-Lucano Apennines (S. Italy): genesis, evolution and tectonic implications*. Tectonophysics, vol. 315, pp. 251-267.

Bechtold, M., Vilardo, G., Battaglia, M., Orsi, G., (2005). *The Campi Flegrei Caldera Database GIS*. http://ipf.ov.ingv.it/cf_gis.html

Benedetti, L., Tapponier, P., King, G. & Piccardi, L. (1998). *Surface rupture of the 1857 Southern Italian earthquake*?.Terra Nova, vol. 10, No. 4, pp. 206-210.

Borraccini, F., De Donatis, M., Di Bucci, D., & Mazzoli, S., 2002. *3D Model of the active extensional fault system of the high Agri River valley, Southern Appennines, Italy*, J. Virtual Expl., 6, 1–6.

Burbank, D.,W., Anderson, R. (2001). *Tectonic geomorphology*. Blackwell, Malden.

Burrato, P., & Valensise, G. (2008). *Rise and fall of a hypothesized seismic gap: source complexity in the M_w 7.0 16 December 1857 Southern Italy Earthquake*. Bull. Seism. Soc. Am., vol. 98, pp. 139-148, doi:10.1785/0120070094.

Cello, G., Tondi E., Micarelli L. & Mattioni L., 2003. *Active tectonics and earthquake sources in the epicentral area of the 1857 Basilicata earthquake (Southern Italy)*, J. Geodyn., Vol. 36, pp. 37–50.

Chiodini, G., Caliro, S., Cardellini, C., Granieri, D., Avino, R., Baldini, A., Donnini, N., & Minopoli, C. (2010). *Long term variations of the Campi Flegrei (Italy) volcanic system as revealed by the monitoring of hydrothermal activity*. J. Geophys. Res., vol. 115, B03205, doi:10.1029/2008JB006258.

Cinque A., Patacca E., Scandone P., Tozzi M. (1993). *Quaternary kinematic evolution of the Southern Apennines. Relationship between surface geological features and deep lithospheric structures*. Ann. Geofis. 36(2), pp. 249-260.

CPTI Working Group (2004). *Catalogo Parametrico dei Terremoti Italiani*, versione 2004, (CPTI04), INGV, Bologna. GNDT- ING-SGA-SSN 1999, pp. 92. (Available at http://emidius.mi.ingv.it/CPTI04/)

Del Gaudio, C., Aquino, I., Ricco, C. & Serio, C., (2009). *Monitoraggio geodetico dell'area vulcanica napoletana: risultati della livellazione geometrica di precisione eseguita ai Campi Flegrei a settembre 2008*. Quaderni di Geofisica, vol. 66, pp. 1-14.

Del Gaudio, C., Aquino, I., Ricciardi, G.P., Ricco, C. & Scandone, R. (2010). *Unrest episodes at Campi Flegrei: A reconstruction of vertical ground movements during 1905-2009*. J. Volcanol. Geotherm. Res., vol. 195, pp. 48-56.doi:10.1016/j.jvolgeores.2010.05.014.

De Vivo, B., Rolandi, G., Gans, P.B., Calvert, A., Bohrson,W.A., Spera, F.J. & Belkin, H.E. (2001). *New constraints on the pyroclastic eruptive history of the Campanian volcanic Plain (Italy)*. Mineral. Petrol. vol. 73, pp. 47-65.

Di Bucci, D., Massa, B. & Zuppetta, A. (2006). *Relay ramps in active normal fault zones: A clue to the identification of seismogenic sources (1688 Sannio earthquake, Italy).* GSA Bullettin, vol. 118 No. 3/4, pp. 430-448.

DISS (Database of Individual Seismogenic Sources) Working Group (2007). Version 3.0.4: *A compilation of potential sources for earthquakes larger than M 5.5 in Italy and surrounding areas,* available at www.ingv.it/DISS/.

Di Vito, M., Lirer, L., Mastrolorenzo, G. & Rolandi, G. (1987). *The Monte Nuovo eruption (Campi Flegrei, Italy).* Bull. Volcanol., vol. 49, pp. 608-615.

Giano, S.I., Lapenna, V., Piscitelli, S. & Schiattarella, M. (1997). *Nuovi dati geologici e geofisici sull'assetto strutturale dei depositi continentali quaternari dell'alta Val d'Agri (Basilicata).* Il Quaternario, vol. 10, No. 2, pp. 591-596.

Giano, S.I., Maschio, L., Alessio, M., Ferranti, L., Improta, S. & Schiattarella, M. (2000). *Radiocarbon dating of active faulting in the Agri High Valley, Southern Italy.* J. Geodyamics., vol. 29, pp. 371-386.

Giano, S. (2011). *Quaternary alluvial fan systems of the Agri intermontane basin (Southern Italy): tectonic and climatic controls. GEOLOGICA CARPATHICA,* (February 2011), vol. 62, No. 1, pp. 65-76 ,doi: 10.2478/v10096-011-0006-y.

Hilley, E. & Arrowsmith, G.R. (2000). *Thrust fault slip rates deduced from coupled geomorphic and tectonic models of active faults and folds in the San Francisco bay area*: collaborative research with Arizona State University, and University of California, Davi, 1434-H-97-GR-03113, Final Technical Report Research NEHPR support, 2000.

Hippolyte, J.C., Angelier, J., Roure, F. (1994). *A major geodynamic change revealed by Quaternary stress patterns in the Southern Apennines (Italy).* Tectonophysics, vol. 230, pp. 199-210.

Improta, L., Ferranti, L., De Martini, P. M., Piscitelli, S., Bruno, P. P., Burrato, P., Civico, R., Giocoli, A. Iorio, M., D'Addezio, G. & Maschio, L. (2010). *Detecting young, slow-slipping active faults by geologic and multidisciplinary high-resolution geophysical investigations: A case study from the Apennine seismic belt, Italy.* J. of Geoph. Research, vol. 115, B11307, doi:10.1029/2010JB000871, 2010.

Jordan, G., Mejninger, B.M.L., van Hinsbergen, D.J.J., Meulenkamp, J.E. & van Dijk, P.M. (2005). *Extraction of morphotectonic features from DEMs: Development and applications for study areas in Hungary and NW Greece.* Int. J. of Applied Earth Observation and Geoinformation, vol. 7, pp. 163-182.

Keller, E.A. & Pinter, N. (2002). *Active Tectonics: earthquakes, uplift and landscape,* 2nd ed. Prentice Hall pp. 359.

Mallet, R. (1862). The great Neapolitan earthquake of 1857. The first principles of observational seismology, London, 1862, In: *Mallet's macroseismic survey on the Neapolitan earthquake of 16th December 1857,* Bologna, 1987, ING-SGA.

Maschio, L., Ferranti L. & Burrato, P. (2005). *Active extension in Val d'Agri area, Southern Apennines, Italy: implications for the geometry of the seismogenic belt.* Geophys. J. Int., vol. 162, pp. 591-609.

Milano, G., Digiovanbattista, R. & Alessio, G., (1999). *Earthquake swarms in the Southern Apennines chain (Italy): the 1997 seismic sequence in the Sannio-Matese mountains.* Tectonophysics, vol. 306, pp. 57-78.

Milano, G., Digiovanbattista, R. & Ventura G., (2005). *The 2001 seismic activity near Isernia (Italy): implications for the seismotectonics of the Central-Southern Apennines.* Tectonophysics, vol. 401, pp. 167-178.

Moore, ID, Lewis, A. &Gallant, JC. (1993). In: Jakeman AJ. (ed) *Terrain properties: estimation methods and scale effects, modeling, change in environmental systems.* Wiley, New York.

Nappi R., Burrato P., Vilardo G., Ventura G. (2002). *Identificazione ed analisi di elementi morfologici connessi alla deformazione di aree sismicamente attive: la Val d'Agri (Basilicata).* Extended Abstract 6ª Conferenza Nazionale ASITA "Geomatica per l'ambiente, il territorio e il patrimonio culturale". Perugia 5-8 novembre 2002, pp. 1647-1653.

Nappi, R., Alessio, G., Bronzino, G., Terranova, C. & Vilardo, G. (2008). *Contribution of the SISCam Web-based GIS to the seismotectonic study of Campania (Southern Apennines): an example of application to the Sannio-area.* Natural Hazards, vol. 45, pp. 73-85.

Nappi, R., Alessio, G., Vilardo, G. & Bellucci Sessa, E. (2009). *Analisi Morfometrica integrata in ambiente GIS applicata ad aree tettonicamente attive.* Memorie della Società Geografica Italiana vol. 87 No.I–II., ISBN 978-88-88692-57-9.

Nappi, R., Alessio, G. & Bellucci Sessa, E. (2010a). *A case study comparing landscape metrics to geologic and seismic data from the Ischia Island (Southern Italy).* Appl. Geomat. Vol. 2, pp. 73–82, DOI 10.1007/s12518-010-0023-z.

Nappi, R., Alessio, G., Aquino, I., Del Gaudio, C. & Ricco, C., (2010b). *Relationships between the active structural lineaments of the Campi Flegrei area (Southern Italy) from morphometric analysis and recent ground deformation.* Poster - 6th Cities on Volcanoes, May 31 to July 4, 2010, Puerto de La Cruz, Tenerife (Canarie Islands, Spain).

Ollier, C. (1991). *Ancient Landforms,* Belhaven Press, (eds).

Orsi, G., Civetta, L., Del Gaudio, C., de Vita, S., Di Vito, M.A., Isaia, R., Petrazzuoli, S.M., Ricciardi, G., & Ricco, C. (1999). *Short-term ground deformation and seismicity in the nested Campi Flegrei caldera (Italy): an example of active block-resurgence in a densely populated area. J. Volcanol.* Geotherm. Res., vol. 91, pp. 415-451.

Pantosti, D., and G. Valensise (1988). *La faglia sud-appenninica: identificazione oggettiva di un lineamento sismogenetico nell'Appennino meridionale,* Proc. VII Meeting Gruppo Nazionale di Geofisica della Terra Solida, Roma 1988, pp. 205–220.

Pondrelli, S., Salimbeni, S., Ekström, G., Morelli, A., Gasperini, P. & Vannucci, G. (2006). *The Italian CMT dataset from 1977 to the present.* Phys. Earth Planet. Int., vol. 159, No. 3–4, pp. 286–303, doi:10.1016/j.pepi. 2006.07.008.

Scandone, R., Bellucci, F., Lirer, L., Rolandi, G., 1991. *The structure of the Campanian plain and the activity of the neapolitan volcanoes.* J. Volcanol. Geotherm. Res., 48 (1-2), 1-3.

Schiattarella, M., Di Leo, P., Beneduce, P. & Giano, S.I. (2003). *Quaternary uplift vs tectonic loading: a case study from the Lucanian Apennine, southern Italy.* Quat. Int., vol. 101-102, pp. 239-251.

Valensise, G. & Pantosti D. eds (2001a). *Database of potential sources for earthquakes larger than 5.5 in Italy,* Annali di Geofisica, 2001a, supplemento al vol. 44 (4) 180, con CD-ROM.

Valensise, G. & Pantosti, D. (2001b). *The investigation of potential earthquake sources in peninsular Italy: A review.* Journal of Seismology, vol. 5 No.3, pp. 287-306.

Valoroso, L., Improta, L., Chiaraluce, L., Di Stefano, R., Ferranti, L., Govoni, A., & Chiarabba, C. (2009). *Active faults and induced seismicity in the Val d' Agri area*

(Southern Apennines, Italy). Geophys. J. Int., vol. 178, pp. 488 - 502, doi:10.1111/j.1365-246X.2009.04166.x.

Vilardo G., Nappi R., Petti P., Ventura G. (2003). *Fault geometries from the space distribution of the 1990–1997 Sannio-Benevento earthquakes: inferences on the active deformation in Southern Apennines*. Tectonophysics vol. 363 No.3–4, pp.259–271.

Westaway, R. (1993). *Quaternary Uplift of Southern Italy*, Journal of Geophysical Research, vol. 98 No. 21, pp. 741-772.

19

The Automated Identification of Volcanic Earthquakes: Concepts, Applications and Challenges

Mauricio Orozco-Alzate[1], Carolina Acosta-Muñoz[2]
and John Makario Londoño-Bonilla[2]
[1]*Depto. de Informática y Computación, Universidad Nacional de Colombia Sede Manizales*
[2]*Observatorio Vulcanológico y Sismológico de Manizales, INGEOMINAS*
Colombia

1. Introduction

Classifying seismic signals into their corresponding types of volcanic earthquakes is among the most important tasks for monitoring volcano activity. Such a duty must be routinely conducted —in a daily basis— and implies, therefore, a significant workload for the personnel. The discipline of pattern recognition (PR) provides volcanic seismology practitioners with theories and methods to design classification systems and, together with digital signal processing (DSP) techniques, has given rise to promising and challenging opportunities for the automated identification of volcanic earthquakes.

A wealth of recently published studies have demonstrated the applicability of PR tools to volcano-seismic monitoring; however, in spite of that, several cutting-edge approaches have not yet been applied to the problem; moreover, there is still a gap between research achievements reported in the literature and the deployment of custom solutions at the volcano observatories. This chapter introduces fundamental concepts regarding seismic volcanic signals and PR systems, reviews research contributions and case studies, and highlights open issues, future directions for research and challenges to bridge the gap in the transfer of prototype academic results into deployed technology.

In this preliminary section, important definitions and concepts from volcano seismology and PR are considered. First, fundamentals of measurement, data acquisition and telemetry are presented. This is followed by an overview of the different types of volcanic earthquakes, including concise explanations of their geophysical origin and importance for monitoring and forecasting volcanic activity. Advantages of using PR tools in the identification of seismic volcanic signals are discussed. Lastly, stages of a PR system —namely detection or segmentation, representation and generalization — are introduced.

1.1 Measurement, data acquisition and data transmission

The foundation of volcano monitoring is the collection of experimental physical data and their subsequent analysis and correlation with the associated underlying phenomena. Measuring volcanic earthquakes is particularly important, since seismic events are a first sign of renewed

Fig. 1. Seismic monitoring station installed by Observatorio Vulcanológico y Sismológico de Manizales (OVSM) at Nevado del Ruiz Volcano, Colombia.

volcanic activity (Chouet, 1996) and reveal processes such as transport of magma and gases or fracture of solid rock. Nowadays, seismic data collection is typically automated and telemetered. Both properties are required in order to guarantee (1) continuous —24 hours a day— records, (2) real time surveillance, and (3) data acquisition in remote areas where frequent visits to collect data are not feasible.

The automated collection of seismic volcanic data can be divided into three stages: measurement, data acquisition and data transmission. Measurement is performed by using sensing devices that convert ground motion into measurable output signals: electrical energies as voltages; data acquisition is composed, in turn, by several substages including signal conditioning, analog to digital (A/D) conversion and further signal processing; data transmission is performed by radio link systems, either analog or digital whether the A/D conversion is carried out after or before transmission. A standard seismic monitoring station —loosely thought of as being composed by a buried sensor, an electronics box, a solar panel and a Yagi antenna— is shown in Fig. 1. Further descriptions regarding sensors and telemetry are given below. For a general introduction to data measurement and analysis, the reader is referred to (Brown & Musil, 2004) and the classic book by Bendat & Piersol (2010).

1.1.1 Seismic sensors

Comprehensive book chapters on seismic instruments have been written by Havskov & Alguacil (2004, Chap. 2), Bormann (2009, Chap. 5) and Havskov & Ottemöller (2010, Chap. 3). In spite of that and for the sake of a self-contained presentation, brief discussions on physical principles, types and technical properties of seismic sensors —also known as seismometers— are given below.

Seismometers are usually categorized into passive short period sensors and active broadband sensors, see Figs. 2(a) and 2(b) respectively. The former consist in a magnetic mass which is suspended in a spring and surrounded by a coil; as a result of the mass movement, an electric current is induced in the coil; the associated voltage is proportional to the velocity of the mass. In these sensors, the relationship between the induced signal and the actual velocity is linear in a bandwidth typically ranging from 1.0 to 100 Hz (Havskov & Ottemöller, 2010).

Active broadband sensors are based on the so-called *force balance accelerometer* principle. It roughly consists in extending the linear bandwidth response, down to about 0.01 Hz, by including a feedback coil that limits the motion of the mass in a desired range. The linear bandwidth of broadband sensors typically ranges from 0.01 to 50 Hz. Both types of sensors require corrections to reflect the actual ground motion in length-related units, namely corrections for the instrument response and phase shift. Such topics are note covered here but are well explained in the above cited references.

(a) Short period sensor. (b) Broadband sensor.

Fig. 2. Examples of seismic sensors installed in the field.

1.1.2 Telemetry, A/D conversion and data storage

Seismic stations may be designed to be either portable or permanent. Portable ones are equipped with on-site data storage devices such as internal memories and external hard drives and are specially deployed for medium time periods. In order to avoid periodic visits to collect data in remote areas and ensure continuity in the historical records, permanent stations are installed by applying telemetry technologies, see Fig. 1. A typical analog radio telemetry system comprises —in the transmitting side— a sensor (see Sec. 1.1.1), a modulator, a radio and an antenna; similarly, in the receiving side, it is composed by an antenna, a radio, a demodulator or discriminator and an A/D system coupled to a storage device. The modulator usually corresponds to a voltage controlled oscillator with frequency modulation (Havskov & Alguacil, 2004, Chap. 8) followed by a second modulation introduced by the radio and aimed to transmit the signals in VHF or UHF bands[1]. When signals are digitized on-site, a digital telemetry system is used with a variety of modulation schemes (Bormann, 2009, Chap. 7). Moreover, recent deployments of seismic arrays have taken advantage of mobile telephone

[1] VHF band: 30 to 300 MHz; UHF band: 300 MHz to 3 GHz.

networks and internet technologies (Vargas-Jimenez & Rincón-Botero, 2003; Werner-Allen et al., 2006). Readers that require a thorough introduction to data transmission are referred to (Temes & Schultz, 1998) and (Eskelinen, 2004) for the analog case and to (Hsu, 2003) for both analog and digital cases.

The digital acquisition of seismic signals involves stages for signal conditioning and A/D conversion. The first one includes amplifiers and antialias filters, required to scale low-level outputs of passive sensors and fulfill the Nyquist criterion[2], respectively. The A/D conversion is carried out by using analog-to-digital converters (ADCs), typically having sampling rates of 50, 100 or 200 Hz and resolutions between 12 and 24 bit. Individual events are extracted from the continuous records by applying segmentation methods, see Sec. 1.3.1. Further details about A/D conversion and filtering can be found in publications by Scherbaum (1994; 2002; 2007).

Segmented seismic events can be stored in a variety of file formats. The choice of a particular format depends on technical convenience for both space and compatibility. Plain text files are simple enough that most programs can read them because they use the ASCII standard to represent characters (Brown & Musil, 2004); however, text files are neither optimized in size according to the number of bits of the corresponding ADC nor suitable to embed codes indicating formatting and additional capabilities. These weaknesses are overcome by special binary formats such as the Seismic Unified Data System (SUDS), the Seismic Analysis Code (SAC), the SEISmic ANalysis system (SEISAN), the Guralp Compressed Format (GCF) and the Standard for the Exchange of Earthquake Data (SEED).

1.2 Seismic waveforms and classes of volcanic earthquakes

Seismic signals reveal the propagation of elastic waves through the ground. An earthquake generates two different types of such waves; namely body waves and surface waves (Kayal, 2008). The former propagate within a body of rock; the latter travel along the ground surface. A further distinction is made in body waves between the primary wave (P-wave) and the secondary or shear wave (S-wave). The P-wave is faster than the S-wave; therefore, it appears before the S-wave in the seismograph record as shown in Fig. 3.

The vibrations following the arrival of a wave are called *coda*. Since the coda of the P-wave is often hidden by the onset of the S-wave, the term coda usually refers to S-coda (i.e. the trailing part of the seismogram) unless indicated otherwise. Refer again to Fig. 3.

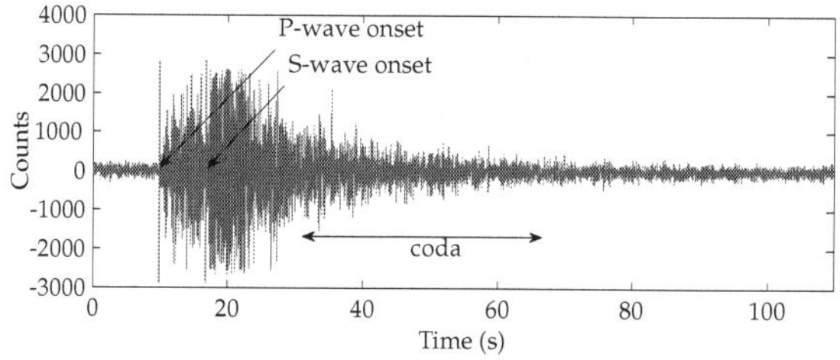

Fig. 3. Parts of a seismic signal.

[2] The sampling rate must be greater than twice the highest frequency component of the signal.

Volcanic earthquakes are typically categorized into four classes according to their mode of generation and the time-frequency behavior of their associated seismic signals. The first criterion —the mode of generation— corresponds to two distinct types of processes occurring either in the solid rock or in the magmatic and hydrothermal fluids within the volcanic edifice. A variety of names have been used to describe the four classes of volcanic earthquakes (McNutt, 2005; Zobin, 2003); however, nowadays, the following denominations are widely accepted: *volcano tectonic* (VT) events, *long period* (LP) events, *tremors* (TR), and *hybrid* (HB) events; see Fig. 4. Concise explanations including their geophysical origin, time-frequency characteristics and importance for monitoring and forecasting volcanic activity are given below. Some special events are observed in particular volcanoes, e.g. multiphase (MP) earthquakes at Mt. Merapi volcano (Hidayat et al., 2000); and flute tremors, spasmodic tremor (Gil-Cruz, 1999) and 'tornillo'-type signals at Galeras volcano (Narváez-M. et al., 1997).

Tectonic earthquakes such as teleseismic (TS), regional (RE) and local (TL) ones are also observed at the seismic volcanic stations. Furthermore, rock falls (RF), explosions (EX), landslides (LS), avalanches, icequakes (IC) and even lightnings are also recorded by the instruments. Descriptions for those non-volcanic events are not given here due to space constraints. Details of the TS, RE and TL classes are available in (Kayal, 2008).

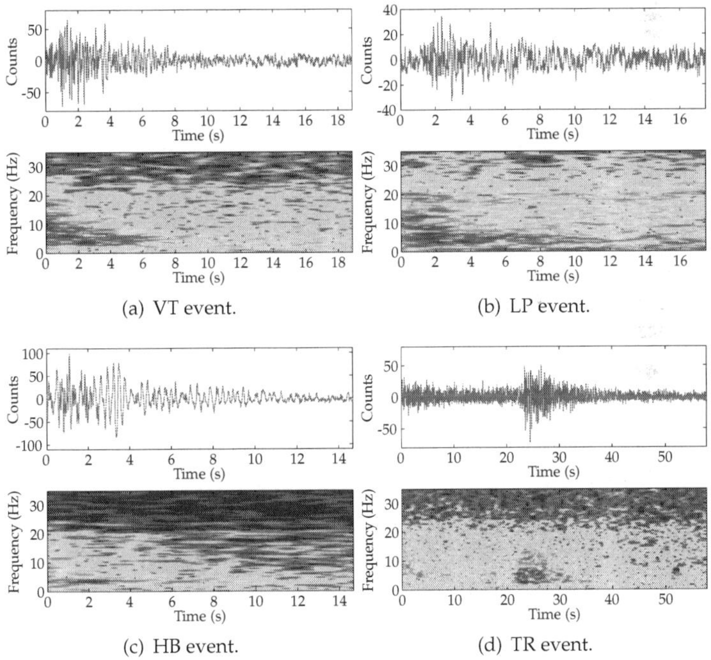

(a) VT event. (b) LP event.

(c) HB event. (d) TR event.

Fig. 4. Examples of seismic volcanic signals observed at Nevado del Ruiz Volcano, together with their associated spectrograms. Events were recorded at Olleta station in 2006. Spectrograms were scaled to highlight the top 50 dB of the signals.

1.2.1 Volcano Tectonic (VT) earthquakes

These earthquakes are indicative of fractures in the solid rock, which are caused by either pressure from magmatic intrusion into the volcano or stress relaxation due to a withdrawal of magma in the crust (Guillier & Chatelain, 2006). VT waveforms are characterized by clear and impulsive arrivals of P and S waves and a short coda typically lasting 7 to 15 s. In the spectral domain, VT events are characterized by a relatively high-frequency content with energy peaking in the band from 6 to 8 or 10 Hz (Chouet, 1996; Guillier & Chatelain, 2006), little energy in the frequencies below 3.5 Hz and significant components up to 15 or 20 Hz, see Fig. 4(a). It is important to monitor VT events because an increase in such seismic activity has been found to be often a first sign of volcanic unrest (Trombley, 2006); nonetheless, their consideration as eruption precursors may not be reliable since the activity may last from days to months or even years (Chouet, 1996). Therefore, VT events must be always correlated with the locations of occurrence and the other classes of volcanic earthquakes (Londoño-Bonilla, 2010).

1.2.2 Long Period (LP) earthquakes

These events are caused by pressure changes in channels filled with magmatic and hydrothermal fluids. Such changes, in turn, are produced by unsteady mass transport and/or thermodynamics of the fluid (Chouet, 1996). The interaction between the surrounding solid and the aforementioned pressure fluctuations constitutes a resonator system (Kumagai & Chouet, 1999) that exhibits decaying harmonic oscillations. LP waveforms are characterized by more or less emergent first arrivals, a lack of clear S waves (Lesage, 2009) and coda waves lasting up to 1.5 minutes (Gil-Cruz & Chouet, 1997). In the spectral domain, energies are concentrated in low frequencies ranging from 0.5 to 3 Hz according to Trombley (2006) or up to 5 Hz according to Chouet (1996). Weak energies at higher frequencies, up to 13 Hz, are only present at the onset. These time and frequency properties can be examined in the sample signal shown in Fig. 4(b).

The forecasting potential of LP events has been pointed out by several studies. They commonly precede and accompany volcanic eruptions (Chouet, 2003) and their analysis may provide an understanding of the dynamic state and mechanical properties of the fluids at their sources.

1.2.3 Tremors (TR)

Tremors are produced by the same phenomena that cause LP earthquakes but their oscillations may last from minutes to days, and sometimes for months or longer (Chouet, 1996). Such an extended manifestation reveals the presence of a sustained excitation. Trombley (2006) claims that such a sustained excitation is caused by extra pushes that the waves of pressure, traveling through the magma, get as a result of pressure changes coming from below.

There is no significant difference between the signal characteristics of LP and TR events, except for the longer duration of the latter. The study of TR earthquakes is considered crucial for the investigation of gas/liquid within a magma conduit (Martinelli, 1997) and also for improving eruption forecasting since, as LP earthquakes, TR events have been frequently observed prior to volcanic eruptions (Lesage et al., 2002).

1.2.4 Hybrid (HB) earthquakes

The occurrence of a VT earthquake may trigger a LP event or vice versa (Trombley, 2006). As a result, a combined event — so-called HB earthquake— appears, containing a mixture of the two former ones. HB earthquakes may be episodic or be related to a steady process as, for instance, the interaction between magmatic heat and underground water systems (Guillier & Chatelain, 2006).

The longest HB events last a few tens of seconds (Neuberg, 2000). Chouet (1996) highlights two particular properties of HB seismic signals: a high-frequency onset and a LP-like coda. The first property is caused by a VT event preceding the LP event. The ambiguous physical origin of HB earthquakes limits their use for forecasting purposes (Harrington & Brodsky, 2007).

1.3 Pattern recognition systems

Duin et al. (2002) define PR as an engineering field that studies theories and methods for designing machines that are able to recognize patterns in noisy data. Many of the techniques and methods in the PR field are borrowed from other fundamental and applied disciplines such as DSP, statistics and machine learning. DSP techniques are mainly applied in the first two stages of the PR system pipeline, see Fig. 5. Statistical and machine learning methods are used in the classification task. The remaining stage —representation— is the focus of interest for PR practitioners and researchers working towards the solution of the following questions: (1) how to represent real-world objects or phenomena in such a way that measurements coming from the sensor stage can be appropriately arranged, e.g. in a vector space, to be provided to the classification methods? and (2) is the representation technically suitable in terms of discriminant power and computational complexity? In addition, the PR community is also devoted to modify classification methods in order to adapt them to the particular technical requirements of the application.

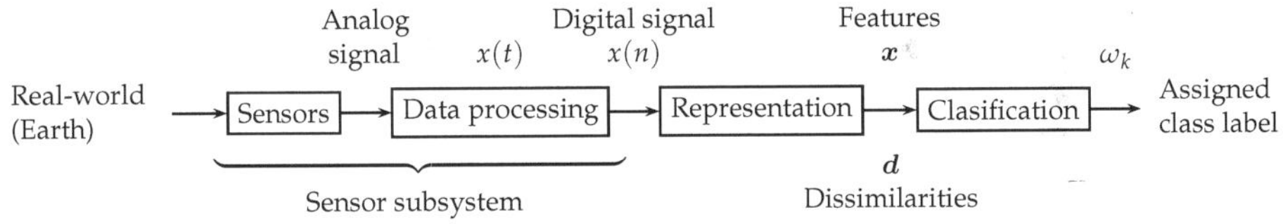

Fig. 5. Building blocks of a PR system.

1.3.1 Sensor subsystem

Consider the particular case of the automated identification of volcanic earthquakes and refer again to Fig. 5. Sensors, as described in Sec. 1.1.1, are seismometers. The subsequent stage —data processing— includes data storage and/or telemetered transmission, A/D conversion (Sec. 1.1.2), and segmentation. This last task in the data processing stage is carried out with a two-fold purpose: (1) to detect the events of interest in the whole continuous raw data; (2) to save space for data storage. In real time implementations, the conventional method for segmenting seismic events is the so-called short-term average - long-term average (STA/LTA) trigger (Havskov & Ottemöller, 2010). Since a detailed discussion of the STA/LTA trigger method is out of the scope of this chapter, the reader is referred to (Havskov & Alguacil, 2004).

1.3.2 Representation approaches

The issue of representation has been traditionally addressed by extracting a set of discriminant features from the segmented sensor measurements. Those features span a vector space which is consequently known as the *feature space*. Good features should allow the building of accurate classifiers to partition the space into decision regions that are associated to the classes to be distinguished —types of volcanic earthquakes in this case. Let $x(t)$ be a segment of the continuous record containing a seismic event and let $x(n)$ be its associated discrete-time sequence. N features extracted from $x(n)$ are arranged in a *feature vector* $\boldsymbol{x} \in \mathbb{R}^N$. Typical features extracted from the morphology of a seismic signal in the time-domain are amplitudes and durations of the waves shown in Fig. 3.

The dissimilarity representation has been proposed as a feasible alternative to represent signals for PR (Pekalska & Duin, 2005). For a given signal $x(n)$, this representation approach consists in computing a dissimilarity measure between either $x(n)$ or some associated transform and a set of M reference signals belonging to a so-called *representation set*. The reference signals are called *prototypes* whenever the set is composed by archetypal examples of each class. Similarly to the feature-based approach, dissimilarities are arranged as a *dissimilarity vector* $\boldsymbol{d} \in \mathbb{R}^M$ in the so-called *dissimilarity space*. Dissimilarity measures typically correspond to metric distances; however, relaxed versions of the metrics are also common in practical applications, e.g. the weighted edit distance and the modified Hausdorff distance which are asymmetric.

Pekalska & Duin (2005) advocate the use of dissimilarity representations instead of classical feature-based ones by presenting several conceptual and practical motivations. Here it is worthwhile to mention the following practical ones: dissimilarities can be derived from raw data such as images, spectra or time samples; dissimilarity-based classifiers outperform the nearest-neighbor rule.

1.3.3 Classification approaches

The last block in Fig. 5 consists in applying classification algorithms to infer a class label $\hat{\omega}(\boldsymbol{x}) \in \Omega$, where $\Omega = \{\omega_1, \ldots, \omega_K\}$ is the set of labels for the K different types of volcanic earthquakes to be identified. According to the nature of the classification algorithms, three different approaches can be distinguished (Jain et al., 2000): similarity-based classification, density-based classifiers and geometric classifiers. These approaches are succinctly described below, including the relatively recent strategy of combining multiple classifiers. A thorough presentation of the classification algorithms can be found in several good textbooks on the subject of PR, such as the ones by Duda et al. (2001), Webb (2002), van der Heijden et al. (2004), Theodoridis & Koutroumbas (2006) and Bishop (2006).

Similarity-based classifiers

This classification approach is based on the elementary rationale of resemblance, i.e. similar events —volcanic earthquakes in our problem— should be identified as belonging to the same class. Among the classifiers in this category, the following two are widely used: the nearest mean classifier (NMC), and the k-nearest neighbor (k-NN) rule. Decision in the first one is taken by examaning the class label of the closest vector among the mean vectors per class; in the second one, the closest event in the vector space defines the assigned class label $\hat{\omega}(\boldsymbol{x})$ for a new incoming event to be identified.

Density-based classifiers

These classifiers are based on the well-known Bayesian decision theory, i.e. on the application of the Bayes decision rule, which consists in the maximization of the posterior probability $P(\hat{\omega}_k|\boldsymbol{x})$ across Ω. $P(\hat{\omega}_k|\boldsymbol{x})$ corresponds, in turn, to the conditional probability density $p(\boldsymbol{x}|\omega_k)$ weighted by the prior probability $P(\omega_k)$. Costs of missclassifications are often included in the rule as an additional weighting parameter.

The key issue in this approach is the estimation of the conditional probability densities, i.e. $\hat{p}(\boldsymbol{x}|\omega_k)$. A distinction between parametric and nonparametric estimates can be made (Jain et al., 2000), where the parametric case corresponds to the assumption of a model for the probability density (e.g. a Gaussian distribution) and the nonparametric one consists in either estimating the probability densities by the standard histogramming technique or by defining window functions in the vector space. Such windows are used to define the contribution of the samples contained in them to the estimation of the probability density. A further division in the window-based nonparametric case is the one between the Parzen window approach and the k-nearest-neighbor method, whether the estimation process is space-invariant or not, respectively.

Consider again the parametric case and the assumption of Gaussian distributions. Parameters to be estimated are the mean vectors and the covariance matrices. According to the assumptions made about the latter, two well-known decision rules result: (1) the Bayes-normal-linear classifier (LDC), when covariance matrices are assumed to be equal; (2) the Bayes-normal-quadratic classifier (QDC), when the covariance matrices are assumed to be different.

Seismic volcanic signals are composed by sequential data, analogously to the case of speech records and time series. A widely used tool for modeling and classifying such sequences is the hidden Markov model (HMM) method. A HMM is composed by a set of states, a matrix of probabilities of transitions between the states, a vector of initial probabilities and an emission model. The HMM-based classification typically consists in training one HMM for each class and, afterwards, using a density-based classifier. Additional details of this method are not given here but can be found in (Rabiner, 1989) as well as in the reviewed studies referenced in Sec. 2.3.

Geometric classifiers

In these classifiers, decision boundaries are built by optimizing a performance criterion instead of considering proximities or densities as in the two previous approaches. Examples of geometric classifiers are the Fisher's linear discriminant, decision trees, single- and multi-layer perceptrons (and, in general, artificial neural networks) and the support vector classifier. Here we only describe the last two classifiers in more detail since they are the most used in volcano seismology applications, as it will be discussed in Sec. 2.

Artificial neural networks (ANNs) are able to implement linear as well as nonlinear classifiers, depending on their architecture (number of layers and number of neurons) and training method. In spite of their tricky tuning procedures, they are still extensively used due to their flexibility and potential good performance. Nonetheless, the emergence of the support vector method has progressively displaced ANNs from their consideration as general solutions for classification and regression; indeed, over the last 15 years, the support vector method have gained a solid theoretical development and an overwhelming number of applications. In few words, the basic principle of the support vector classifier (SVM) is to maximize the margin between two classes, which is defined by the so-called *support vectors*: the closest training examples to the decision boundary. The SVM is extended to nonlinear and multiclass problems by using strategies called the kernel trick and the one-against-rest approach. Further details can be found in some of the PR textbooks cited above as well as in the original work by Vapnik (1998).

Combination of multiple classifiers

The strategy of combining multiple classifiers aims to exploit (1) the availability of multiple sources of data from different sensors or representations, and (2) the possibilities of training several classifiers for the same training set and performing different tuning sessions for the same classifier. Data mentioned in item 1 may belong to either the same events or to different ones. Most seismic volcanic data sets are multiple in nature since they are acquired at multiple recording stations and across several months or years; thereby, multiple sources —stations— for the same events are often available and different sets of examples can be arranged by date of acquisition.

Several strategies for combining classifiers have been proposed. They are typically categorized according to their architecture into parallel, serial and hierarchical; or according to the combination rule into static and trainable (Kuncheva, 2004). PR systems that include these strategies are called *multiple classifier systems*. There has been a sustained interest in this field during the last decade as evidenced by the series of workshops started by Kittler & Roli (2000) and recently organized by Gayar et al. (2010).

2. A Review of research on automated identification of volcanic earthquakes

This section is meant to be a compact but comprehensive survey of research efforts, achievements and case studies on automatic classification of seismic volcanic signals. Reviewed studies are grouped into categories according to the various approaches and methods discussed in Secs. 1.3.2 and 1.3.3.

2.1 Research teams and study sites

A literature search was performed in the main technical databases. Most of the applications on the automated identification of volcanic earthquakes have been undertaken through the inter-institutional and international research collaboration of four teams composed by: (1) Departamento de Teoría de la Señal Telemática y Comunicaciones, Universidad de Granada, Spain; and Instituto Andaluz de Geofísica, Universidad de Granada, Spain; (2) Dipartimento di Fisica, Università di Salerno, Italy; and Osservatorio Vesuviano, Istituto Nazionale di Geofisica e Vulcanologia, Italy; (3) Departamentos de Ingeniería Eléctrica y Física, Universidad de La Frontera, Temuco, Chile; and Observatorio Volcanológico de los Andes del Sur, Servicio Nacional de Geología y Minería, Chile; and (4) Departamento de Informática y Computación, Universidad Nacional de Colombia Sede Manizales, Colombia; Observatorio Vulcanológico y Sismológico de Manizales, INGEOMINAS, Colombia; and Pattern Recognition Lab, Delft University of Technology, The Netherlands. In these collaborative studies, it seems that spatial proximity between volcano observatories and at least one expert in DSP and/or PR encourages collaboration, probably due to the possibility of establishing informal communication as pointed out by Katz & Martin (1997). Other active teams are composed by personnel from Istituto Nazionale di Geofisica e Vulcanologia,

Catania, Italy; and Institut für Erd- und Umweltwissenschaften, Universität Potsdam, Germany.

The found studies have been applied to data sets of the following volcanoes: Ambrym volcano, Vanuatu (AMV); Deception Island Volcano, Antarctica (DIV); Etna Volcano, Italy (ETV); Las Cañadas Volcano, Tenerife, Spain (LCV); Llaima Volcano, Chile (LLV); Mt. Merapi Volcano, Indonesia (MMV); Mt. Vesuvius Volcano, Italy (MVV); Nevado del Ruiz Volcano, Colombia (NRV); Phlegraean Fields, Italy (PFV); San Cristóbal Volcano, Nicaragua (SCV); Soufrière Hills Volcano, Montserrat (SHV); Stromboli Volcano, Italy (STV); and Villarica Volcano, Chile (VRV). Other studies are not applied to signals of volcanic origin but to tectonic seismic events. In spite of that and considering the affinity between these two problems, such studies have also been reviewed here. Data considered in those studies come from the European Broadband Network (EBN), the Mediterranean Seismic Network (MSN), the Hyblean Plateau network (HPN), the Marmara Region Network (MRN) and the Bavarian Earthquake Service Network (BEN). See Table 1 for associations between study sites and publications.

Publication	Data set	Classes	Representation	Classification
(Avossa et al., 2003)	STV	Two types of EX	Time samples + PCA	ANN
(Benítez et al., 2007)	DIV	VT, LP, TR and HB	MFCCs	HMM
(Beyreuther et al., 2008)	LCV	VT, RE, day noise and night noise	Morphological attributes of waveforms and spectrograms	HMM
(Beyreuther & Wassermann, 2008)	BEN	VT, RE, TL, day noise and night noise	Wavefield parameters	HMM
(Chu-Salgado et al., 2010)	NRV	VT and LP	1-D spectra + feature selection	NMC, *k*-NN and LDC together with fixed combining rules
(Curilem et al., 2009)	VRV	LP, TR and energetic TR	Morphological attributes of waveforms and spectra	ANN
(Del Pezzo et al., 2003)	PFV	VT and fishermen's EX	LPC coefficients	ANN
(Duin et al., 2010)	NRV	VT, LP and IC	1-D spectra and spectrograms + PCA	QDC and a trained combining rule: decision templates.
(Esposito et al., 2006)	STV	LS, EX and micro-TR	LPC coefficients and morphological waveform atttributes	ANN
(Ezin et al., 2002)	MVV	Earthquakes and false events (thunders, quarry blasts and man-made undersea explosions)	LPC coefficients, and morphological waveform atttributes	ANN

Continued on next page

Continuation of Table 1

Publication	Data set	Classes	Representation	Classification
(Falsaperla et al., 1996)	STV	Four classes of EX	Auto-correlation function, envelope function and spectra	ANN
(Gutiérrez et al., 2009)	SCV	LP, EX, TR and background noise	MFCCs	HMM
(Gutiérrez et al., 2006)	STV and ETV	EX and tremor bursts	MFCCs	HMM
(Hoogenboezem, 2010)	NRV	LP, RE,TL and VT	Spectrograms + PCA and Fisher mapping	Parzen classifier, NMC, HMM, 1-NN, SVM, ANN and combining rules
(Ibáñez et al., 2009)	STV and ETV	Strombolian EX and background noise (STV). TR and tremor bursts (ETV)	MFCCs	HMM
(Langer & Falsaperla, 2003)	STV	Four classes of EX	Auto-correlation function, envelope function and spectra	ANN
(Langer et al., 2006)	SHV	VT, RE, LP, HB, RF and LP+ RF	Autocorrelation function; morphological and statistical attributes of the waveforms	ANN
(Ohrnberger, 2001)	MMV	VT, MP and RF	Wavefield parameters	HMM
(Orozco-Alzate et al., 2006)	NRV	VT, LP and IC	1-D spectra + Dissimilarities	1-NN in feature space. LDC and QDC in dissimilarity space
(Orozco-Alzate et al., 2008)	NRV	VT, LP	1-D spectra + Band selection	Regularized LDC
(Porro-Muñoz et al., 2010a)	NRV	VT and LP	1-D spectra and spectrograms + Dissimilarities (multiway approach)	Fisher linear classifier
(Porro-Muñoz et al., 2010b)	NRV	VT and LP	Spectrograms and scalograms + Dissimilarities (multiway approach)	Fisher linear classifier
(Porro-Muñoz et al., 2011)	NRV	VT and LP	Spectrograms + Dissimilarities (multiway approach)	Regularized LDC
(Riggelsen et al., 2007)	EBN	Tectonic earthquakes and noise	Continuous wavelet transform	Dynamic Bayesian networks
(Romeo, 1994)	MSN	RE, TL, TS, sausage-like and spike-like	Morphological spectral attributes	ANN
(Romeo et al., 1995)	MSN	TL, RE, TS, spike-like and noise	Morphological spectral attributes	ANN

Continued on next page

Continuation of Table 1

Publication	Data set	Classes	Representation	Classification
(Rouland et al., 2009)	AMV	TR and tectonic earthquakes	Morphological waveform attributes	Decision based on the presence or absence of S-waves
(Scarpetta et al., 2005)	MVV	Two problems: VT and quarry blast; underwater EX and thunder	LPC coefficients and morphological waveform atttributes	ANN
(San-Martín et al., 2010)	LLV	LP, TR and VT	Hilbert and wavelet transforms	LDC
(Ursino et al., 2001)	HPN	Tectonic earthquakes and quarry blasts	Autocorrelation functions and spectra	ANN
(Yıldırım et al., 2011)	MRN	Tectonic earthquakes and quarry blasts	Morphological waveform attributes	ANN

Table 1. Summary of reviewed studies and their associated experimental setups.

2.2 Applications and representation approaches

Raw seismic signals are the simplest and straightforward representation to be provided to a classifier. That option exempts designers from the need to find good features and may be convenient if sufficient training examples are available. However, building a vector space by using the original time samples yields to the following drawbacks: (1) it is mandatory to have equal-length and aligned signals, which is often not possible due to the intrinsic variable duration of seismic events; and (2) high dimensional vector spaces are spanned by the samples and, thereby, large training sets are required in order to avoid the "curse of dimensionality" phenomenon. The second drawback can be overcome by applying dimensionality reduction techniques such as principal component analysis (PCA) and feature selection methods. Avossa et al. (2003) adopted this approach, reducing the dimension from 240 to 15. Langer & Falsaperla (2003); Ursino et al. (2001); and Langer et al. (2006) used the autocorrelation function instead of the original waveforms in order to avoid the phase alignment problem.

Morphological features can be extracted directly from the examination of the waveforms. Curilem et al. (2009) measured the following values from the absolute value of the signals: standard deviation, mean, median and maximum value, as well as kurtosis and skewness from a histogram of the signal amplitudes. Scarpetta et al. (2005) and Esposito et al. (2006) extracted time-domain information by computing differences, properly normalized, between the maximum and minimum signal amplitudes. Similarly, Ezin et al. (2002) measured maximum and minimum signal amplitudes, Yıldırım et al. (2011) obtained peak S-to-P amplitude ratios and complexity values and Rouland et al. (2009) detected the presence or absence of S-waves. Signal envelopes, that are smoothed versions of the original waveforms, were also tested for data representation by Falsaperla et al. (1996), Langer & Falsaperla (2003) and Beyreuther et al. (2008). A collection of morphological and statistical attributes of the waveforms were considered in the study by Langer et al. (2006). The most specialized representation is that reported in (Ohrnberger, 2001, Chap. 7) and (Beyreuther & Wassermann, 2008), which includes several wavefield parameters.

An alternative consists in computing intermediate representations, usually spectra and spectrograms because differences in spectral content allow a visual discriminating of different types of volcanic earthquakes (Zobin, 2003, Chap. 9). This approach was followed by Orozco-Alzate et al. (2008); Chu-Salgado et al. (2010); Duin et al. (2010); Hoogenboezem (2010); Orozco-Alzate et al. (2006); and Porro-Muñoz et al. (2010a;b; 2011). In the first four studies, the computation of spectra was followed by dimensionality reduction techniques such as sequential feature selection, PCA and Fisher mapping. In the remaining ones, dissimilarity representations were computed after transforms to the frequency or the time-frequency domain. Porro-Muñoz et al. (2010a;b; 2011) included multiway data analysis techniques, see Sec. 3.5.

Additional features can be extracted from spectral representations by measuring morphological attributes such as the mean frequencies of the five highest peaks, energies in given frequency bands (Curilem et al., 2009; Romeo, 1994; Romeo et al., 1995) and the instantaneous frequency (Beyreuther et al., 2008), or by computing variables such as the Mel-frequency cepstral coefficients (MFCCs), their associated log-energies and the so-called delta and delta-delta coefficients (Benítez et al., 2007; Gutiérrez et al., 2009; 2006). Spectra and spectrograms are typically computed by using the Fourier or the cosine transforms. Other ones, such as the Hilbert and wavelet transforms have been applied for representation; e.g. by Riggelsen et al. (2007), San-Martín et al. (2010), and Porro-Muñoz et al. (2010b).

The linear predictive coding (LPC) coefficients have been widely used in speech recognition and, by extension, also chosen for representation in several projects of seismic signal classification (Del Pezzo et al., 2003; Esposito et al., 2007; 2006; 2005; Ezin et al., 2002; Scarpetta et al., 2005). They are aimed to predict samples as linear combinations of several previous ones, based on the correlation between successive samples in a seismic signal.

2.3 Applications and classification approaches

In the majority of the reviewed applications, ANNs have been used for classification; particularly multilayer perceptrons (MLPs). Summarized descriptions of publication references, input-hidden-output architecture (number of neurons per layer) and training method are shown in Table 2. Architecture and training method, in almost all the studies, were selected either by trial and error or by agreement with a previous publication. An exception is the study by Curilem et al. (2009), who optimized the size of the hidden layer and selected the training process by means of a genetic algorithm, finding that 14 hidden neurons and the Levenberg-Marquardt training algorithm were the optimal choice.

Publication	Architecture	Training method
(Avossa et al., 2003)	15-3-1	Quasi-Newton
(Curilem et al., 2009)	8-14-1	Levenberg-Marquardt
(Del Pezzo et al., 2003)	105-6-1	Quasi-Newton
(Esposito et al., 2006)	71-5-3	Quasi-Newton
(Ezin et al., 2002)	174-6-1	Quasi-Newton and scaled gradient descent
(Falsaperla et al., 1996)	600-8-4	Gradient descent
(Langer & Falsaperla, 2003)	600-8-4	Backpropagation algorithm
(Langer et al., 2006)	103-20-6	—
(Romeo, 1994)	40-12-40	—
(Romeo et al., 1995)	10-9-9-5	—
(Scarpetta et al., 2005)	[70,79]-[4,5]-1	Scaled conjugate gradient and Quasi-Newton
(Ursino et al., 2001)	100-5-2	Back propagation

Continued on next page

Continuation of Table 2

Publication	Architecture	Training method
(Yıldırım et al., 2011)	2-5-1	Levenberg-Marquardt

Table 2. MLP architecture and training methods used in several ANN-based applications.

HMMs have been widely used in the speech recognition framework. Given the analogous nature of speech and seismic signals, authors have also successfully applied them to the automated classification of volcanic earthquakes. Similarly to the case of ANNs, the performance of HMMs is controlled by several free parameters, namely: the topology of the models, the number of states for the models, the number of multivariate Gaussian probability density functions and the number of iterations of the Baum-Welch algorithm for training. Topology usually corresponds to a left-to-right configuration. Values used for the second parameter —the number of states— in the reviewed applications are listed in Table 3.

Publication	Number of states
(Benítez et al., 2007)	11
(Beyreuther et al., 2008)	—
(Beyreuther & Wassermann, 2008)	3,4,5
(Alasonati et al., 2006)	2
(Gutiérrez et al., 2006)	13,20
(Gutiérrez et al., 2009)	13
(Hoogenboezem, 2010)	20
(Ibáñez et al., 2009)	17,22
(Ohrnberger, 2001)	6,12

Table 3. Configurations of HMMs applied in the reviewed publications.

A conceptual discussion on the use of wavelet-based HMMs to the classification of seismic volcanic signals is presented in (Alasonati et al., 2006). Several reasons have motivated researchers to prefer a left-to-right HMM topology instead of an ergodic one; Ohrnberger (2001, Chap. 7) points out the following reasons: (1) seismic signals are causal in time; (2) seismic signals are analogous to speech signals, for which left-to-right models are widely used; and (3) the degree of freedom of a model —with equal number of states— is lower for a left-to-right topology than that for an ergodic one. Readers are referred again to (Rabiner, 1989) for details on the difference between these two topologies. A generalization of HMMs are the so-called dynamic Bayesian networks. Riggelsen et al. (2007) applied them to the real-time identification of seismic signals.

Less complex classifiers were applied by San-Martín et al. (2010); Chu-Salgado et al. (2010); Orozco-Alzate et al. (2006); Porro-Muñoz et al. (2010a;b); and Duin et al. (2010). Authors of the first study built classifiers on top of a classical feature representation while the others employed simple ones, either in the dissimilarity space or to be combined in a second step of the classification process as explained at the end of Sec. 1.3.3. The reader is referred again to Table 1 to associate studies and classifiers.

Hoogenboezem (2010) presented a compendious survey of classifiers and representations applied to signals from NRV. However, more rigorous experimentations and statistical comparisons are a must when a comprehensive study is planned to be conducted. Recommendations such as those made by Demšar (2006); Duin (1996); Salzberg (1997) and in Sec. 2.4 should be taken into account. An additional concern is the methodological rigor in the evaluation of performances for multiclass problems; even though most of the studies report confusion matrices, others draw conclusions from overall accuracies that are likely to be unreliable for multiclass and/or unbalanced data sets.

This subsection is concluded with a mention to the following studies dealing with the unsupervised classification problem: (Ansari et al., 2009; Esposito et al., 2007; 2008; 2005; Orozco-Alzate & Castellanos-Domínguez, 2007). They are aimed at finding clusters in seismic volcanic data and understanding their structure. A separate chapter would be required to properly discuss them.

2.4 The need of a benchmarking data set

Classification accuracies and other performance measures reported in the literature are not comparable across the reviewed studies because, unfortunately, there are no standard and publicly-available data sets of seismic volcanic signals. Furthermore, authors have used different sets even when they performed studies for the same volcano. Thus, the need for a benchmarking data set is evident. Researchers in this field are encouraged to define such a reference set to be made available for rigorous comparative studies. Ultimately, it is the only reliable way of measuring relative system performance.

3. Open issues and research opportunities

The area of PR has developed itself into a mature engineering field (Duin & Pekalska, 2005). As a result, in practical applications and particularly in volcano seismology, a number of recent approaches and techniques have not yet been explored. This section is concerned with future directions for research, considering not just the state-of-the-art in PR but also possibilities offered by the development of sensors and computer resources. Prospective projects are briefly outlined, considering novel approaches such as multiple instance learning, one-class classification, adaptive single and multiple classifiers, classifier optimization and multi-way representations.

3.1 Multiple instance learning

A multiple instance problem occurs when training objects are naturally organized into bags of feature vectors, also known as *multisets*, instead of being composed by individually labeled ones (Ray & Craven, 2005). It happens, for instance, when objects are too rich and contain too many details and information that can not be easily represented by a single feature vector (Tax & Duin, 2008), e.g. images that depict several objects —in addition to the one of interest, also known as *concept*— in the same picture. Feature vectors (called in this framework as *instances*) in the bag are assumed to be independent and are not individually labeled since the class labels are only assigned to the complete bags. In a two-class case, with a positive class and a negative class to be distinguished, a negative bag only contains vectors that are not members of the concept; whereas a positive bag contains at least one vector that is member of the concept and, consequently, may contain other vectors that are not.

A prospective application of multiple instance learning to the automated identification of volcanic earthquakes would consider waveforms and spectrograms as bags of feature vectors. In such a way, labels might be more accurately assigned to those segments in the signals or patches in the spectrograms clearly belonging to the concept class. Moreover, ill-defined classes might be more properly treated, e.g. the HB events.

3.2 One-class classification

Seismic signal classification problems are unbalanced. Events of some classes are very common and, therefore, a lot of examples are available. In contrast, other classes are rare and just a few examples of them can be collected. Based on the given examples, only a boundary descriptor of the most frequent classes can be accurately built. Considering a rare type of seismic events as the outliers and the rest of the events as the target class clearly follows the definition of a one-class classification problem (Juszczak, 2006; Tax, 2001).

One-class classifiers are sound alternatives to multi-class ones for cases when rare or abnormal states are very infrequent, costly to be forced (e.g. faults in machinery) or impossible to obtain upon request: a person can not be asked to get sick with particular symptoms and a volcano can not be artificially induced to exhibit particular rare seismic events. This approach, to the best knowledge of the authors, has not yet been applied to the automated identification of earthquakes.

3.3 Adaptive single and multiple classifiers

Seismic signals of the same events may look completely different across seismic stations, waveforms of the same classes of events differ among volcanoes and; moreover, volcano geophysical conditions change over time. These dynamic nature motivates the application of classifier adaptation strategies, either for single or multiple classifiers (Aksela, 2007), that allow the possibility of learning from the test set to adapt or modify the decision regions.

Individually adaptive classifiers have been employed in optical character recognition (OCR) in order to prevent accuracy deterioration due to the statistical dissimilarity between the training and test data (Veeramachaneni & Nagy, 2003). Such a dissimilarity is introduced in OCR by the proliferation of fonts and typefaces. Similarly, in speech recognition, adaptation has been extensively applied to deal with unseen conditions or time-variant speakers (Herbig et al., 2011). In summary, undertaking an exploratory study on the application of adaptive single and multiple classifiers may provide a convenient solution for seismic signal classification under the varying conditions mentioned above. It might be indeed an alternative to re-training or entirely re-designing deployed PR systems.

3.4 Classifier optimization

The relative importance of different classification outcomes must be taken into account when optimizing and evaluating the design of a PR system. Such differences are reflected in a trade-off between the values of true positive rate and false positive rate and can be represented in receiver operator characteristic (ROC) curves, whose examination gives the designer insights to tune the classifiers. Classical ROC curves are restricted to two-class problems, in which one class is designated as positive (target) and the other one is assumed as negative.

The automated classification of seismic volcanic signals is a multiclass PR problem. Therefore, the application of classical ROC analysis is only possible under a one-against-rest approach. Nonetheless, recent research efforts have extended ROC analysis to multiclass cases while overcoming restrictive computational complexity issues that limit straightforward multiclass generalizations; see for instance (Landgrebe, 2007; Landgrebe & Paclík, 2010; Paclík et al.,

2010). Optimal classification systems for the automated identification of volcanic earthquakes might be designed by using those novel ROC approaches.

3.5 Multiway representations

Multiway data analysis has been extensively used in chemometrics and psychometrics. It extends classical multivariate statistical techniques such as component analysis, factor analysis, cluster analysis, correspondence analysis, and multidimensional scaling to multiway data (Kroonenberg, 2008). Multiway means that data are arranged in high-order arrays instead of the usual two-dimensional matrices, in which each row represents an object and each column is associated to a feature or measurement. Data collected at different times, conditions or locations are suitable to be considered as multiway data sets (Porro-Muñoz et al., 2009).

Porro-Muñoz et al. (2010a;b; 2011) derived intuitive multiway representations for classifying seismic volcanic signals. Spectrograms and scalograms are computed for each segmented seismic signal and, afterwards, the whole set is arranged by stacking those initial two-dimensional representations. As a result, a so-called profile-data configuration is obtained, where the three dimensions are associated to signals, time and frequency; respectively. Further studies on the design of custom classifiers for multiway data sets are needed. Moreover, other multiway arrangements might be created by considering, for instance, the recording stations or the sensor components (vertical, North-South, and East-West) as additional ways, i.e. dimensions.

4. Challenges and constraints in deploying automated systems

This section is devoted to a discussion on the difficulties and challenges for the design and deployment of custom solutions at the volcano observatories. Technical challenges and non-technical constraints are summarized. Lastly, a few remarks concerning industrial and commercial implementation alternatives are made.

4.1 Technical challenges and non-technical constraints

Technical challenges in the deployment of PR systems for the automated recognition of seismic volcanic signals are mainly related to the following issues: (1) computational aspects and (2) local conditions. The first issue depends on the actual computational requirements of classification algorithms and their associated demands for data storage. The latter is becoming less relevant since disk storage capacity has grown exponentially and hardware prices have declined. In spite of that, processing the stored data may still be cumbersome, especially when dealing with continuous recording as commented by Langer & Falsaperla (2003). Classification speed is of crucial importance for real-time applications. Computational complexities of all stages in the PR pipeline (see Fig. 5) must be carefully estimated in terms of orders or FLOPS[3] in order to guarantee fast execution. Such a condition implies a reasonable trade-off between complexity and classification performance.

The second issue —local conditions— includes the consideration of several volcano-specific factors as those mentioned at the beginning of Sec. 3.3. In addition, the so-called source, path and local site effects require special attention. They cause that waveforms of the same seismic

[3] Floating point operations per second.

event but recorded at different stations exhibit distinct characteristics; for instance, time delays introduced by the physical distance between stations and amplifications or attenuations of signal components at certain frequencies due to geophysical properties that act as filters. See (Havskov & Alguacil, 2004, Chap. 9) and (McNutt, 2005) for further details about these effects, their characterizations and corrections.

Non-technical constraints are mainly related to budget limitations to undertake R&D projects at volcano observatories. Even though the research stage can be achieved in association with universities and institutes, as reflected in the discussion in Sec. 2.1, the development and implementation of in-house solutions is subject to organizational practices and policies at the observatories. Therefore, formalizing high-level collaboration is needed, in such a way that isolated partnership between individuals become supported by inter-institutional cooperation agreements.

4.2 Industrial and commercial alternatives

Almost all the above-reported applications were developed in mathematical scripting languages, such as MATLAB and its free clones; see e.g. (Lesage, 2009). They certainly offer unparalleled advantages in the design of academic prototypes but are often not well-suited to deliver tools for real-world applications. Main constraints include the inherent slowness of interpreted languages, external dependencies with other third-party toolboxes and prohibitive licensing or pricing terms.

Two alternatives can be identified when economic constraints are critical: (1) developing the entire application from scratch in compiled languages such as Fortran and C, probably incorporating freely available numerical and graphical libraries, see for instance (Ottemöller et al., 2011); and (2) developing programs in high-level and free numerical languages such as OCTAVE and SCILAB, from where stand-alone compiled routines can be invoked, see (Laverde & Manzo, 2009) for an example.

A reasonable trade-off between affordability and performance is offered by a number of commercial software packages that also maintain the advantage of a faster development time in a scripting environment. PERCLASS —formerly PRSD STUDIO— allows the design and easy deployment of PR systems and has proved to be a successful solution in a variety of industrial applications[4]. It is based on the MATLAB platform and follows the style of PRTOOLS (Duin et al., 2007), an academic toolbox for PR, but is not dependent on it. Another possibility is using the SIGNAL PROCESSING TOOLBOX together with the STATISTICS TOOLBOX, both by MATHWORKS, and translating the codes to platform-independent files such as DLLs.

5. Conclusion

Multiple research studies have shown that PR tools can be successfully used in the volcano-seismic monitoring task. Several data representations have been explored, including raw and processed signals in the time- and/or frequency-domain as well as other measurements related to geophysical wave properties. ANNs and HMMs have been preferred to be used in the classification stage, thanks to their flexibility and in spite of being heavily parameterized. Other classifiers, on the contrary, do not demand much

[4] `http://perclass.com/index.php/html/applications/`

parameter adjustments and have being used in combination with novel representations such as dissimilarities and multiway configurations.

The state-of-the-art in PR offers a number of new techniques and methods that might be suitably applied to the automated recognition of volcanic earthquakes. Such technological trends and research directions could effectively incorporate inherent properties of the problem, e.g. multiple channels (stations and components), variations over time and multiclass unbalanced nature. Results obtained by different research teams are unfortunately not comparable because different data sets were used across the studies. A rigorous and comprehensive comparison has not yet been made. If undertaken, defining a benchmark set of problems would be mandatory.

Transferring research achievements to the seismological practice demands careful feasibility evaluations of implementation alternatives and would greatly benefit from working cooperations agreements between volcano observatories and universities. One of the ways to achieve an effective technology transfer is the provision of grants and scholarships.

6. Acknowledgments

This chapter was carried out within the research program "Fortalecimiento de capacidades conjuntas para el procesamiento y análisis de información ambiental" (code Hermes 12677), which is funded by Universidad Nacional de Colombia.

7. References

Aksela, M. (2007). *Adaptive Combinations of Classifiers with Application to On-Line Handwritten Character Recognition*, PhD thesis, Helsinki University of Technology, Espoo, Finland.

Alasonati, P., Wassermann, J. & Ohrnberger, M. (2006). Signal classification by wavelet-based hidden Markov models: application to seismic signals of volcanic origin, *in* H. Mader, C. Connor & S. Coles (eds), *Statistics in Volcanology*, number 1 in *Special Publications of IAVCEI*, Geological Society of London, Trowbridge, UK, chapter 13, pp. 161–174.

Ansari, A., Noorzad, A. & Zafarani, H. (2009). Clustering analysis of the seismic catalog of Iran, *Computers & Geosciences* 35(3): 475–486.

Avossa, C., Giudicepietro, F., Marinaro, M. & Scarpetta, S. (2003). Supervised and unsupervised analysis applied to Strombolian E.Q., *in* B. Apolloni, M. Marinaro & R. Tagliaferri (eds), *14th Italian Workshop on Neural Nets, WIRN VIETRI 2003*, Vol. 2859 of *Lecture Notes in Computer Science*, Springer, pp. 173–178.

Bendat, J. S. & Piersol, A. G. (2010). *Random Data: Analysis and Measurement Procedures*, Wiley Series in Probability and Statistics, 4 edn, John Wiley & Sons, Hoboken, New Jersey.

Benítez, M. C., Ramírez, J., Segura, J. C., Ibáñez, J. M., Almendros, J., García-Yeguas, A. & Cortés, G. (2007). Continuous HMM-based seismic-event classification at Deception Island, Antarctica, *IEEE Transactions on Geoscience and Remote Sensing* 45(1): 138–146.

Beyreuther, M., Carniel, R. & Wassermann, J. (2008). Continuous hidden Markov models: Application to automatic earthquake detection and classification at Las Cañadas caldera, Tenerife, *Journal of Volcanology and Geothermal Research* 176(4): 513 – 518.

Beyreuther, M. & Wassermann, J. (2008). Continuous earthquake detection and classification using discrete hidden Markov models, *Geophysical Journal International* 175(3): 1055–1066.

Bishop, C. M. (2006). *Pattern Recognition and Machine Learning*, Information Science and Statistics, Springer, New York, NY.

Bormann, P. (ed.) (2009). *New Manual of Seismological Observatory Practice NMSOP*, revised edn, IASPEI, Potsdam, Germany.

Brown, P. W. & Musil, S. A. (2004). Automated data acquisition and processing, *in* J. F. Artiola, I. L. Pepper & M. Brusseau (eds), *Environmental Monitoring and Characterization*, Elsevier Academic Press, Burlington, USA, chapter 4, pp. 49–67.

Chouet, B. A. (1996). Long-period volcano seismicity: its source and use in eruption forecasting, *Nature* 380(6572): 309 – 316.

Chouet, B. A. (2003). Volcano seismology, *Pure and Applied Geophysics* 160(3–4): 739–788.

Chu-Salgado, C. A., Orozco-Alzate, M. & Londoño-Bonilla, J. M. (2010). Combinación fija de clasificadores para la discriminación de señales sísmicas volcánicas, *Boletín de Ciencias de la Tierra* (27): 37–48.

Curilem, G., Vergara, J., Fuentealba, G., Acuña, G. & Chacón, M. (2009). Classification of seismic signals at Villarrica volcano (chile) using neural networks and genetic algorithms, *Journal of Volcanology and Geothermal Research* 180(1): 1 – 8.

Del Pezzo, E., Esposito, A., Giudicepietro, F., Marinaro, M., Martini, M. & Scarpetta, S. (2003). Discrimination of earthquakes and underwater explosions using neural networks, *Bulletin of the Seismological Society of America* 93(1): 215–223.

Demšar, J. (2006). Statistical comparisons of classifiers over multiple data sets, *Journal of Machine Learning Research* 7: 1–30.

Duda, R. O., Hart, P. E. & Stork, D. G. (2001). *Pattern Classification*, 2nd edn, John Wiley & Sons, Inc.

Duin, R. P. W. (1996). A note on comparing classifiers, *Pattern Recognition Letters* 17(5): 529 – 536.

Duin, R. P. W., Juszczak, P., Pekalska, E., de Ridder, D., Tax, D. M. J. & Verzakov, S. (2007). PRTools 4.1: A Matlab toolbox for pattern recognition. Delft University of Technology, The Netherlands.

Duin, R. P. W., Orozco-Alzate, M. & Londoño-Bonilla, J. M. (2010). Classification of volcano events observed by multiple seismic stations, *Proc. of the 20th Int. Conf. on Pattern Recognition (ICPR2010)*, IEEE Computer Society, pp. 1052–1055.

Duin, R. P. W. & Pekalska, E. (2005). Open issues in pattern recognition, *in* M. Kurzyński, E. Puchała, M. Woźniak & A. Żołnierek (eds), *Computer Recognition Systems: Proceedings of the 4th International Conference on Computer Recognition Systems CORES '05*, Vol. 30 of *Advances in Soft Computing*, Springer Verlag, Berlin, Germany, pp. 27–42.

Duin, R. P. W., Roli, F. & de Ridder, D. (2002). A note on core research issues for statistical pattern recognition, *Pattern Recognition Letters* 23(4): 493 – 499.

Eskelinen, P. (2004). *Introduction to RF Equipment and System Design*, Artech House, Norwood, MA.

Esposito, A., Esposito, A. M., Giudicepietro, F., Marinaro, M. & Scarpetta, S. (2007). Models for identifying structures in the data: A performance comparison, *in* J. G. Carbonell & J. Siekmann (eds), *11th International Conference on Knowledge-Based Intelligent Information and Engineering Systems KES 2007, XVII Italian Workshop on Neural Networks*, Vol. 4694 of *Lecture Notes in Artificial Intelligence*, Springer, Berlin, pp. 275–283.

Esposito, A. M., Giudicepietro, F., D'Auria, L., Scarpetta, S., Martini, M. G., Coltelli, M. & Marinaro, M. (2008). Unsupervised neural analysis of very-long-period events at Stromboli Volcano using the self-organizing maps, *Bulletin of the Seismological Society of America* 98(5): 2449–2459.

Esposito, A. M., Giudicepietro, F., Scarpetta, S., D'Auria, L., Marinaro, M. & Martini, M. (2006). Automatic discrimination among landslide, explosion-quake, and microtremor seismic signals at Stromboli Volcano using neural networks, *Bulletin of the Seismological Society of America* 96(4): 1230–1240.

Esposito, A. M., Scarpetta, S., Giudicepietro, F., Masiello, S., Pugliese, L. & Esposito, A. (2005). Nonlinear exploratory data analysis applied to seismic signals, *in* B. Apolloni, M. Marinaro, G. Nicosia & R. Tagliaferri (eds), *16th Italian Workshop on Neural Nets and International Workshop on Natural and Artificial Immune Systems, WIRN/NAIS 2005*, Vol. 3931 of *Lecture Notes in Computer Science*, Springer, Berlin Heidelberg, pp. 70–77.

Ezin, E. C., Giudicepietro, F., Petrosino, S., Scarpetta, S. & Vanacore, A. (2002). Automatic discrimination of earthquakes and false events in seismological recording for volcanic monitoring, *WIRN VIETRI 2002: Proceedings of the 13th Italian Workshop on Neural Nets-Revised Papers*, Vol. 2486/2002 of *Lecture Notes in Computer Science*, Springer-Verlag, London, UK, pp. 140–145.

Falsaperla, S., Graziani, S., Nunnari, G. & Spampinato, S. (1996). Automatic classification of volcanic earthquakes by using multi-layered neural networks, *Natural Hazards* 13(3): 205–208.

Gayar, N. E., Kittler, J. & Roli, F. (eds) (2010). *Multiple Classifier Systems: 9th International Workshop, MCS 2010 (Cairo, Egypt)*, Vol. 5997 of *Lecture Notes in Computer Science*, Springer, Berlin, Germany.

Gil-Cruz, F. (1999). Observations of two special kinds of tremor at Galeras volcano, Colombia (1989-1991), *Annali di Geofisica* 42(3): 437–449.

Gil-Cruz, F. & Chouet, B. A. (1997). Long-period events, the most characteristic seismicity accompanying the emplacement and extrusion of a lava dome in Galeras Volcano, Colombia, in 1991, *Journal of Volcanology and Geothermal Research* 77(1-4): 121 – 158. Galeras Volcano, Colombia: Interdisciplinary Study of a Decade Volcano.

Guillier, B. & Chatelain, J.-L. (2006). Evidence for a seismic activity mainly constituted of hybrid events at Cayambe volcano, Ecuador. Interpretation in a iced-domes volcano context, *Comptes Rendus Geosciences* 338(8): 499 – 506.

Gutiérrez, L., Ibañez, J., Cortés, G., Ramírez, J., Benítez, C., Tenorio, V. & Álvarez, I. (2009). Volcano-seismic signal detection and classification processing using hidden Markov models. application to San Cristóbal volcano, Nicaragua, *Proceedings of the IEEE International Geoscience and Remote Sensing Symposium, IGARSS 2009*, Vol. 4, pp. IV–522 –IV–525.

Gutiérrez, L., Ramírez, J., Benítez, C., Ibañez, J., Almendros, J. & García-Yeguas, A. (2006). HMM-based classification of seismic events recorded at Stromboli and Etna volcanoes, *IEEE International Geoscience & Remote Sensing Symposium, IGARSS 2006*, pp. 2765–2768.

Harrington, R. M. & Brodsky, E. E. (2007). Volcanic hybrid earthquakes that are brittle-failure events, *Geophysical Research Letters* 34(L06308): 1–5.

Havskov, J. & Alguacil, G. (2004). *Instrumentation in Earthquake Seismology*, Vol. 22 of *Modern Approaches in Geophysics*, 1 edn, Springer, Dordrecht, the Netherlands.

Havskov, J. & Ottemöller, L. (2010). *Routine Data Processing in Earthquake Seismology with Sample Data, Exercises and Software*, Springer, Dordrecht, The Netherlands.

Herbig, T., Gerl, F. & Minker, W. (2011). *Self-Learning Speaker Identification: A System for Enhanced Speech Recognition*, Signals and Communication Technology, Springer, Berlin, Germany.

Hidayat, D., Voight, B., Langston, C., Ratdomopurbo, A. & Ebeling, C. (2000). Broadband seismic experiment at Merapi Volcano, Java, Indonesia: very-long-period pulses embedded in multiphase earthquakes, *Journal of Volcanology and Geothermal Research* 100(1-4): 215 – 231.

Hoogenboezem, R. M. (2010). *Automatic classification of segmented seismic recordings at the Nevado del Ruiz volcano, Columbia*, Master's thesis, Delft University of Technology, Delft, The Netherlands.

Hsu, H. (2003). *Schaum's Outline of Analog and Digital Communications*, Schaum's Outlines series, 2 edn, McGraw-Hill.

Ibáñez, J. M., Benítez, C., Gutiérrez, L. A., Cortés, G., García-Yeguas, A. & Alguacil, G. (2009). The classification of seismo-volcanic signals using Hidden Markov Models as applied to the Stromboli and Etna volcanoes, *Journal of Volcanology and Geothermal Research* 187(3-4): 218 – 226.

Jain, A. K., Duin, R. P. W. & Mao, J. (2000). Statistical pattern recognition: A review, *IEEE Transactions on Pattern Analysis and Machine Intelligence* 22(1): 4–37.

Juszczak, P. (2006). *Learning to recognise: A study on one-class classification and active learning*, PhD thesis, Delft University of Technology, Delft, The Netherlands.

Katz, J. S. & Martin, B. R. (1997). What is research collaboration?, *Research Policy* 26(1): 1 – 18.

Kayal, J. R. (2008). *Earthquakes and Seismic Waves*, Springer, chapter 2 in Microearthquake Seismology and Seismotectonics of South Asia, pp. 6–80.

Kittler, J. & Roli, F. (eds) (2000). *Multiple Classifier Systems: First International Workshop, MCS 2000 (Cagliari, Italy)*, Vol. 1857 of *Lecture Notes in Computer Science*, Springer, Berlin, Germany.

Kroonenberg, P. M. (2008). *Applied Multiway Data Analysis*, Wiley, Hoboken, New Jersey.

Kumagai, H. & Chouet, B. A. (1999). The complex frequencies of long-period seismic events as probes of fluid composition beneath volcanoes, *Geophysical Journal International* 138(2): F7–F12.

Kuncheva, L. I. (2004). *Combining Pattern Classifiers: Methods and Algorithms*, Wiley-Interscience, Hoboken, New Jersey.

Landgrebe, T. C. W. (2007). *Operating characteristics for the design and optimisation of classification systems*, PhD thesis, Delft University of Technology, Delft, The Netherlands.

Landgrebe, T. C. W. & Paclík, P. (2010). The ROC skeleton for multiclass ROC estimation, *Pattern Recognition Letters* 31(9): 949 – 958.

Langer, H. & Falsaperla, S. (2003). Seismic monitoring at Stromboli volcano (Italy): a case study for data reduction and parameter extraction, *Journal of Volcanology and Geothermal Research* 128(1-3): 233 – 245.

Langer, H., Falsaperla, S., Powell, T. & Thompson, G. (2006). Automatic classification and a-posteriori analysis of seismic event identification at Soufrière Hills volcano, Montserrat, *Journal of Volcanology and Geothermal Research* 153(1-2): 1 – 10.

Laverde, C. A. & Manzo, Ó. (2009). Manual del procesamiento de la información sismológica: Sistema PROVIG - INGEOMINAS, *Technical report*, Instituto Colombiano de Geología y Minería - INGEOMINAS.

Lesage, P. (2009). Interactive Matlab software for the analysis of seismic volcanic signals, *Computers & Geosciences* 35(10): 2137–2144.

Lesage, P., Glangeaud, F. & Mars, J. (2002). Applications of autoregressive models and time-frequency analysis to the study of volcanic tremor and long-period events, *Journal of Volcanology and Geothermal Research* 114(3-4): 391 – 417.

Londoño-Bonilla, J. M. (2010). Activity and Vp/Vs ratio of volcano-tectonic seismic swarm zones at Nevado del Ruiz volcano, Colombia, *Earth Sciences Research Journal* 14(1): 111–124.

Martinelli, B. (1997). Volcanic tremor and short-term prediction of eruptions, *Journal of Volcanology and Geothermal Research* 77(1-4): 305 – 311. Galeras Volcano, Colombia: Interdisciplinary Study of a Decade Volcano.

McNutt, S. R. (2005). Volcanic seismology, *Annual Review of Earth and Planetary Sciences* 33(1): 461–491.

Narváez-M., L., Torres C, R. A., Gómez M., D. M., Cortés J., G. P., Cepeda V., H. & Stix, J. (1997). 'Tornillo'-type seismic signals at Galeras volcano, Colombia, 1992-1993, *Journal of Volcanology and Geothermal Research* 77(1-4): 159 – 171. Galeras Volcano, Colombia: Interdisciplinary Study of a Decade Volcano.

Neuberg, J. (2000). Characteristics and causes of shallow seismicity in andesite volcanoes, *Philosophical Transactions of the Royal Society* 358(1770): 1533–1546.

Ohrnberger, M. (2001). *Continuous Automatic Classification of Seismic Signals of Volcanic Origin at Mt. Merapi, Java, Indonesia*, PhD thesis, University of Potsdam, Postdam, Germany.

Orozco-Alzate, M. & Castellanos-Domínguez, C. G. (2007). Clustering on dissimilarity representations for detecting mislabelled seismic signals at Nevado del Ruiz volcano, *Earth Sciences Research Journal* 11(2): 131–138.

Orozco-Alzate, M., García-Ocampo, M. E., Duin, R. P. W. & Castellanos-Domínguez, C. G. (2006). Dissimilarity-based classification of seismic volcanic signals at Nevado del Ruiz volcano, *Earth Sciences Research Journal* 10(2): 57–65.

Orozco-Alzate, M., Skurichina, M. & Duin, R. P. W. (2008). Spectral characterization of volcanic earthquakes at Nevado del Ruiz Volcano using spectral band selection/extraction techniques, *in* J. Ruiz-Shulcloper & W. G. Kropatsch (eds), *Progress in Pattern Recognition, Image Analysis and Applications. Proceedings of the 13th Iberoamerican Congress on Pattern Recognition CIARP 2008*, Vol. 5197 of *Lecture Notes in Computer Science*, IAPR, Springer, pp. 708–715.

Ottemöller, L., Voss, P. & Havskov, J. (2011). SEISAN: Earthquake analysis software for Windows, Solaris, Linux and MacOSX. Version 9.0.1. Department of Earth Science, University of Bergen, Norway.

Paclík, P., , Lai, C., Landgrebe, T. C. W. & Duin, R. P. W. (2010). ROC analysis and cost-sensitive optimization for hierarchical classifiers, *Proc. of the 20th Int. Conf. on Pattern Recognition (ICPR2010)*, IEEE Computer Society, pp. 2977 –2980.

Pekalska, E. & Duin, R. P. W. (2005). *The Dissimilarity Representation for Pattern Recognition: Foundations and Applications*, Vol. 64 of *Machine Perception and Artificial Intelligence*, World Scientific, Singapore.

Porro-Muñoz, D., Duin, R. P. W., Orozco-Alzate, M., Talavera-Bustamante, I. & Londoño-Bonilla, J. M. (2010a). Classifying three-way seismic volcanic data by dissimilarity representation, *Proc. of the 20th Int. Conf. on Pattern Recognition (ICPR2010)*, IEEE, pp. 814–817.

Porro-Muñoz, D., Duin, R. P. W., Orozco-Alzate, M., Talavera-Bustamante, I. & Londoño-Bonilla, J. M. (2010b). The dissimilarity representation as a tool for three-way data classification: a 2D measure, *in* E. R. Hancock, R. C. Wilson, T. Windeatt, I. Ulusoy & F. Escolano (eds), *Structural, Syntactic and Statistical Pattern Recognition: Proceedings of the Joint IAPR International Workshop, SSPR and SPR 2010*, Vol. 6218 of *Lecture Notes in Computer Science*, IAPR, Springer, Berlin Heidelberg, pp. 569–578.

Porro-Muñoz, D., Duin, R. P. W., Talavera, I. & Orozco-Alzate, M. (2011). Classification of three-way data by the dissimilarity representation, *Signal Processing* 91(11): 2520–2529.

Porro-Muñoz, D., Talavera-Bustamante, I. & Duin, R. P. W. (2009). Multi-way data analysis, *Technical Report 14*, Centro de Aplicaciones de Tecnologías de Avanzada, La Habana, Cuba. Reporte técnico: Reconocimiento de patrones - Serie azul. RNPS: 2142. ISSN: 2072-6287.

Rabiner, L. R. (1989). A tutorial on hidden Markov models and selected applications in speech recognition, *Proceedings of the IEEE* 77(2): 257 –286.

Ray, S. & Craven, M. (2005). Supervised versus multiple instance learning: an empirical comparison, *Proceedings of the 22nd international conference on Machine learning*, ICML '05, ACM, New York, NY, USA, pp. 697–704.

Riggelsen, C., Ohrnberger, M. & Scherbaum, F. (2007). Dynamic bayesian networks for real-time classification of seismic signals, *PKDD 2007: Proceedings of the 11th European conference on Principles and Practice of Knowledge Discovery in Databases*, Lecture Notes in Computer Science, Springer-Verlag, Berlin, Heidelberg, pp. 565–572.

Romeo, G. (1994). Seismic signal detection and classification using artificial neural networks, *Annali di Geofisica* XXXVII(3): 343–353.

Romeo, G., Mele, F. & Morelli, A. (1995). Neural networks and discrimination of seismic signals, *Computers & Geosciences* 21(2): 279 – 288.

Rouland, D., Legrand, D., Zhizhin, M. & Vergniolle, S. (2009). Automatic detection and discrimination of volcanic tremors and tectonic earthquakes: An application to Ambrym volcano, Vanuatu, *Journal of Volcanology and Geothermal Research* 181(3-4): 196 – 206.

Salzberg, S. (1997). On comparing classifiers: Pitfalls to avoid and a recommended approach, *Data Mining and Knowledge Discovery* 1(3): 317–328.

San-Martín, C., Melgarejo, C., Gallegos, C., Soto, G., Curilem, M. & Fuentealba, G. (2010). Feature extraction using circular statistics applied to volcano monitoring, *in* I. Bloch & R. M. Cesar (eds), *Progress in Pattern Recognition, Image Analysis, Computer Vision, and Applications. Proceedings of the 15th Iberoamerican Congress on Pattern Recognition CIARP 2010*, Vol. 6419 of *Lecture Notes in Computer Science*, Springer, Berlin, pp. 458–466.

Scarpetta, S., Giudicepietro, F., Ezin, E. C., Petrosino, S., Pezzo, E. D., Martini, M. & Marinaro, M. (2005). Automatic classification of seismic signals at Mt. Vesuvius volcano, Italy, using neural networks, *Bulletin of the Seismological Society of America* 95(1): 185–196.

Scherbaum, F. (1994). *Basic Concepts in Digital Signal Processing for Seismologists*, Vol. 53 of *Lecture Notes in Earth Sciences*, Springer, Berlin / Heidelberg.

Scherbaum, F. (2002). Analysis of digital earthquake signals, *in* W. H. K. Lee, H. Kanamori, P. C. Jennings & C. Kisslinger (eds), *International Handbook of Earthquake and Engineering Seismology*, Vol. 81, Part A of *International Geophysics Series*, Academic Press, San Diego, CA, chapter 22, pp. 349 – 355.

Scherbaum, F. (2007). *Of Poles and Zeros: Fundamentals of Digital Seismology*, Vol. 15 of *Modern Approaches in Geophysics*, revised 2nd edn, Springer, Dordrecht, The Netherlands.

Tax, D. & Duin, R. P. W. (2008). Learning curves for the analysis of multiple instance classifiers, *in* N. da Vitoria Lobo, T. Kasparis, F. Roli, J. Kwok, M. Georgiopoulos, G. Anagnostopoulos & M. Loog (eds), *Structural, Syntactic, and Statistical Pattern Recognition. Proceedings of the Joint IAPR International Workshop, SSPR & SPR 2008*, Vol. 5342 of *Lecture Notes in Computer Science*, Springer, Berlin, Germany, pp. 724–733.

Tax, D. M. J. (2001). *One-class classification*, PhD thesis, Delft University of Technology, Delft, The Netherlands.

Temes, L. & Schultz, M. E. (1998). *Electronic Communication*, Schaum's Outlines series, 2 edn, McGraw-Hill.

Theodoridis, S. & Koutroumbas, K. (2006). *Pattern Recognition*, 3 edn, Academic Press, London, UK.

Trombley, R. B. (2006). *Probability Contribution Due To Seismic Analysis: Types of Volcanic Earthquakes*, iUniverse, chapter 3 in The Forecasting of Volcanic Eruptions, pp. 14–17.

Ursino, A., Langer, H., Scarfì, L. & Giuseppe di Grazia, S. G. (2001). Discrimination of quarry blasts from tectonic microearthquakes in the Hyblean Plateau (Southeastern Sicily), *Annali di Geofisica* 44(4): 703–722.

van der Heijden, F., Duin, R. P. W., de Ridder, D. & Tax, D. M. J. (2004). *Classification, Parameter Estimation and State Estimation: An Engineering Approach Using MATLAB*, Wiley, Chichester, UK.

Vapnik, V. N. (1998). *Statistical Learning Theory*, Adaptive and Learning Systems for Signal Processing, Communication and Control, John Wiley & Sons, New York, NY.

Vargas-Jimenez, C. A. & Rincón-Botero, S. (2003). Portable digital seismological AC station over mobile telephone network and internet, *Computers & Geosciences* 29: 685–694.

Veeramachaneni, S. & Nagy, G. (2003). Adaptive classifiers for multisource OCR, *International Journal on Document Analysis and Recognition* 6: 154–166.

Webb, A. R. (2002). *Statistical Pattern Recognition*, 2nd edn, John Wiley & Sons, West Sussex, England.

Werner-Allen, G., Lorincz, K., Ruiz, M., Marcillo, O., Johnson, J., Lees, J. & Welsh, M. (2006). Deploying a wireless sensor network on an active volcano, *IEEE Internet Computing* 10(2): 18 – 25.

Yıldırım, E., Gülbag, A., Horasan, G. & Dogan, E. (2011). Discrimination of quarry blasts and earthquakes in the vicinity of Istanbul using soft computing techniques, *Computers & Geosciences* 37(9): 1209 – 1217.

Zobin, V. M. (2003). *Introduction to Volcanic Seismology*, Vol. 6 of *Developments in Volcanology*, Elsevier, Amsterdam.